ROBOTICS AND INDUSTRIAL AUTOMATION

ROBOTICS
AND
INDUSTRIAL AUTOMATION

For Engineering Students of B.Tech./B.E. Courses

Er. R.K. RAJPUT

M.E. (Hons.)-Gold Medalist; Grad. (Mech. Engg. & Elect. Engg.); MIE (India);
MSESI; MISTE; CE (India)

Recipient of :

Best Teacher (Academic) Award

Distinguished Author Award

Jawahar Lal Nehru Memorial Gold Medal
for an Outstanding Research Paper
(Institution of Engineers–India)

PRINCIPAL *(Formerly)*

- *Thapar Polytechnic College*
- *Punjab College of Information Technology*

Patiala

S. CHAND
PUBLISHING

S Chand And Company Limited

(ISO 9001 Certified Company)

S Chand And Company Limited

(ISO 9001 Certified Company)

Head Office: Block B-1, House No. D-1, Ground Floor, Mohan Co-operative Industrial Estate, New Delhi – 110 044 | Phone: 011-66672000

Registered Office: A-27, 2nd Floor, Mohan Co-operative Industrial Estate, New Delhi – 110 044 Phone: 011-49731800

www.**schandpublishing.com**; e-mail: **info@schandpublishing.com**

Branches

Ahmedabad	:	Ph: 27542369, 27541965; ahmedabad@schandpublishing.com
Bengaluru	:	Ph: 22354008, 22268048; bangalore@schandpublishing.com
Bhopal	:	Ph: 4274723, 4209587; bhopal@schandpublishing.com
Bhubaneshwar	:	Ph: 2951580; bhubaneshwar@schandpublishing.com
Chennai	:	Ph: 23632120; chennai@schandpublishing.com
Guwahati	:	Ph: 2738811, 2735640; guwahati@schandpublishing.com
Hyderabad	:	Ph: 40186018; hyderabad@schandpublishing.com
Jaipur	:	Ph: 2291317, 2291318; jaipur@schandpublishing.com
Jalandhar	:	Ph: 4645630; jalandhar@schandpublishing.com
Kochi	:	Ph: 2576207, 2576208; cochin@schandpublishing.com
Kolkata	:	Ph: 23357458, 23353914; kolkata@schandpublishing.com
Lucknow	:	Ph: 4003633; lucknow@schandpublishing.com
Mumbai	:	Ph: 25000297; mumbai@schandpublishing.com
Nagpur	:	Ph: 2250230; nagpur@schandpublishing.com
Patna	:	Ph: 2260011; patna@schandpublishing.com
Ranchi	:	Ph: 2361178; ranchi@schandpublishing.com
Sahibabad	:	Ph: 2771238; info@schandpublishing.com

First Edition 2008
Reprints 2010, 2011, 2012 (Twice), 2013
Second Revised Edition 2014
Reprint 2018
Reprint 2020 (Twice)

ISBN : 978-81-219-2997-4 **Product Code :** H3RIA62ETRN10ENAB14R

PRINTED IN INDIA

By Vikas Publishing House Private Limited, Plot 20/4, Site-IV, Industrial Area Sahibabad, Ghaziabad – 201 010 and Published by S Chand And Company Limited, A-27, 2nd Floor, Mohan Co-operative Industrial Estate, New Delhi – 110 044.

PREFACE TO THE SECOND EDITION

I take the pleasure in presenting the Second Edition of this book. The warm reception which the previous editions and reprints have received is a matter of great satisfaction to me.

In this edition, the book has been thoroughly revised and a new section on Short Answer Questions has been added to make the book still more useful to the students.

The constructive suggestions for the improvement of the book are most welcome.

Er. R.K. Rajput

PREFACE TO THE FIRST EDITION

This treatise on the subject **Robotics and Industrial Automation** contains a comprehensive treatment of the subject matter in a simple, lucid and direct language. It covers the syllabi of the various Indian Universities exhaustively.

The book contains ten chapters in all, namely :

1. Fundamentals of Robot, 2. Robot End-effectors, 3. Robotic Sensors and Vision, 4. Robot Programming Languages, 5. Robot Drives, Actuators and Controls, 6. Hydraulic and Pneumatic Circuits, 7. Electrical and Electronic Controls, 8. Fluidics, 9. Industrial Automation, and 10. Robot Applications.

All these chapters are saturated with much needed text, supported by simple and self-explanatory figures, and worked examples, wherever required. At the end of each chapter *Highlights, Objective Type Questions,* and *Theoretical Questions* have been added to make the book a comprehensive and complete unit in all respects.

The author's thanks are due to his wife Ramesh Rajput for extending all cooperation during preparation and proofreading of the manuscript.

Finally, the author wishes to express his gratitude to the Managing Director and all the concerned staff members of S. Chand & Company Ltd. who were involved so dedicatedly in the publication of this book.

Although every care has been taken to make the book free of errors both in text as well as in solved examples, yet the author shall feel obliged if any errors present are brought to his notice. Constructive criticism of the book will be warmly received.

Er. R.K. Rajput

CONTENTS

CHAPTER

1 Fundamentals of Robot

1.1 ROBOTICS – PHILOSOPHICAL CONSIDERATIONS

During the last century, the modern science grew at a fast pace in "*stages*" given below.

First stage – Invention of tools.

Second stage – Invention of steam and I.C. engines.

Third stage – Technological development; a brand of numerically controlled (NC) machines came into existence in the early 1950s.

Fourth stage – Invention of computers/microprocessors etc.; development of computer numerically controlled (CNC) machines.

Final stage – Development of concept of *Robotics*.

Robotics, conceptually, differs from *conventional automation* in the following respects:

(*i*) Its ability to perform on its own, going to the extent of unmanning several operations.

(*ii*) To relieve man of tedious and mindless, repetitive and hazardous jobs.

- Robotics is an exciting, dynamic interdisciplinary field of study.

Important areas of robotics :

The important areas of robotics are :

1. *Prosthesis* :

— It deals with artificial replacements for parts of the human body.

2. *Exoskeletons :*
   — It deals with frames which surround the human limbs or frames and devices and means for amplification of available power.
3. *Telecherics :*
   — It is concerned with remote manipulation.
4. *Locomotive devices :*
   — This area deals with robots designed to walk on legs.

1.2 DEFINITION AND ADVANTAGES OF ROBOTICS

Definition :

Robotics may be defined as follows :

*"**Robotics** is the science of designing and building robots suitable for real-life applications in automated manufacturing and other non-manufacturing environments."*

Or

*"**Robotics** is the art, knowledge base and know how of designing, applying and using robots in the human endeavors."*

- *Robotics* is an interdisciplinary subject that benefits from mechanical engineering, electrical and electronic engineering, computer science, and several other disciplines.

Advantages :

The main *advantages* of robotics are :

1. Reliability.
2. Increased flexibility.
3. Low cost in the long run.

1.3 LAWS OF ROBOTICS

Following are the laws (philosophical in nature) of robotics (Sir Issac Asimov proposed three basic laws; 'zeroth law' was added later on):

Zeroth law. A robot must not injure humanity or, through inaction, allow humanity to come to harm.

First law. A robot must not harm a human being or, through inaction, allow one to come to harm.

Second law. A robot must always obey human beings unless it is in conflict with a higher order law.

Third law. A robot must protect itself from harm unless that is in conflict with a higher order law.

1.4 MOTIVATING FACTORS FOR THE INTRODUCTION OF ROBOTICS SYSTEMS TO THE INDUSTRIAL WORLD

The following are the factors which vouch for the introduction of robotics systems to the industrial world :

1. Improved quality of products.
2. Lesser preparation time than hard automation.
3. Lower rejects and less waste than labour intensive production.
4. Higher flexibility of product type and variation.

} Technical factors

5. Skilled labour shortage.
6. Constant demand for improvement in quality.
7. Rising costs.
8. Pressure to increase production rates to compete the market.

} Technical factors

1.5 ROBOT

— The origin of word *'robot'* is in the Czech word 'robota' meaning either a slave or a mechanical item that would help its master. A robot therefore carries out the task done by a human being.

— The word **robot* always refers to an automated multifunctional manipulator that works by energy, to perform a variety of tasks.

— A robot is any mechanical device operated automatically to perform in a seemingly human way. By this definition, a garage door opener, which automatically opens the door by remote control is also a robot; obviously this is not an industrial robot.

— Robot, once a creature of science fiction, is today a reality. It is the off-shoot of the second industrial revolution.

- *A robot by virtue of its reprogrammability and versatility is productive, dynamic and flexible to an extent.*
- Robots range from toys to automated assembly lines.

1.5.1. Definition of a Robot

A robot may be defined as follows :

The definition adopted by International Standards Organisation (ISO) and agreed upon by most of the users and manufacturers is :

*"An industrial **robot** is a reprogrammable multifunctional manipulator designed to move materials, parts, tools or specialized devices through various programmed motions for the performance of a variety of tasks".*

Or

According to the Robotics Industries Association (formerly the Robotics Institute of America) :

*"An industrial **robot** is a reprogrammable multifunctional manipulator designed to move material, parts, tools or specialized devices through various programmed motions for the performance of a variety of tasks".*

This definition is compatible with the classification of programmable automation, but robots are also used in flexible automation and fixed automation systems. *Example.* An automation line using many robots to perform spot welds in which robot programs are downloaded from a computer to a specific robot controller could be considered a high-production flexible automation system.

Or

*"A **robot** is an automatic device that performs functions normally ascribed to humans or a machine in the form of a human."*

Or

*"A **robot** is a software-controllable mechanical device that uses sensors to guide one or more end-effectors through programmed motions in a workpiece in order to manipulate physical objects.*

---

* The word robot was introduced in 1920 in a play by Karel Capek called R.U.R. or Rossum's Universal Robots.

Or

*"Industrial **robot** may also be defined as a number of rigid links connected by joints of different types that are controlled and monitored by computer.*

— A robot (industrial) is also called a *robotic manipulator* or a *robotic arm.*

• So long as the evolution of robotics continues, the definition of robot will change from time to time depending on the technological advances in its sensory capability and intelligence level.

1.5.2. Functions of a Robot

The functions of a robot can be classified into *three* areas :

1. "*Sensing*" the environment by external sensors.

   Example : Vision, voice, touch, proximity and so on.
2. "*Decision making*" based on the information received from the sensors.
3. "*Performing*" the task decided.

1.5.3. Difference between a Robot and an Automated Machine

— In automation (hard), the machine produces a job following a set operational sequence, while a robot can be made to do different jobs at different times and in different sequences. This can be achieved by *programming*.

A robot *can* be programmed to change the sequence of tasks while a fixed machine set to perform certain tasks in sequence *cannot* be programmed. An automated machine does not have sensory feedback to reprogram the predetermined path.

— An automated machine has neither a "knowledge base" nor "intelligence"; at best it can be made to adopt to slight changes in a known environment, as in the case of a few numerically controlled (NC) machines.

Thus, *a robot is more than an automated machine/equipment*.

— A *machine can qualify as a robot* subject to the following *conditions* :

(*i*) Reprogrammable.

(*ii*) Sensing and perception.

(*iii*) Function autonomously and/or interact with human beings.

(*iv*) Carry out different tasks.

1.5.4. Robot v/s NC Technology

The similarities and dissimilarities between Robots and NC machines are as follows:

Similarities :

(*i*) Power drive technology.

(*ii*) Feedback systems.

(*iii*) Trend of increased computer control.

(*iv*) Actuating mechanisms.

(*v*) Several industrial applications.

Dissimilarities :

(*i*) A robot is a lighter and more portable equipment comparatively.

(*ii*) A robot's programming is different from the part programming used in NC machine tools.

(*iii*) The technology of robots differs from NC technology in that robots effect higher velocity and movements in more axes of motions.

(*iv*) The applications of robots are more general in nature as compared to a NC machine tool.

1.5.5. Capabilities of Industrial Robots

Industrial robots (tough and tireless) are capable of doing the following jobs :

1. To handle a variety of jobs right from material handling to complex assembly tasks.
2. To perform hazardous and monotonous tasks with tireless precesion.
3. To improve productivity and reduce manufacturing costs.
4. To perform complex jobs.
5. To even cope with changing conditions in the workplace, when fitted with sensors and adaptive controls.

1.6 HISTORY OF ROBOTS

The history of robots is as under :

Year	Event
• *1958*	Devol entered into a license agreement of an industrial robot with consolidated Control Corporation, a subsidiary of Condec. Corp.
• *1959*	The planet corporation introduced a pick and place robot.
• *1961*	— The first industrial robot was commercialised by Unimation Inc.
	— Microprocessor technology was brought by INTEL.
• *1963*	Robot models of historical interest developed by AMF – Versatran.
• *1969*	General motors installed the first robot for spot welding automobiles.
• *1974*	Cincinnati Milacron T^3 indroduced – the first computer controlled industrial robot.
• *1978*	ASEA introduced its model IRB-6
• *1983-84*	— Bhabha Atomic Research Centre developed a 6-axes multipurpose robot.
	— A microprocessor-based pick and place robot designed in the Military college of Electronics and Mechanical Engineering, Hyderabad.
	— Southerland's hexapod of Carnegie Mellon by University – the first man-carrying computer-controlled walking machine by Raibert (1983) and Southerland and Ullner (1984)
	— BIPER-4 – A two-legged walking machine designed by Miura-Shimoyana.

Some of the pioneering institutions propagating the ideas and ideologies of the robotics to the profession are :

— Japan Industrial Robot Association (JIRA, 1971)

— Robot Institute of America (RIA, 1975)

— British Robot Association (BRA, 1977)

— Robotics International/SME, 1981.

Robot generations :

Based on characteristic breakthroughs in robot's capabilities, the growth of robots can be grouped into four robot generations as follows :

First generation (1960–onwards) :

— The robots belonging to this generation are *repeating, non-servo, pick-and-place, or point-to-point kind.*

— At present 80 percent robots in use in the industry are of this kind.

— It is predicted that their use will continue for a long period of time.

Second generation (1980–onwards) :

— These robots exhibit *path-control capabilities.*

— The technological breakthrough which came around 1980s is yet not mature.

Third generation (1990–onwards) :

The robots belonging to this generation have human like intelligence.

— The growth in computers led to high-speed processing of information and, thus, robots also acquired *artificial intelligence, self learning, and conclusion-drawing capabilities by past experiences.* The significant characteristics of these robots are : (*i*) On-line computations and control ; (*ii*) Artificial vision ; (*iii*) Active force/torque interaction with the environment.

— The technology is still in infancy and has to go a long way.

Fourth generation (2000–onwards) :

— This is futuristic and may be reality only during this millenium.

— Prediction about its features is difficult, if not impossible. It may be a true android or an artificial biological robot or super humanoid capable of producing its own clones. This might provide for fifth and higher generation robots.

1.7 OBJECTIVES OF USING INDUSTRIAL ROBOTS

The use of industrial robots is increasing day-by-day with a view to achieve the following main *objectives* :

1. To reduce production time.
2. To minimise the labour requirement.
3. To raise the quality level of products.
4. To increase productivity.
5. To improve existing manufacturing processes.
6. To enhance the life of production machines.
7. To minimise the loss of man-hours on account of accidents and diseases.
8. To make the reliability and applicability of new high-speed production processes and their related machinery possible.
9. To take advantage of fatigue-free continuous deployment of robots, because the human beings are always bound to experience fatigue when put to continuous working.

1.8 ADVANTAGES AND DISADVANTAGES OF ROBOTS

Following are the *advantages and disadvantages* of employing robots :

Advantages :

1. Lifting and moving heavy objects.
2. Working in hostile environments.
3. Providing repeatability and consistency.
4. Working during unfavourable hours.
5. Performing dull or monotonous jobs.
6. Increasing productivity, safety, efficiency and quality of products.
7. Achieving more accuracy than human beings.

Disadvantages :

1. The robots lack capability to respond in emergencies.
2. The initial and installation costs of equipments of robots are quite high.
3. They replace human workers, thus causing resentment among workers.

1.9 TYPES OF INDUSTRIAL ROBOTS

Industrial robots can be broadly *divided* into two main groups as follows :

1. General purpose robots.
2. Special purpose robots.

1. General purpose robots :

- These robots carry standard designs and parts and are readily available.
- They can be easily adapted to the users' requirements by attaching suitable end-effectors or fingers to them according to the requirement of the work, such as a part picking operation, welding operation, spray painting, etc.
- Since such robots are mass produced, they are cheaper.

2. Special purpose robots :

- These robots are tailor made to specific job requirements. The ultimate user has to feed his requirements and, based on them, these robots are specially designed and built to cater to such specific needs. Obviously, their designing and manufacturing consumes a lot of time. As such, they can not be readily available in market.
- Since they can not be manufactured on mass scale, their *prices* are bound to be *higher*.

Robots may also be categorised as follows :

1. *Tele-robots* :

   — These are guided by human operators through remote control.

2. *Telepresence robots* :

   — This type of robot is similar to a tele-robot with a difference that it is provided with a feedback of video, sound and other data.

3. *Mobile robots* :

   These robots are so designed as to navigate and carry out the task with the intervention of human beings.

4. *Autonomous robots* :

   These robots perform their tasks independently and receive their power from their environment.

5. *Androids* :

   These robots are built to mimic humans.

6. *Static + industrial robots* :

   These are the widely used arms employed in factories and laboratories all over the world.

1.10 ROBOTIC SYSTEMS

A **system** *is an integral whole of parts or subsystems.* It has a specific goal or output for a given set of inputs; a system may have many goals as well.

• A "robot" is a system as it combines several subsystems that interact themselves as well as with the environment in which the robot works.

— A robot has some specific objective. It may be designed for the following jobs/ assignments :

(*i*) To simply pick up and place the workpieces;

(*ii*) To interact with and work load a lathe, a milling machine or any equipment;

(*iii*) To perform some assembly work.

To accomplish these assignments a robot should have the following *"components"* :

— A suitable *manipulator arm* with specified coordinate systems to attain a designed reach in the working space.

— A suitable *gripper* to match the geometry of the workpiece to be handled.

— A suitable *control system* with or without servo-mechanisms for sending signals to the drives, or permitting storage of programmes and data for desired path planning with adequate speed and good accuracy.

— Some *sensors* to feed back information for modifying the motion or path.

— A *controller* is provided with interfacing units connected to external equipment in the outside world.

• Fig. 1.1 shows a scheme of robotic system.

Robot components : Refer to Fig. 1.1

The various components of a robot are enumerated and discussed below :

1. Base.
2. Manipulator arm.
3. End-effector.
4. Actuators and transmissions.
5. Controller.
6. Sensors.

1. Base :

— The base may be fixed or mobile.

2. Manipulator arm :

— The most obvious mechanical configuration of the robot is the manipulator arm.

— There are *several designs of the arm* to facilitate movement within the work envelope with maximum possible load and speed with high precision and repeatability.

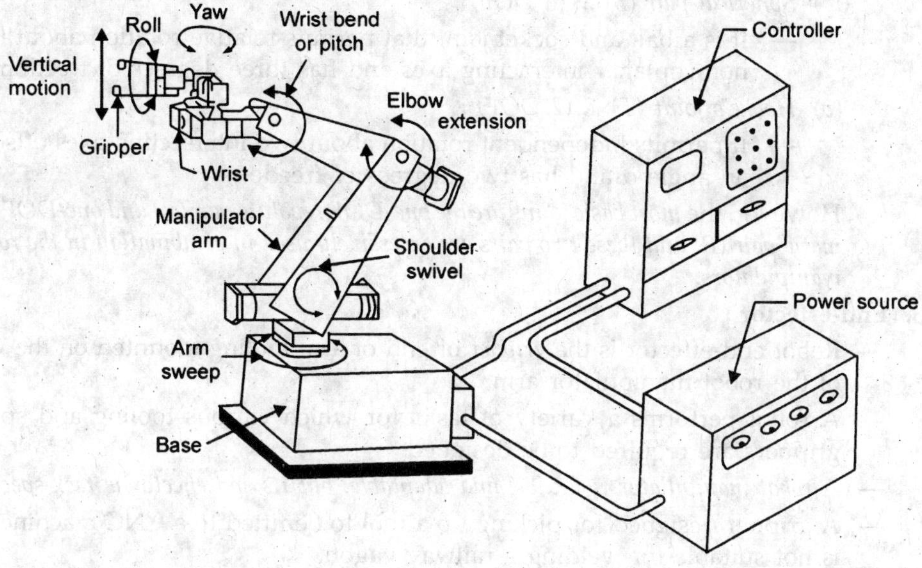

Fig. 1.1. Robotic system — Main components of a robot and the basic motions.

— The simplest robot may be a two or three-axes arm. *The axis is meant to understand independent movement or degree of freedom* (**DOF*).

— A robotic manipulator arm consists of several separate links making a *chain*. The arm is located relative to the ground on either a fixed base or a movable base. It has a free-end where an end-deflector or gripper or sometimes a specialised tool holder (for holding, say, a welding gun) or any powered device (say, a drill) is attached.

— In a *fixed base, six* degrees of freedom robot, the first three links of the manipulator constitute the *body* and they help to place the end-effector at a desired location inside its work environment or working volume. The remaining three links make up the *wrist* of the manipulator and are used to define the orientation of the manipulator end points.

- A robot is essentially a movable open chain of successively coupled bodies with one end fixed to the ground and the free end containing an end-effector. The bodies of the open chain are usually links which are joined together by some lower pair connectors. The most common types of lower pair connectors are :

  (*i*) *Revolute pair (R) ... (1 DOF) :*

  — It permits relative rotation about a unique pair axis and has a single degree of freedom.

  (*ii*) *Prismatic pair (P) ... (1 DOF) :*

  — It allows relative sliding parallel with a unique pair axis and has one degree of freedom.

  (*iii*) *Cylindric pair (C) ... (2 DOF) :*

  — It permits independent relative rotation about and relative sliding parallel to a pair axis and it has two degrees of freedom.

* The number of independent movements that an object can perform in a 3-D space is called the number of degrees of freedom (DOF).

(*iv*) *Spherical pair (S) ... (3 DOF) :*

— It is a ball and socket joint that permits relative rotation about three non-coplanar interacting axes and has three degrees of freedom.

(*v*) *Hooke's joint (T) ... (2 DOF) :*

— It permits independent rotation about two intersecting axes offset by an angle α and has two degrees of freedom.

However, the *most basic joints are the one-DOF revolute pair (R) and one-DOF prismatic pair (P) and these two pairs are excessively used in combination in the robotic manipulators.*

3. End-effector :

— Robot end-effector is the *gripper* or end of arm tooling mounted on the wrist of the robot manipulator arm.

— A robot performs a variety of tasks for which various tooling and special grippers are required to be designed.

— *A robot manipulator is flexible and adaptable, but its end-effector is task-specific.*

— A gripper designed for picking up a tool to be fitted to a CNC machine tool is not suitable for welding a railway vagon.

The wide range of gripping methods include :

(*i*) Mechanical clamping ;

(*ii*) Magnetic gripping ;

(*iii*) Vacuum (suction) gripping.

4. Actuators and Transmissions :

Actuators :

The robot arm can be put to a desired motion with its payload if actuator modules are fitted in to provide *power drives* to the systems.

There are three different types of power drives in common use. They are :

(*i*) ***Pneumatic drives :***

— These systems *use compressed air to move the robot arm.*

— The pneumatic systems may employ a *linear actuator, i.e.,* double acting cushioned cylinders or it may *employ rotary* actuators like vane motors. However, linear actuators are more popular.

— The *"advantages"* of pneumatic actuators are : *Simple construction,* relatively inexpensive, fast and reliable. The *"disadvantages"* of pneumatic system are smaller payloads, the *mass inertia and delayed response of the robot arm* due to the sponginess and reduced repeatability.

- *Non-servo robots* can be built up with pneumatically powered actuators.

(*ii*) ***Hydraulic drives :***

— In a hydraulic system, the electric motor pumps fluid (oil) from a reserve tank to the hydraulic actuators which are, in general, double acting piston-cylinder assemblies. Fluid at a higher pressure passes through control valves before its entry into the *linear actuators.* On the other hand, *rotary actuator* comprising some motors or hydraulic motors which rotate continuously may also be employed.

- The hydraulic drives have *high payload capacities* and are relatively *easy to maintain.* They are, however, rather *expensive* and *not* as *accurate* as either the pneumatic or electric drives.

(*iii*) ***Electrical drives :***

— These drives are *clean and quiet with a high degree of accuracy and reliability*. They also offer a *wide range of payload capacity*, accompanied by an equally *wide range of costs*.

- *D.C. servo motors, Brushless D.C. motors, Reversible A.C. servo motors and Stepper motors* are important electrical drives.

Transmissions :

"Transmissions" are elements between the actuators and the joints of the mechanical linkage.

They are generally used for the following three *reasons* :

(*i*) Often the actuator output is not directly suitable for driving the robot linkage.

Example : The high speed D.C. motor running at 3000 r.p.m. (say) may not be suitable for running a robot at slower speeds. However, with appropriate gearing or transmission, the speed may be reduced to 30 r.p.m. (*i.e.* $\frac{1}{2}$ rotation per second) which is reasonably fast. In addition, the rated torque at 3000 r.p.m. is amplified by 100 (assuming a highly efficient gearbox).

(*ii*) The output of the actuator may be kinematically different from the joint motion.

(*iii*) The actuators are usually big and heavy and often it is not practical to locate the actuator at the joint.

— Firstly, big actuators have large inertias and they are harder to move around in space than the links that comprise the mechanical linkage. So it is desirable to locate them at a fixed base.

— Secondly, because of their size, they can impede the motion of one or more links of the robot.

Thus, it is not uncommon to find linkages or gear train that transmits the power from the actuator over a large distance to the joint.

5. Controller :

The "controller" provides the intelligence that is necessary to control the manipulator system.

It looks at the sensory information and *computes the control commands* that must be sent to the actuators to carry out the specified tasks. It generally includes :

(*i*) *Memory* to store the control program and the state of the robot system obtained from the sensors.

(*ii*) A *computational unit* that computes the control commands.

(*iii*) The appropriate *hardware* to interface with the external world (sensors and actuators)

(*iv*) The *hardware for a user interface.*

- The *"user interface" allows the use of a human operator to monitor or control the operation of the robot.*

— It must have a display that shows the status of the system.

— It must also have an input device that allows the human to enter commands to the robot.

The user interface may be a *'personal computer'* with the *'appropriate software'* or a *'teach pendant'*.

6. Sensors :

The sensors perform the following *functions* :

(*i*) To act as feedback devices to direct further actions of the manipulator arm and the end effector (gripper), and (*ii*) To interact with the robot's working environment.

Usually there are two basic types of sensors. These are :

(*i*) **Tactile sensors :**

These are *"contact sensors" that must be brought in contact with the object to obtain signals to measure the necessary qualities.*

— When the tactile sensors make physical contact with the object, an electrical analog or digital signal is generated and sent to the robot controller. Electrical signals may be obtained through the contacts of *microswitches.* Signals may also be obtained through mechanical pressures which change resistances of electrical strain gauges or general electrical potentials in piezoelectric crystals.

— Typical contact type robotic sensors include :

(*a*) Force sensors ; (*b*) Torque sensors ;

(*c*) Touch sensors ; (*d*) Position sensors

(*ii*) **Non-tactile sensors :**

These are *"contactless sensors" which sense the signals remotely,* but only within the specified range of distance from the object.

— They detect and measure magnetic fields, infrared and ultraviolet light, X-rays, electrical fields, ultrasonic sound waves or electromagnetic waves.

— Typical non-contact robotic sensors include :

(*a*) Proximity sensors ;

(*b*) Electro-optical sensors ;

(*c*) Range imaging sensors

Basic Motions : Refer to **Fig. 1.1.**

The *six basic motions* or degrees of freedom (DOFs) are as follows : Refer to Fig. 1.1.

1. ***Vertical motion.*** The entire manipulator arm can be moved up and down vertically either by means of the *shoulder swivel, i.e.,* turning it about a horizontal axis, or by sliding it in a vertical slide.

2. ***Radial motion.*** Radial movement, *i.e., in* and *out* movements, to the manipulator arm is provided by *Elbow extension* by extending it and drawing back.

3. ***Rotational motion.*** Clockwise or anticlockwise rotation about the vertical axis to the manipulator arm is provided through *Arm sweep.*

4. ***Pitch motion.*** It enables up and down movement of the wrist and involves rotational movement as well. It is also known as *wrist bend.*

5. ***Roll motion.*** Also known as *wrist swivel,* it enables rotation of wrist.

6. ***Yaw.*** Also called *wrist yaw,* it facilitates rightward or leftward swivelling movement of the wrist.

The most versatile robots can have following degrees of feedom (DOFs) :

(*i*) Horizontal travel.

(*ii*) Rotary movement.

(*iii*) Radial arm movement.

(*iv*) Vertical arm movement.

(*v*) Rotary wrist movement.

(*vi*) Wrist bend.

(*vii*) Wrist sweep.

These axes of movement enable movements to be programmed that duplicate those of a human operator in performing a job.

- **Fig. 1.2.** shows the *basic components of microprocessor-based pneumatic robotic system.*

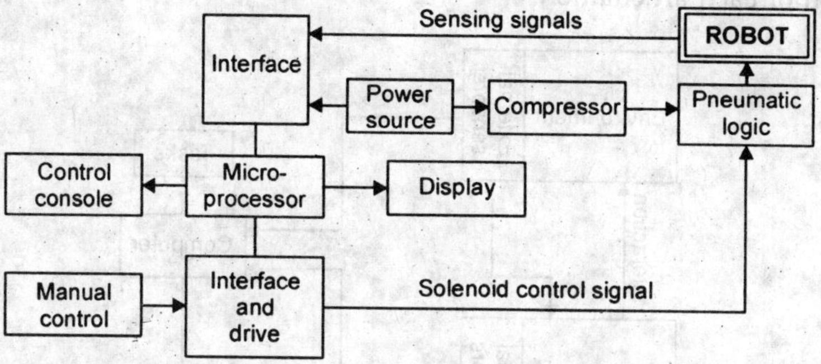

Fig. 1.2. Basic components of a microprocessor-based robotic system.

- **Fig. 1.3(*a*)** *shows a six axes puma manipulator and* **Fig. 1.3(*b*)** *shows a scheme of computer-controlled puma robotic system.*

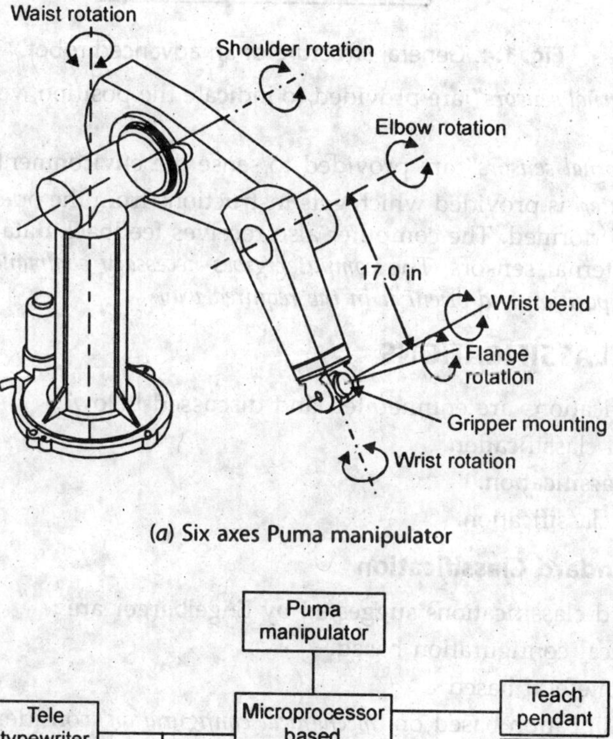

(*a*) Six axes Puma manipulator

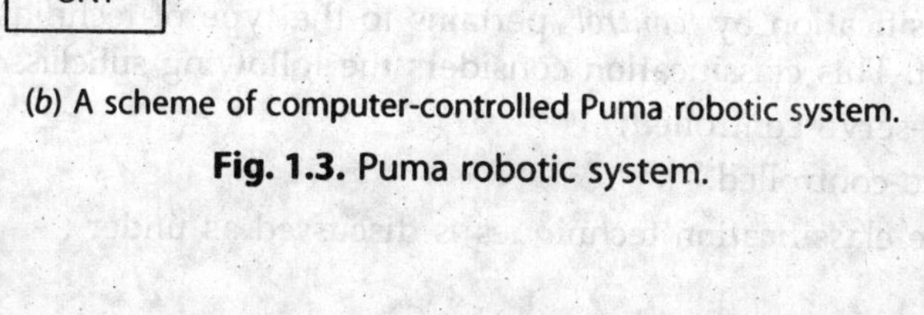

(*b*) A scheme of computer-controlled Puma robotic system.

Fig. 1.3. Puma robotic system.

● **Fig. 1.4.** *Shows a general structure of an advanced robot.*

— The operational unit consists of *"articulated mechanical system"* (AMS – comprising of rigid links and kinetic joints), transmission system and actuators which control the configuration of each articulation.

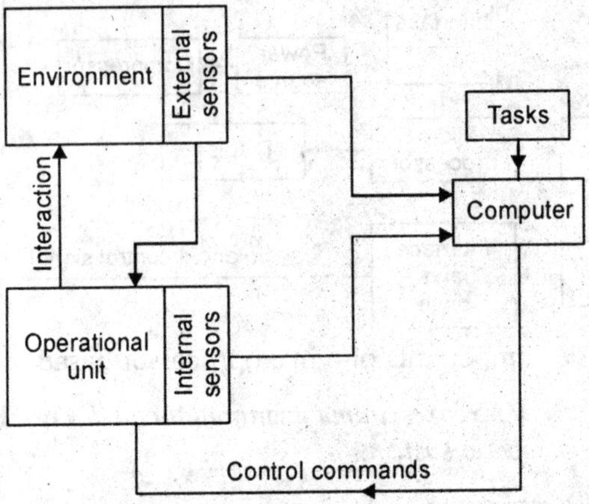

Fig. 1.4. General structure of an advanced robot.

— The *"internal sensors"* are provided to indicate the position, velocity and forces of the end-effector.

— The *"external sensors"* are provided to sense the environment.

— A *'computer'* is provided which gets instructions from the operator depending on the tasks to be performed. The computer also receives feedback data regarding the joint variable from internal sensors. The *computer gives necessary commands to locate the end-effector at desired position and orient it in the required way.*

1.11 ROBOT CLASSIFICATIONS

Robot classifications are enumerated and discussed below :

1. Standard classification.
2. Broad classification.
3. General classification.

1.11.1. Standard Classification

Two standard classifications suggested by Engelberger are :

1. Mechanical configuration based.
2. Control method based.

● The classification based on *'mechanical configuration'* considers the various joints and links comprising the physical structure of the robot and their relationship to each other.

● The classification by *'control'* pertains to the type of technique implemented to control the robot. This classification considers the following subclasses :

— Non-servo controlled;

— Servo-controlled.

Ease of these classification techniques is discussed as under :

1.11.1.1. Mechanical configuration

The majority of commercially available robots can be grouped into four basic configurations :

(*i*) Cartesian coordinate configuration.

(*ii*) Cylindrical configuration.

(*iii*) Spherical configuration.

(*iv*) Jointed-arm configuration (Revolute).

(*i*) ***Cartesian coordinate configuration*** : Refer to Fig. 1.5.

- The cartesian or rectilinear robot also termed as gantry robot, has three mutually perpendicular axes which define a rectangular work volume.
- In this, the simplest of configurations, the links of the manipulator are constrained to move in a linear manner. Axes of a robotic device that behave in this way are referred to as "*prismatic*".
- The cartesian devices may be of two types :

(*a*) *Cantilevered cartesian* :

— Such devices tend to have a limited extension from the support frame, are less rigid, but have a less restricted workspace than other robots.

— They have good repeatability and accuracy (even better than SCARA types) and are easier to program because of the "more natural" coordinate system.

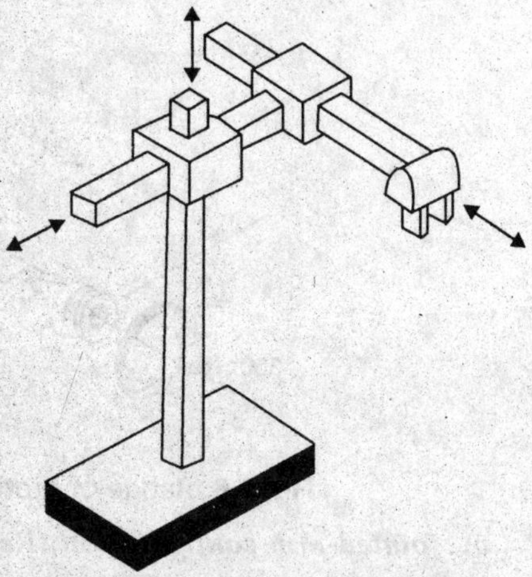

Fig. 1.5. Cartesian configuration.

(*b*) *Gantry-style cartesian* :

— Such robots are used when extremely heavy loads must be precisely moved.

— They are often mounted on the ceiling.

— They are generally more rigid but may provide less access to the workspace.

(*ii*) ***Cylindrical configuration*** : Refer to Fig. 1.6.

— Cylindrical configured robots use a vertical column with the robot arm attached to a side which can move up and down the column. Simultaneously, the arm can move radially with respect to the column.

— Usually, a full 360° rotation in θ is not permitted, due to restrictions imposed by hydraulic, electrical, or pneumatic connections or lines. Also, there is minimum, as well as maximum extension (*i.e.* *R*), due to mechanical requirements. Consequently, the overall volume or work envelope is a portion of a cylinder.

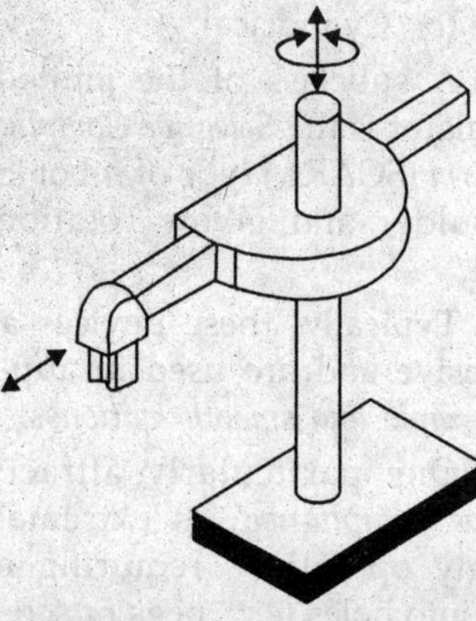

Fig. 1.6. Cylindrical configuration.

(*iii*) ***Spherical (or polar) configuration*** : Refer to Fig. 1.7.

— This configuration has a telescopic arm which pivots about a horizontal axis and also rotates about a vertical axis.

— Owing to mechanical and/or actuator connection limitations, the work envelope of such a robot is a portion of sphere.

Fig. 1.8 shows the working volume of spherical coordinate robot.

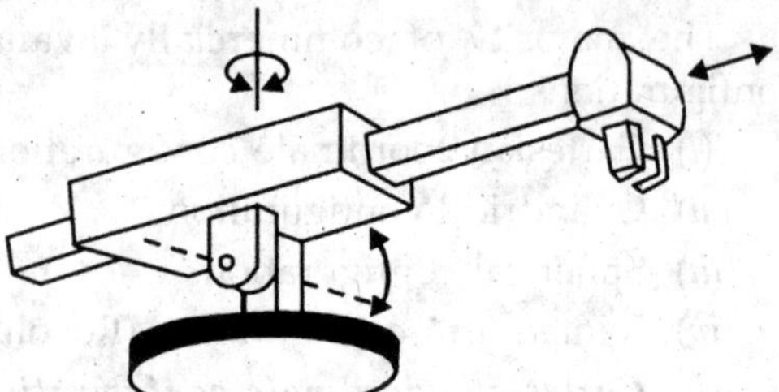

Fig. 1.7. Spherical configuration.

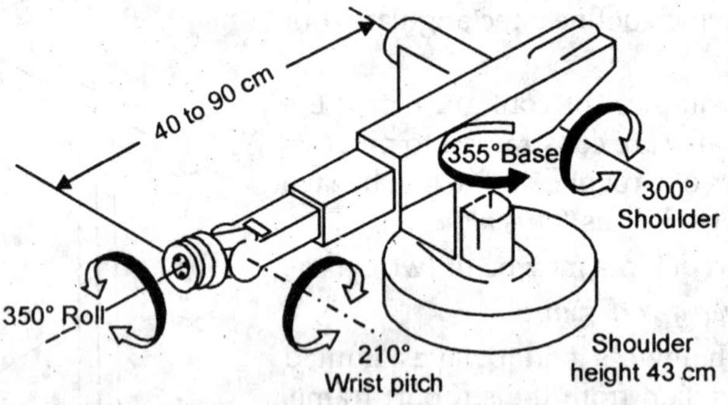

Fig. 1.8. Range of motion of a spherical coordinate robot.

(*iv*) ***Jointed-arm configuration (Revolute)*** : Refer to Fig. 1.9.

— The jointed-arm robot most resembles an human arm and consists of a series of links connected by rotary joints which when referenced from base are referred to as the shoulder, arm and wrist joints.

— There are actually three different types of jointed arm robots :

(*a*) Pure spherical.

(*b*) Parallelogram spherical.

(*c*) Cylindrical.

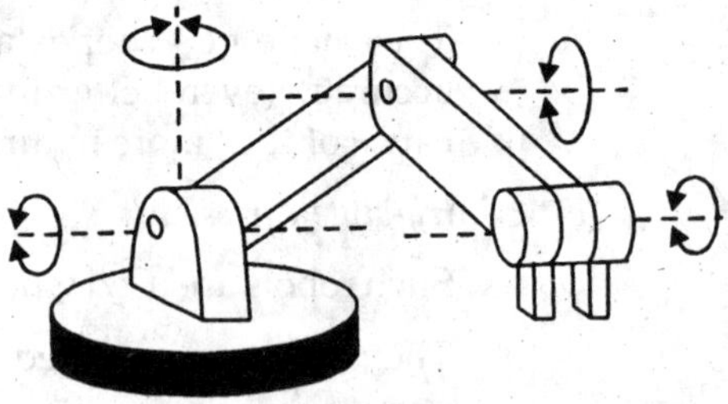

Fig. 1.9. Jointed-arm configuration.

• A subclass of the jointed cylindrical manipulator is the *Selective Compliance Assembly Robot Arm* (SCARA) type of robot (See Fig. 1.10); its shoulder and elbow rotational axes are vertical.

— Typically, these devices are relatively inexpensive and are used in applications that require *rapid and smooth motions*.

— One particularly altractive feature, '*selective compliance*', is extremely useful in assembly operations requiring insertions of objects into holes (*e.g.*, pegs or screws). Because of its construction, the SCARA is *extremely stiff*

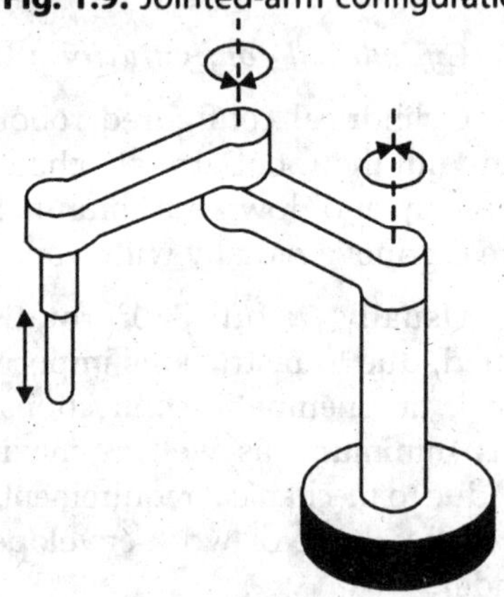

Fig. 1.10. SCARA body-and-arm assembly configuration.

in the vertical direction but has some lateral "give" (*i.e.* compliance), thereby facilitating the insertion process.

— Some SCARAs even permit the lateral compliance to be increased during an operation by reducing appropriate electronic amplifier gains.

1.11.1.2. Control classification

Several techniques have been developed to control the various axes of a robot simultaneously. The two general classes are :

(*i*) ***Non-servo controlled robots :***

— Non-servo-controlled robots are also called *limited-sequence robots, end-point robots, pick and place robots* or *bang-bang robots.*

— Such a robot is controlled by setting mechanical stops or limit switches to establish end points of travel of each joint. The mechanical set-up to give the proper position and sequence of stops serves as a rudimentary programming approach rather than a computer-intensive robot programming language.

(*ii*) ***Servo-controlled robots* :**

— These robots are normally sub-divided into either 'continuous-path' (CP) or 'point-to-point' (PTP) devices. In either case, however, information about the position and velocity (and perhaps other physical quantities) is continuously monitored and fed back to the control system associated with each of the joints of the robot. Consequently, each axis loop is *"closed"*.

— Use of closed-loop control permits the manipulator's members to be commanded to move and stop *anywhere* within the limits of travel for the individual axes.

— In addition, it is possible to control the velocity, acceleration, deceleration, and jerk (*i.e.*, time derivative of acceleration) for various axes between the end points. Manipulator vibration, as a consequence, is reduced significantly.

- Regardless of configuration, the function of the robot arm configuration is to position a wrist assembly which orients an end-effector to the proper position and orientation (jointly referred to as the POSE) dictated by the task at hand.

Fig. 1.11 shows the degrees of freedom (DOFs) associated with a robot *wrist*.

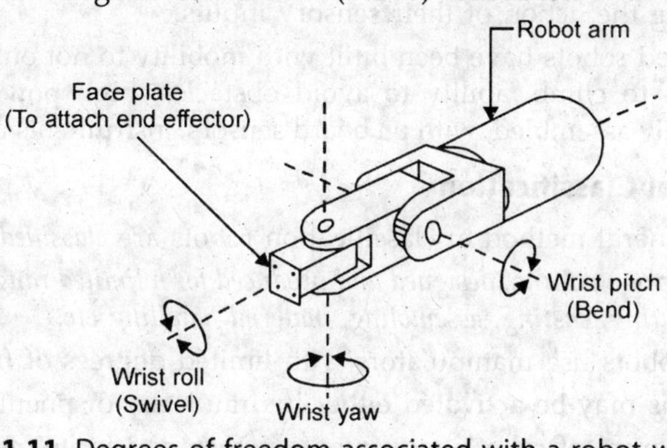

Fig. 1.11. Degrees of freedom associated with a robot wrist.

- ***Mechanical inaccuracies*** result from the various errors that can occur in a robot's construction or operation. These include *gear backlash, stretching of pulley cords, leakage of fluids* or *materials and structural imperfections*. Such errors degrade the overall accuracy of the robot and are magnified when the arm is fully extended. Mechanical inaccuracies are principally responsible for repeatability errors. A robot manufacturer typically specifies repeatability as the radius of an idealised sphere and expresses the specification as a plus or minus value. The majority of robots have extremely high repeatability (can be ± 0.0025 cm or less) but many of these robots can have comparatively greater inaccuracy values.

1.11.2. Broad Classification

Robots can be broadly *classified* as follows :

1. ***Programmable/reprogrammable general purpose robots :***
   — These robots operate fully by programmed computer control.
   — The robot is taught before hand to perform the necessary action in the teach mode. It can then take over and execute the operation repetitively such as in welding, painting, assembly of components for mass manufacture, loading/ unloading of jobs into and from machine tools etc.
   —These are most useful for all structured operations.
2. ***Tele-operated, man-controlled robots or man-in-the-loop manipulator :***
   — These robots differ from totally machine-controlled robots in the sense that the advantage of presence of man is taken in situations where it is not possible to anticipate all the motion and handling requirements in such details as to render them programmable or teachable for machine control. This type of requirement is found in *hazardous locations*.
   — The servodriven master-slave manipulator with force feedback, or vehicle mounted heavy duty multi-axis power manipulator performs the necessary work in hazardous environment, taking commands from a human controller who can manipulate the slave arms at the scene of operation from safe location, relying for viewing on closed circuit television.
3. ***Intelligent robots :***
   — These are very advanced state of art robots and *possess adequate artificial or machine intelligence*, somewhat analogous to the sensory perception of the neuro-muscular coordination that human beings are capable of.
   — They can not only explore the environment on their own machine perceptions and evaluate them in real time, but also execute the necessary motor functions matching the action of their sensory inputs.
   — Advanced robots have been built with mobility to not only move over floors but also to climb, ability to avoid obstacles, high power-to-weight ratios, compactly assembled, with on board sensors, instruments and power supplies.

1.11.3. General Classification

According to general method of classification robots are *classified* as given below :

1. **Special purpose robots**, *designed and produced for a limited range of specific jobs, like welding, painting, casting, assembling, material handling etc.*
   — These robots use manipulators with limited degrees of freedom.
   — The arms may be activated either hydraulically or pneumatically.
   — The program of robot's movements is recorded in its memory in sequence.
2. **General purpose robots**, *designed and produced to form a wide variety of jobs.*
   — These may be non-servo-controlled, servo-controlled or sensory type depending on sophistication.

(*i*) Non-servo-controlled robots :
   — These robots are *simplest type* and have *inherent limited control sequence characteristics.*

— Their arms are operated by an open loop between exact end point positions on each axis. In *point-to-point control*, only end points are defined and how these are achieved, i.e., how the arm moves to reach end points is of no concern. However, in *trajectory control*, all points between start and end points are predefined.

— Arm's motion may be required along straight lines, circles, spirals, or arbitrary curves or combination of these to reach from point to another.

(*ii*) ***Servo-controlled robots*** :

— These robots utilise the position velocity, acceleration, force and torque sensors additionally in order to estimate the internal states of the manipulator. If the states at any time are in deviation to predetermined operational parameters of the control program, then the corrective action is taken to reduce the deviation to zero. It is possible to operate them with precision of the order of 0.1 to 1 mm movement.

— Robots capable of approximating the motions of a human arm with high lifting capacity and variable lifting speed are available.

— Robots capable of calculating automatically the shortest route between two points with given constraints are also available.

(*iii*) ***Sensory robots*** :

— These robots in *addition to internal control sensors, employ external sensors* like TV cameras, pressure detectors, magnetic sensors, force torque sensors, laser range finders etc.

— The sensors find out the robots' relation to its environment and determine the location of the parts to be handled.

— The computer based on information from sensors, effect appropriate program modifications thus coping with changing requirements and unpredictable situations.

— Robots with arms having eight degrees of freedom and capable of identifying a given object from a random collection of objects have been produced.

— Robots working automatically under the direction of human voice have also been developed.

— Robots with vision system are used for inspecting and sorting parts for automatic assembly at high speed.

— Some robots work on pattern recognition principle by selecting distinctive parts of the object, and operating on this information using advanced software algorithms.

— Robots with *proximity sensing* (sense without functioning) and *tactile sensing* (sensing by grasping gently) have also been developed.

- One of the most fundamental classification schemes is based upon the source of power used to drive the joints of the robot. The two most popular "drive technologies" are :

  (*i*) Electric drives;

  (*ii*) Hydraulic drives.

  — Many robotic manipulators today use electric drives in the form of either D.C. servomotors or D.C. stepper motors. However, *when high-speed manipulation of substantial loads is required, such as in molten steel handling or auto body part handling "hydraulic drive robots" are preferred.*

One serious drawback of hydraulic-drive robots lies in their *lack of cleanliness*, a characteristic that is important for many assembly applications.

— Both electric-drive robots and hydraulic-drive robots often use *pneumatic tools* or end-effectors, particularly when the only gripping action required is a simple open-close type of operation.

— An important characteristic of an *"air-activated tools"* is that they exhibit *built-in compliance in grasping objects*, since air is compressible fluid. This is in contrast to sensorless rigid mechanical grippers, which can easily damage a delicate object by squeezing too hard.

Selection of a robot :

The following factors should be considered while selecting a robot for a particular application :

1. Load capacity required.
2. Adequate number of degrees of freedom.
3. Speed of movement.
4. Reach.

● The robots may be powered electrically, hydraulically or pneumatically. The electrically powered robots have the advantage of accurate movements and quiet operation.

1.12 MECHANICAL DESIGN OF A ROBOT

— The mechanical design of a robot is an *iterative process* involving evaluation and choice among a large number of engineering and technical considerations in several disciplines.

— A purely static, rigid-body approach to design is not sufficient and factors like mechanical system stiffness, natural frequencies, control system compatibility also need to be considered.

Design considerations :

The various design considerations include the following :

1. *System specification* :
   — Reach;
   — Range;
   — Work envelope;
   — Load capacity.
2. *System configuration* :
   — Number of degrees of freedom;
   — Joint configuration;
   — Joint travel range;
   — Drive configuration.
3. *System performance* :
   — System velocity and acceleration;
   — Repeatability;
   — Resolution;
   — Accuracy;
   — Component life and duty cycle.

- Detailed design of major components relates to :
  — Robot structures:
  — Robot joints;
  — Actuators;
  — Transmission;
  — Wiring and routing of cables and hoses.

It is advantageous to evaluate the possible flexibility of the robot, grippers, tools, and peripheral units and integrate all components to one system.

1.13 TYPES OF MECHANICAL JOINTS

The function of a **joint** *is to permit relative motion between two links or arms of a robot*. It provides controlled relative movement between two links (input and output).

Generally one joint provides the robot with one degree of freedom. The robots are usually classified according to number of degrees of freedom possessed by them.

The various types of mechanical joints are discussed below : Refer to Fig. 1.12.

1. **Linear joint – L-joint** [See Fig. 1.12(*i*)]
   — This joint permits linear sliding motion between two links whose axes are parallel.

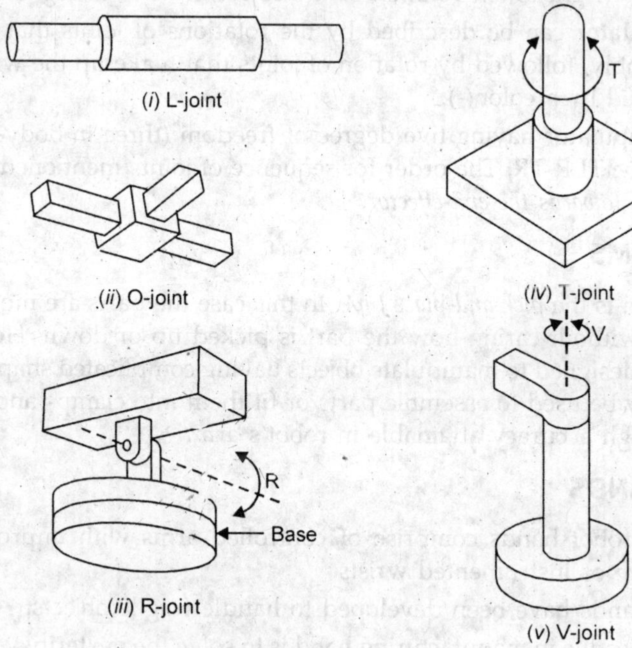

Fig. 1.12. Types of mechanical joints.

2. **Orthogonal joint – O-joint** [See Fig. 1.12(*ii*)]
   — In this type of joint the two links are *perpendicular* to each other but motion between them at the joint is *linear sliding type*.
3. **Rotational joint – R-joint** [See Fig. 1.12(*iii*)]
   — This joint provides *rotational relative motion* of the joints, with the axis of rotation *perpendicular* to the axes of the two links.

4. **Twisting joint – T-joint** [See Fig. 1.12(*iv*)]
   — This joint permits *rotary motion* between two links, the axis of rotation being *parallel* to the axes of the two links.

5. **Revolving joint – V-joint** [See Fig. 1.12(*v*)]
   — This joint also provides rotary motion, but the axis of the input link is parallel to the axis of rotation of the joint, and the axis of output link is perpendicular to the axis of rotation.

- A robot, in a very general way, could be considered to be made up of the following two sections :

(*i*) Body-and-arm. (*ii*) Wrist assembly.

— The object to be moved is handled by the manipulator's wrist with the help of end-effector.

— The *body-and-arm* help to position the object (by having three degrees of freedom of moving the object in vertical motion, *i.e.*, "*Z-axis motion*", radial or in-and-out, *i.e.*, "*Y-axis motion*", and right-to-left motion, *i.e.*, "*X-axis motion*" or a swivel about a vertical axis).

— The *wrist* helps to orient the object by having *three* degrees of freedom, *viz.* "*roll*" (rotation of object about the arm axis), "*pitch*" (up-and-down rotation of object about the arm axis), and "*yaw*" (right-to-left rotation of the object) ...Refer to Fig. 1.11.

- The **manipulator** can be described by the rotations of joints that make up the body-and-arm assembly, followed by rotation of joints that make up the wrist assembly. The two are separated by a colon(-).

Example. A manipulator having five degree of freedom (three in body-and-arm and two in wrist) could be TLR-TR. The order for sequence of joints mentioned is by *starting from base and moving towards the end-effector.*

1.14 ROBOT ARMS

The simplest arm is the *pick-and-place type*. In this case the parts are moved from one location to another without caring how the part is picked up or down. However, these days robot arms are designed to manipulate objects having complicated shapes and fragile in nature. These may be used to assemble parts or fit them into clamps and fixture. This is possible due to high accuracy attainable in robot's arm.

1.15 ROBOT HANDS

— These days robot hands comprise of controlled arms with improved grippers backed up by passive or instrumented wrists.

— Dexterous hands have been developed to handle fragile objects.

— The main objective in manufacturing hand is to solve the dexterity versus strength trade off in the context of machine tools and flexible manufacturing/assembly systems. *The aim is to build a hand with adequate manipulating abilities without unduly sucrificing power.*

— The wide range of manipulation tasks to be performed calls for an array of special purpose grippers. These days *three* types of gripper are used :

(*i*) For precision grasps with opposed fingers;
(*ii*) For power wrap grasps;
(*iii*) For pinch grasps.

These three basic designs can be modified to fit a wide variety of part shapes, and finally, finger adapters may be used for specific geometric constraints encountered with any unusual parts.

— *'Dexterous hands with active fingers'* are also available; these can be used for tasks like screwing a large nut into a bolt.

— 'Compliant wrist units' that can be mounted between the gripper and the robot arm are available. These possess the ability to deform in response to forces and torques and return to equilibrium position after the deflecting forces are removed.

1.16 ROBOTS QUALITIES

The following qualities are expected in robots :

1. **Tactile sensing :**
   — Robot with tactile sensor can identify an object and perform the function based on the referenced data.
   — Grippers have been developed which can pick up any shape of objects and at the same time not exert enough force to crush them.
2. **Vision :**
   — By incorporating vision systems, the utility of robots is increased several folds.
   — Vision systems capable of identifying the part for pick up by pattern recognition data based on object's silhoutte have been developed. Such systems can *transform the position and orientation of the object into robot coordinates enabling the robot to acquire the object in a known manner.*
   — Other types of vision systems can recognise different objects.
   — In another vision system, a fiber sensor is used to look at a seam to be welded and automatically adjusts the weld path of the robot.
3. **Mobility :**
   — Generally the robot stands in a single station for the bulk of factory requirements. However, to handle intermittent and asynchronous demands, compact mobile device which would move in complex paths and access large areas economically has been developed.
4. **Other important qualities :**
   — General purpose hands ;
   — Computer interpretation of the visual and tactile data ;
   — Multiple appendage hand-to-hand coordination ;
   — Minimised spatial intrusion ;
   — Man-robot voice communication ;
   — Inherent safety ;
   — Interaction with other techologies ;
   — Total self-diagnostic fault tracing etc.

1.17 ROBOT SPECIFICATIONS

The various *specifications* of robot are enumerated and discussed below :

1. Work envelope.
2. Load carrying capacity.
3. Speed of movement.
4. Repeatability.
5. Control resolution.
6. Spatial resolution.

7. Mechanical errors.
8. Accuracy.
9. Stability.

1. Work envelope :

"Work envelope" (or work volume of a manipulator) is defined as the envelope or space within which the robot can manipulate the end of the wrist.

— It depends on the number of types of joints, physical size of the joints and links and the ranges of various joints.

— Since the arm movements in different robot configurations are different, the work volumes or work envelopes of different coordinate systems are also different.

Example : The work envelopes of cartesian, cylindrical, spherical and jointed arm coordinate robots respectively will be as shown in Fig. 1.13 to 1.16.

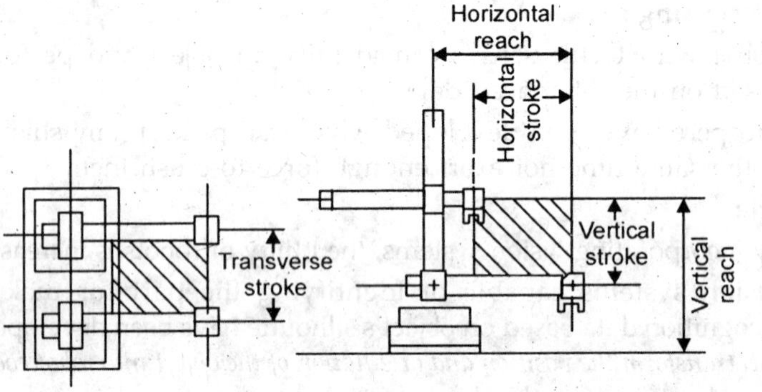

Fig. 1.13. Work envelope of a cartesian coordinate robot.

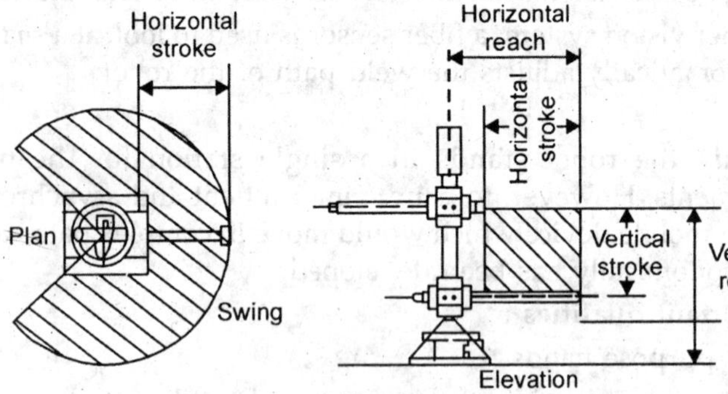

Fig. 1.14. Work envelope of a cylindrical coordinate robot.

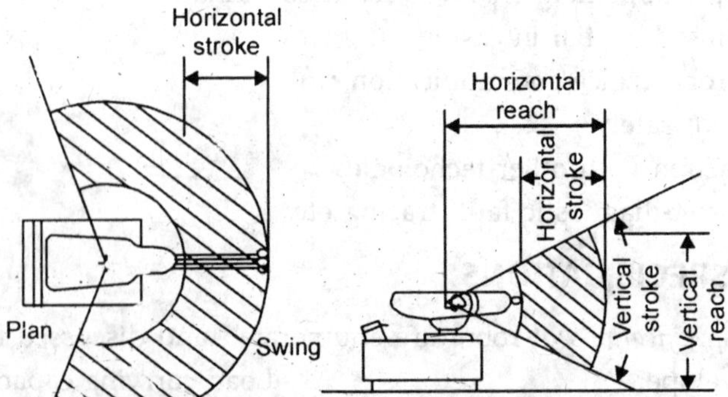

Fig. 1.15. Work envelope of a spherical coordinate robot.

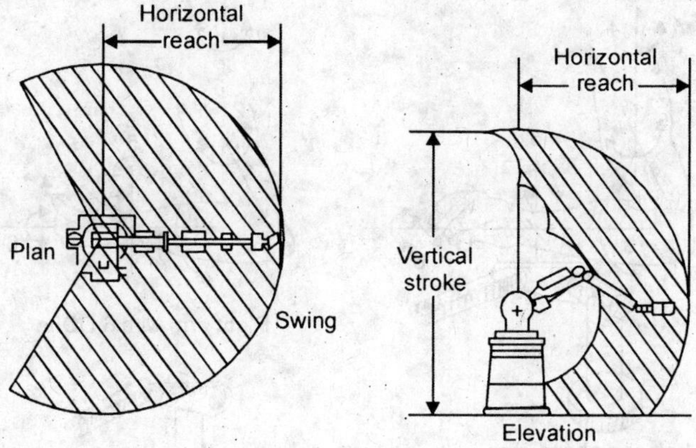

Fig. 1.16. Work envelope of a jointed-arm coordinate robot.

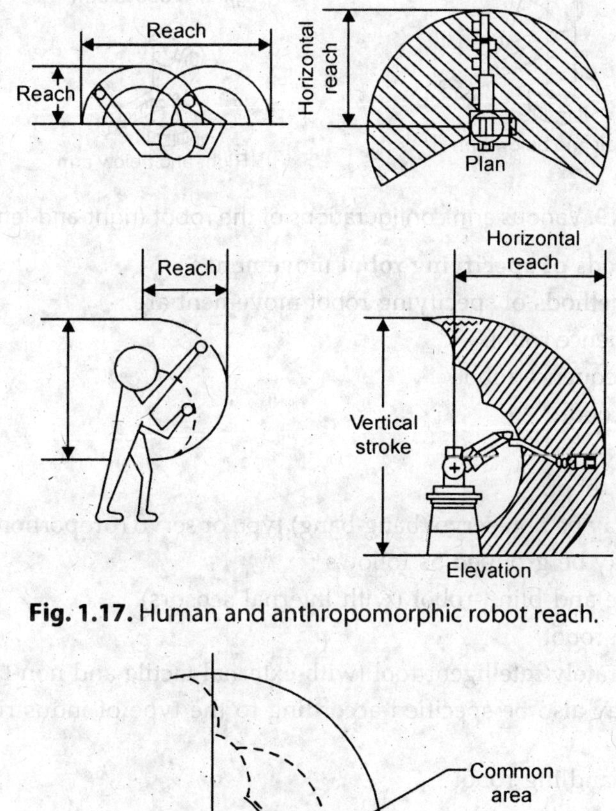

Fig. 1.17. Human and anthropomorphic robot reach.

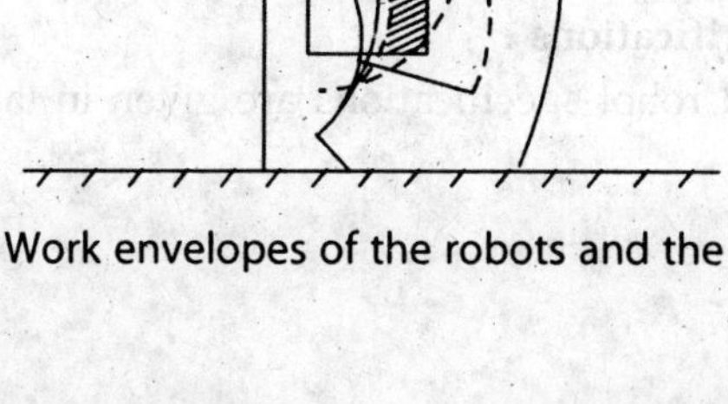

Fig. 1.18. Work envelopes of the robots and the human operative.

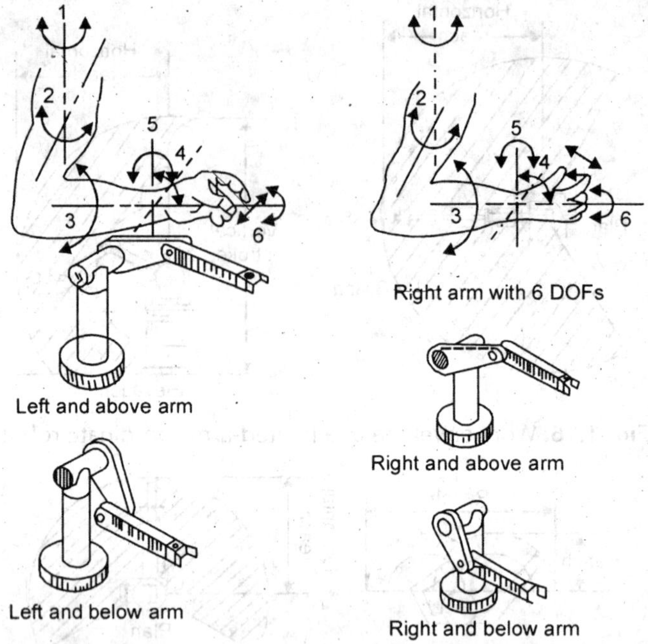

Fig. 1.19. Various arm configurations of the robot (right-and-left arms)

Various methods of specifying robot movement :

The various methods of specifying robot movement are :

(*i*) Fixed sequence robot.

(*ii*) Variable sequence robot.

(*iii*) Playback robot.

(*iv*) NC robot.

(*v*) Intelligent robot.

- A robot may be non-servo (bang-bang) type or servo (proportional feedback) type.
- Robots may be grouped as follows :
  (*i*) Simple and blind robot (with internal sensors).
  (*ii*) Vision robot.
  (*iii*) Moderately intelligent tool (with external tactile and non-tactile sensors).
- Robots may also be specified according to the type of industrial applications as follows :
  (*i*) Part handling robot ;
  (*ii*) Tool operating robot ;
  (*iii*) Assembly robot etc.

Typical broad robot specifications :

Some of the typical broad robot specifications are given in tabular form on the next page (Table 1.1) :

Table 1.1 : Typical Broad Robot Specifications

S.No.	Specifications	ROBOTS		
		Cincinnai Milacron	Puma (Unimation)	Seiko Instruments
1.	*Model*	T3 586	550	M 700
2.	*Manipulator end-effector*	6 DOF, RRR-3A	6 DOF, RRR-3A	3 DOF RP-1A
3.	*Speed*	900 m/s	1000 m/s	312 mm/s
4.	*Actuator*	Hydraulic	Electrical	Pneumatic
5.	*Payload*	100 kg	3 kg	1 kg
6.	*Repeatability*	± 1.25 mm	± 0.10 mm	± 0.03 mm
7.	*Applications*	Forging, investment casting, machine tool loading, welding, machining, inspection etc.	Machine tool loading, part transfer, assembly, welding, inspection, education.	Machine tool loading/unloading, material handling, assembly, inspection.

2. Loading carrying capacity :

— The load carrying capacity of a robot depends on :

(*i*) Its physical size and construction ;

(*ii*) Its capability to transmit force and torque to the end effector in the wrist.

— While this capacity for very light models can be as low as 1.5 kg (including the weight of end-effector), the heavier class of robots can have their capacities as high as 1000 kg.

3. Speed of movement :

"Speed of movement" is the speed at which a robot is capable of manipulating its end-effector. It is governed by :

(*i*) The distance to be moved ;

(*ii*) The weight of the part to be moved ;

(*iii*) The accuracy required in placement of the part in position.

— Heavier parts and higher placement accuracy demand slower movement while the lighter parts can be moved at faster speeds.

— The '*speed*' varies from one point to another.

4. Repeatability :

"Repeatability" is the measure of robot's ability to position an object at a previously taught point in the work envelope.

— Owing to inherent errors present, particularly due to mechanical sources, the robot will not be able to return to exact programmed point.

5. Control resolution :

It relates to the *system's capability* (both controller and the positioning device) *to divide the range of the total movement into closely spaced points that can be identified*. Thus it would represent the minimum noticeable movement achievable.

It may be added that the controller can generate pulses of very small duration but the positioning device should be able to respond and change its position accordingly. In such a case :

$$\textit{Control resolution} = \frac{\text{Range of movement}}{2^n}$$

where, n = Number of bits devoted to a joint, and

2^n = Number of addressable points.

6. Spatial resolution :

Whereas control resolution concerns the resolution for only one link and one motion, "spatial resolution" combines the control resolution of all motions and also considers the mechanical errors in the points and associated links.

— The spatial resolution varies depending on the exact position of wrist end because certain joint combinations would tend to magnify the effect of the control resolution and the mechanical errors.

7. Mechanical errors :

The mechanical errors arise from :

- Backlash in gears ;
- Hysteresis ;
- Deflection of links ;
- Hydraulic leaks etc.

These errors are characterised by a normal distribution.

8. Accuracy :

"Accuracy" is the measure of the ability of robot to position the end of wrist at a desired location in the work envelope.

Let us consider the accuracy treatment for a single link and single motion (the worst case would occur when the desired location lies directly between two adjacent control points – points set by control resolution). The inaccuracies in mechanical positioning system can be considered to have normal distribution with a constant variance over the range of movement. With these data it is possible to create a mathematical model for further design improvements.

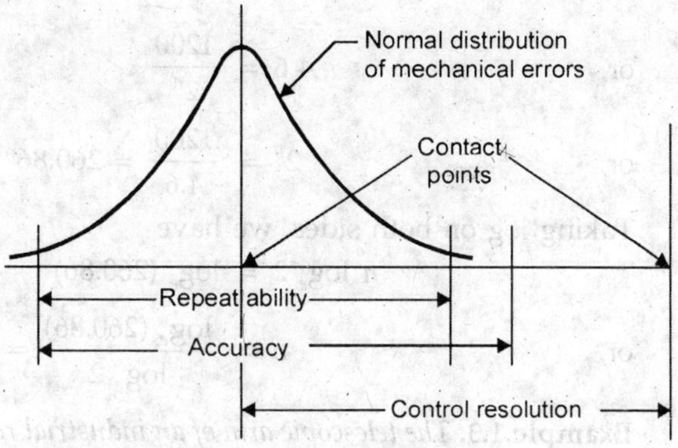

Fig. 1.20

Refer to Fig. 1.20. The following definition can be established :

$$\textit{Accuracy} = \frac{\text{Control resolution}}{2} + 3\sigma$$

(where, σ = standard deviation of mechanical error)

$$\text{Repeatability} = \pm 3\sigma = 6\sigma$$

$$\text{Spatial resolution} = \text{Control resolution} + 6\sigma$$

$$\text{Accuracy, in terms of spatial resolution} = \frac{\text{Spatial resolution}}{2}$$

— As robots move in 3-dimensional space, the distribution of all above items is also 3-dimensional. The normal distribution in 3-D can be conceptualised as a *sphere* whose mean is the programmed point and radius is equal to *three-times* standard deviation of the repeatability error distribution.

- For modern robots the repeatability values are of the order of ± 0.05 mm.

9. Stability :

It relates to the *amount of overshoot and oscillations in robot motion as it is about to reach a certain location.*

— A stable system has less oscillation but it becomes *inherently slower in response.*

Example 1.1. *One of the links of a robot has a telescoping arm with a stroke of 768 mm. The control memory of the robot has 8-bit storage capacity for this axis. Determine the control resolution for the same.*

Solution. *Given* : Stroke length (or range of movement) = 768 mm

Bit storage capacity for the axis, $n = 8$

Now, $\quad \textit{control resolution} = \frac{\text{Stroke length}}{2^n} = \frac{768}{(2)^8} = \mathbf{3\ mm}$ **(Ans.)**

Example 1.2. *A cartesian robot has a slide with a total range of 1.2 m and it is desired that it will have a control resolution of 4.6 mm on this axis. Determine the bit storage capacity which the control memory must possess to accommodate this level of precision.* **(VTU- 2003)**

Solution. *Given* : Total range = 1.2 m or 1200 mm ; Control resolution = 4.6 mm

Bit storage capacity, *n* :

$$\text{Control resolution} = \frac{\text{Total range}}{2^n}$$

or, $$4.6 = \frac{1200}{2^n}$$

or, $$2^n = \frac{1200}{4.6} = 260.86$$

Taking log on both sides, we have

$$n \log_e 2 = \log_e (260.86)$$

or, $$n = \frac{\log_e (260.86)}{\log_e 2} = 8.027 \text{ say } \mathbf{8} \quad \textbf{(Ans.)}$$

Example 1.3. *The telescopic arm of an industrial robot obtains total range of rotation of 120°. The robot has a 12-bit storage capacity for the axis. The arm fully extends to 1500 mm and fully retracts to 750 mm from the pivot point. Determine the robot's control resolution for the axis (i) in degrees of rotation, and (ii) on linear scale in fully extended and retracted position.* **(VTU-2004)**

Solution. *Given* : Range of rotation = 120° ; Number of bit storage capacity = 12 ; Maximum reach of extension, H_{max} = 1500 mm ; Minimum reach retraction, H_{min} = 750 mm

(*i*) *The control resolution of "rotation"*

$$= \frac{\text{Range of rotation}}{2^n} = \frac{120}{(2)^{12}} = \mathbf{0.0293°} \quad \textbf{(Ans.)}$$

(*ii*) *The control resolution for "translation"*

$$= \frac{\text{Range of movement } (i.e.\ H_{max} - H_{min})}{2^n} = \frac{1500 - 750}{(2)^{12}}$$

$$= \mathbf{0.183\ mm} \quad \textbf{(Ans.)}$$

Example 1.4. *The mechanism connecting the wrist assembly is a twisting joint which can be rotated through 7.5 revolutions from the start to end position. It is desired to have control resolution of rotation of ± 0.34° at the least. What is the number of bit storage capacity to achieve this resolution ?*

Solution. *Given* : Range of rotation = 7.5 × 360° = 2700° ; Total control resolution = ± 0.34° *i.e.*, 0.68°.

Bit storage capacity, *n* :

We know that,

$$\text{Control resolution} = \frac{\text{Range of rotation}}{2^n}$$

or, $$0.68° = \frac{2700°}{2^n}$$

or, $$2^n = \frac{2700°}{0.68°} = 3970.6$$

Taking log on both sides, we get

$$n \log_e 2 = \log_e (3970.6)$$

or, $$n = \frac{\log_e (3970.6)}{\log_e 2} = 11.955 \text{ say } \mathbf{12} \quad \textbf{(Ans.)}$$

Example 1.5. *The angular position of a high speed motor shaft is monitored by an incremental shaft encoder with 2 emitter detector pairs and 15 slots around the circumference. The precision of the load shaft is measured and found to be 0.06 degree per count. What is the gear ratio between high speed shaft and the load shaft ?*

Solution. *Given* : Number of slots, $S = 15$; Number of emitter detector pairs, $n = 2$; The angular control resolution, $A_c = 0.06°$ (precision) ; Range of rotation = $2\pi = 360°$.

The gear reduction ratio, G :

$$\text{Gear reduction ratio,} \quad G = \frac{360°}{S \times A_c \times (2)^n} = \frac{360°}{15 \times 0.06° \times (2)^2} = 100$$

i.e. The *gear reduction ratio* between high speed shaft and the load shaft = **100 : 1 (Ans.)**

Example 1.6. *A cylindrical robot has a prismatic joint with a range of travel of 900 mm. The control memory for this joint has 10 bit capacity. It has been recorded that the associated mechanical accuracies with the said arm show a random distribution of random variable of the robot position about the mean position of the taught point gives a standard deviation of 0.08 mm. The standard deviation is equal in all directions.*

Calculate : (*i*) *The control resolution for the axis.*

(*ii*) *The spatial resolution for the prismatic joint.*

(*iii*) *The accuracy defined.*

(*iv*) *The repeatability of the robot link.*

Solution. *Given* : Range of travel = 900 mm ; n = 10 bits ; The standard deviation of the mechanical inaccuracies, σ = 0.08 mm.

(*i*) **The control resolution for the axis :**

$$\text{Control resolution} = \frac{\text{Range of travel}}{2^n} = \frac{900}{(2)^{10}} = \mathbf{0.8789 \text{ mm}} \quad \textbf{(Ans.)}$$

(*ii*) **The spatial resolution :**

$$\text{Spatial resolution} = \text{Control resolution} + 6\sigma$$
$$= 0.8789 + 6 \times 0.08 = \mathbf{1.3589 \text{ mm}} \quad \textbf{(Ans.)}$$

(*iii*) **The accuracy :**

$$\text{Accuracy} = \frac{\text{Control resolution}}{2} + 3\sigma$$
$$= \frac{0.8789}{2} + 3 \times 0.08 = \mathbf{0.6794 \text{ mm}} \quad \textbf{(Ans.)}$$

$$(\textit{Alternatively} : \text{Accuracy} = \frac{\text{Spatial resolution}}{2} = \frac{1.3589}{2} = 0.6794 \text{ mm})$$

(*iv*) **The repeatability of the robot link :**

$$\text{The repeatability of the robot link} = \pm 3\sigma$$
$$= \pm (3 \times 0.08) = \pm \mathbf{0.24 \text{ mm}} \quad \textbf{(Ans.)}$$

1.18 ROBOTS PERFORMANCE TESTING

In order to judge the suitability of robots the following tests are performed on them :

1. **Geometric values :**

The geometric values include the following :

(*i*) *Workspace* : It is the envelope reached by the centre of the interface between the wrist and the tool, using all available axis motions.

(*ii*) *Static behaviour* : It indicates the deformation of a fixed robot structure under different load cases.

(*iii*) *Position accuracy* : It is the repeatable accuracy that can be achieved at nominal load and normal operating temperature.

— Position accuracy is based on two types of errors :

(*a*) Repeatability error ; (*b*) Reversal error.

(*iv*) *Path accuracy* : It indicates at what level of accuracy programmed path curves can be followed at nominal load.

- Following are the typical errors in path accuracy of a robot :
  — Path accuracy or mean-path dispersion error ;
  — Trailing error or mean-path deviation ;
  — Overshoot during acceleration/deceleration.

(*v*) *Reproduction of smallest steps* : The slip-stick effect may become serious when the velocities are very low.

(*vi*) *Synchronous travel accuracy* : It relates to cases where robot has to perform tasks sychronous to a moving conveyor as in spray painting and assembly.

(*vii*) *Long-term behaviour* : It provides information on the time required to achieve thermal stability.

2. **Kinematic values :**

— The kinematic values include cycle time, speed, acceleration.

— It involves measuring of attainable cycle times for a defined sequence in different areas of the working place.

3. **Dynamic values :**

— It involves determination of dynamic behaviour of simple components and the total structure.

— The response of the robot structure is elicited by the following excitation methods:

(*a*) Shaker (Sinus, random) ; (*b*) Hammer (impact) ;
(*c*) Snapback (impact) ; (*d*) Drives (Sinus, random).

4. **Power noise values :**

— These values are usually measured in decibel at a distance of one metre from the working space.

5. **Thermal values :**

— The changes in temperature affect deviation of the structure.

1.19 ROBOTS KINEMATIC CONTROL

1.19.1. Robot Arm Kinematics

"Robot arm kinematics" deals with analytical study of the geometry of motion of a robot arm with respect to a fixed reference coordinate system without regard to the forces/moments that cause

the motion. Thus kinematics deals with the analytical description of the *spatial displacement* of the robot as a *function of time*, in particular the relations between the joint-variable space and the position and orientation of the end-effector of a robot arm.

Forward and reverse kinematics :

● When the position and orientation of end-effector, for a manipulator, are derived from the given joint angles and link parameters, the scheme is called the *"**Forward kinematics**"* (Fig. 1.21) problem.

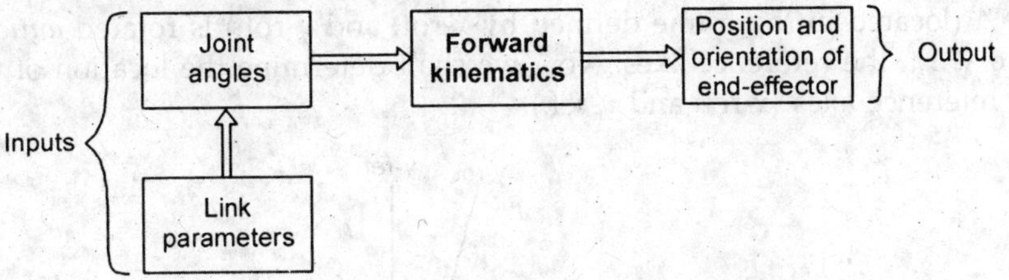

Fig. 1.21. Forward kinematics scheme.

● When the joint angles and the different configurations of the manipulator are derived from the position and orientation of the end-effector, the scheme is known as the *"**Inverse/Reverse kinematics**"* problem.

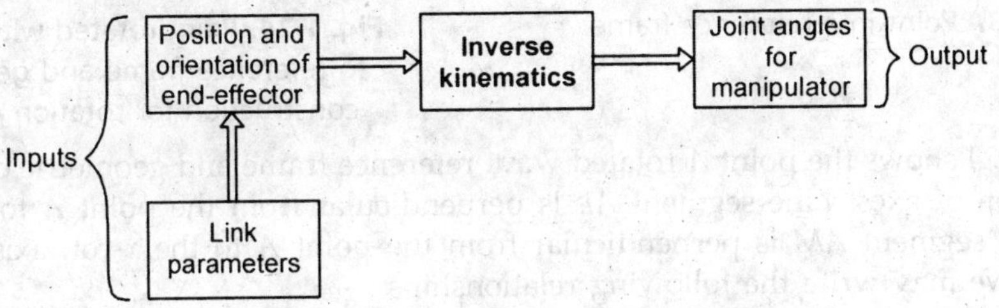

Fig. 1.22. Inverse kinematics scheme.

— Since the independent variables in robot arm are the joint variables, and a task is usually stated in terms of the reference coordinate frame, the *inverse kinematics problem is used more frequently*.

— The *inverse kinematics problem*, in general, can be solved by several techiques ; the most commonly used are :

(*i*) Matrix algebraic ;

(*ii*) Iteratine, or ;

(*iii*) Geometric approach.

1.19.1.1. Transformations

● For describing the position and orientation of the tool with respect to the base frame it is imperative to know and formulate the body attached coordinate frame along the joint axis for each link in the manipulator chain of the robot.

● *The relation between the body attached frame with the base frame of reference is described by* **transformation matrix.**

● The transformation is represented by the following component transformations:

(*i*) Rotation matrix ; (*ii*) Translation or position vector ;

(*iii*) Perspective transformation ; (*iv*) Scaling or stretching.

1.19.1.2. Rotational transformations

Besides moving points along vectors, it is possible to generate new positions in space by rotating them about an axis.

Let us derive the rotation formula for a set of orthogonal axes rotated in a plane. Consider a point *A* located at x_1, y_1 with respect to the *x-y* reference axis (called as *x*-ref and *y*-ref, Fig. 1.23). In addition, let us assume initially that a second set of axes is lying on top of the reference axes. This set will be called *x*-rot. and *y*-rot. and the relationship of the point *A* w.r.t. these axes will remain *fixed*. Now assume that the plane containing the point *A* (located in the frame defined by *x*-rot. and *y*-rot.) is rotated *anticlockwise* by an angle α w.r.t. the reference axes. Now we shall determine the location of the point *P* w.r.t. the reference axes (*x*-ref. and *y*-ref.).

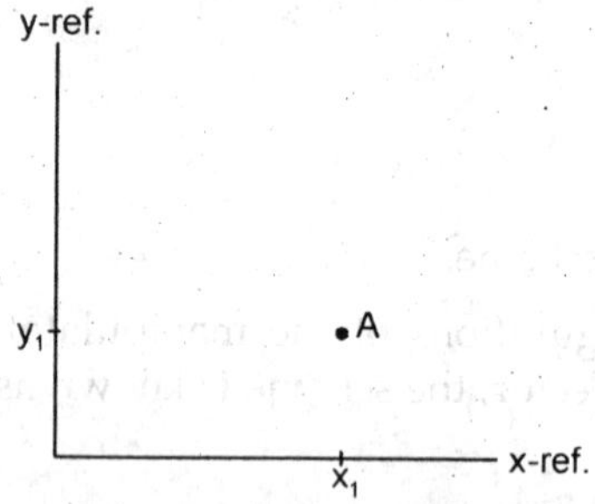

Fig. 1.23. Point in a reference frame.

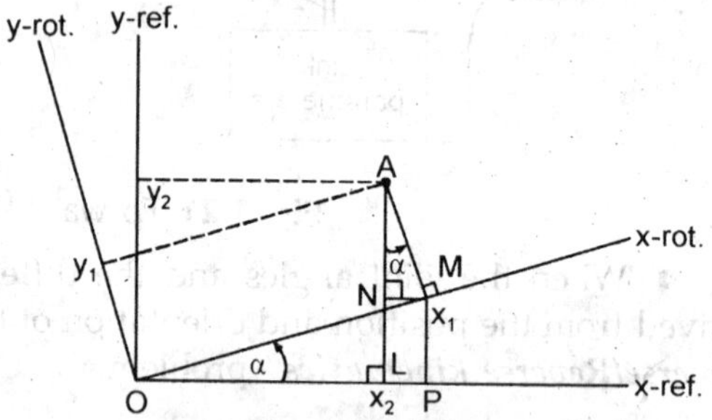

Fig. 1.24. Point rotated with respect to reference frame and geometric construction for rotation of axes.

Fig. 1.24 shows the point *A* rotated w.r.t. reference frame and geometric construction for rotation of axes. Line segment *AL* is perpendicular from the point *A* to *x*-ref. axis, while line segment *AM* is perpendicular from the point *A* to the *x*-rot. axis. From the diagram we may write the following relationships :

$$x_2 = |OL| = |OP| - |LP| \quad ...(i)$$

$$y_2 = |LA| = |LN| + |NA| \quad ...(ii)$$

$$|OP| = |OM| \cos\alpha$$

$$= x_1 \cos\alpha \quad ...(iii)$$

$$|LN| = |PM| = |OM| \sin\alpha$$

$$= x_1 \sin\alpha \quad ...(iv)$$

$$|LP| = |NM| = |MA| \sin\alpha$$

$$= y_1 \sin\alpha \quad ...(v)$$

$$|NA| = |MA| \cos\alpha$$

$$= y_1 \cos\alpha \quad ...(vi)$$

Combining these eqns., we get

$$\left.\begin{aligned} x_2 &= x_1 \cos\alpha - y_1 \sin\alpha \\ y_2 &= x_1 \sin\alpha + y_1 \cos\alpha \end{aligned}\right\} \quad ...(1.1)$$

The above relationships allow us to determine the location of points affixed to movable frames of reference in terms of a fixed reference frame. The derivation in the *x-y* plane is easily extended to either *y-z* or *z-x* plane.

Geometric interpretation of rotation matrix :

Rotation	*Rotation matrix*	p_{abc}	*Geometric representation*
$R(x, \alpha)$	$\begin{bmatrix} 1 & 0 & 0 \\ 0 & \cos\alpha & -\sin\alpha \\ 0 & \sin\alpha & \cos\alpha \end{bmatrix}$ The first column/row is same as the unit vector $p_{abc} = (1, 0, 0)^T$ in *OABC*	$\begin{bmatrix} 0 \\ 1 \\ 0 \end{bmatrix}$	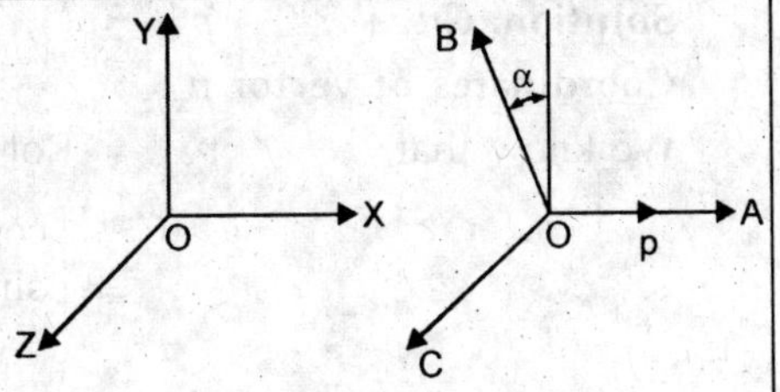
$R(y, \alpha)$	$\begin{bmatrix} \cos\alpha & 0 & \sin\alpha \\ 0 & 1 & 0 \\ -\sin\alpha & 0 & \cos\alpha \end{bmatrix}$ The second row/column of *R* matrix is same as $p_{abc} = (0, 1, 0)^T$ in *OABC*	$\begin{bmatrix} 0 \\ 1 \\ 0 \end{bmatrix}$	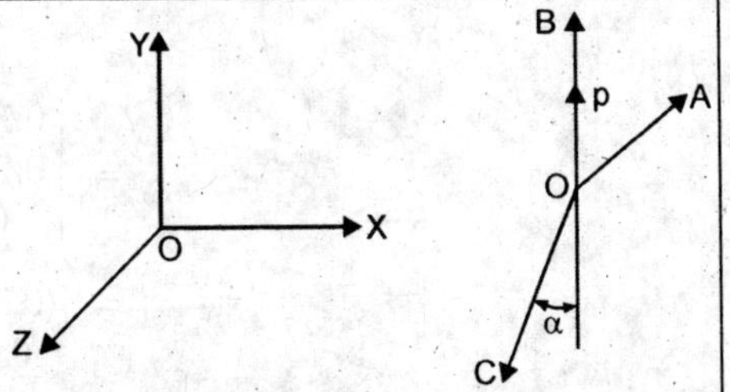
$R(z, \alpha)$	$\begin{bmatrix} \cos\alpha & -\sin\alpha & 0 \\ \sin\alpha & \cos\alpha & 0 \\ 0 & 0 & 1 \end{bmatrix}$ The third column/row of *R* is same as the unit vector $p_{abc} = (0, 0, 1)^T$ in *OABC*	$\begin{bmatrix} 0 \\ 0 \\ 1 \end{bmatrix}$	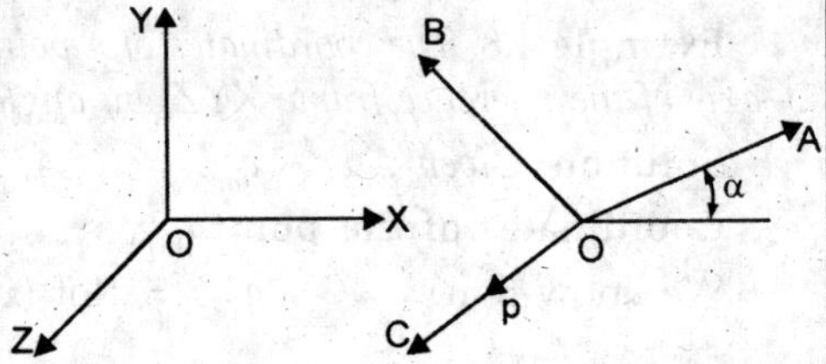

Inverse transformations (*Rotation matrix*) **:**

If two co-ordinate frames are orthonormal and are given by *X* and *A*, and *R* to be the transformation that maps *X* co-ordinate into *A* co-ordinate,

Then, $X = R.A,$

and *A* maps with *X* by the relation

$$A = R^{-1}X = R^T X. \quad \text{...(1.2)}$$

Rotation	*R matrix*	$R^{-1} = R^T$
1. $R(x, \alpha)$	$\begin{bmatrix} 1 & 0 & 0 \\ 0 & \cos\alpha & -\sin\alpha \\ 0 & \sin\alpha & \cos\alpha \end{bmatrix}$	$\begin{bmatrix} 1 & 0 & 0 \\ 0 & \cos\alpha & \sin\alpha \\ 0 & -\sin\alpha & \cos\alpha \end{bmatrix}$
2. $R(y, \alpha)$	$\begin{bmatrix} \cos\alpha & 0 & \sin\alpha \\ 0 & 1 & 0 \\ -\sin\alpha & 0 & \cos\alpha \end{bmatrix}$	$\begin{bmatrix} \cos\alpha & 0 & -\sin\alpha \\ 0 & 1 & 0 \\ \sin\alpha & 0 & \cos\alpha \end{bmatrix}$
3. $R(z, \alpha)$	$\begin{bmatrix} \cos\alpha & -\sin\alpha & 0 \\ \sin\alpha & \cos\alpha & 0 \\ 0 & 0 & 1 \end{bmatrix}$	$\begin{bmatrix} \cos\alpha & \sin\alpha & 0 \\ -\sin\alpha & \cos\alpha & 0 \\ 0 & 0 & 1 \end{bmatrix}$

Example 1.7. *The coordinates of a point $p_{abc} = (4, 3, 2)^T$ in the body coordinate frame OABC are rotated through 45° about OZ-axis. Determine the coordinates of the vector p_{xyz} with respect to base reference coordinate frame.*

Solution. *Given :* $p_{abc} = (4, 3, 2)^T$; $\alpha = 45°$

Coordinates of vector p_{xyz} :

We know that,

$$p_{xyx} = \text{Rot}(z, \alpha) \cdot p_{abc}$$

$$= \begin{bmatrix} \cos\alpha & -\sin\alpha & 0 \\ \sin\alpha & \cos\alpha & 0 \\ 0 & 0 & 1 \end{bmatrix} \begin{bmatrix} 4 \\ 3 \\ 2 \end{bmatrix}$$

$$= \begin{bmatrix} \cos 45° & -\sin 45° & 0 \\ \sin 45° & \cos 45° & 0 \\ 0 & 0 & 1 \end{bmatrix} \begin{bmatrix} 4 \\ 3 \\ 2 \end{bmatrix}$$

$$= \begin{bmatrix} 0.707 & -0.707 & 0 \\ 0.707 & 0.707 & 0 \\ 0 & 0 & 1 \end{bmatrix} \begin{bmatrix} 4 \\ 3 \\ 2 \end{bmatrix}$$

$$= [(0.707 \times 4 - 0.707 \times 3 + 0 \times 2), (0.707 \times 4 + 0.707 \times 3 + 0 \times 2), (0 \times 4 + 0 \times 3 + 1 \times 2)]$$

$$= [0.707, 4.949, 2]^T$$

Hence, $p_{xyz} = \mathbf{[0.707, 4.949, 2]^T}$ **(Ans.)**

Example 1.8. *The coordinates of a point q_{abc} are given by $(6, 4, 2)^T$ which are rotated about X-axis of the reference frame XYZ, by angle of 30°. Determine the coordinates of the point q_{xyz}.*

Solution. *Given :* $q_{abc} = (6, 4, 2)^T$; $\alpha = 30°$

Coordinates of the point q_{xyz} :

We know that,

$$q_{xyz} = \text{Rot}(x, 30°) \cdot q_{abc}$$

$$= \begin{bmatrix} 1 & 0 & 0 \\ 0 & \cos\alpha & -\sin\alpha \\ 0 & \sin\alpha & \cos\alpha \end{bmatrix} \begin{bmatrix} 6 \\ 4 \\ 2 \end{bmatrix}$$

$$= \begin{bmatrix} 1 & 0 & 0 \\ 0 & \cos 30° & -\sin 30° \\ 0 & \sin 30° & \cos 30° \end{bmatrix} \begin{bmatrix} 6 \\ 4 \\ 2 \end{bmatrix}$$

$$= \begin{bmatrix} 1 & 0 & 0 \\ 0 & 0.866 & -0.5 \\ 0 & 0.5 & 0.866 \end{bmatrix} \begin{bmatrix} 6 \\ 4 \\ 2 \end{bmatrix}$$

$$= [(1 \times 6 + 0 \times 4 + 0 \times 2), (0 \times 6 + 0.866 \times 4 - 0.5 \times 2), (0 \times 6 + 0.5 \times 4 + 0.866 \times 2)]$$

$$= [6, 2.464, 3.732]^T$$

Hence the coordinates of the vector $q_{xyz} = \begin{bmatrix} 6 \\ 2.464 \\ 3.732 \end{bmatrix}$

$$= \mathbf{[6, 2.464, 3.732]^T} \quad \textbf{(Ans.)}$$

Example 1.9. *For the following rotation matrix determine the axis of rotation and the angle of the rotation about the same.*

$$R = \begin{bmatrix} 0.5 & 0 & 0.866 \\ 0 & 1 & 0 \\ -0.866 & 0 & 0.5 \end{bmatrix}$$

Solution. *Given* :
$$R = \begin{bmatrix} 0.5 & 0 & 0.866 \\ 0 & 1 & 0 \\ -0.866 & 0 & 0.5 \end{bmatrix}$$

Axis and angle of rotation :

From geometric interpretation of rotation matrix table, by comparing the rotation matrix R, we find that

$$R = \text{Rot}(y, \alpha)$$

Hence the *axis of rotation is* **OY-axis (Ans.)**

Also,
$$R = \text{Rot}(y, \alpha) = \begin{bmatrix} \cos\alpha & 0 & \sin\alpha \\ 0 & 1 & 0 \\ -\sin\alpha & 0 & \cos\alpha \end{bmatrix} = \begin{bmatrix} 0.5 & 0 & 0.866 \\ 0 & 1 & 0 \\ -0.866 & 0 & 0.5 \end{bmatrix}$$

i.e.,
$$\cos\alpha = 0.5, \text{ and}$$
$$\sin\alpha = 0.866$$

$\therefore$
$$\tan\alpha = \frac{0.866}{0.5} = 1.732$$

or,
$$\alpha = 60° \quad \textbf{(Ans.)}$$

1.19.1.3. Homogeneous transformations

A general method for solving the kinematic equations of a robot manipulator with many joints is the use of *"homogeneous transformations"*. A generalised transformation is now described by a *single matrix that combines the effects of translation and rotation.*

Rotational transformations which operate on *homogeneous coordinates* and perform rotations about a given axis of the reference coordinate system can be shown to be :

(*i*) *Rotation α degrees about the x-axis* :

$$\text{Rot}(x, \alpha) = \begin{bmatrix} 1 & 0 & 0 & 0 \\ 0 & \cos\alpha & -\sin\alpha & 0 \\ 0 & \sin\alpha & \cos\alpha & 0 \\ 0 & 0 & 0 & 1 \end{bmatrix} \qquad ...(1.3)$$

(*ii*) *Rotation α degrees about the y-axis* :

$$\text{Rot}(y, \alpha) = \begin{bmatrix} \cos\alpha & 0 & \sin\alpha & 0 \\ 0 & 1 & 0 & 0 \\ -\sin\alpha & 0 & \cos\alpha & 0 \\ 0 & 0 & 0 & 1 \end{bmatrix} \qquad ...(1.4)$$

(*iii*) *Rotation α degrees about z-axis* :

$$\text{Rot}(z, \alpha) = \begin{bmatrix} \cos\alpha & -\sin\alpha & 0 & 0 \\ \sin\alpha & \cos\alpha & 0 & 0 \\ 0 & 0 & 1 & 0 \\ 0 & 0 & 0 & 1 \end{bmatrix} \qquad ...(1.5)$$

- The general form of the homogeneous transformation can be partitioned into submatrices as shown in a tabular form below :

Rotation matrix (3 × 3)	Position vector (3 × 1)
Perspective transform (1 × 3)	Scaling factor (1 × 1)

The entries consist of a 3 × 3 submatrix reponsible for the *rotation* and 3 × 1 submatrix responsible for *translation*. In addition, two other partitions may be identified as the perspective transformation and scaling factor. Typically we set the scaling factor to unity. The perspective transformation is set to (0, 0, 0) so that the columns of rotation matrix taken along with the appropriate entry of the perspective transformation define direction vectors.

The real power of this matrix is seen by interpreting its columns. With the perspective transform and scaling factor set to the values mentioned above, the homogeneous transformation may be interpreted as defining a relationship between two frames.

- *Vector and position nomenclature for homogeneous transformation matrix* :

$$H = \left[\begin{array}{ccc|c} n_x & 0_x & a_x & p_x \\ n_y & 0_y & a_y & p_y \\ n_z & 0_z & a_z & p_z \\ \hline 0 & 0 & 0 & 1 \end{array}\right] \quad \text{...(1.6)}$$

Eqn. (1.6) shows the *labeling of the individual vector components and positions*. We may think of this matrix as containing the direction vectors for the x, y, and z axes of frame-2 written in terms of the directions of the x, y, and z axes of frame-1. *For example,* the elements of the first column (n_x, n_y, n_z) are the components of the unit vector defining x-axis of frame-2, written in terms of the three unit vectors defining the axes of frame-1. The *second column* defines the y-axis of frame-2, while the *third column* defines the direction of the z-axis. The fourth column is the location of the origin of frame-2 w.r.t. frame-1 and may be thought of as a vector (having coordinates p_x, p_y, p_z) drawn from the origin of frame-1 to the point defining the origin of frame-2.

1.19.1.4. Forward and reverse kinetics (transformation) of 3 DOFs (degrees of freedom) robot arm

In a revolute type of joint only, the position of the end-effector can be represented by the joint angles α_1, α_2, α_3, ... α_n as :

$$P_{\text{Joint}} = (\alpha_1, \alpha_2, \alpha_3, \ldots \alpha_n) \quad \text{...(1.7)}$$

The end-effector's position can also be defined in world space, as :

$$P_{\text{World}} = (x, y, z) \quad \text{...(1.8)}$$

Fig. 1.25 shows a 3 DOFs 2-D (dimensional) manipulator.

Forward transformation :

The end-effector's position and orientation in world space can be determined from the joint angles and link parameters by the equations given below :

$$x_3 = l_1 \cos \alpha_1 + l_2 \cos (\alpha_1 + \alpha_2) + l_3 \cos (\alpha_1 + \alpha_2 + \alpha_3) \quad \text{...(1.9)}$$

$$y_3 = l_1 \sin \alpha_1 + l_2 \sin (\alpha_1 + \alpha_2) + l_3 \sin (\alpha_1 + \alpha_2 + \alpha_3) \quad \text{...(1.10)}$$

$$\gamma = \alpha_1 + \alpha_2 + \alpha_3 \quad \text{...(1.11)}$$

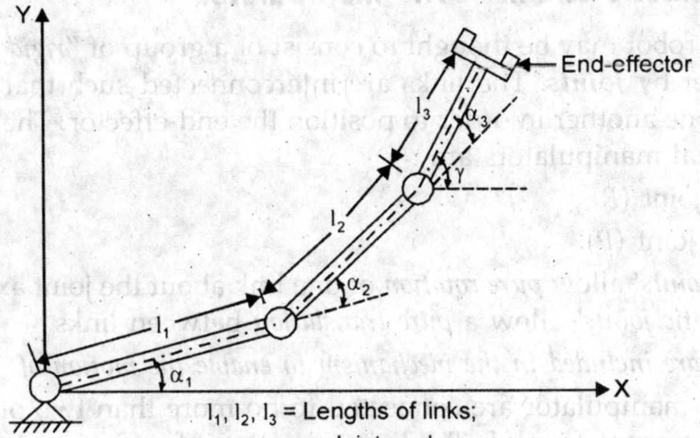

Fig. 1.25. 3 DOFs 2-D manipulator.

Reverse transformation :

By using reverse transformation, we can determine joint angles also from the end-effector's position (x_3, y_3) and the orientation (γ) as follows :

$$x_2 = x_3 - l_3 \cos \gamma \quad \text{...(1.12)}$$

$$y_2 = y_3 - l_3 \sin \gamma \quad \text{...(1.13)}$$

From the given geometry, we have :

$$x_2 = l_1 \cos \alpha_1 + l_2 \cos \alpha_1 \cos \alpha_2 - l_2 \sin \alpha_1 \sin \alpha_2 \quad \text{...(1.14)}$$

$$y_2 = l_1 \sin \alpha_1 + l_2 \sin \alpha_1 \cos \alpha_2 + l_2 \cos \alpha_1 \sin \alpha_2 \quad \text{...(1.15)}$$

By squaring and adding eqns. (1.13) and (1.14), we get :

$$\cos \alpha_2 = \frac{x_2^2 + y_2^2 - l_1^2 - l_2^2}{2 l_1 l_2} \quad \text{...(1.16)}$$

By inserting the value of α_2 in eqns (1.13) and (1.14), we obtain the value of α_1.

The value of α_3 can be obtained from the following relation :

$$\alpha_3 = \gamma - (\alpha_1 + \alpha_2) \quad \text{...(1.17)}$$

1.19.2. Robot Arm Dynamics

"Robot arm dynamics" deals with the mathematical formulation of the equations of the robot arm motion.

— The dynamic equations of motion of a manipulator are a set of mathematical equations describing the dynamic behaviour of the manipulator. Such equations of motion are useful for computer simulation of the robot arm motion, the design of suitable control equations for a robot arm and the evaluation of the kinematic design and structure of a robot arm.

— The actual dynamic model of an arm can be obtained from known physical laws such as the laws of newtonian and lagrangian mechanics. This leads to the development of dynamic equations of motion for various articulated joints of the manipulator in terms of specified geometric and inertial parameters of the links. Conventional approaches like the Lagrange–Euler and the Newton-Euler formulations can then be applied systematically to develop the actual robot arm motion equations.

1.19.3. Kinematic Chains – The Manipulator

An industrial robot may be thought to consist of a group of *"rigid bodies"* called ***links*** connected together by ***joints***. The links are interconnected such that they are forced to move relative to one another in order to position the end-effector. The two types of joints used in commercial manipulators are :

(*i*) Revolute joint (*R*)

(*ii*) Prismatic joint (*P*).

— *'Revolute joints'* allow *pure rotation* of one link about the joint axis of the preceding link, while *'prismatic joints'* allow a *pure translation* between links.

— *Actuators are included in the mechanism to enable the motion of the joints.*

The links of a manipulator are connected to no more than two others (via joints) so that closed loops are not formed. *The links and joints of a manipulator form a **kinematic chain** which is open at one end and connected to ground at the other.* The end-effector or hand is connected to the free end and the control objective of the robot system is to position the end-effector at a desired location in space.

- The degrees of freedom (DOFs) of a system depend on the number of variables or coordinates that are needed to describe its position.

Graphic symbols for kinematic chains :

Fig. 1.26, shows the graphic symbols for kinematic chains.

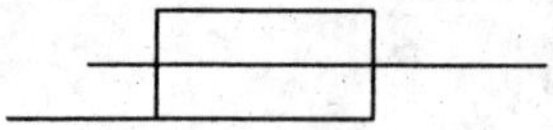

(*i*) Prismatic joint with its axis in plane of the paper.

(*ii*) Prismatic joint with its slider perpendicular to plane of the paper.

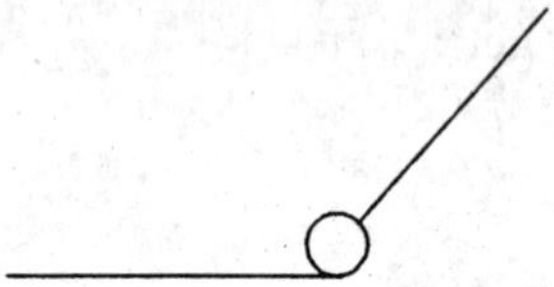

(*iii*) Revolute joint with its axis perpendicular to plane of the paper.

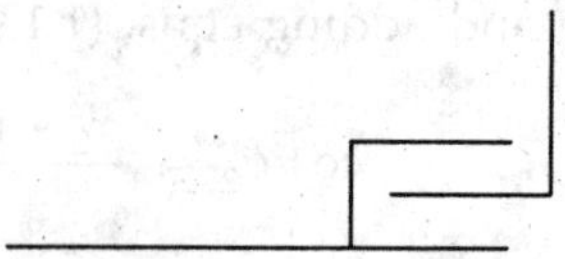

(*iv*) Revolute joint with its axis in the plane of the paper.

Fig. 1.26. Graphical symbols for kinematic chains.

Useful major linkages :

Fig. 1.27, shows *twelve* useful major linkages :

(P = Prismatic joint ; R = Revolute joint)

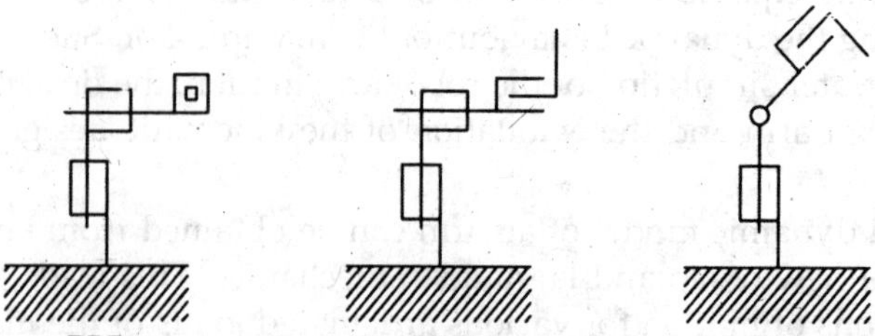

PPP PPR PRR

(i) Rectangular

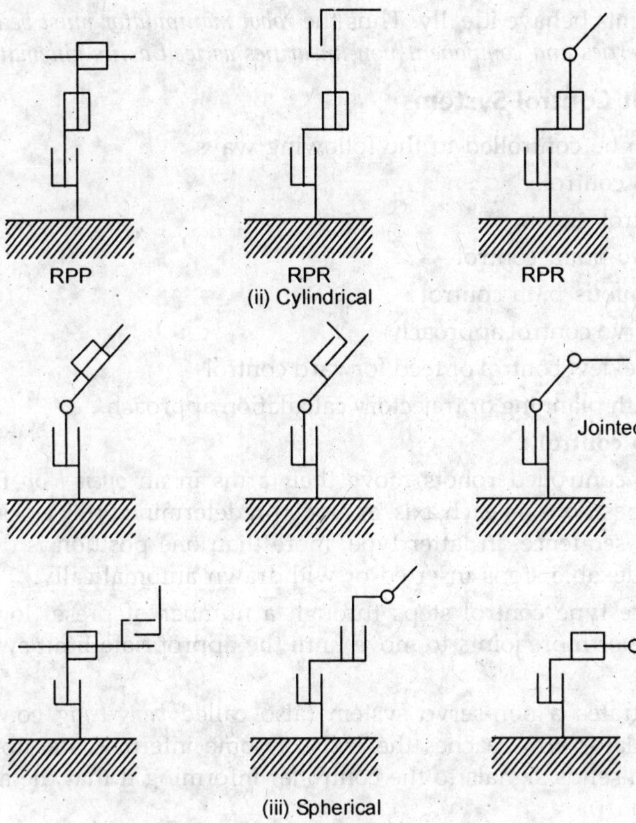

Fig. 1.27. Twelve useful major linkages.

Wrist configurations :

Fig. 1.28 shows some possible wrist configurations.

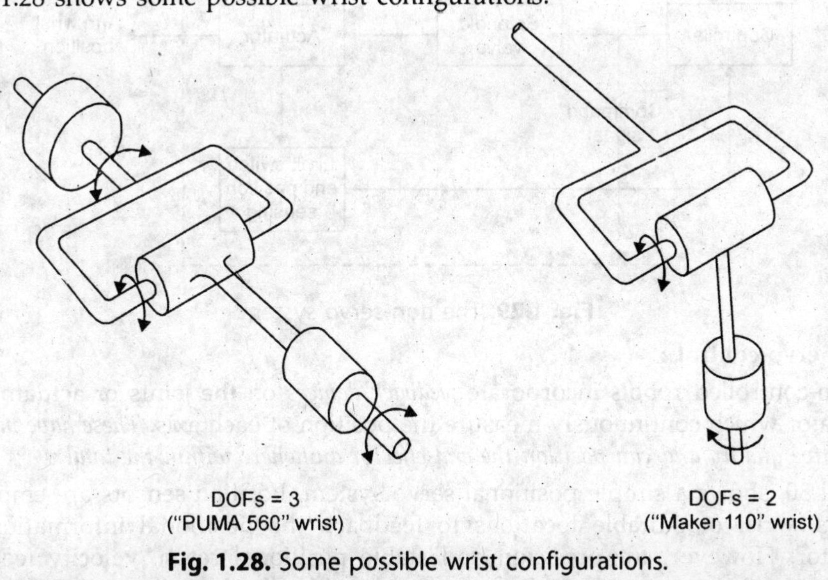

Fig. 1.28. Some possible wrist configurations.

- *'Kinematic formulation'* does *not* consider the weights or inertia of the links and the actuators for the joints that may be housed inside the links. In addition, no link is perfectly rigid, nor do all joints behave ideally. Thus the *robot manipulator must be analysed in terms of its dynamic properties and component non-linearities as well as its kinematic properties.*

1.19.4. Robot Control Systems

The robots can be controlled in the following ways :

1. Non-servo control.
2. Servo-control.
   (*i*) Point-to-point control
   (*ii*) Continuous path control :
      (*a*) Servo control approach
      (*b*) Preview control or feed forward control
      (*c*) Path planning or trajectory calculation approach.

1. **Non-servo control :**

— Non-servo-controlled robots move their arms in an *open loop* fashion between exact end positions on each axis, or along predetermined trajectories in accordance with fixed sequence. In latter type, more than one position is defined along an axis by indexable stops inserted or withdrawn automatically.

— A sequence type control steps through a number of preset logic steps, which causes one or more joints to move until the appropriate limit switch on the axis is reached.

Fig. 1.29 illustrates a non-servo system (also called *bang-bang* controlled system). When the manipulator arm reaches the ends of some intermediate stops, it touches a microswitch which sends signals to the controller informing it that the arm has reached the desired position.

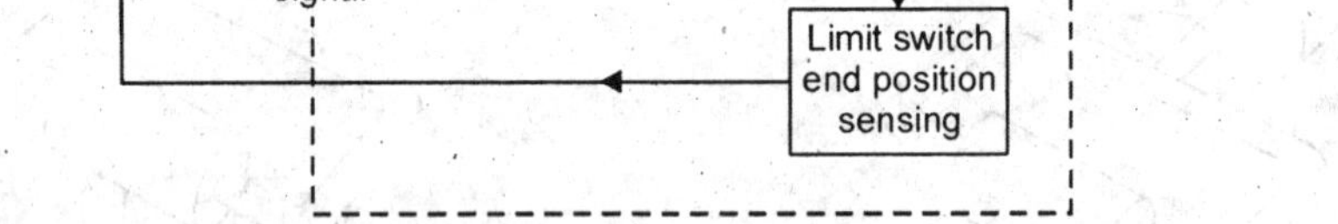

Fig. 1.29. The non-servo system.

2. **Servo-control :**

Servo-controlled robots incorporate *feedback devices* on the joints or actuators of the manipulator which continuously measure the position of each axis. *These have much more manipulative quality and can position the end-effector anywhere within the total work envelope.*

Fig. 1.30, shows a simple positional servo system. Position sensors are employed at the joints, wrist or suitable locations to feedback the positional information to the comparator. However, for precision and stable positional reach, velocity feedback is provided by tachogenerator in addition to positional feedback as shown in Fig. 1.31. Thus *servo system with position and velocity feedback ensures controlled arm position.*

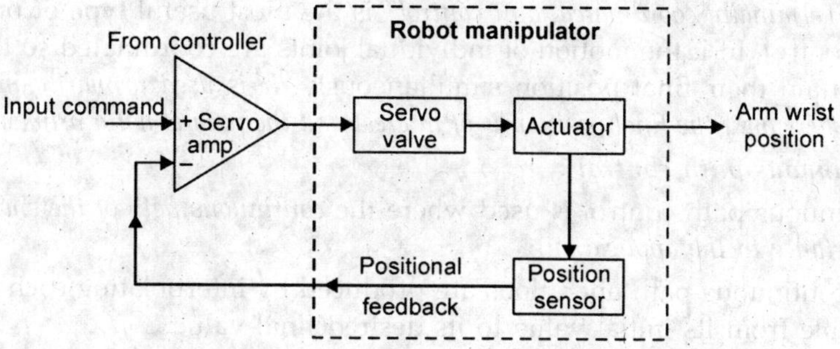

Fig. 1.30. Simple positional servo system.

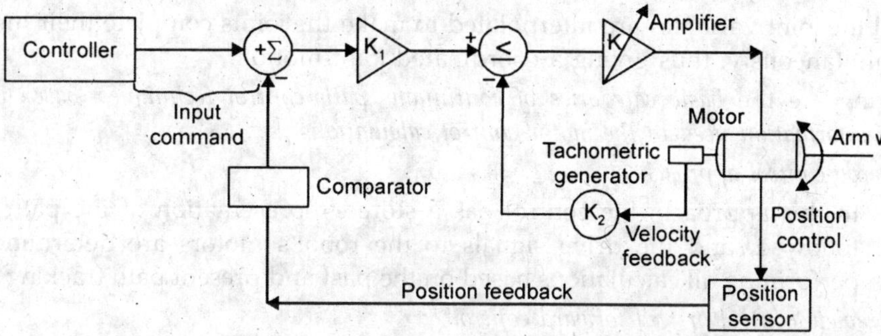

Fig. 1.31. The servo control loop with positional and velocity feedback.

Further classification of these robots is discussed below :

(*i*) ***Point-to-point control :***

- In this system each joint is controlled by an independent position servo with all joints moving from position to position *independently*.
- Here, each joint or axis of the robot is moved individually until the combination of joint position yields the desired position of the end-effector.
- The way each joint is to move to achieve final position is practiced before hand and stored in a memory device. As per this stored information each joint runs freely at its maximum or limited rate until it reaches its final position.
- Point-to-point motion could be controlled independently in *sequence joint control, uncoordinated joint control,* or *terminally coordinated joint control.*
  - — In "*sequential joint operation*" *one joint is activated at a time,* while all other axes are immobilised. A single joint may operate more than once in a sequence associated with such a motion. The resulting path of the manipulator end-effector will thus have a zig-zag form associated with the motion directions of the manipulator joints. *It results in immediate simplification in the control.* However, it causes *longer* point-to-point motion time.
  - — In "*uncoordinated joint control*" *the motions are not coordinated,* in the sense that if one joint has made some fraction of its motion it does not imply that all other joints will have made the same fractions of their respective motions. When each joint reaches its final position, it holds and waits until all joints have completed their motions. Due to non-coordination of motion between joints, the path and velocity of end-effector between points is *not* easily produced.

— *"Terminally coordinated joint control"* is the most useful type of point-to-point control. In it the motion of individual joints are coordinated so that all joints attain their final position simultaneously. *It is used primarily in applications where only the final position is of interest and the path is not a prime consideration.*

(*ii*) ***Continuous path control*** **:**

The continuous path control is used where the *continuous path of the end-effector is of primary importance to the application.*

— The continuous path operations are produced by interpolating each joint control variable from its initial value to its desired final value.

— Each joint is moved the maximum amount required to achieve the desired final positions to *give the robot tool a controlled predicted path.*

— All the joint variables are interpolated to make the joints complete their motions simultaneously, thus giving a coordinated joint motion.

Following are the *basic categories of continuous path control techniques, based on the quantum of information used in the motor control calculations* :

(*a*) *"Servo-control approach"* :

— In this approach, the control has a stored representation of the path to be followed, and the drive signals to the robot's motors are determined by performing all calculations based on the past and present path tracking error.

(*b*) *"Preview control or feed forward control"* :

— This technique uses some knowledge about how the path changes immediately ahead of robot's current location, in addition to the past and present tracking error used by the servo-controller.

(*c*) *"Path planning or trajectory calculation approach"* :

— In this technique, the controller is fed with a complete description of the manipulator from one point to another.

— It uses a mathematical physical "model" of the arm and its load, and precomputes an acceleration profile for every joint, predicting the nominal motor signals that should cause the arms to follow the desired path.

- *Continuous path control requires a lot of memory space to store all the axis positions needed to smoothly record the desired path.* In practice, the device is *moved actually through the desired path manually* and the position of each axis is recorded on a time base, thus generating continuous time history of each axis position.

Servo-robots' hierarchy :

The robot controller may employ a processor which accomplishes the tasks by breaking them into several subtasks for which, a hierarchical level of control may be organised :

Example **:**

— A vision robot may use a high level vision processor that provides a knowledge of position as a component identity including its position, and orientation and avoidance of obstacles; the robot controller through its other sensory devices feeds back the data to accomplish the tasks at various levels including the movement of joints or axes of the robot. Thus a *complex task may be divided into simple tasks.*

— A *simple task*, say 'load', may be broken down to elemental moves, say 'move' and 'grasp' and the elemental moves may further be decomposed into action commands in X, Y, Z primitives of trajectory in the world coordinates. The world coordinate commands are then transformed into joint actuator commands, $\alpha_1, \alpha_2, \alpha_3, ...$ through transformation coordinates and finally the drive signals are sent through positional and velocity servo of various joint axes of the robot.

1.19.5. Types of Controllers used in Robotics

The various *controllers* used in robotics are enumerated and discussed as under :

1. Drum controller
2. Air logic controller
3. Programmable controller
4. Microprocessor-based controller
5. Minicomputer-based controller.

1. Drum controller :

— In this type of controller, as the drum rotates, it actuates those switches which are wired to hydraulic or pneumatic valves. Thus, the manipulator movements are controlled by the rotational advancement of the drum.

— This controller is now obsolete.

2. Air logic controller :

— It employs a number of *pneumatic valves* which in turn control the opening and closing of the main valves of the robot manipulator in *close synchronisation with the timers.*

3. Programmable controller :

— In this type of controller, the *sequential order* in which the switches are to be operated is kept in the *memory*. It can be entered into the controller with the help of a keyboard.

— The programme can also be displayed on the CRT screen.

— This controller may be *used to control and coordinate various tasks to be done by the peripheral equipment including robots.*

4. Microprocessor-based controller :

— The microprocessor-based control is the *most popular robot control system.*

— The various types of microcomputers may be employed to program the sequential tasks or motions and store them in its *memory*.

— *It contains special circuitry to interpret the programs kept in its memory and at the same time it sends drive signals to various actuators of the robot manipulator.*

— It can also count the number of sequential events or tasks accomplished.

- A microprocessor-based controller is a *versatile, programmable and has good memory.*
- *Point-to-point, continuous path and controlled path motions can be easily programmed in microprocessor-based robotic systems.*

5. Minicomputer-based controller :

— A minicomputer-based controller is employed to manipulate robots having *higher payload.*

1.19.6. Trajectory Planning and Motion Control of Manipulator

- For accomplishing a specified task, the end-effector of the manipulator is required to *move in a particular fashion.*

- The *goal of trajectory planning* is to describe the requisite motion of the manipulator at a time sequence of point/link/end-effector locations and derivatives of locations, which are generated by *"interpolating"* or *"approximating"* the desired path by a polynomial function [The space curve that the manipulator hand moves along from an initial location (position and orientation) to final position is called the *"path"*]. These time base sequence locations, also called time law or time history obtained from the trajectory planning serve as reference input or "control set points" to the manipulator's control system. *The control system, in turn, assures that the manipulator executes the planned trajectories.*

- The kinematics and dynamics of the manipulator govern the transition of posture with actuators exerting the desired torques. Because no limits must be violated and no unmodelled behaviour should be executed, it is imperative to devise *planning algorithms* that generate smooth trajectories. One of the primary objectives of any algorithm is *to achieve a smooth motion of the manipulator*. The smooth motion results in reduced vibrations and wear of the mechanical system.

A systematic approach to the trajectory problem is to view the trajectory planner as a black box, as shown in Fig. 1.32.

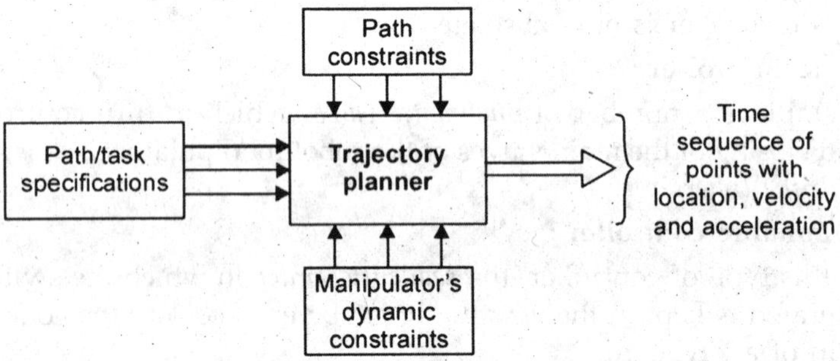

Fig. 1.32. Trajectory planning block diagram.

— *The trajectory planner accepts input variables which indicate the constraints of the path and outputs a sequence of time-based intermediate configurations of the manipulator hand (position and orientation, velocity, and acceleration).*

Planning of manipulation trajectories :

For planning manipulator trajectories, the following *two approaches* are used :

(*i*) The *first approach* requires the user to explicitly *specify a set of constraints* (*e.g.* continuity and smoothness) on position, velocity, and acceleration of the manipulator's generalised coordinates at selected locations (called "knot points" or interpolation points") along the trajectory. The trajectory planner then selects a parameterised trajectory from a class of functions that *interpolates* and satisfies the constraints at the interpolation points.

In this approach, the constraint specification and the planning of the manipulator trajectory are performed in *joint coordinates.* Since no constraints are imposed on the manipulator hand, it is difficult for the user to trace the path that the manipulator hand traverses. Hence, manipulator hand may hit obstacles with no prior warning.

(*ii*) In the *second approach,* the user explicitly specifies the path that the manipulator must traverse by an analytical function.

In this approach, the path constraints are specified in *cartesian coordinates, and the joint actuators are servoed in joint coordinates.* Hence, to find a trajectory that approximates the desired path closely, *one must convert cartesian path constraints to joint path constraints by some functional approximations and then find a parameterised trajectory that satisfies the joint path constraints.*

Motion control problem :

In general, the motion control problem consists of :

(*i*) Obtaining dynamic model of the manipulator ;

(*ii*) Using these models to determine control laws or strategies to achieve the desired system response and performance.

- From the control analysis point of view, the movements of a robot arm is usually performed in two distinct control phases :

— The *first* is the "*gross motion control*" in which the arm moves from an initial position/orientation to the vicinity of the desired target position/orientation along a planned trajectory.

— The *second* is the "*first motion control*" in which the end-effector of the arm dynamically interacts with the object using sensory feedback information from the sensors to complete the task.

- Now-a-days, industrial approaches to robot arm control *treat each joint of the robot as a simple joint servomechanism.*

— The "*servomechanism approach*" models the varying dynamics of a manipulator *inadequately* because it neglects the motion and configuration of the whole arm mechanism. These changes in the parameters of the controlled system sometimes are significant enough to render conventional feedback control strategies ineffective. The result is reduced servo response speed and damping limiting the precision and speed of the end-effector and making it appropriate only for limited-precision tasks. Manipulators controlled in this manner move at slow speeds with unnecessary vibrations. Any significant performance gain in this and other areas of robot arm control require the consideration of more efficient dynamic models, sophisticated control approaches and the use of dedicated computer architectures and parallel processing techniques.

1.20 CONVERSION OF MOTION

— The *joints of a robotic manipulator* may be *rotary or linear in nature.*

— The components employed to implement a practical joint usually consists of an 'actuator' (e.g. motor) coupled to the physical joint by a mechanical transmission. This transmission is used to take the actuator motion and direct it to the joint in order to provide such characteristics as a change of rotational direction, a change of axes, torque multiplication (or reduction), and speed reduction (or multiplication).

— The "*transmission*" *may be used to convert rotary motion to linear motion or to provide a "match" between the actuator and load in order that the maximum energy is transferred to the load.*

— Some *devices* produce linear relationships between input and output motions and thus are well suited for use in closed-loop control. However, some of the more traditional mechanical schemes have highly nonlinear input/output relationships, which can create several control problems.

The various types of motion conversions are given in the following text.

1. Rotary-to-rotary motion conversion :

Such type of conversion is associated with the following "*components*" :

(*i*) *Gear trains* ;

(*ii*) *Harmonic drives* ;

(*iii*) *Belts and pulleys* ;

(*iv*) *Sprockets and chains.*

— Each of the above components will take the torque and speed provided by a rotary actuator connected to the input shaft and provide rotary motion on the output shaft.

— Depending on the type of component and its characteristics, the motion on the output shaft may be *reversed*, have an increased torque (with a reduced speed), or have an increased speed (with a reduced torque).

2. Rotary-to-linear motion conversion :

Such a conversion is concerned with taking the *rotational motion* and *torque* from an actuator and producing a *linear motion* and *force* on the output. "*Components*" whose ideal model produces a linear relationship between the input shaft's rotation and output's linear displacement are :

(*i*) *Lead screw ;*

(*ii*) *Rack and pinion ;*

(*iii*) *Rohlix* – This device is based on the *principle of rolling contact.*

- Non-linear relationships are available by "*slider-crank assemblies*" and "*cams*".

3. Linkages :

Before the advent of computer-controlled machinery the complete cycle of an automated machine was controlled by cam profiles and/or linkages. The linkages may be :

(*i*) Four-bar linkage ;

(*ii*) Four-bar linkage using discs for two links ;

(*iii*) Crank and rocker ;

(*iv*) Slider crank from four-bar linkage.

4. Couplers :

A coupler is a device used to connect shafts so that one can drive the other. Typically it is used to *connect an actuator or motor to the rest of the joint mechanism.*

- "*Oldham coupling*" is used to connect two shafts that are parallel but not collinear.
- "*The universal joint*" is used for *connecting two shafts whose axes are not in line but intersect at a point.*

The "*flexible shaft coupler*" is a feasible alternative to both the oldham coupler and the universal joint.

1.21 TECHNIQUES OF ROBOT CALIBRATION

The basic methods of improving robot accuracy are enumerated and discussed below:

1. Joint calibration.
2. Open-loop calibration.
3. Closed-loop calibration.

1. Joint calibration :

— It relates to the procedure to establish the zero-reference position of the robot joints.

— It determines the relationship between encoder values and the actual joint angles or position.

— This type of calibration improves the robot accuracy at one particular point only in the robot workspace. Thus, this is of limited use as most of the applications require that the robot be accurate at more than one point.

- The majority of industrial robots follow some form of initial calibration, yet still have larger inaccuracies.

2. Open-loop calibration :

— In this type of calibration, measurements are taken off-line and then some form of compensation is applied retrospectively; no further monitoring of the accuracy is carried out while the robot is in use. An on-line application requires access to robot controller, which is not widely available.

— Whereas on-line application provides the advantage of trajectory control, off-line solutions are adequate for the purpose of some high accuracy tasks such as assembly.

Following are the *two* distinct approaches to open-loop calibration :

(*i*) ***Forward calibration*** **:**

— It involves producing an improved kinematic model of the robot.

— In this approach the measurements of end-effector pose are taken at several known joint configurations. The parameters of the model are identified using these measurements.

— The inverse of this kinematic model is used in the compensator.

Advantages :

(*a*) The calibration covers the entire work volume.

(*b*) Both the position and orientation inaccuracies are corrected automatically.

(*c*) The physical sources of error can be identified.

Limitations :

(*a*) An iterative procedure is needed to solve for the correct joint.

(*b*) A good understanding of the specific robot kinematics is required.

(*c*) Other possible contributions to error are ignored.

(*ii*) ***Inverse calibration*** **:**

— This approach considers methods of modelling the error between the existing nominal inverse model and the actual inverse model.

— These error models have no physical significance, but rather are based on observed inaccuracies.

Advantages :

(*a*) All sources of error may not be modelled.

(*b*) The method is independent of robot construction and can be easily applied to any system.

(*c*) There is no need to solve for joint angles by an iterative process.

Limitations :

(*a*) The inaccuracy mapping is good only for specific test conditions (e.g. arm configuration and load).

(*b*) The inaccuracies are not attributed to any particular defect and cannot be used in diagnosis as with forward calibration.

(*c*) Compensation can only realistically be achieved in a localised area/volume.

3. Closed-loop calibration :

— In this method, sensors are used to provide *feedback* to the robot controller on-line, thus enabling compensation as inaccuracies occur. An additional control loop modifies the robot paths.

Presently this approach is not available to robots with full six DOFs as autonomous sensors that can detect the full pose of the robot are not yet available.

— This type of calibration can reduce the system response time.

1.22 ROBOT SENSING AND VISION

Robot sensing :

- A robot can interact with its environment in a flexible manner with the use of external sensing mechanisms. This is in contrast to preprogrammed operations in which a robot is 'taught' to perform repetitive tasks via a set of programmed functions. Although the latter is by far the most predominant form of operation of present industrial robots, the use of sensing technology to endow machines with a greater degree of intelligence in dealing with their environment is indeed an active topic of research and development in the robotics field.
- The two principal categories of robot sensors are :

(*i*) *"Internal state sensers"* :

— These sensors deal with the detection of variables such as arm joint position, which are used for robot control.

(*ii*) *"External state sensors"* :

— These sensors deal with the detection of variables such as range, proximity and touch.

— External sensing is used for robot guidance, as well as for object identification and handling.

Robot vision :

Although proximity, touch, and force sensing play a significant role in the improvement of robot performance, *'vision'* is recognised as the most powerful of robot sensory capabilities.

Robot vision *may be defined as a process of extracting, characterizing, and interpreting information from images of a three-dimensional world*. This process, also commonly referred to as *machine or computer vision* may be subdivided into following six areas :

(*i*) Sensing ; (*ii*) Preprocessing ;
(*iii*) Segmentation ; (*iv*) Description ;
(*v*) Recognition ; (*vi*) Interpretation.

1.23 ROBOT PROGRAMMING LANGUAGES

There are several ways to communicate with a robot, and the three major approaches to achieve it are :

1. Discrete word recognition.
2. Teach and playback.
3. High level programming languages.

1. Discrete word recognition :

- Current state-of-the art speech recognition is quite primitive and generally speaker-dependent. It can recognize a set of discrete words from a limited vocabulary and usually requires the user to pause between words.
- Although it is now possible to recognise words in real time due to faster computer components and efficient processing algorithms, the usefulness of

discrete word recognition to describe a task is *limited*. Moreover, it requires a large memory *space* to store speech data, and it usually requires a training period to build up speech templates for recognition.

2. Teach and playback :

- This method involves teaching the robot by leading it through the motions to be performed. This is usually accomplish in the following *steps* :
  - (*i*) Leading the robot in slow motion usual manual control through the entire assembly task, with the joint angles of the robot at appropriate locations being recording in order to replay the motion ;
  - (*ii*) Editing and playing back the taught motion ; and
  - (*iii*) If the taught motion is correct, then the robot is run at an appropriate speed in a repetitive motion.

This method is also known as '***guiding***" and is the most commonly used approach in present-day industrial robots.

3. High level programming languages :

- The use of high-level programming is a more general approach to solve the human-robot communication problem.
- The robots are commonly used in areas such as welding, spot welding, and paint spraying. *These tasks require no interaction between the robot and the environment and can be easily programmed by guiding*. However, *the use of robots to perform assembly tasks generally requires high-level programming techniques.*

QUESTIONS WITH ANSWERS

Q. 1. Define the following robotic terms :

(*i*) Degree of freedom (DOF) ;	**(*ii*) Articulated structure ;**
(*iii*) Active accommodation ;	**(*iv*) Balance interface ;**
(*v*) Compliance ;	**(*vi*) Controller ;**
(*vii*) Cybernetics ;	**(*viii*) Flexible automation ;**
(*ix*) Interface ;	**(*x*) Joystick ;**
(*xi*) On-line programming ;	**(*xii*) Off-line programming ;**
(*xiii*) Play back robot ;	**(*xiv*) Pose-to-pose control ;**
(*xv*) Rack indexing ;	**(*xvi*) Response time ;**
(*xvii*) Wrist ;	**(*xviii*) Pendant.**

Ans. (*i*) **Degree of freedom (DOF).** It is defined as one of the variables (maximum number of six) required to define the motion of a body in space.

(*ii*) **Articulated structure.** It is a set of links and joints that constitute arm and the wrist of a robot.

(*iii*) **Active accommodation.** It is an integration of sensors, control, and robot motion to achieve alteration of a robot's preprogrammed motions in response to sensed forces.

It is used to stop a robot when forces reach set levels, or to perform force feedback tasks like insertions, door opening and edge tracing.

(*iv*) **Balance interface.** It is the electronic hardware and software necessary to utilise an electronic balance as an automated workstation.

(*v*) **Compliance.** It is the flexible behaviour of a robot or any associated tool in response to external forces exerted on it.

(*vi*) **Controller.** It is an information processing device whose inputs are both desired and measured position, velocity or other pertinent variables in a process and whose outputs are drive signals to a controlling motor or actuator.

(*vii*) **Cybernetics.** It is the study of control and communication in, and particularly between, humans and machines or robots.

— A human machine cybernetic system is a functional synthesis of a human system and a technological system or machine.

(*viii*) **Flexible automation.** It refers to the multitasking capability of robots; multipurpose, adaptable, reprogrammable.

(*ix*) **Interface.** It refers to those connections of one system that are matched to another system that is distinctly different because of the basic nature of each system.

(*x*) **Joystick.** It is a mannually controlled device whose variable position and orientation or applied forces are measured and results in commands to the robot control system.

(*xi*) **On-line programming.** It is the robotic programming that makes use of the manipulator. It utilizes the actual robot in order to develop procedures and define the values of data items in a task program.

Example. The record-playback method, which is dependent on an actual robot for testing and demonstration.

(*xii*) **Off-line programming.** It is the programming of robotic controllers and computers that involves the writing of task programs, the running of simulations, and the collection and organisation of data either away from the robot or when it is not in operation.

(*xiii*) **Play back robot.** It is a robot that can repeat a task program which is entered through teach programme.

(*xiv*) **Pose-to-pose control.** It is a control scheme whereby the inputs or commands specify only a limited number of points along a desired path of motion. The control system determines the intervening path segments.

(*xv*) **Rack indexing.** It is the ability to define and access all positions in any rectangular array or rack by teaching a few positions.

(*xvi*) **Response time.** It is the period of time that lapses from the moment an order to start a robotic operation is given and the moment the actual operation begins. This takes into account data transmission and reception, memory access time, and computer interfacing.

(*xvii*) **Wrist.** It is the set of rotary joints to which a robot's end-effector is attached.

— It may exhibit compliance, overload protection and strength.

(*xviii*) **Workstation.** It is an apparatus that performs some automated function or with which a laboratory unit operation is performed.

Q. 1.2. With the help of a neat sketch shows the basic components of a robot conneted as a system.

Ans. Fig. 1.33, shows the four basic components of a robot system, these are :

1. Manipulator ;
2. Sensory devices ;
3. Controller ;
4. Power conversion unit.

It may be noted that the *sensory devices are spread throughout the system*. For example, in addition the TV camera (a visual sensor), each joint contains sensors for position, velocity, and/or acceleration.

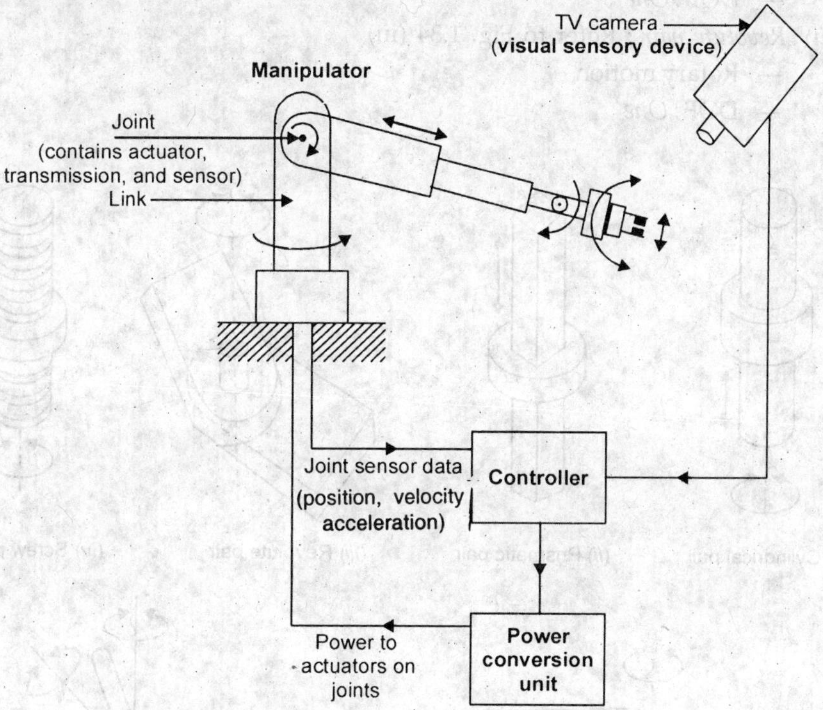

Fig. 1.33. Components of a robot system.

- The *manipulator* is composed of the following three divisions : (*i*) The major linkages ; (*ii*) The minor linkages (wrist components) ; (*iii*) The end-effector (gripper or tool).
- *Sensory* devices may monitor position, speed, acceleration or torque.
- *Controller* provides the "intelligence" to cause the manipulator to perform in the manner described by its tranier/user.
- *Power conversion unit* contains the components necessary to take a signal from the sequencer (either digital or low-level analog) and convert it into a meaningful power level so that the actuators can move.

Q. 1.3. Explain briefly the various types of joints/pairs used in robots.

Ans. A **joint** is a *'*lower pair'* formed between two links.

- The motion in the joint can be translatory (linear/sliding) or rotary/rotational, about or along the cartesian axes.
- The joints can exhibit one or more relative motion(s) at a time, depending on that they are *classified* as follows :

(*i*) *Cylindrical pair* : Refer to Fig. 1.34 (*i*)

— One sliding and one rotary motions.

* The relative motion featured by the sliding action between the two surfaces describes the characteristic connection known as the *lower pair*.

— DOF (Degree of freedom) : *Two*

(*ii*) *Prismatic pair* : Refer to Fig. 1.34 (*ii*)

— Sliding motion

— DOF : *One*

(*iii*) *Revolute pair* : Refer to Fig. 1.34 (iii)

— Rotary motion

— DOF : *One*

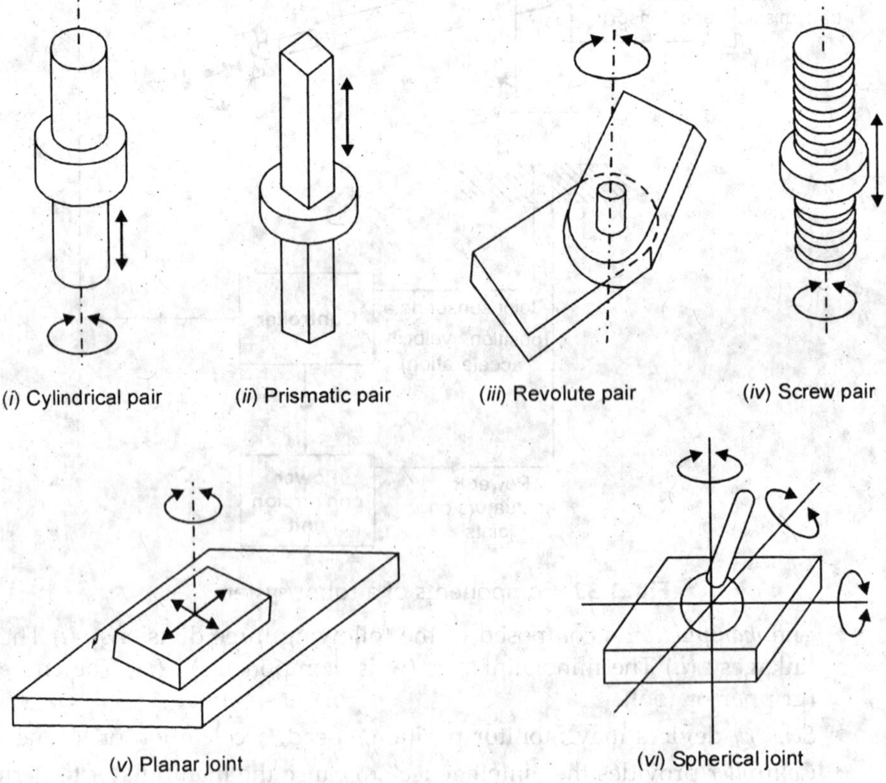

Fig. 1.34. Joints /pairs used in robots.

(*iv*) *Screw pair* : Refer to Fig. 1.34 (*iv*)

— One translatory and one rotary motions

— DOF : *Two*

(*v*) *Planar joint* : Refer to Fig. 1.34 (*v*)

— Two sliding and one rotary motions

— DOF : *Three*

(*vi*) *Spherical joint* : Refer to Fig. 1.34 (*vi*)

— Three rotary motions

— DOF : *Three*

Q. 1.4. What is an industrial robot ? Discuss briefly.

Ans. *An **industrial robot** is a general purpose, computer-controlled manipulator consisting of several rigid links connected in series by revolute or prismatic joints.*

— One end of the chain is attached to a supporting base while the other end is free and equipped with a tool to manipulate objects or perform assembly tasks. The motion of the joints results in relative motion of links.

— Mechanically, a robot is composed of an *arm* (or mainframe) and a *wrist subassembly plus a tool*. It is designed to reach a workpiece located within its work volume. The work volume is the sphere of influence of a robot whose arm can deliver the wrist assembly unit to any point within the sphere. The arm assembly generally can move with 3 DOFs (degrees of freedom). The combination of the movements positions the wrist unit at the workpiece.

— The *wrist unit* usually consists of *three* rotary motions. The combination of these motions orients the tool according to the configuration of the object for each in pickup. These three motions are often called *"pitch"*, *"yaw"*, and *"roll"*.

Hence, for a *6-jointed robot, the arm subassembly is the* ***positioning mechanism****, while the wrist subassembly is the* ***orientation mechanism****.*

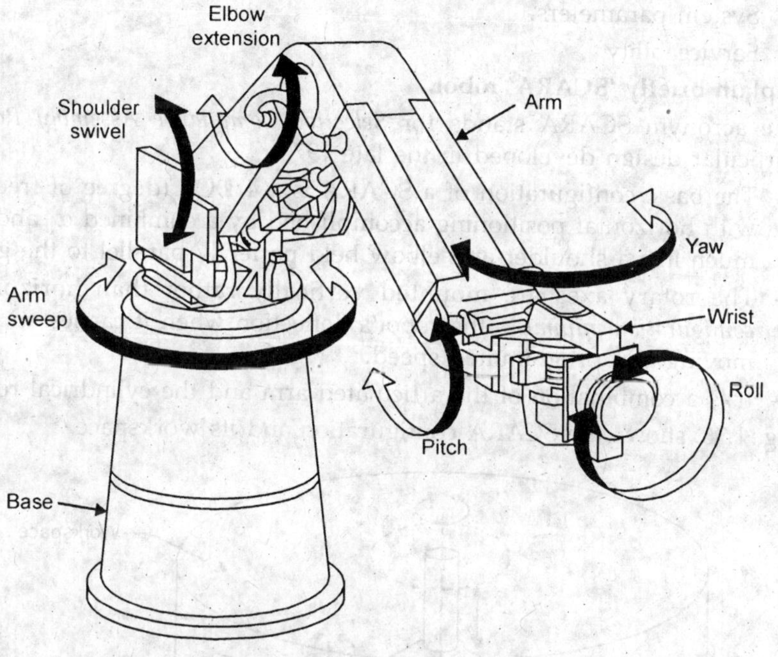

Fig. 1.35. Cincinnati Milacron.

These concepts are illustrated by Cincinnati Milacron T^3 robot (Fig. 1.35) and th Unimation PUMA robot arm [Fig. 1.3(*a*)]

Q. 1.5. What are the applications of industrial robots ?

Ans. Many commercially available industrial robots are widely used in manufacturing and assembly tasks such as :

- Spot/arc welding.
- Material handling.
- Parts assembly.
- Paint spraying.
- Loading and unloading NC (numerically controlled) machines.

- Space and undersea exploration.
- Handling hazardous materials.
- Prosthetic arm research.

Q. 1.6. List the desirable capabilities of a robot system.

Ans. The desirable capabilities of a robot system are as under :

1. Ability to move between points in various ways.
2. Ability to define points (or locations) in space to which the robot is to go.
3. Program control.
4. Control for the end-effector.
5. Provide interfaces to ancillary equipment such as parts, feeders, other robots, and/or vision systems.
6. External robot control and communications.
7. Housekeeping features.
8. Program debug and simulation.
9. System parameters.
10. Serviceability.

Q. 1.7. Explain briefly 'SCARA' robot.

Ans. The acronym SCARA stands for *Selective Compliance Assembly Robot Arm*, a particular design developed in the late 1970s.

— The basic configuration of a SCARA is a 4 DOF (degree of freedom) robot with horizontal positioning accomplished by a combined α_1 and α_2 motion, much like a shoulder and elbow held perfectly parallel to the ground.

— The rotary axes are mounted vertically rather than horizontally. This *configuration minimises* the robot's deflection when it carries an object while moving at a programmed speed.

— It is a combination of the articulated arm and the cylindrical robot.

Fig. 1.36, shows the SCARA configuration and its workspace.

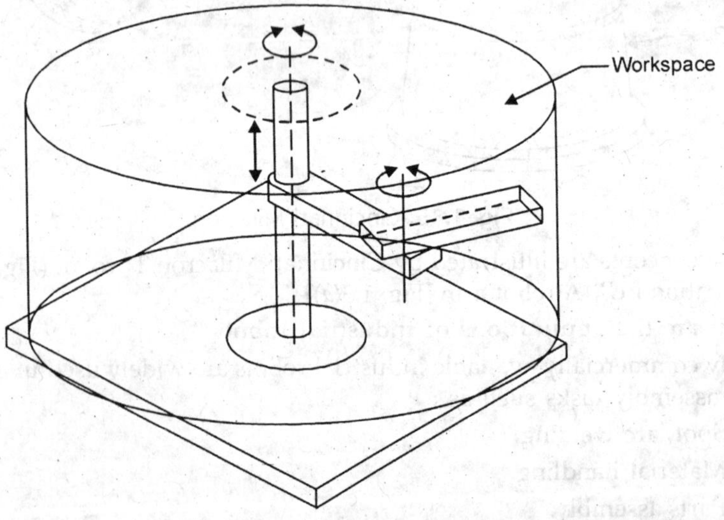

Fig. 1.36. The SCARA configuration and its workspace.

— SCARA robots are shaped somewhat like a donut and possess the following characteristics :

(*i*) A large work space.

(*ii*) Fast cycle times.

(*iii*) Excellent repeatability.

(*iv*) Good payload capacity.

(*v*) Sophisticated motion control with full programmability.

- The *"applications"* of these robots include :

(*i*) Pick and place work.

(*ii*) Application of sealout.

(*iii*) Assembly operations.

(*iv*) Handling mchine tools.

Q. 1.8. With the help of a neat sketch give the description of 'Articulated robot'.

Ans. Fig. 1.37. shows an articulated robot :

— It consists of three constant parts (*links*) that are joined with revolute joints and are placed on a *"rotating base"*.

— The kinematics layout is similar to human arm.

— The *"tool"* (gripper) is similar to a palm and is adjusted to the lower part of the arm through the wrist.

— The *"elbow"* connects lower and upper parts of the arm and the *"shoulder"* connects upper part of arm with base. Several times the shoulder joint has a rotational motion in horizontal plane.

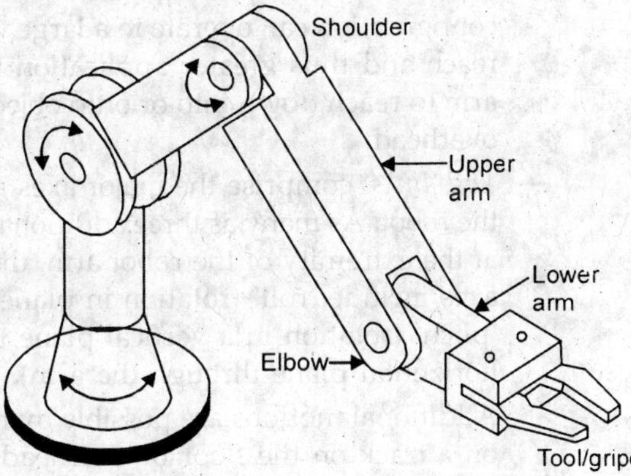

Fig. 1.37. Articulated robot.

— This robot has all three axes revolute, *so position resolution is completely dependent on arm's position*.

- The total accuracy of this type of robot is small, because joint errors are accumulated at the end of the arm, which is at wrist position.
- The *advantages* of these robots are :

(*i*) High mechanical flexibility.

(*ii*) Can move with high speed at the three degrees of freedom.

- These robots are used for :

(*i*) Gas welding ;

(*ii*) Arc welding ;

(*iii*) Assembly operations ;

(*iv*) Die casting ;

(*v*) Fettling machines ;

(*vi*) Spray painting.

Q. 1.9. Describe briefly robots with jointed-spherical or jointed arm coordinate system.

Ans. Fig. 1.38, shows the schematic form of robot with jointed-spherical coordinated system which is frequently used in several applications :

— This configuration consists of a base or trunk, an upper arm and a forearm which move in a vertical plane through the trunk. Further, an elbow joint is located between the fore arm and upper arm and the trunk. Rotary motion in the horizontal plane is also provided at the shoulder joint.

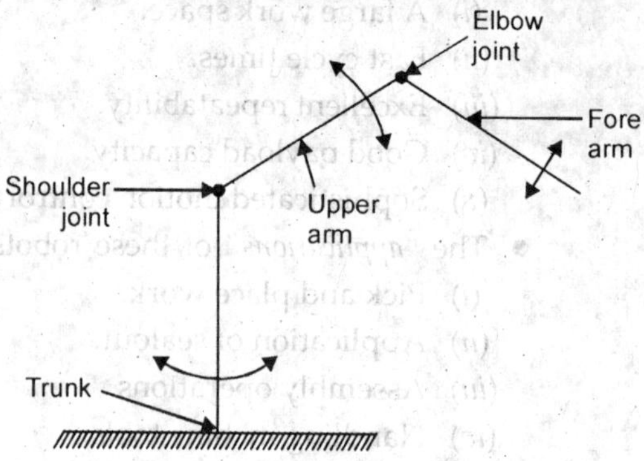

Fig. 1.38. Schematic form of robot with jointed-spherical coordinated system.

— The *work envelope approximates a portion of a sphere*. This configuration can operate in a large working volume for its size giving greater reach and thus greater application flexibility. It is also possible for the fore arm to reach down into or onto objects as easily as if the robot were mounted overhead.

— The *'links'* comprise the major axes and offer desired 'degrees of freedom' of the robot. As many as three additional DOFs (degrees of freedom) are provided at the extremity of the robot arm, the unit commonly called the *'wrist'*. Wrist axes include 'roll' (rotation in plane perpendicular to the plane of the arm), 'pitch' (rotation in a vertical plane through the arm), and 'yaw' (rotation in horizontal plane through the arm).

Additional motions are possible by mounting the robot on a two-axis table or on a track on the floor or overhead.

—Many of the robots available are 'modular' in design and the user may select as few as two or as many as seven or eight degrees of freedom, depending upon his needs.

Q.1.10. Draw workspaces for the following kinematic configurations :
(*i*) Rectangular ; (*ii*) Cylindrical ; (*iii*) Spherical.

Ans. Fig. 1.39, shows the workspaces for rectangular, cylindrical and spherical configurations.

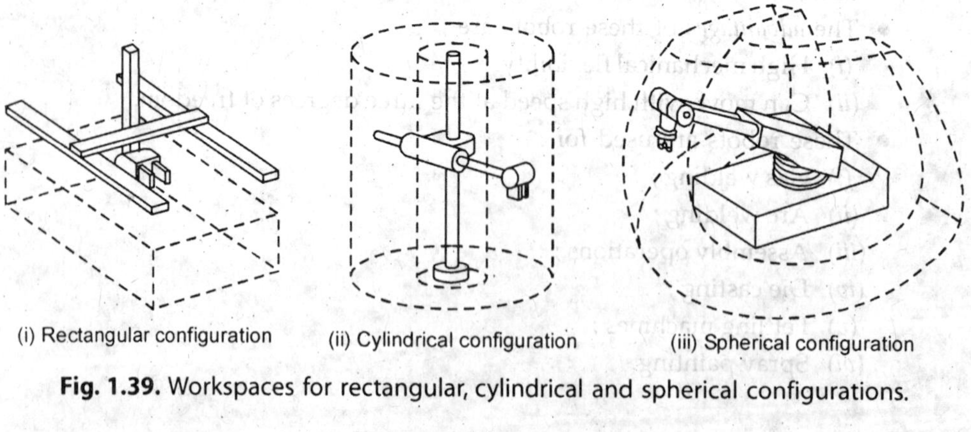

(i) Rectangular configuration (ii) Cylindrical configuration (iii) Spherical configuration

Fig. 1.39. Workspaces for rectangular, cylindrical and spherical configurations.

Q.1.11. Give the comparison of robots on the basis of coordinate systems.

Ans. The comparison of robots on the basis of co-ordinate systems is given in tabular form below :

S.No.	Type of robot	Axes	Merits	Demerits
1.	*Cartesian robot* (*e.g.* IBM's RS-1 robot and the Sigma robot from Olivetti)	3 linear axes	• Rigid structure. • Linear axes make for easy mechanical stops. • Easy to visualise. • Easy to program offline.	• Axes hard to seal. • Requires large-floor space for size of work envelope. • Can only reach in front of itself.
2.	*Cylindrical robot* (*e.g.* Versatran 600 robot from Prab)	2 linear axes; 1 rotary axis	• Rotational axis easy to seal. • Can reach all around itself. • Reach and height axes rigid.	• Cannot reach around obstacles. • Cannot reach above itself. • Base rotation axis is less rigid than a linear axis. • Horizontal motion is circular.
3.	*Spherical (polar) robot* (*e.g.* Unimate 2000B from Unimation Inc.)	1 linear axis; 2 rotary axes.	• Long horizontal reach.	• Generally has short vertical reach. • Cannot reach around obstacles.
4.	*Revolute or Articulated robot* (*e.g.* T^3 from Cincinnati Milacron and PUMA from Unimation Inc.)	3 rotary axes.	• Largest work area for least floor space. • Can reach above or below obstacles.	• Two or four ways to reach a point. • Most complex manipulator. • Difficult to program off-line.
5.	*SCARA*	1 linear axis ; 2 rotating axes.	• Can reach around obstacles. • Height axis is rigid. • Large work area for floor space.	• Highly complex arm. • Two ways to reach a point. • Difficult to program off-line.

Q.1.12. Enlist some advanced technological features of modern robots.

Ans. Some advanced technological features of modern robots are :

(*i*) *Multiple modes with controller* assisted by temperature gauge, pressure gauge and position sensors.

(*ii*) *Multiple adaptable robotic arms with modular construction.*

(*iii*) *Motion control* by mechanical couplings (coupled motion control) and coordinated kinematics and dynamics.

(*iv*) *Large work envelopes* and *higher payioaas* managed and controlled with servo tuning to avoid resonance and vibration.

(*v*) *Usage of micro-controllers and embedded systems* for less power requirement, compactness in size, changeable functions, less movable parts for longer life, chip forming the brain.

(*vi*) *Controller Area Network connections* :
- Good price-performance ratio ;
- Efficient control of the distributed intelligence ;
- Adequate reliability through error detection and error handling system ;
- Immunity against electromagnetic interference ;
- Provide dynamic connection and disconnection of nodes for flexibility ;
- Provide real time capability for better repeatability, accuracy and precision;
- Radio frequency and infrared links for digital communications.

(*vii*) *Programmable Automation Control* (PAC) for rapid advancement in capability for which re-engineering is needed.

(*viii*) *Robot vision.* Machine vision replaces human vision through video cameras, special computer hardware and software.

Q.1.13. What are the factors which should be considered while installing a robot or robot system ?

Ans.
- The installation of a robot or robot system should be carried out by the users in accordance with the manufacturer's recommendations and in conformation to acceptable industry standards.
- Temporary safeguarding devices and practices should be used to minimise the hazards associated with the installation of new equipment. The facilities, peripheral equipment, and operating conditions that should be considered are :

(*i*) Installation specifications ; (*ii*) Physical facilities ;
(*iii*) Electrical facilities ; (*iv*) Indentification requirements ;
(*v*) Special robot operating procedures or conditions; (*vi*) Control and emergency spot requirements ;
(*vii*) Action of peripheral equipment integrated with the robot.

Q.1.14. List the characteristics which should be considered while selecting a robot.

Ans. The following characteristics should be considered while selecting a robot :

(*i*) Degrees of freedom ; (*ii*) Size of class ;
(*iii*) Velocity ; (*iv*) Drive type ;
(*v*) Control mode ; (*vi*) Repeatability ;
(*vii*) Lift capacity ; (*viii*) Right-left traverse ;
(*ix*) Up-down traverse ; (*x*) In-out traverse ;
(*xi*) Yaw (*xii*) Pitch ;
(*xiii*) Roll (*xiv*) Weight of the robot.

Q.1.15. What do you understand by the terms "Workcell" and "Interlocks" ?

Ans. Workcell :

The term *"workcell" represents a collection of automated equipment and controls that act as a mean to coordinate all the activities within the robot workstation.* It is required because the robot system by itself has a very limited utility.

— In order to utilise it for performing some really useful tasks, it has to be *integrated* with many other things like workparts, part feeders, conveyors, control devices, process machinery, tools, machine tools, etc.

— A *workcell* need not essentially cover a single robot only, but may contain several robots.

— In a robot *workstation*, some of the activities take place in a sequence and some others occur simultaneously. A controlling device, called a ***workcell controller*** or ***workstation controller***, is, therefore suitably incorporated, which coordinates and regulates these different activities and ensures that they are performed in a proper sequence.

Interlocks :

There has to be some means in *workcell control* to check the continuation of sequence of the work cycle whenever needed. For example, after a certain sequence, suppose the next cycle sequence is required to be performed only under some specified conditions. If these conditions are not met, the sequence need to be discontinued until the required conditions are available. The means used to apply this check is known as an ***Interlock***.

—The workcell carries two types of interlocks—one *outgoing* and the other *incoming*. The signal sent by the workstation controller to an external equipment or machinery to operate or not operate is known as *external interlock*. In the same way, an *internal interlock* is the signal sent by the external machinery (or device) to the workstation controller to allow or discontinue further sequencial operation of the work cycle. Thus, the *provision of interlocks facilitates an effective control over the workcycle by permitting the sequential activities to take place only when they should and preventing them when they should not*.

Q.1.16. What is the significance of "Finite element techniques" in robots ? Explain briefly.

Ans. Industrial robots known for positioning flexibility have an undesired characteristic of elastic flexibility. The elastic flexibility impairs the positional accuracy in assembly tasks.

The beam-like structure used in most of the robots leads to considerable bending and deflection. *Finite element method (FEM) technique permits static and dynamic analysis of structure and can be used for estimating deflection and stresses at various locations, optimisation of structure, for estimating natural frequencies etc.*

Q.1.17. Draw the basic control block diagram for robot manipulators. Considering the robot arm control as a path-trajectory problem, give the classification of motion control.

Ans.
- Fig. 1.40, shows basic control block diagram for robot controllers.
- Considering the robot arm control as a path-trajectory tracking problem, motion control can be classified as follows :

  (*i*) *Joint motion controls* :

  — Computed torque technique ;
  — Joint servomechanism ;
  — Minimum-time control ;
  — Variable structural control ;
  — Non-linear decoupled control.

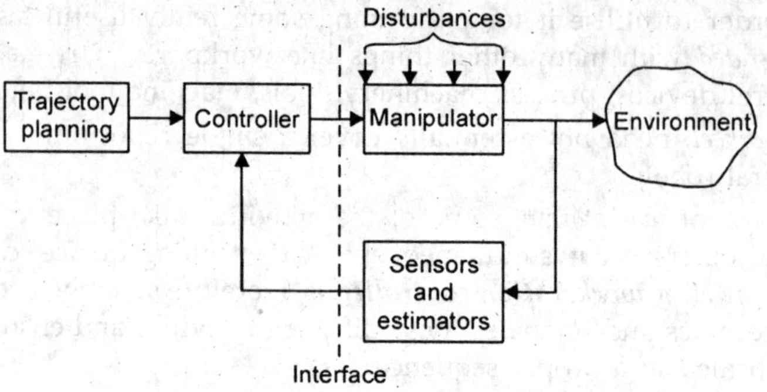

Fig. 1.40. Basic control block diagram for robot controllers.

(*ii*) *Resolved motion controls* (cartesian space control) :

— Resolved motion rate control ;

— Resolved motion acceleration control ;

— Resolved motion force control.

(*iii*) *Adaptive controls* :

— Self-tuning adaptive control ;

— Model referenced adaptive control ;

— Resolved motion adaptive control ;

— Adaptive perturbation control with feedforward compensation.

Q.1.18. Explain briefly the use of a robot in a palletizing operation.

Ans. Fig. 1.41. shows the use of a robot in a *palletizing operation* :

—When a part moving on a conveyor interrupts the light beam from a photo emitter, the controller commands the robot to acquire the part.

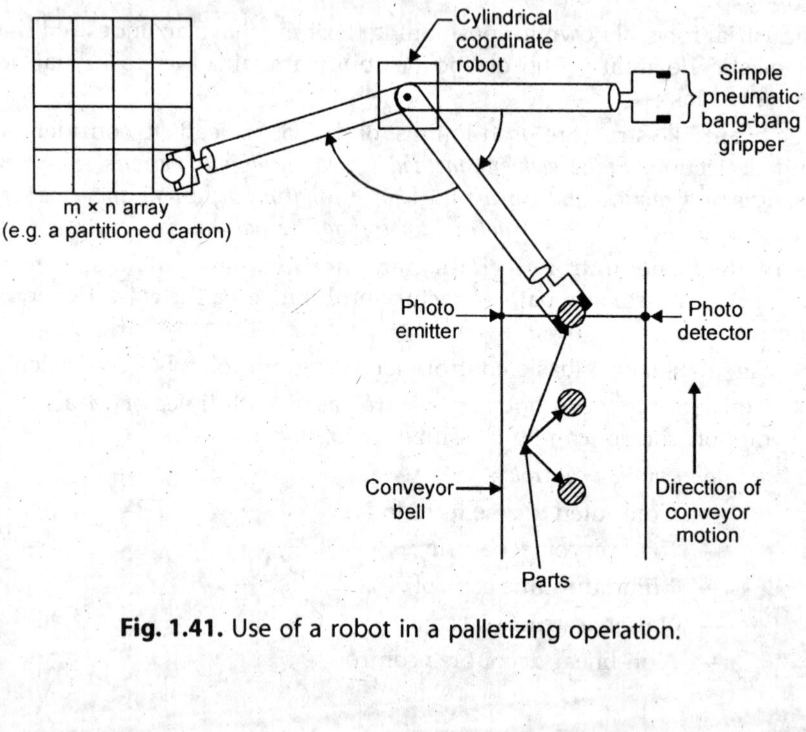

Fig. 1.41. Use of a robot in a palletizing operation.

— This part is then moved to and placed in one of the locations (empty) in a partitioned carton.

— This process is repeated until all such locations are filled, at which time the carton is removed from the loading station, any empty one replaces it, and the operation is repeated.

HIGHLIGHTS

1. An *industrial robot* is a reprogrammable multifunctional manipulator designed to move material, parts, tools or specialised devices through various programmed moitons for performance of a variety of tasks.
2. The various *components of a robot* are :

   Base; Manipulator arm; End-effector; Drives; Controller; Sensors.
3. Regardless of configuration, the function of the robot arm configuration is to position a wrist assembly which orients an end-effector to the proper position and orientation (jointly referred to as POSE) dictated by the task at hand.
4. The function of a *"joint"* is to permit relative motion between two links or arms of a robot.
5. *"Work envelope"* is defined as the envelope or space within which the robot can manipulate the end of the wrist.
6. The degrees of freedom (DOFs) of a system depend on the number of variables or coordinates that are needed to describe its position.
7. *"Robot arm kinematics"* deals with analytical study of the geometry of motion of a robot arm with respect to a fixed reference coordinate system without regard to the forces/moments that cause the motion.
8. Types of controllers used in robotics are :

   Drum controller; Air logic controller; Programmable controller; Microprocessor-based controller; Minicomputer-based controller.

OBJECTIVE TYPE QUESTIONS

Choose the Correct Answers :

1. Prosthesis area of robotics deals with
   (*a*) artificial replacements for parts of the human body
   (*b*) remote manipulation
   (*c*) robot designed to walk on legs
   (*d*) none of these.
2. The main advantages of robotics include :
   (*a*) Reliability (*b*) Increased flexibility
   (*c*) Low cost in the long run (*d*) All of the above.
3. The main function of a robot is
   (*a*) 'sensing' the environment by external sensors
   (*b*) 'decision making' based on the information received from the sensor
   (*c*) 'performing' the task decided
   (*d*) all of the above.

4. A machine can qualify as a robot subject to which of the following conditions ?
   (*a*) Reprogrammble.
   (*b*) Sensing and perception.
   (*c*) Function autonomously and/or interact with human being.
   (*d*) Carry out different tasks.
   (*e*) All of the above.
5. Tele-robots are
   (*a*) so designed as to navigate and carry out the task with the intervention of human beings.
   (*b*) built to mimic humans.
   (*c*) provided with a feedback of video, sound and other data.
   (*d*) guided by human operators through remote control.
6. A robot may be designed for which of the following jobs ?
   (*a*) To simply pick up and place the workpieces.
   (*b*) To interact with and work load a lathe, a milling machine or any equipment.
   (*c*) To perform some assembly work.
   (*d*) All of the above.
7. Which of the following methods of gripping is used in a robot ?
   (*a*) Mechanical gripping. (*b*) Magnetic gripping.
   (*c*) Vacuum gripping. (*d*) All of the above.
8. Which of the following drives are clean and quiet with a high degree of accuracy and reliability ?
   (*a*) Pneumatic drives. (*b*) Hydraulic drives.
   (*c*) Electrical drives. (*d*) All of the above.
9. A controller includes which of the following components ?
   (*a*) Memory
   (*b*) Computational unit
   (*c*) Hardware to interface with external world (sensors and actuators)
   (*d*) Hardware for a user interface.
   (*e*) All of the above.
10. Which of the following sensors are non-contact type robotic sensors ?
    (*a*) Force sensors. (*b*) Touch sensors.
    (*c*) Torque sensors. (*d*) Position sensors.
    (*e*) None of the above.
11. Which of the following are contact robotic sensors ?
    (*a*) Proximity sensors. (*b*) Electro-optical sensors.
    (*c*) Range imaging sensors. (*d*) None of the above.
12. 'Pitch motion' enables
    (*a*) rotation of wrist.
    (*b*) rightward or leftward swivelling movement of the wrist.
    (*c*) up and down movement of the wrist and involves rotational movements as well.
    (*d*) none of these.
13. Which of the following configurations has three mutually perpendicular axes ?
    (*a*) Cartesian coordinate configuration. (*b*) Cylindrical configuration.
    (*c*) Spherical configuration. (*d*) None of the above.

14. In which of the following configurations, there is a telescopic arm which pivots about a horizontal axis and also rotates about a vertical axis ?
    (*a*) Cylindrical configuration. (*b*) Spherical configuration.
    (*c*) Jointed-arm configuration. (*d*) None of above.
15. Mechanical errors arise due to which of the following causes ?
    (*a*) Hysteresis. (*b*) Deflection of links.
    (*c*) Hydraulic leaks. (*d*) All of the above.
16. Which of the following controllers is most popular ?
    (*a*) Drum controller. (*b*) Air logic controller.
    (*c*) Programmable controller. (*d*) Microprocessor-based controller.

ANSWERS

1. (*a*)	2. (*d*)	3. (*d*)	4. (*e*)	5. (*d*)	4. (*d*)	7. (*d*)
8. (*c*)	9. (*e*)	10. (*e*)	11. (*d*)	12. (*c*)	13. (*a*)	14. (*b*)
15. (*d*)	14. (*d*).					

THEORETICAL QUESTIONS

1. Define the term 'Robotics' ?
2. What are the important areas of robotics ?
3. What are the main advantages of robotics ?
4. State the laws of robotics.
5. List the motivating factors for the introduction of robotics systems to the industrial world.
6. What is a robot ?
7. What are the functions of a robot ?
8. How does a robot differ from an automated machine ?
9. What are the conditions which qualify a machine as a robot ?
10. What are the similarities and dissimilarities between Robots and NC machines ?
11. Enumerate the capabilities of industrial robots.
12. What are the objectives of using industrial robots ?
13. State the advantages and disadvantages of robots.
14. What are the basic components of a robotic system ? State the main function of each of the components.
15. Explain briefly the following robot components :
    (*i*) Manipulator arm ; (*ii*) End-effector ;
    (*iii*) Controller ; (*iv*) Sensors.
16. Explain briefly tactile and non-tactile sensors.
17. Discuss briefly the six basic motions or degrees of freedom (DOFs).
18. How are robots classified ? Explain briefly.
19. Explain briefly the following configurations of robots :
    (*i*) Cartesian coordinate configuration.
    (*ii*) Cylindrical configuration.
    (*iii*) Spherical configuration.
20. Explain briefly non-servo controlled and servo-controlled robots.
21. Discuss briefly the following types of robots :
    (*i*) Programmable/reprogrammable general purpose robots.

(*ii*) Tele-operated robots.

(*iii*) Intelligent robots.

22. Differentiate between general purpose and special purpose robots.
23. List the factors which should be considered while selecting a robot for a particular application.
24. Explain briefly various types of mechanical joints used in robots.
25. Explain briefly the following :

    (*i*) Robot arms.

    (*ii*) Robot hands.
26. List the qualities which are expected in robots.
27. Explain briefly the following robot specifications :

    (*i*) Work envelope ; (*ii*) Load carrying capacity ;

    (*iii*) Speed of movement ; (*iv*) Repeatability ;

    (*v*) Spatial resolution ; (*vi*) Accuracy.
28. Explain briefly the various tests which are performed on robots to judge their suitability.
29. What do you understand by forward and reverse kinematics ? Explain briefly.
30. Draw the graphic symbols for kinematic chains.
31. How can robots be controlled ? Explain briefly.
32. Explain briefly the following controls :

    (*i*) Point-to-point control.

    (*ii*) Continuous path control.
33. Discuss briefly the following controllers used in robotics :

    (*i*) Air logic controller.

    (*ii*) Programmable controller.

    (*iii*) Microprocessor-based controller.
34. Explain briefly the following basic methods of improving robot accuracy :

    (*i*) Joint calibration.

    (*ii*) Open-loop calibration.

    (*iii*) Closed-loop calibration.
35. Explain briefly the following :

    (*i*) Robot sensing and vision.

    (*ii*) Robot programming languages.
36. Discuss briefly "Trajectory planning and motion control of manipulator".

CHAPTER

2

Robot End-Effectors

2.1 Introduction; 2.2 Types of end-effectors; 2.3 Classification of end-effectors; 2.4 Mechanical grippers; 2.5 Vacuum grippers; 2.6 Magnetic grippers; 2.7 Adhesive grippers; 2.8 Design of grippers – Requirements for the grippers design – tips for designing a gripper – Gripper force analysis. **2.9 Remote Center Compliance Devices (RCCs).** ***Questions with Answers*** – Highlights – Objective Type Questions – Theoretical Questions.

2.1 INTRODUCTION

2.1.1. Definition

End-effector *is gripper or end-of-arm tooling attached to the wrist of manipulator to accomplish the desired task.*

Its design depends on the shape and size of part to be held. In some applications end-effector may be required *to hold tool* and the robot must be able to manipulate the tool also in addition to control the relative position of the tool with respect to workpiece.

The end-effectors are also called ***grippers***

Or

End-effector *is any object attached to the robot flange ("wrist") that serves a function.* This includes :

— Robotic grippers;
— Robotic collision sensors;
— Robotic press tooling;
— Robotic point gun;
— Robotic tool changers;
— Robotic rotary joint;
— Compliance device;
— Robotic deburring tool etc.

2.1.2. General Aspects

- End-effectors range from commercial devices such as *pneumatic grippers to special tools* for applications ranging from arc welding and spray painting to the handling of silicon wafers in a corrosive acid environment.
- Due to diverse applications of robots, end-effectors are usually customized for a particular application and may include additional components such as force measuring devices or mechanical linkages that can be rocked in various positions. They may be multifuntional; that is, the same tool may be used to handle a part as it changes shape during a machining or assembly process.
- While considering the design of the end-of-arm tooling for a manipulator, certain parameters of the manipulator and the surrounding environment of the workcell must also be addressed. The *end-effector itself has weight and moments of inertia; thus these properties, along with the weight and shape of the payload, must be considered to ensure that the specifications of the robot are not violated.*

- The end-of-arm tool not *only has to interface to the robot, but must not interfere with any of the peripheral devices in the workspace.*
- In today's automated factories, reliability is of paramount importance. Since reliability decreases with increasing complexity, it is better to try to *keep the end-effector simple in both its design and functions.*
- In addition to gripping a part, the *end-effector can include a sensor to determine if a part is present.* The addition of a simple sensor can *make a gripper a relatively intelligent device.*
- More sophisticated end-effectors are also possible that include such features as *servo-controlled axes with force control, and sensors to measure the dimensions or weight of an object.*
- The *"structure" of an end-effector, and the nature of the programming and hardware that drives it, depends on the intended task.*

— If a robot is designed to set a table and serve a meal, then robotic hands, more commonly called grippers, are the most functional end-effectors. The same or similar gripper might be used with greater force, as a plier or wrench for tightening nuts or crimping wire. In a robot designed to tighten screws, however, a driver-head end-effector is more appropriate.

2.2 TYPES OF END-EFFECTORS

The end-effector is the special propose tooling which enables the robot to perform a particular job. Various types of end-efforts are designed for the same robot to make it more flexible and versatile.

End-effectors are categorised into two major types :

1. Grippers. 2. Tools.

2.2.1. Grippers

Grippers are used to hold either workparts (in pick-and-place operations, machine loading, or assembly work) or tools.

These are further *classified* as :

(*i*) *Mechanical grippers* — Here friction or the physical configuration of the gripper retains the object.

(*ii*) *Suction or vacuum cups* — Used for flat objects.

(*iii*) *Magnetised gripper* — These devices are used for ferrous objects.

(*iv*) *Hookes* — Used to lift parts off conveyors.

(*v*) *Scoopes or ladles* — Used for fluids, powders, pellets etc.

(*vi*) *Adhesive or electrostatic grippers.*

Fig. 2.1. shows sample gripper designs.

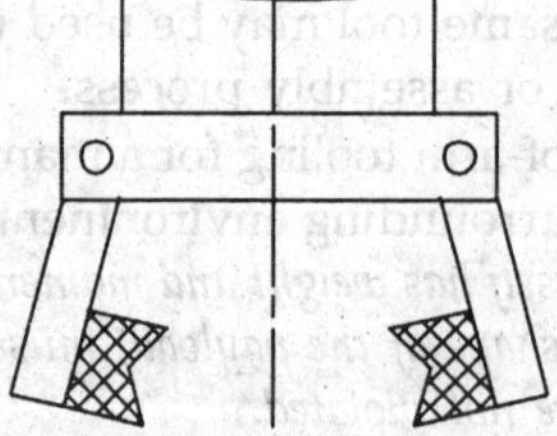

(a) Pivot-action gripper

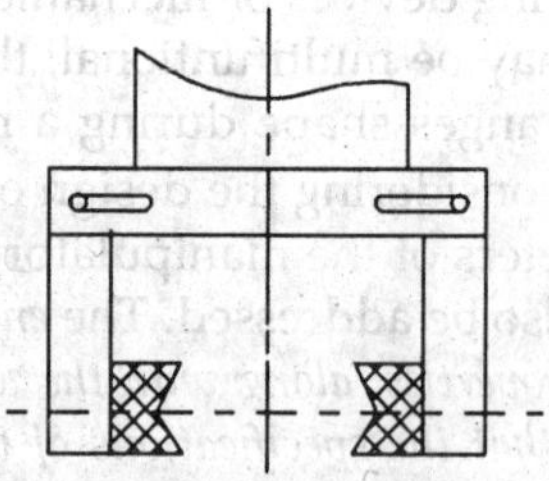

(b) Slide action gripper

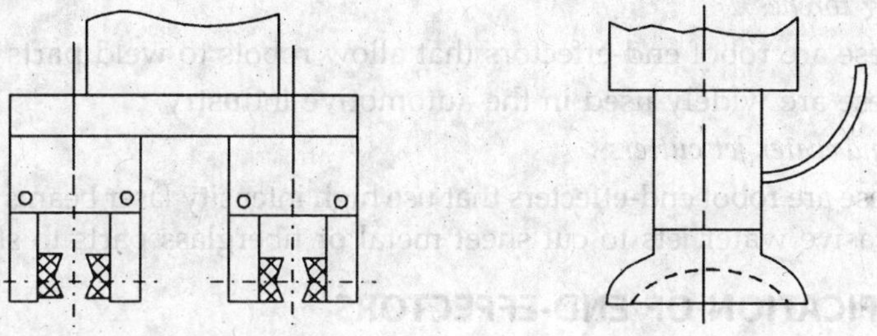

Fig. 2.1. Sample gripper designs.

- The grippers *may also be classified* as follows :

  (*i*) *Part handling grippers* :

  — These grippers are used to grasp and hold objects that are required to be transported from one point to another or placed for some assembly operations.

  — The part handling applications include :

  (*a*) Machine loading and unloading;

  (*b*) Picking parts from a conveyor and moving parts etc.

  (*ii*) *Tool handling grippers* :

  — These grippers are used to hold tools like welding gun or spray painting gun to perform a specific task.

  — The robot hand may hold a deburring tool.

  (*iii*) *Specialised grippers* :

  — These grippers may be specialised devices like Remote Centre Compliance (RCC) to insert an external mating component into an internal member, viz. inserting a plug into hole.

  — The other types of end-effectors employ some physical principle like magnetism or vacuum technology to hold the object securely.

2.2.2. Tools

"Tools" are fastened directly to the robot wist and become the end-effectors.

Invariably, a robot is required to manipulate a tool to perform an operation on a work part. In such applications end-effector is used as a gripper that can grasp and handle a variety of tools and the robot has multitool handling function. However, in most robot applications in which only one tool is to be manipulated, the *tool is directly mounted on the wrist, here the tool itself acts as the end-effector.*

Example :

(*i*) *Machine tools :*

Robot end-effectors can also be machine tools *e.g.*,

— Drills; — Grinding wheels;

— Cutting wheels; — Sanders.

(*ii*) *Measuring instruments :*

— These are end-effectors that allow the robot to precisely measure parts by running the arm lightly over the part using a measuring probe or gauge.

(*iii*) *Welding torches :*

— These are robot end-effectors that allow robots to weld parts together.

— These are widely used in the automotive industry.

(*iv*) *Laser and water jet cutters :*

— These are robot end-effecters that use high intensity laser beams or high pressure abrasive water jets to cut sheet metal or fiberglass parts to shape.

2.3 CLASSIFICATION OF END-EFFECTORS

The end-effectors can be *classified* in the following ways :

1. According to the types of grasping modality :

(*i*) *Mechanical fingers.*

(*ii*) *Special tools.*

(*iii*) *Universal fingers.*

- *"Mechanical Fingers"* are used to perform *special tasks.*

— Gripping by mechanical type fingers is *less versatile* and less *dextrous* than holding by universal fingers as the grippers with mechanical fingers have fewer number of joints and lesser flexibility. However, they economise the device cost.

— The grippers can also be subgrouped according to the number of fingers, typically two-, three-, and five-finger types. Out of these *two-finger gripper is the most popular.*

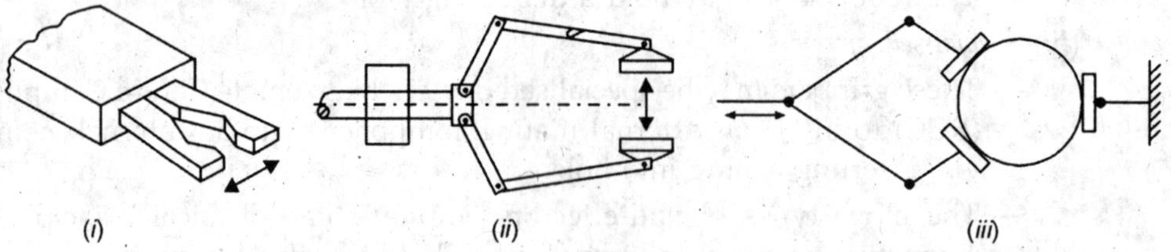

Fig. 2.2. Some fingered grippers for holding different types of jobs.

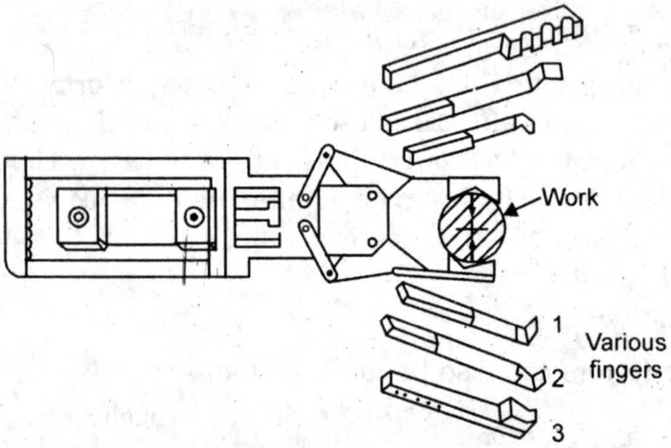

Fig. 2.3. Two fingered end-effector with changeable finger-tips.

Fig. 2.2. shows some fingered grippers for holding different types of jobs.

Fig. 2.3, shows a typical end-effector in which various designs of *interchangeable finger-tips can be adopted*.

— This is a two linkage gripper and it is actuated by a linear drive system-a *pneumatic hydraulic piston* that pulls or pushes the linkage causing rotation of the gripper linkages to close on or open free an object.

— *Electrical motors* may also be used to open or close the gripper with suitable linkage systems. However, for gripping rotational components suitable locations can be arranged.

- *Universal fingers* usually comprise multipurpose grippers of more than these fingers and or more than one joint on each finger which provide the capacity to perform a wide variety of grasping and manipulating assignments.

2. According to number of grippers mounted on the wrist :

(*i*) *Single grippers* mounted on the wrist.

(*ii*) *Multiple grippers* mounted on the wrist.

- Fig. 2.4. shows a multi gripper system which enables simultaneous execution of more than two different jobs.

— In this system, design methods for each individual gripper are subject ot those of single grippers.

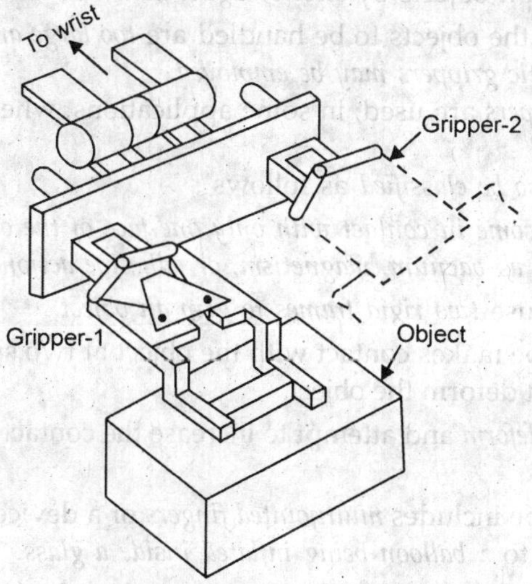

Fig. 2.4. Multiple grippers system.

3. According to mode of gripping :

(*i*) *Internal gripping.* (*ii*) *External gripping.*

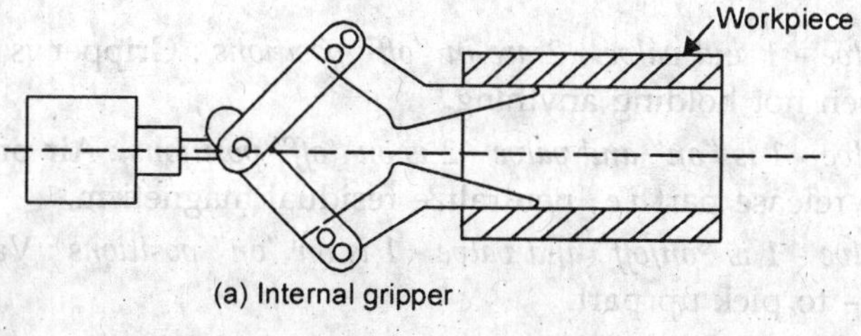

(a) Internal gripper

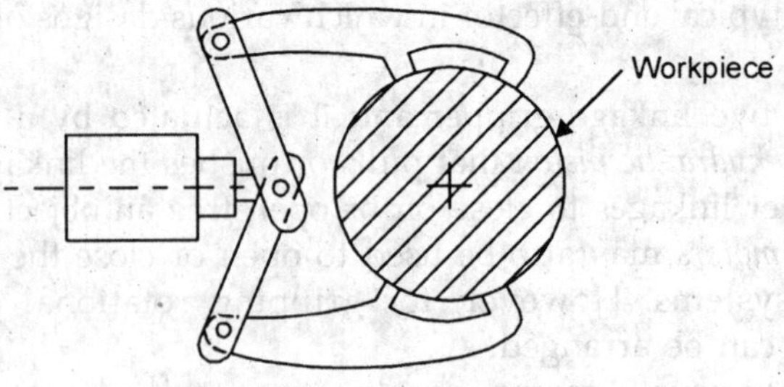

(b) Extenal gripper

Fig. 2.5. Internal and external gripping systems.

- The internal gripping system shown in Fig. 2.5 (*a*) grips the internal surface of object with open fingers whereas the external gripper system shown in Fig. 2.5 (*b*) grips the exterior surface of object with closed fingers.

4. According to the number of degrees of freedom (DOFs) :

— Typical mechanical grippers belong to the class of 1 DOF.

— A few grippers can be found with more than 2 DOFs.

- The robot grippers, using some special tooling action, robot grippers can be designed to retain objects by *electromagnet action* or under the action of *vacuum*.
  — Generally if the objects to be handled are *too large and ferromagnetic in nature, electromagnetic grippers may be employed.*
  — *Vacuum grippers* are used, in some applications, where the objects are *too thin to be handled.*
- Grippers *can also be classified* as follows :
  1. Those that *come in contact with only one face of the object to be lifted and use a method such as vacuum, magnetism, or adhesive action to capture the object.*
  2. Those that use *two rigid frames to grip an object.*
     — This type makes contact with the object at two specific points and may or may not deform the object.
  3. Those that *deform* and attempt to increase the contact area between the gripper and object.
     — This type includes *multijointed fingers* or a device operating on a principle similar to a *balloon being inflated inside a glass.*

     The above three classifications are referred to as systems using :

     (*i*) Unilateral action; (*ii*) Bilateral action; (*iii*) Multilateral action.

     While using above devices, the following ***considerations*** must be made :

(*i*) ***Unilateral action devices.*** Fig. 2.6. shows end effector with unilateral gripping action :

- *When valve - 1 and valve - 2 are in "off" positions* : Gripper is ported to free air – state when not holding anything.
- *When valve - 1 is "on" and valve - 2 is on "off" positions* : Air pressure is applied to cup – to release part *i.e.*, neutralize residual magnetism.
- *When valve - 1 is "on/off" and valve - 1 is on "on" positions* : Vacuum is applied to the cup – to pick up part.

The following *considerations* must be made while using these devices.

— The end effector must also be capable of releasing the part as well as picking it up.

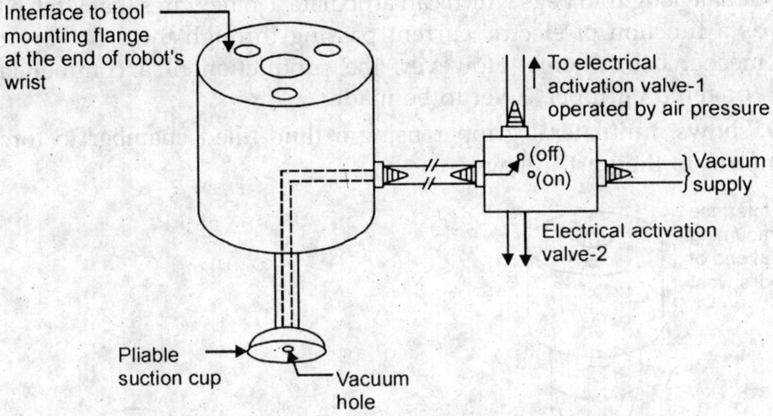

Fig. 2.6. End effectors with unilateral gripping action.

— When vacuum tool is used, the vacuum may have to be plugged with air to release the part; otherwise, if the seal between the part and vacuum cup is sufficient the part may stay on the tool even when the vacuum is shut off until some external force overcomes the residual vacuum.

— In case an adhesive or magnetic approach is used it may be necessary to have another device to 'kick' the part from the pickup tool.

(*ii*) **Bilateral action devices :** Fig. 2.7 shows end-effectors with bilateral gripping action :

— In this case, it may be necessary to include a material such as a piece of rubber on both points of contact *to increase the fiction*; otherwise, the part may fall out of the gripper.

— In such system, the devices such as rack or pinions or mechanical linkages may be required to achieve the required motions.

— The bilateral devices are binary in nature *i.e.*, either opened or closed. In this simple case, a pneumatic actuator is commonly used,.

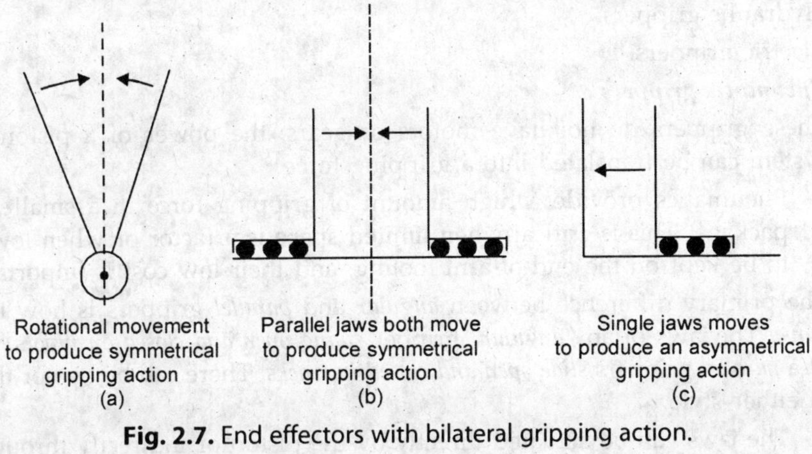

Fig. 2.7. End effectors with bilateral gripping action.

(*iii*) **Multilateral action devices :** Fig. 2.8, shows end-effectors with multilateral gripping action.

— The implementation of multilateral devices uses pneumatics to inflate and deflate long thin bags which can articulate a joint, wires that stretch or compress as a function of electric current passing through them or very complicated mechanical systems. However, the production of a commercially feasible "fingered" gripper is yet to be made.

Fig. 2.9. shows multilateral gripper using a fluid-filled chamber to force the outer layer to conform to the part in jaws.

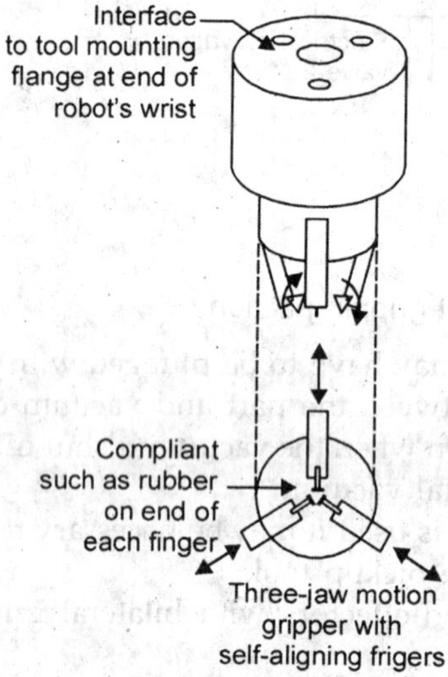

Fig. 2.8. End-effectors with multilateral gripping action.

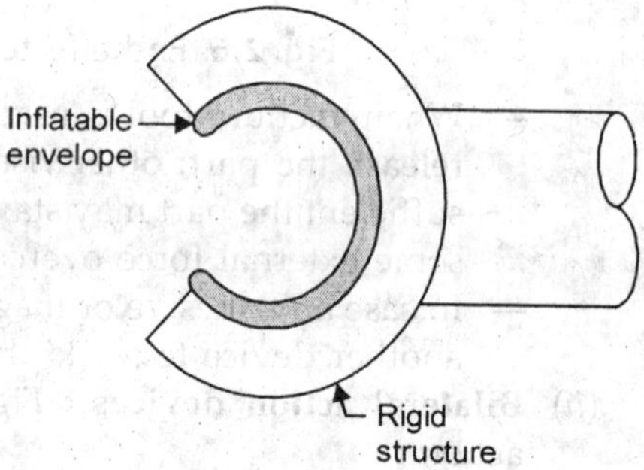

Fig. 2.9. Multilateral gripper.

5. According to source of power :

(*i*) Pneumatic grippers.

(*a*) Angular grippers

(*b*) Parallel grippers

(*ii*) Hydraulic grippers.

(*iii*) Electric grippers.

(*i*) *Pneumatic grippers :*

- These grippers do not have motors or gears; the power of a piston cylinder system can be translated into a gripping force.

  — Pneumatics provide a high amount of gripping force in a small and light package. This is critical when limited space is a factor or when low mass is to be kept on the end-of-arm tooling, and their low cost is important.

- The primary difference between *angular* and *parallel* grippers is how their *jaws move.* The jaws of an *"angular"* gripper *swing open and closed on pivots.* The jaws of a *parallal* grippers *slide open and closed in tracks.* There can be two or three jaws in either style.

  — The jaws can be actuated directly by the piston or indirectly through a cam or wedge mechanism. The former have longer stroke lengths, but the latter produce more gripping force.

— Angular grippers are less costly than parallel grippers (these require precise machining and more expensive bearings for the jaws). Further, the jaws on angular grippers can swing out of the way of incoming parts; that can save time when picking parts off a conveyor, because the actuator does not have to move as much.

However, the angular grippers, as compared with parallel grippers, *cannot handle as wide a range of part sizes, and they do not resist side loads as well.*

(*ii*) *Hydraulic grippers :*

They provide better gripping power (since hydraulic fluid is incompressible). However their *disadvantages* are :

(*i*) They are more costly and are generally less accurate than pneumatic or electric grippers.

(*ii*) They are messy and much more costly to maintain.

(*iii*) They are not suited for clean room applications (pneumatic grippers are common because they are clean especially when compared to hydraulic grippers).

(*iii*) *Electric grippers :*

These grippers are suitable for applications that require *high speeds* and for those requiring *light* or *moderate grip forces.*

These grippers have the following *advantages* :

(*i*) They do not put any dirt or particulate (as such are suitable for clean room operation).

(*ii*) Better control (it is relatively easy to add a stepper motor to control the gripper as to how far it opens and closes)

Its *disadvantages* are :

(*i*) An electric gripper tends to be a bit larger.

(*ii*) These grippers provide less force than pneumatic grippers.

2.4 MECHANICAL GRIPPERS

*A **mechanical gripper** is an end-effector that uses mechanical fingers actuated by mechanism to grasp an object.*

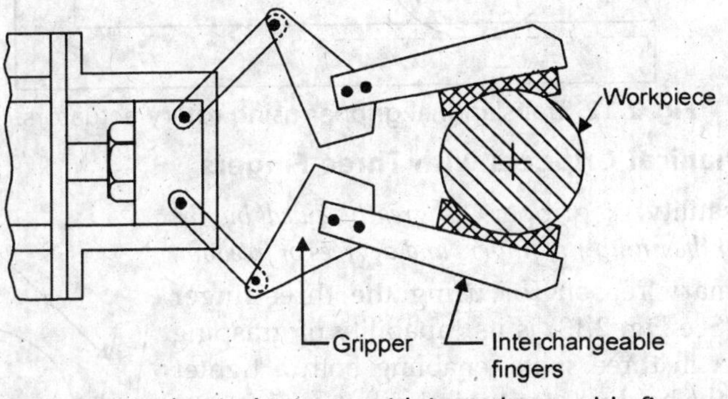

Fig. 2.10. Mechanical gripper with interchangeable fingers.

Fig. 2.10. shows a mechanical gripper with interchangeable fingers. In most cases two fingers are enough to grip the workpiece in good condition.

The various gripping mechanisms are discussed in the following articles.

2.4.1. Mechanical Grippers with Two Fingers

This type of gripper is the *most popular mechanical gripper for industrial robots.*

— *It can be designed for limited shapes of an object, especially cylindrical workpiece.*

— If actuators that produce linear movement are used, like pneumatic piston-cylinders, the device contains a pair of slider-crank mechanisms.

These grippers may have the following *mechanisms* :

(*i*) Pivoting or swinging gripper mechanisms :

— Using slider-crank mechanism;

— Using swing block mechanism;

— Using rotary actuator;

— Using cam actuator (See Fig. 2.11); the advantages of this arrangement is that the spring action would accommodate objects of different sizes.

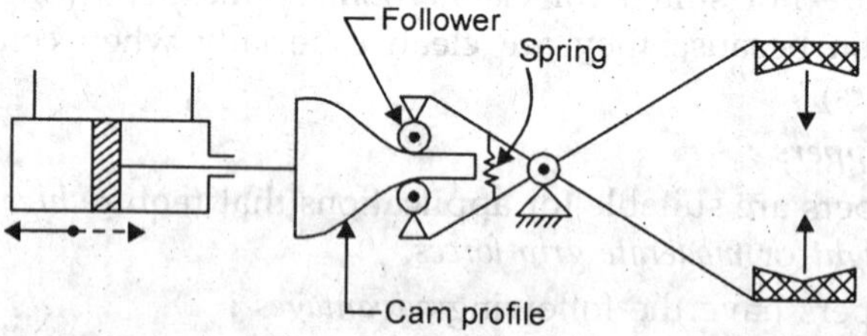

Fig. 2.11. Cam activated gripper.

(*ii*) Translation gripper mechanisms :

— Using screw;

— Using parallel bar linkages;

— Using cylinder-piston;

— Using rotary actuators (See Fig. 2.12)

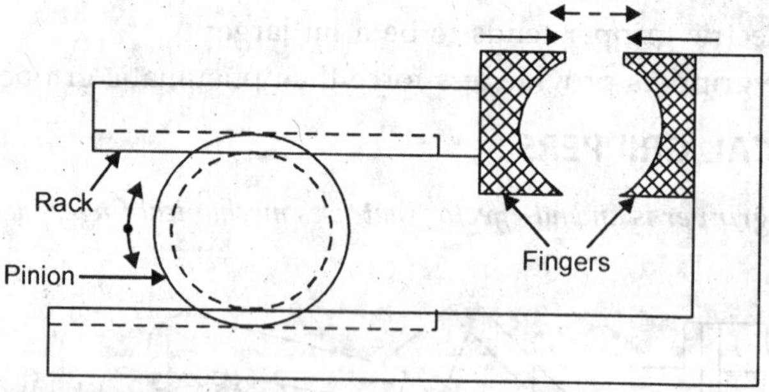

Fig. 2.12. Translational gripper using rotary actuators.

2.4.2. Mechanical Grippers with Three Fingers

— The versatility of grippers is *greatly aided by the increase of the number of fingers and degrees of freedom.*

— The primary reason for using the three-finger gripper (See Fig. 2.13) is its capability of grasping the object in three spots, enabling both a tighter grip and the holding of spherical objects of different sizes keeping the centre of the object at a specified position. Three point chuck mechanisms are typically used for this purpose; each finger motion is performed using ball-screw mechanism.

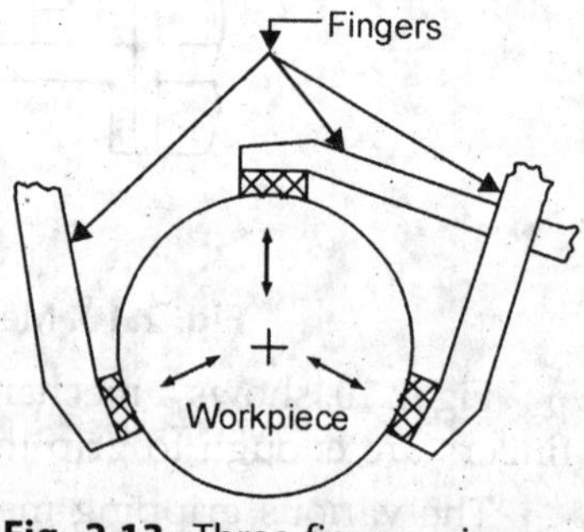

Fig. 2.13. Three-finger gripper.

2.5 VACUUM GRIPPERS

It is often difficult to grasp large flat objects; however, it can be done by the use of vacuum gripper.

- These grippers are mainly *"suction grippers"* that use a *vacuum pump or a venturi jet to generate vacuum* between a suction cup and the gripping plane.
  - — The suction cups are made of rubber like materials (silicone, neoprene). Based on the design and the material, suction cups are suitable for workpiece temperatures upto 200°C. These cups are really important because the efficiency depends on the air-pressure source.
  - — *Number of grippers/cups determines the size and weight of object to be grasped.*
  - — *For handling soft materials cups made of harder material are used.*
- The vacuum gripper is tolerant of errors in the orientation of the part and is *especially suited for pick-and-place work*

Uses. Vacuum grippers are used for picking up *metal plates, pans of glass, or large light weight boxes.*

Fig. 2.14. shows a typical vacuum gripper (using venturi device) widely employed for lifting fragile materials.

- Instead of a venturi, a vacuum pump driven by an electric motor may also be used.

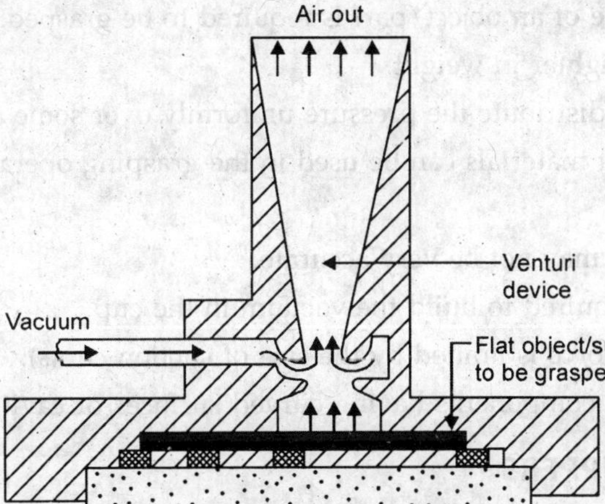

Fig. 2.14. Vacuum gripper for flat surface.

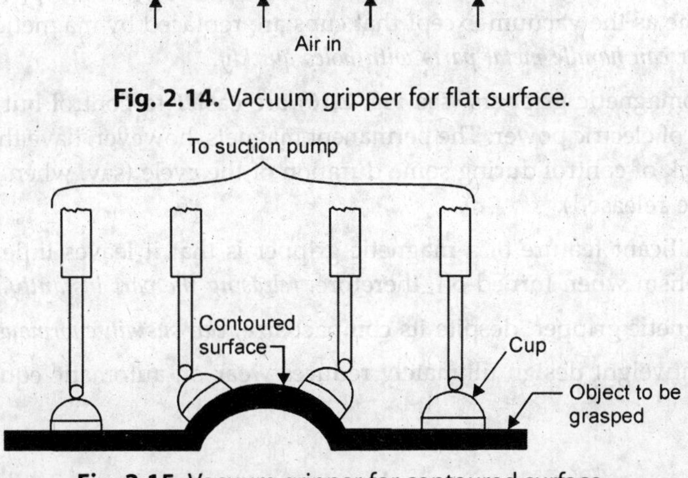

Fig. 2.15. Vacuum gripper for contoured surface.

Lift capacity of suction pump : Refer to Fig. 2.15.

The lift capacity of the suction pump depends on the effective area of the cup and negative air pressure (vacuum) between the cup and the object.

The lift capacity/force is given by the following equation :

$$F = k\,p\,A_{\text{eff.}} = k\,A_{\text{eff.}}\,(p_a - p_{\text{res.}}) \quad ...(2.1)$$

where, F = Force/lift capacity, N,

k = A coefficient which depends on atmospheric pressure and seal conditions,

$A_{\text{eff.}}$ = Effective area (total) of the suction cup(s) used to create the vacuum,

p = *Negative pressure, N/cm^2,

p_a = Atmospheric pressure, and

p_{res} = Residual pressure in vacuum cup.

(*Negative pressure is the difference of pressure between the inside and the outside of the vacuum pump.)

Advantages and disadvantages of vacuum grippers :

***Advantages* :**

(*i*) Only one surface of an object/part is required to be grasped.

(*ii*) The gripper is lighter in weight.

(*iii*) It is possible to distribute the pressure uniformly over some area.

(*iv*) Several different materials can be used in the grasping operation.

***Disadvantages* :**

(*i*) The positioning may not be very accurate.

(*ii*) Some time is required to build the vacuum in the cup.

(*iii*) The maximum force is limited by the size of suction cup(*s*).

(*iv*) Not suitable for components having curved surfaces or cavities and holes.

2.6 MAGNETIC GRIPPERS

A simple method to grasp ferrous materials is to use a magnetic gripper. The principle used is the same as the vacuum except that cups are replaced by magnetic grippers. This type of gripper *can handle metal parts with holes quickly*.

- Electromagnetic grippers (see Fig. 2.16) are easier to control but they require a source of electric power. The permanent magnets, however, have the disadvantage of a lack of control during some duration of the cycle (say, when the object/part is to be released).
- A significant feature of a magnetic gripper is that it leaves little to no residual magnetism when turned off, therefore, *releasing the part instantly*.
- A magnetic gripper, despite its compact size, *can sustain enormous holding power*.
- Its lightweight design ultimately reduces wear on automatic equipment.

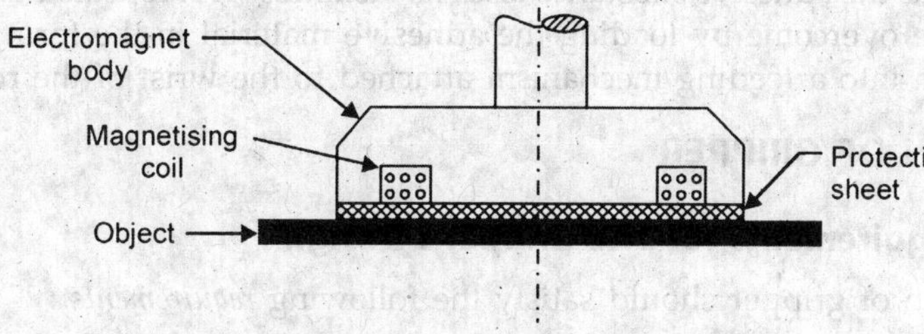

Fig. 2.16. Electromagnetic gripper.

Attractive force of an electromagnet :

The attractive force, P of an electromagnet can be found by using Maxwell's equation, given below.

$$P = \frac{(IN)^2}{25 A_{cont.} (S_a + S_m)} \qquad ...(2.2)$$

where, IN = Number of amp-turns of magnetising coil (here, I = current; N = no. of coil-turns),

$A_{cont.}$ = Area of contact of an object with magnet, and

S_a, S_m = Reluctances of magnetic paths through air and metal respectively.

Also, $$P \geq (a + g)m \times \text{F.O.S.} \qquad ...(2.3)$$

where, a = Acceleration of gripper,

g = Gravitational constant,

m = Mass of the object, and

F.O.S. = Factor of safety.

- Since in case of permanent magnets external power is not required, therefore, they can be used in *hazardous and explosive environments*.
  — In case, the part is to be released at the end of the handling cycle, some means of separating the part from the magnet must be provided.

Advantages and disadvantages of magnetic grippers :

Advantages :
(*i*) They can handle metal parts with holes.
(*ii*) Only one surface is required for gripping.
(*iii*) Pickup times are very fast.
(*iv*) Variations in part size can be tolerated.

Disadvantages :
(*i*) Residual magnetism remaining in the workpiece may cause problems.
(*ii*) While picking up a sheet from the stock, it is possible that more than a single sheet is lifted (since magnetic attraction tends to penetrate beyond the top sheet in the stock.

2.7 ADHESIVE GRIPPERS

- In a gripper design, an adhesive substance can be used for grasping action. The primary requirement in such an arrangement is that the items to be handled must be gripped on one side only.

- The reliability of an adhesive gripper is diminished with each successive operation cycle as the adhesive substance loses its tackiness with repeated use. This difficulty can be overcome by loading the adhesive material in the form of a continuous ribbon into a feeding mechanism attached to the wrist of the robot.

2.8 DESIGN OF GRIPPER

2.8.1. Requirements for The Gripper's Design

The design of gripper should satisfy the following *requirements* :

1. To match the abilities of the arm and controller.
2. To grasp and hold the object securely.
3. To complete the task.

- Different grippers are needed for different applications, which are of highly varied types. These may be : *To hold complicated shapes which may be fragile, assembling parts and fitting them into clamps/fixtures etc.* (Once a part is picked up it must be securely held in such a way that its position and orientation are as desired).

2.8.2. Tips for Designing a Gripper

Following are some tips for designing a gripper :

A. Grippers with passive fingers :

1. Select the grip choice, based on the task to be performed and the geometry and characteristics of parts to be grasped (for example, parts way be fragile, slippery, irregularly spaced, may require contact forces to be distributed etc.; each needs different suitable gripper).
2. Consider the local environment, working space, conveyor position etc. which also influence the design.
3. Determine the number of joint links and their configuration geometry by the part geometry and task movements.
4. Decide material of construction based on type of duty, corrosion resistance, or heat resistance required.
5. Select proper type of actuator as follows :
   — Hydraulic for high forces;
   — Electric motors for best control;
   — Pneumatic for less cost and ease etc.
6. Ensure compliance and proper safety sensors for assembly work and for good part handling.
7. Examine whether the design is serviceable, robust, economical, and whether it is possible to use less moving parts.
8. Examine whether it is desirable to incorporate two or three grippers at the end of arm.

B. Grippers with active feeback :

1. Use sensor based robots to handle heavy parts.
2. Use active wrist with passive fingers to handle forces greater than 1 to 2-5 kg.
3. Group the manufacturing tasks according to the trade off among arms, wrists, and fingers.

2.8.3 Force Analysis of Grippers Mechanism

The primary aim of the gripper mechanism is to convert input power into required motion and force to grasp and retain an object. Following are the *two ways of constraining an object in the gripper* :

1. The gripper fingers may enclose the object partly, thereby constraining its motion. This can be accomplished by designing the contacting surfaces of the fingers as shown in Fig. 2.17.
2. To hold the object between fingers and the object *by friction*. This approach demands that the finger must apply a force that must be adequate to retain the object against gravity, acceleration and any other force that might arise during holding operation.

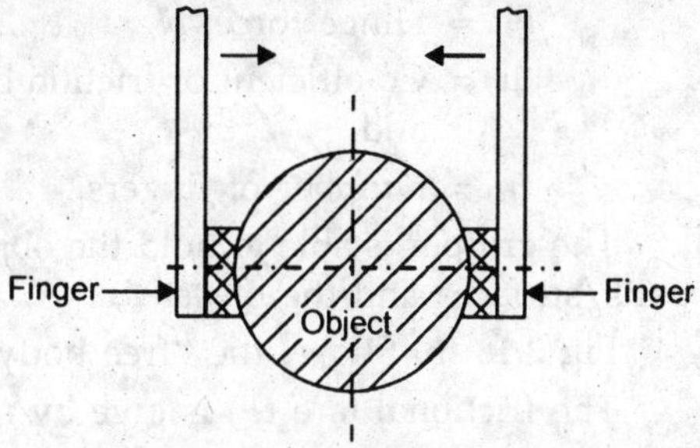

Fig. 2.17. Gripper enclosing the envelope of an object.

Fig. 2.18 (*a*) shows a *"gripper mechanism"* consisting of *fingers, linkages frame* and a *pneumatic cylinder*.

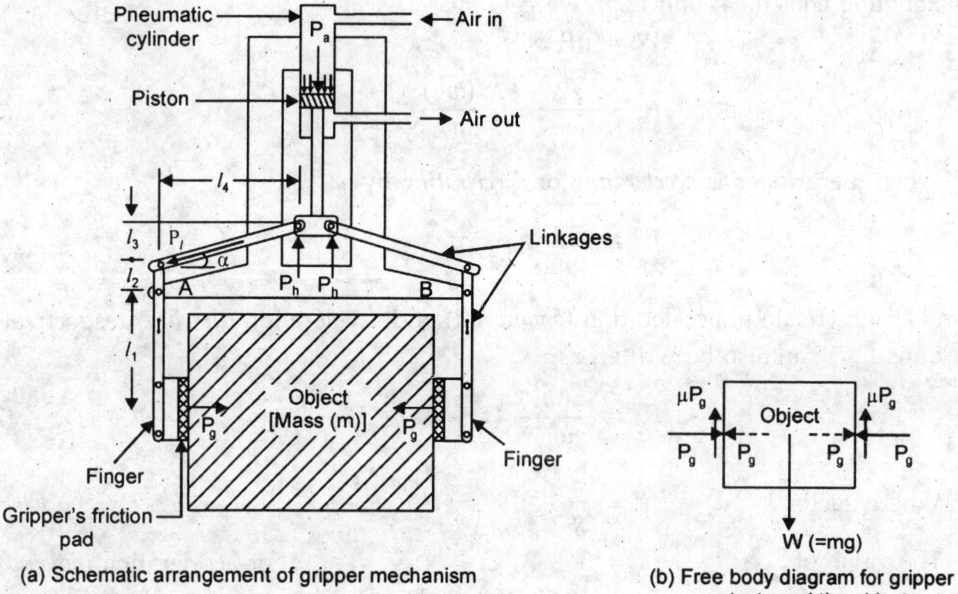

Fig. 2.18. Force analysis of Gripper mechanism, using friction.

When high pressure air is supplied (Air in) to the cyclinder, it actuates the fingers (through linkages) which in turn grasp the object.

Let, m = Mass of the object, kg,

g = Acceleraton due to gravity, m/s^2

W = Weight of the object (= mg), N,

P_a = Force exerted by air on the piston, N,

$= p \times \frac{\pi}{4} d_p^2$, where p is pressure of air needed (in N/m^2) and d_p is the diameter of the piston of the actuator (in m),

P_g = Gripper force, N,

P_l = Linkage force, N,

P_h = Hinge force, N,

μ = Co-efficient of friction between the gripper's pads/surface and the object, and

n_f = Number of fingers.

The gripper is able to hold the object (W = mg) due to friction between the gripper's pads/surfaces and the object.

Fig 2.18 (*b*) Shows the "free body diagram" for gripper contacts and the object.

The frictional fare (F) is give by :

$$F = \mu P_g \times n_f \qquad \text{...(2.4)}$$

The design force (P_d) using a factor of safet z (taking into account uncertain circumstance), is given by :

$$P_d = zW \qquad \text{...(2.5)}$$

Equating eqns. (2.4) and (2.5), we get

$$zW = \mu P_g \times n_f$$

or,
$$P_g = \frac{zW}{\mu n_f} = \frac{z.(mg)}{\mu n_f} \qquad \text{...(2.6)}$$

When the gripper is *accelerating* or *decelerating* by '*a*',

$$P_g = \frac{z.m}{\mu n_f}(g \pm a) \qquad \text{...(2.7)}$$

+ and - signs relate to acceleration (*downward*) and acceleration (*upward*) respectively.

Eqns (2.7) can also be written as :

$$P_g = \frac{z(mg)}{\mu n_f}(1 \pm k) \qquad \text{...(2.8)}$$

Where, factor $k = \frac{a}{g}$

The value of $k = 3$ ...For acceleration (*downward*)

$= 1$ ...For acceleration (*upward*)

$= 2$ ...Acceleration in *horizontal* direction.

Linkage analysis : Refer to Fig. 2.18

Taking moments about point A, we have :

$$P_g \times l_1 = P_l \cdot \cos\alpha \times l_2$$

or,
$$P_l = \frac{P_g \times l_1}{l_2 \cos\alpha} \qquad \text{...(2.9)}$$

Where, $\alpha = \tan^{-1}\left(\frac{l_3}{l_4}\right)$

By resolving P_l, we have :

Hinge force, $P_h = P_l \sin\alpha$

$\therefore$ Force exerted by the piston on the linkage,

$$P_a = 2P_h$$

Using eqns. (2.8) and (2.9), we get

$$P_a = 2 \times P_l \sin\alpha$$

$$= 2 \times \frac{P_g \times l_1}{l_2 \times \cos\alpha} . \sin\alpha = \frac{2P_g \times l_1}{l_2} . \tan\alpha$$

or, $$P_a = \frac{2zW}{\mu n_f}\left(\frac{l_1}{l_2}\right)(l \pm k) . \tan\alpha \quad ...(2.10)$$

Inserting, $\tan\alpha = \frac{l_3}{l_4}$ in the above equation, we have :

$$P_a = \frac{2zW}{\mu n_f}\left(\frac{l_1 . l_3}{l_2 . l_4}\right)(l \pm k) \quad ...(2.11)$$

- **Power Required** to produce air pressure p (in N/m^2) to operate gripper

$$= \frac{p \times Q}{1000} \ kW \quad ...(2.12)$$

where, Q = Discharge (volume flow rate) of air (in m^3/s).

Example 2.1. *The following data relate to a mechanical gripper (See Fig. 2.18) using friction to grasp an object :*

Weight of the part = 30 N

Coefficient of friction between the gripper pad and the object = 0.3

Lengths : l_1= 72 mm; l_2 = 48 mm; l_3 = 18 mm; l_4 = 54 mm

Diameter of the piston of the pneumatic cyclinder = 78 mm

Factor of safety = 1.4

If the gripper is accelerating down with an acceleration = 9.81 m/s^2, calculate :

(i) *The gripping force to retain the part ;*

(ii) *Actuation force required to obtain this gripping force ;*

(iii) *The pressure needed to operate the piston ;*

(iv) *The power required if the discharge is 0.018 m^3/s.*

Solution. *Given :* W = 30 N; μ = 0.3; l_1 = 72 mm; l_2 = 48 mm; l_3 = 18 mm; l_4 = 54 mm; d_p = 78 mm; z = 1.4; a = 9.81 m/s² (= g); Q = 0.018 m³/s.

(i) The gripping force to retain the part, P_g :

Using the relation : $$P_g = \frac{zW}{\mu n_f}(1 + k) \quad ...[\text{Eqn. (2.8)}]$$

Here. $$k = \frac{a}{g} = \frac{9.81}{9.81} = 1; \ n_f = 2$$

$\therefore$ $$P_g = \frac{1.4 \times 30}{0.3 \times 2}(1 + 1) = \mathbf{140\ N} \quad \textbf{(Ans.)}$$

(*ii*) Actuating force required to obtain the gripping force, P_a :

We know that, $$P_a = \frac{2zW}{\mu n_f}\left(\frac{l_1 . l_3}{l_2 . l_4}\right)(1+k) \qquad \text{...[Eqn. (2.11)]}$$

$$= \frac{2 \times 1.4 \times 30}{0.3 \times 2}\left(\frac{72 \times 18}{48 \times 54}\right)(1+1) = \mathbf{140\ N} \quad \textbf{(Ans.)}$$

(*iii*) The pressure needed to operate the piston, p :

$$P_a = p \times \frac{\pi}{4} d_p^2$$

or, $$140 = p \times \frac{\pi}{4} \times \left(\frac{78}{1000}\right)^2 = 4.78 \times 10^{-3} p$$

or, $$p = \frac{140}{4.78 \times 10^{-3}} \text{ N/m}^2 \text{ or } 29298 \text{ N/m}^2 \text{ or}$$

$$\mathbf{29.298\ kN/m^2} \quad \textbf{(Ans.)}$$

(*iv*) The power required :

$$\text{Power required} = \frac{P \times Q}{1000} kW$$

$$= \frac{29298 \times 0.018}{1000} = \mathbf{0.527\ kN} \quad \textbf{(Ans.)}$$

2.9 REMOTE CENTER COMPLIANCE DEVICES (RCCs)

These devices are commonly used for applications requiring the assembly of close-fitting parts, for example, the insertion of a peg into the hole. The task which appears to be simple is however quite difficult :

When the centerlines of the peg and hole are *not coaxial, the parts will not mate;* as a consequence the following two possibilities arise :

(*i*) The centerlines are parallel but not coaxial ..... *A translational misalignment;*

(*ii*) The centerlines interect but are not parallel ..... *A rotational misalignment.*

It is possible that both translational and rotational misalignments may occur simultaneously.

Let us assume there is a rotational misalignment (*i.e.*, a peg goes partially into the hole). If a force is used to attempt to push the peg into the hole, jamming may occur and use of still more excessive force may cause damage to the parts or to the robot.

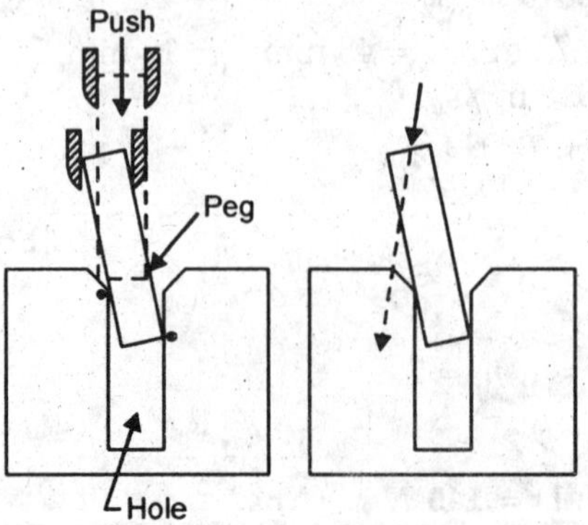

Fig 2.19. Peg being pushed into hole.

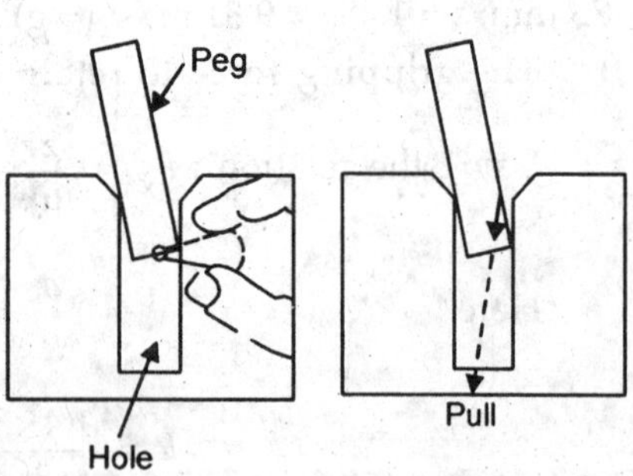

Fig 2.20. Peg being pulled into hole.

— **Fig. 2.19** shows the case of a peg being *pushed* into a hole.

— **Fig. 2.20**, shows one way of solving the problem; this involves *"pulling" the peg into the hole rather than pushing*. The pulling *allows the peg to rotate and align itself rather than jam against the hole, as in the case of pushing*.

— ***RCC device*** solves the peg-in-hole problem by effectively causing the peg to be pulled into the hole (See Fig. 2.20). The rotation is about a point called the *'center of compliance '* or *'elastic center'*; this center is a point that is remote from the unit.

— The RCC device provides the ability to compensate for both translational and rotational errors.

— A typical RCC device consists of two parallel plates which are separated by two rigid rods firmly attached to one plate but with a ball joint on the other. In addition, three elastic members are also placed between the plates which keep them separated and parallel. The rods and plates are so arranged that one plate is fixed and the other one has limited rotation and deflection.

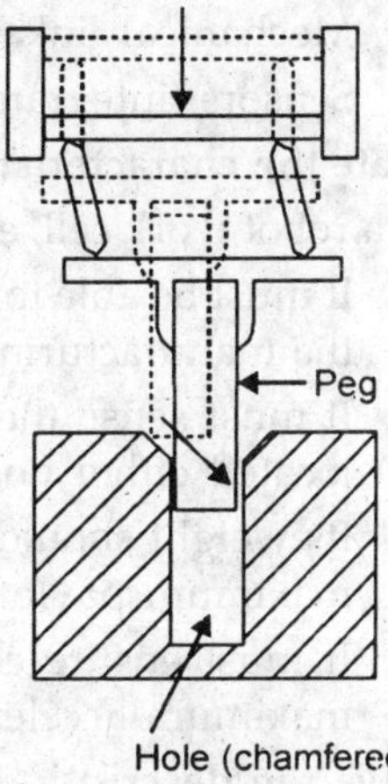

Fig 2.21. RCC device for inserting a peg into chamfered hole

Fig. 2.21. indicates the principle of an RCC device. The gripper with compound compliance can have both lateral as well angular movement to compensate for locational error or orientational error while a peg is inserted into a hole for the purpose of small assembly.

QUESTIONS WITH ANSWERS

Q. 2.1. What is a robotic end-effector ? Explain briefly :

Ans. *A "robotic end-effector" is any object attached to the robot flange/wrist that serves a function.* It includes the following :

— Robotic grippers; — Robotic tool changers;
— Robotic collision sensers; — Robotic rotary joint;
— Robotic press tooling; — Robotic paint gun;
— Robotic deburring tool; — Compliance device; etc.

● Robotic end-effectors are known as :

— Robot tools or robotic tools; — Robotic peripherals;
— Robotic accessories — End-of-arm devices
— End-of-arm tooling (EOA) ..... It is the subsystem of an industrial robot system that links the mechanical option of the robot (manipulator) to the workpiece being handled or worked on.

Q. 2.2. Name five basic types of end-of-arm tooling.

Ans. The five basic types of end-of-arm tooling, *including grippers and tools are :*

1. Attachment devices.
2. Support and containment devices.
3. Pneumatic pick-up devices.

4. Magnetic pick-up devices.
5. Mechanical grip devices.

Q. 2.3. Name the four distinct elements of end-of-arm tooling.

Ans. The four distinct *elements* of end-of-arm tooling are :

1. Mounting plate.
2. Power—for actuation of tooling motions.
3. Mechanical linkages.
4. Sensors (integrated into the tooling).

Q. 2.4. State the characteristics of end-of-arm tooling.

Ans. In a robot work cell, end-of-arm tooling should have the following characteristics :

1. It must be able to grasp, lift and release the part or family of parts required by the manufacturing operation/process.
2. It must sense the presence of a part in the gripper with the help of sensors located either on the tooling or at a fixed position in the work cell.
3. Its weight should be minimum since it is added to part weight to determine maximum payload..
4. It must ensure containment of the part in the gripper under conditions of maximum acceleration and the tool plate and loss of gripper power.

Q. 2.5. Give the description of commonly used gripper (end-effector) mechanisms.

Ans. Commonly used gripper (end-effector) mechanisms are described below :

1. **Linkage actuation** : Refer to Fig. 2.22.
   — Opening width and the speed of gripping depends on the linkage configuration.
   — Higher gripping force at less actuating force due to linkage.
   — Many design possibilities using linkages.

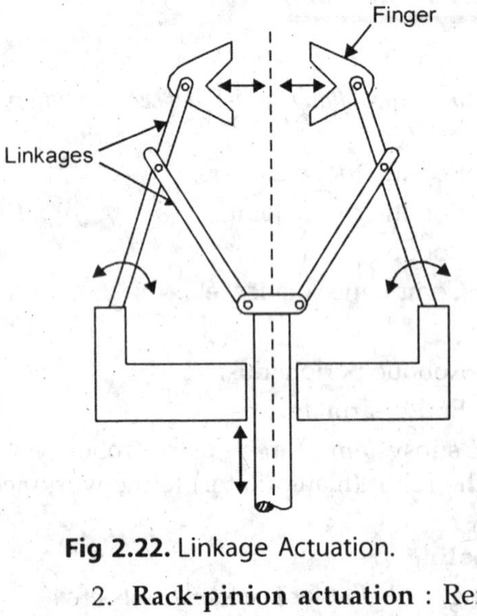

Fig 2.22. Linkage Actuation.

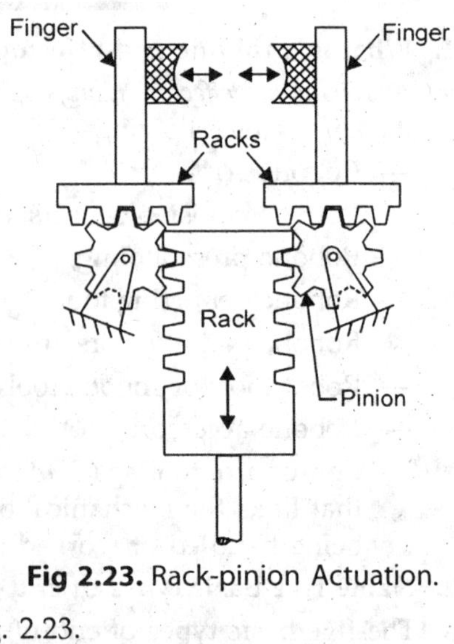

Fig 2.23. Rack-pinion Actuation.

2. **Rack-pinion actuation** : Refer to Fig. 2.23.
   — Two pivoted pinions are operated by a rack which is operated by a hydraulic pneumatic cylinder.
   — Pinions oscillate to move the racks (two).

— Two fingers attached to the racks open and close to grab and release the object.

3. **Cam actuation :** Refer to Fig. 2.24.
   — A cam attached to a cylinder moves to and fro.
   — Spring-locked followers move on the cam profile.
   — The finger pads (attached to the pivoted lever, connected to the followers) open and close to release and grab the object.

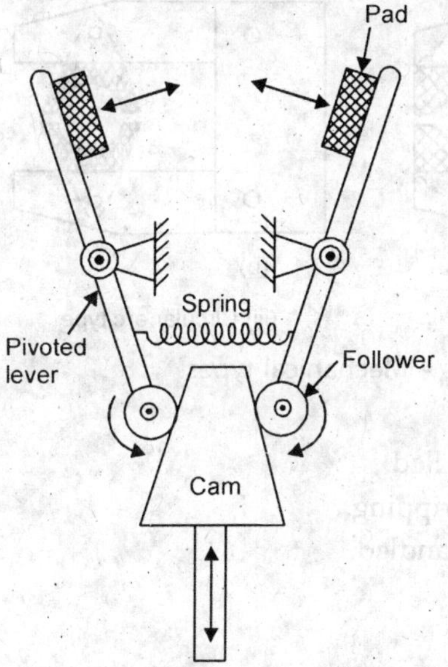

Fig 2.24. Cam actuation.

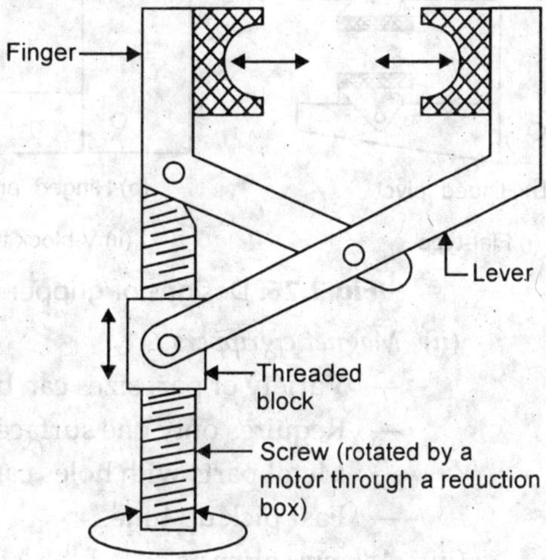

Fig 2.25. Screw actuation.

4. **Screw actuation** : Refer to Fig. 2.25.
   — A threaded block engages with a screw and moves forward and backward; the fingers attached to levers connected to the threaded block by a hinged joint, open and close providing releasing and gripping action.
   — The screw is rotated by a motor through reduction box.

Q. 2.6. Discuss briefly the various designs of gripper fingers.

Ans. Gripper fingers may be of the following types :

1. Mechanical types.
2. Non-mechanical types.

1. **Mechanical types :**

   This category of gripper fingers include the following :

   (*i*) Flat type [Fig. 2.26 (*i*)].
   (*ii*) V-block type [Fig. 2.26 (*ii*)].
   (*iii*) Circular arc type [Fig. 2.26 (*iii*)].

2. **Non-mechanical types :**

   (*i*) *Vacuum cup type grippers :*
   — The objects need to be flat, smooth and clean.
   — Cups are made of elastic materials round in shape.
   — Vacuum is created between cup and the object.

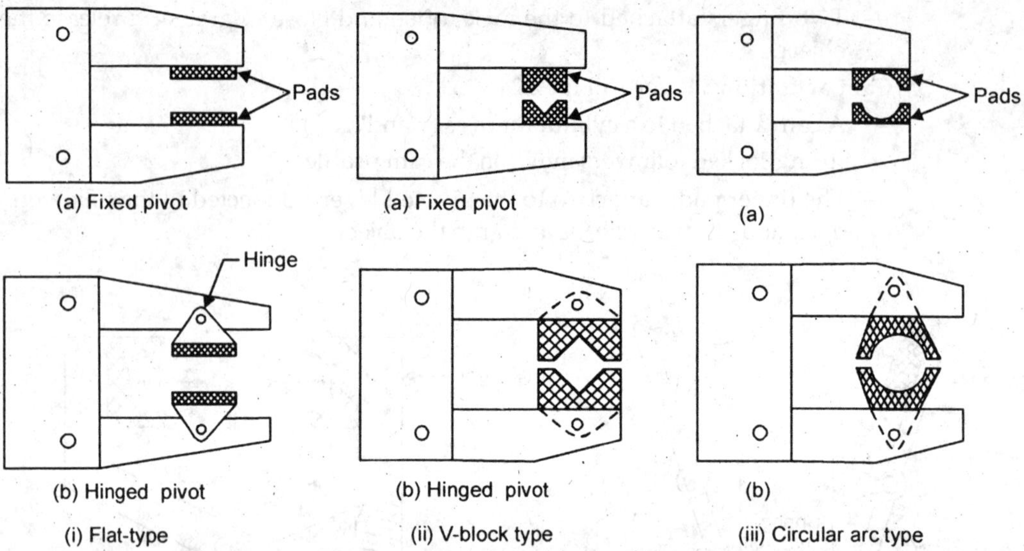

Fig 2.26. Designs of gripper fingers – Mechanical types.

(*ii*) *Magnetic grippers :*

— Variety of part sizes can be handled.

— Requires only one surface for gripping.

— Metal parts with holes can be handled.

— Fast pick up time.

(*iii*) *Adhesive grippers :*

— Gripping ability diminishes with successive operations.

— Adhesive ribbon is alternative.

— Employed to handle fabrics and light weight materials.

(*iv*) *Expandable bladder type :*

— Bladder expands to grasp/grab the object.

— Bladder is made of elastic material, which applies uniform pressure.

— Employed to handle fragile parts.

Q. 2.7. Explain briefly the following end-effectors :

(*a*) Passive end-effectors; (*b*) Active end-effectors.

Ans. (*a*) Passive end-effectors : These are classified as follows :

(*i*) *Non-prehensible end-effectors.* They neither enclose parts nor apply grasp forces across them.

— Vacuum;

— Electromagnetic;

— Bernoulli effect.

(*ii*) *Wrap end effectors* :

— Blades and linkages

(*iii*) *Pinch end effectors* :

— Two or three fingers;

— Parallel or angular motion;

— Finger tip styles like flat, V etc.

These days, mostly passive end effectors are used. They *emulate the grasps that people use for holding a heavy object or tool, without manipulating it in the fingers.* However, these end-effectors *may be equipped with sensors,* and the information from these sensors may be used in controlling the robot arm.

- The "*vaccum grippers*" (perhaps most commonly used gripping device in industry now-a-days) are *ideal for handlig large and delicate items such as glass panels.* They are *easily adapted* to a wide variety of parts from surface mounted microprocessor chips and other small items that require precise placement to large, bulky items such as automobile windshields and panels of aircraft.
- The "*wrap grippers*" hold a part in the same way that a person might hold a heavy hammer or a grape fruit.
- The "*Two-fingered grippers*" employ a strong punch force between two fingers, in the same way that a person might grasp a key when opening a lock. These grippers typically use a single air cylinder or motor that operates both fingers in unison, they will tend to center parts that they grip/grasp.

(*b*) **Active end-effectors :** These end-effectors include the following :

(*i*) Active servo grippers
(*ii*) Dexterous robot hands } *Used in research laboratories and tele-operated applications.*

In the above end-effectors, the distinctions depend largely on the number of fingers and number of joints or degrees of freedom per finger.

— The fingers, when equipped with force sensors such as strain gauges, can provide force sensing and control, typically with better accuracy that can be obtained with robot wrist-or joint-mounted sensors.

— A servo gripper can be programmed either to control the position of an uncontrolled part or to accommodate the position of a constrained part.

Q. 2.8. Explain briefly various drive methods used for robot gripper systems.

Ans. The following three kinds of drive methods are employed in typical robot gripper systems :

1. ***Pneumatic drive :***

- Being less expensive than other methods, this type of drive is used in most of the industrial robots.
- The advantage of the pneumatic system is the low-degree of stiffness of the air-drive system. This feature can be used effectively to achieve compliant grasping which is necessary for one of the most important functions of grippers; to grasp objects with delicate surfaces carefully. However, the relatively limited stiffness of the system makes precise control difficult.

2. ***Hydraulic drive :***

- These drives are usually electrohydraulic drive systems. They have almost the same configuration as pneumatic systems, though their features are different from each other.
- A typical hydraulic drive system consists of the following *components* :

(i) *Actuators*:
— Piston cylinder;
— Swing motor;
— Hydraulic motor.
(ii) *Control valves*.
(iii) *Power units*.

- In order to achieve positional control using electrical signals, electrohydraulic conversion drives are used.

3. **Electric drives :**

- In electric drive system, the following two types of actuators are used :
  (i) *D. C. motors.*
  (ii) *Stepper motors.*
- Normally, each motor requires appropriate reduction gear systems to provide power output force or torque.
- In this system a servo power amplifier is also needed to provide a complete actuation system.

Q. 2.9. Explain briefly translational gripper using : (*a*) cylinder-piston; (*b*) Parallel bar linkages.

Ans. • Fig. 2.27 (*a*) shows the simplest translational gripper using the direct motion of piston-cylinder :

— The finger motion corresponds to the piston movement without any connecting mechanisms between them.

— The *drawback* of this arrangement is that sometimes it is difficult to design the desired size of the gripper, because here the gripper size is decided by the actuator size.

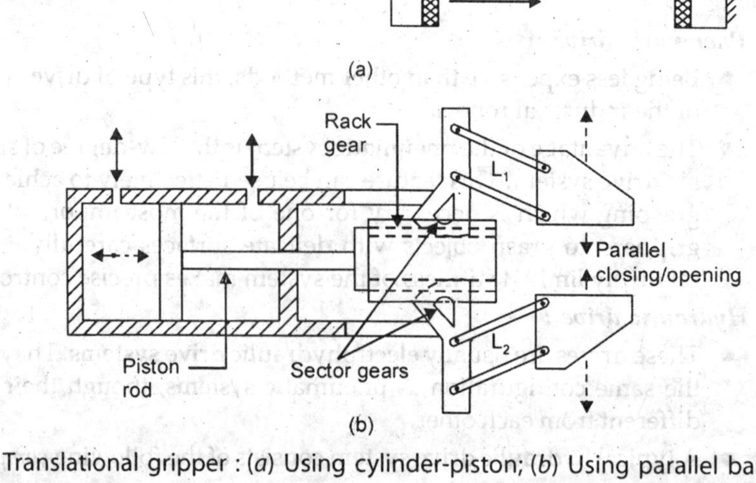

Fig 2.27. Translational gripper : (*a*) Using cylinder-piston; (*b*) Using parallel bar linkages.

- **Fig. 2.27 (*b*)** shows a translational gripper using a hydraulic or pneumatic piston-cylinder, which includes a dual-rack gear mechanism and two pairs of symmetrically arranged parallel closing linkages :
  - — The piston and sector gears are connected to the elements L_1 and L_2 respectively.
  - — When the piston rod moves towards left, the translation of the rack causes the two pinions to rotate clockwise and anticlockwise respectively and produces the grasping action, keeping each finger direction constant. The release action occurs when the piston rod moves to the right in the same way.

Q.2.10. Describe briefly a modular gripper system.

Ans. A modular gripper system, equipped with force/torque sensor consists of the following :

1. Force/torque sensor with online data processing that evaluates signals from additional sensors in the gripper.
2. Gripper with drive control, gripping force sensor, and presence check (light barrier) for workpieces.
3. Collision protection with emergency stop.
4. Flexible coupling.

- The modular gripping system *can apply predetermined gripping forces.*
- The distinguishing feature of *'adaptive gripping' is the ability to handle work-pieces within the minimum necessary gripping force and manipulate them flexibly during assembly tasks on occasions when external forces act on them momentarily.*
- The modular gripper system for industrial robots is *suitable for adaptive gripping.* In addition to extra sources of information, such as visual sensors, previous knowledge of the weight, size, and coefficient of friction of the object are more important and are stored in the system. The gripping force sensor plus other sensors to detect weights and joining forces are integrated in the gripper fingers. But even without the support of extra sensors, the concept of measurement with load cells has yielded good results.

Q.2.11. What is a "universal gripper"? Explain briefly.

Ans. Efforts have always been initiated by man to imitate hand which is most universal type of gripper as it has *many degrees of freedom.* The main difficulty associated with large number of degrees of freedom is with regard to installation of actuators to actuate all degrees of freedom. This requires *miniature actuators that can produce enough power to drive the gripper joint.* Frequently used actuator consist of a *cable and pulleys mechanism that enables a motor to be placed at an appropriate position away from the joint.*

Fig. 2.28. shows an example of the universal gripper :

- — It includes three fingers : a *thumb,* an *index finger,* and a *middle finger.*
- — Each finger contains two or three *segments* made of a 17 mm brass rods bored to be cylindrical. The tip of each segment is truncated at a slope of 30° so that

the finger can be kept both inward and outward, at the maximum angle of 45° at each joint.

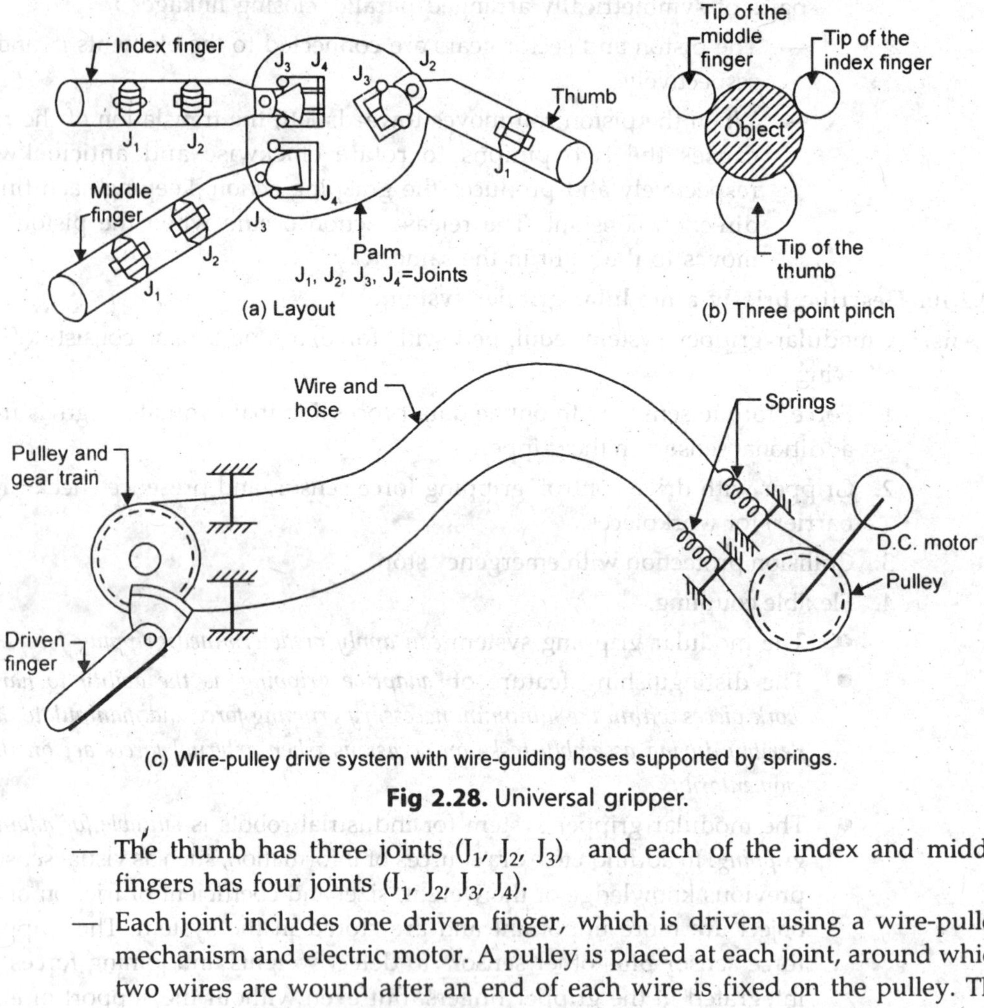

Fig 2.28. Universal gripper.

— The thumb has three joints (J_1, J_2, J_3) and each of the index and middle fingers has four joints (J_1, J_2, J_3, J_4).

— Each joint includes one driven finger, which is driven using a wire-pulley mechanism and electric motor. A pulley is placed at each joint, around which two wires are wound after an end of each wire is fixed on the pulley. The wire is guided through coil like hoses so that it can not interfere with the complicated finger motion.

— In order to make the motion of the fingers flexible and to make the gripper system *more compact*, the wire and hoses are installed through the finger tubes.

— Motors for driving the respective joints are located outside separate from the gripper system.

— The problem of friction between hoses and wires is overcome by supporting the hoses with springs.

Q.2.12. Enlist considerations for selection of gripper (end-effector).

1. *Actuation selection* :

(*i*) Mechanical or friction gripping methods

(*iv*) Magnetic grasping

(*ii*) Pad shape selection
(*iii*) Vacuum actuation
(*v*) Adhesive gripping
(*vi*) Expendable bladder type actuation

2. *Drive selection* :
   (*i*) Pneumatic drive systems
   (*ii*) Hydraulic drive systems for heavy duty operation
   (*iii*) Electrical drive for light duty applications
   (*iv*) Speed reduction of mechanical transmission
3. *Pretection selection* :
   (*i*) Heat shield for sensors and actuators
   (*ii*) Forced cooling by air or water cooling to take away the heat
   (*iii*) Selection of heat resistance material for fingers and components of gripper.
   (*iv*) Shield from hazardous chemicals.
4. *Process selection* :
   (*i*) Accurate processing methods for fingers.
   (*ii*) Leak prevention for pneumatic / hydraulic actuators.
   (*iii*) Interchangeability standards for the fingers.
   (*iv*) Shape compatible processing methods for fingers.
   (*v*) Ease of assembly of fingers and linkages.

Q.2.13. What are the advantages of multigripper systems?

Ans. The *advantages* of multiple gripper systems are :

1. Reduced work-in-process time for part.
2. The robot arm and controls are used more efficiently.
3. Increased production capability of work cell.

Q.2.14. Enlist "Gripper design considerations".

Ans. "Gripper design considerations" include the following *specifications* :

1. *Material specifications* :
   — Strength and rigidity;
   — Durability and fatigue factors;
   — Friction properties;
   — Factor of safety;
   — Compatibility with the environment.
2. *Part specifications* :
   — Weight of the part;
   — Size and shape of the component;
   — Tolerance on the part size;
   — Change of shape and size during processing;
   — Surface finish of the part;
   — Care for delicacy of the part to be handled.
3. *Performance specifications* :
   — Co-efficient of friction between part and the gripper pads;
   — Speed and acceleration during motion;
   — Accuracy and repeatability of the robot;

— Interchangeability of fingers;
— Memory capacity of the controllers;
— Control resolution of the robot;
— Lead time and cycle time.

4. *Source specifications* :
— Pneumatic : Air pressure and discharge, and cylinder size;
— Electrical : The power rating and specifications of actuator;
— Hydraulic : The oil pressure, volume flow rate and power pack specification;
— Mechanical : Linkage design and transmission design specifications.

5. *Position specifications* :
— Holding methods;
— Physical difficulty;
— Length of fingers;
— The object and tool orientation;
— Product design changes;
— Spare parts specifications.

6. *Environmental specifications* :
— Heat and temperature of object and atmosphere;
— Humidity and moisture;
— Dirt and safety;
— Hazardness of chemicals used.

HIGHLIGHTS

1. *End-effector* is gripper attached to the wrist of manipulator to accomplish the desired task.
2. Grippers are classified as :
Mechanical grippers; suction or vacuum cups; magnetised gripper; Hookes; Scoopes or ladles; Adhesive or electrostatic grippers.
3. A mechanical gripper is an end-effector that uses mechanical fingers actuated by mechanism to grasp an object.
4. The lift capacity of the suction pump depends on the effective area of the cup and negative air pressure (vacuum) between the cup and the object.
5. Electromagnetic grippers are easier to control but they require a source of electric power. The permanent magnets, however, have the disadvantage of a lack of control during some duration of the cycle (say, when the object is to be released).
6. RCC device provides the ability to compensate for both translational and rotational errors.

OBJECTIVE TYPE QUESTIONS

Fill in the Blank or Say "Yes" or "No".

1. The end-effector is also called ....
2. The structure of an end-effector, and the nature of the programming and hardware that drives it, depends on the intended task.

3. Suction or vacuum cups are used for .... objects.
4. Tools are fastened directly to the robot wrist and become the end effectors.
5. .... fingers are used to perform special tasks.
6. When the objects to be handled are too large and ferromagnetic in nature, ... grippers may be employed.
7. In .... grippers, the power of a piston/cylinder system can be translated into a gripping force.
8. The suction cups are made of rubber like materials.
9. A magnetic gripper, despite its compact size, can sustain enormous holding power.
10. RCCs are rarely used for the assembly of close-fitting parts.

ANSWERS

1. gripper;	2. Yes;	3. flat;	4. Yes;	5. Mechanical;
6. electromagnetic;	7. pneumatic;	8. Yes;	9. Yes;	10. Yes.

THEORETICAL QUESTIONS

1. What is an end-effector ? Explain briefly.
2. How are end-effectors categorised ?
3. How are grippers classified ? Give neat sketches for four sample gripper designs.
4. Discuss briefly the following grippers :
   (*i*) Part handling grippers; (*ii*) Tool handling grippers;
   (*iii*) Specialised grippers.
5. Give the classification of end-effectors.
6. Explain briefly the following end-effectors :
   (*i*) Mechanical fingers; (*ii*) Universal fingers;
   (*iii*) Internal and external grippers.
7. Describe briefly the following :
   (*i*) Unilateral-action devices; (*ii*) Bilateral-action devices;
   (*iii*) Multilateral-action devices.
8. Explain briefly any two of the following :
   (*i*) Pneumatic grippers.
   (*ii*) Hydraulic grippers.
   (*iii*) Electric grippers.
9. Explain with the help of a neat sketch the mechanical gripper with interchangeable fingers.
10. With the help of a neat sketch, explain the working of a vacuum gripper for flat surface.
11. State the advantages and disadvantages of vacuum grippers.
12. Explain briefly "magnetic grippers".
13. Distinguish between magnetic and vacuum grippers
14. List the advantages and disadvantages of magnetic grippers.
15. Write short note on "Adhesial grippers".
16. What are the requirements for the gripper's design?
17. List some important tips for designing gripper.
18. Explain briefly Remote Center Compliance devices (RCCs).

CHAPTER

3 Robotic Sensors and Vision

ROBOTIC SENSORS

3.1 INTRODUCTION TO TO ROBOTIC SENSORS

A **sensor** *is a *transducer used to make a measurement of a physical variable.*

The *basic function* of an electric sensor is *to measure some feature of the world, such as light, sound or pressure and convert that measurement into an electrical signal, usually a voltage or current.*

- For a robotic manipulator to operate effectively and intelligently and enable it to work in unstructured environment, it must be equipped with sensors, which give information about itself and its environment. The computer-cum-controller operating the robot must receive information from the sensors and process it in real-time for decision making. *Robots capable of such decision making are called* ***intelligent robots***, the *environment sensing is called* ***intelligent sensing****, and the sensors used for such operations* are called ***intelligent sensors***.
  - — The intelligent sensors need associated hardware and software to process the signals from the robot's *data-acquisition system*. The performance, capability, and flexibility of a robot are greatly dependent on its data-acquisition system. A robot without intelligent sensors is handicapped and can only do very specific tasks, while a robot that is able to ***see*** and ***feel*** is easier to perform complex tasks.
- *The most common and minimal use of sensors is to provide information about the status of links and joints of the manipulator and about the working environment of the robot.*

---

* A *transducer* is a device which converts the energy from one form to another.

In addition, *sensors may be deployed to provide data for inspection and quality control, safety monitoring, and to detect and resolve interlocks in the workcell.*

- Properly designing or choosing and using sensors is an essential part of robotics, and sensors of all types are used in many different kinds of robots for different tasks. For example, some robots use only simple position sensors, while some may have combination of special cameras and advanced software to gather information about their surroundings, and orient themselves based on the behaviour of other robots.

3.2 CHARACTERISTICS OF A SENSING DEVICE

A sensory device should possess the following *characteristics* :

1. **Range :**
   - It refers to the *minimum and maximum change in input signal to which the sensor can respond.*
   - The sensor should have a *wide operating range.*
2. **Response :**
   - The sensor should be capable of responding to changes in the sensed variable in *minimum time.*
   - Ideally, the response should be *instantaneous.*
3. **Accuracy :**
   - The measurement accuracy should be *as high as possible.*
   - The sensor's output should properly reflect the input quantity being measured or sensed.
4. **Sensitivity :**
   - The sensitivity relates to the *change in the output exhibited by the sensor for the change in input.*
   - It should be *as high as possible.*
5. **Linearity :**
   - The sensory device should exhibit the same sensitivity over its entire operating range.
6. It should be suitable for the environment in which it is to be employed.
7. It should have a suitable physical size, cost and ease of operation.
8. It should not disturb or have any effect upon the quantity it senses or measures.
9. It should include isolation from receiving excess signals or electrical noise that could give rise to the possibility of misoperation or damage of the sensor, circuit or system.

3.3 TYPES OF SENSERS

A. The sensors can broadly be *classified* as:

1. **Internal state sensors.** There are used *to measure position, velocity or acceleration of robot joints and/or the end-effector.* The following devices fall into this class :

   (*i*) Potentiometers ("pots"); (*ii*) Synchros

   (*iii*) Resolvers; (*iv*) Linear inductive scales;

   (*v*) Differential transformers (i.e. LVDTs and RVDTs); (*vi*) Optical interrupters;

   (*vii*) Optical encoders; (*viii*) Tachometer;

   (*ix*) Accelerometers, etc.

2. External state sensors. These are used *to monitor the robot's geometric and/or dynamic relation to its task, environment, or the objects that it is handling*. Such devices can be of either the *visual or nonvisual variety*. The following devices are include in this class :

(*i*) Strain gauges; (*ii*) Pressure transducers;
(*iii*) Proximity devices; (*iv*) Ultrasonic sensors;
(*v*) Electromagnetic sensors etc.

B. The sensors can also be *classified* as follows :

1. Tactile sensors. *These are contact sensors that must be brought in contact with the object to obtain signals to measure the necessary quantities.*

— When the tactile sensors make contact with the object, an electrical analog or digital signal is generated and sent to the robot controller. Electrical signals may be obtained through the contact of microswitches. Signal may also be obtained through mechanical pressures which change resistances of electrical strain gauges or generate electrical potentials in piezoelectric crystals.

Typical *contact type* robotic sensors include :

(*i*) Force sensors; (*ii*) Torque sensors;
(*iii*) Touch sensors; (*iv*) Position sensors.

Tactile sensors :

(*i*) *Microswitches* .... Signalling the current under rated voltage.
(*ii*) *Piezoelectric crystals* .... Generating voltage under pressure.
(*iii*) *Potentiometers* .... Measuring current or voltage drop.
— Linear;
— Rotary
(*iv*) *Electric strain gauges* .... Changing electrical resistance.
(*v*) *Linear voltage differential transformer (LVDT)* .... Measuring output voltage.
(*vi*) *Resolvers* .... Counting position.
(*vii*) *Encoders* .... Informing angular position.
— Absolute;
— Incremental.

- Tactile sensors can be used to *sense a diverse range of stimulus ranging from detecting the presence or absence of a grasped object to complete tactile image.*
- A tactile sensor consists of an *array of touch sensitive sites*, the sites may be capable of *measuring more than one property*.
- The contact forces measured by the sensor are able to convey a large amount of information about the state of a grip. Texture, slip, impact and other contact conditions generate force and position signatures that can be used to identify the state of manipulation. This information can be determined by examination of the frequency domain.

2. Non-tactile sensors. *These are contactless sensors which sense the signals remotely, but only within the specified range of distance from the object.*

— These sensors detect and measure magnetic fields, infrared and ultraviolet light, X-rays, electircal fields, ultrasonic sound waves or electromagnetic waves.

Non-tactile/non-contact sensor :

(*i*) *Proxmity sensors* .... Knowing the presence of the object.
(*ii*) *Range imaging sensors* .... Measuring the range (distance).

(*iii*) *Ultrasonic sensors* .... Measuring range by sound waves.

(*iv*) *Electro-optical vison sensors* .... Computer vision or image processing for contour search.

(*v*) *Magnetic sensor* .... Measuring Hall voltage.

C. Robotic sensors can also be *classified* as follows :

Robotic sensors :

1. *Status sensors* .... To detect position, velocity, acceleration, torque, force etc.
   — Potentiometer;
   — Tachometer;
   — Optical encoder;
   — Microswitch etc.
2. *Environment sensor* .... To detect presence, motion, force, torque, touch/tactile, etc.

(*i*) Contact sensors :

— Pressure, force, slip, torque, surface finish, temperature, pH etc.

(*ii*) Non-contact sensors :

— Visional, optical, accoustic, infrared, proximity, range, temperature, chemical etc.

- *"Contact" or "touch" sensors* are one of the most common sensors in robotics. These are generally used to detect a change in position, velocity, acceleration, force, or torque at the manipulator joints and/or the end-effector. These are mainly of two types : *"Bumper"* and *"tactile"*

— *"Bumper type sensors"* detect whether they are touching anything, the information is either *"Yes"* or *"No"*. They cannot give information about *how hard is the contact* or *what they are touching.*

☞ — ***"Tactile sensors"*** *are more complex and provide information on how hard the sensor is touched, or what is the direction and rate of relative movement.*

The technologies used for the status sensors are *electric, electromagnetic, electronic,* and *optic.*

- *"Non-contact sensors"* are used to get parameteric information about the enviorment, task, objects in the workspace and to detect presence, distance or features of the workpiece.

  Various technologies used in these sensors are : *Optic, acoustic, infrared, electromagnetic* etc.

- **Sensors and controls for I/O interfaces :**

(*i*) Switches :

— Push button switch
— Flush head switch
— Mushroom head switch
— Key lock switch
— Selector switch.

(*ii*) Indicating lamps.

(*iii*) Vacuum switch.

(*iv*) Control relays.

(*v*) Time relay delays.

(*vi*) Limit switches.

(*vii*) Proximity switches :

— Capacitance type

— Magnetic type.

(*viii*) Photodetectors :

— Photo conductive cell: photoresistors; photodiodes; photo transistors

— Photovoltaic cell.

(*ix*) Electrical gauges.

3.4 TOUCH OR TACTILE SENSORS

In robotics, touch or contact sensors are used to obtain information associated with the contact between a manipulator hand and objects in the workspace.

- Several physical principles have been exploited in the development of tactile sensors. In most cases, the developments in tactile sensing technologies are *application driven*. It should be recognised that the operation of a touch or tactile senor is *amply dependent on the material of the object being gripped*. Although the sensors discussed here are capable of working with rigid objects, yet work has shown that conventional sensors can be modified to operate with non-rigid materials.

  Touch/Contact sensors can be subdivided into the following two principal categories :

  1. Binary sensors.
  2. Analog sensors.

3.4.1. Binary Sensors

A binary sensor is a contact device *e.g. microswitch*. In the simplest arrangement, a switch is placed on the inner surface of each finger of a manipulator hand as shown in Fig. 3.1 :

— This type of sensing is useful for determining whether a part is present between the fingers.

— Furthermore, with the movement of the hand over the object and sequentially making contact with it surface, it is posssible *to center the hand over the object for grasping and manipulation.*

— For providing further tactile information, *multiple binary touch sensors* can be used on the inner surface of each finger.

— Besides this, the binary sensors are often mounted on the external surfaces of a manipulator hand *to provide control signals useful for guiding the hand throughout the workspace* (similar to what human do in feeling their way in a total dark room)

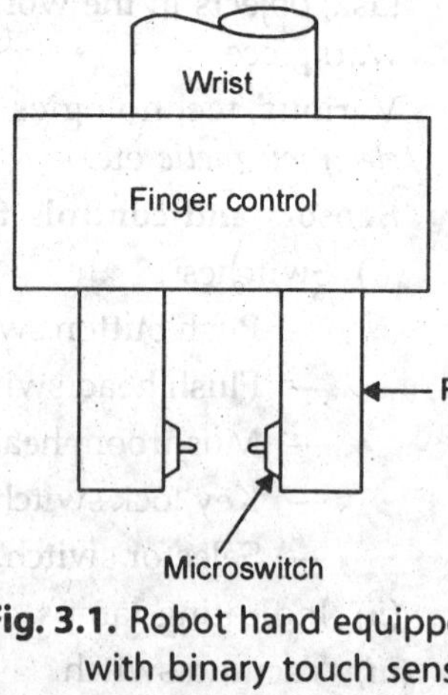

Fig. 3.1. Robot hand equipped with binary touch sensors.

3.4.2. Analog Sensors

The *output of an analog sensor (compliant device) is proportional to a total force.*

- In its simplest arrangement, an analog device consists of a spring–loaded rod which is mechanically linked to a rotating shaft in such a way that the *displacement of the rod due to a lateral force results in a proportional rotation of the shaft*. The rotation is then measured continuously using a potentiometer or digitally using a code wheel. Knowledge of spring constant yields the force corresponding to a given displacement.
- In the past ample efforts have been initiated for the development of *tactile sensing arrays* capable of yielding touch information over a wider area than that afforded by a single sensor. In a better approach involving sensing arrays, in the device often called *"artificial skins"*, an object pressing against the surface causes local deformations which are measured as continuous resistance variations. The latter are easily transformed into electrical signals whose amplitude is proportional to the force being applied at any given point on the surface of the material.

Fig. 3.2 shows a robot hand equipped with tactile sensing arrays.

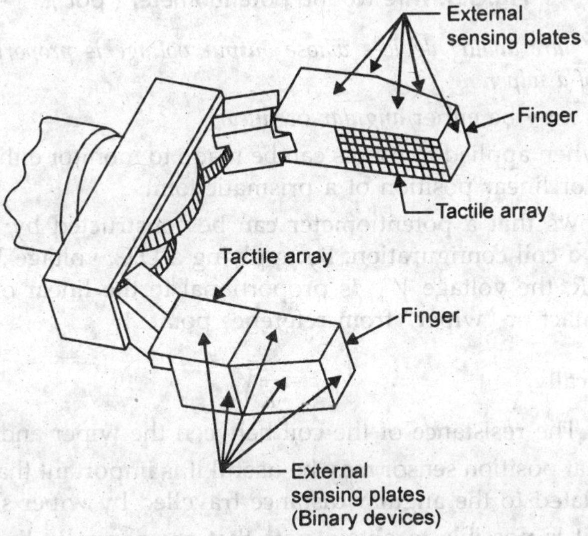

Fig. 3.2. Robot hand equipped with tactile sensing arrays.

3.5 POSITION AND DISPLACEMENT SENSORS

These sensors are employed as components of the robot control system.

The robot's control structure needs to know the position of each joint in order to calculate the position of the end-effector thus enabling the successful completion of the programmed task. The movements of the joint can be either linear or angular/rotary depending on the type of robot .

The two main types of position sensors are :

1. Absolute; 2. Incremental.

Some common devices which are used as position sensors are :

1. Potentiometers (or Pot)
2. Encoders(optical)
3. Linear variable differential transformer (LVDT).

3.5.1. Potentiometers

A potentiometer ("pot") is the simplest device that can be used to measure position.

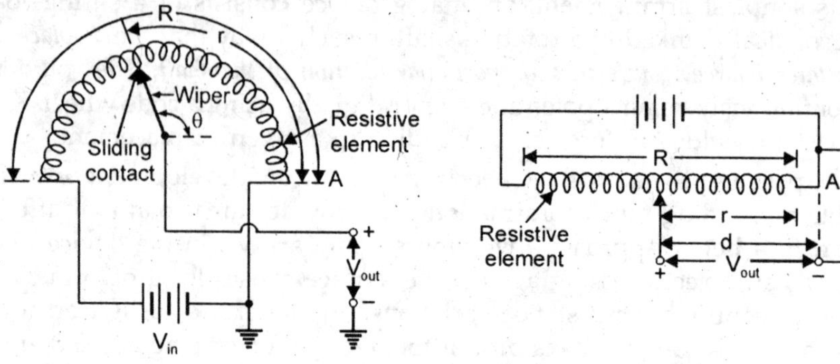

Fig. 3.3. Wire wound potentiometer ("pot").

"Potentiometers" are analog devices whose output voltage is proportional to the position (angular or linear) of a wiper.

Potentiometers may be either *angular* or *linear*.

Such devices when applied to robots can be made to monitor either angular position of a revolute joint or linear position of a prismatic joint.

- Fig 3.3 shows that a potentiometer can be constructed by winding a resistive element in a coil configuration. By applying a D.C. voltage V_{in} across the entire resistance R, the voltage V_{out} is proportional to the linear or rotary distance of sliding contact or "wiper" from reference point A.

  Mathematically, $$V_{out} = \frac{r}{R} V_{in} \quad ...(3.1)$$

  where, r = The resistance of the coil between the wiper and the reference.

  In order that position sensor may be useful it is important that the resistance r be linearly related to the angular distance travelled by wiper shaft.
- Although it is possible to obtain pots that are normally linear, there is always some deviation from linearity. Thus, except in the case of robots where extreme accuracy is not needed (such as educational purposes), the potentiometer is *not* generally used as a primary position-monitoring sensor.
- Despite the fact that there are several characteristics that limits its usefulness in high-accuracy position-measurement applications, *it is possible to use "pots" to provide a limited amount of feedback control in robots where high positional resolution and accuracy are not essential to the operation and where a certain amount of jitter caused by noisy contacts is tolerable.*

3.5.2. Optical Encoders

- Unlike potentiometers which give analog signals, *"encoders"* (non-contact devices) *give digital signals directly.*
- *"Optical encoder"* is one of the most widely used position sensors, due to the following ***merits*** :

  (*i*) Simple construction.

(ii) Low cost.

(iii) Ease of application.

(iv) Versatility.

It converts linear or angular displacement into digital code or pulse signals.

- Capable of resolutions that are more than adequate for robotic applications these *non-contact* sensory devices come in two distinct classes :

(i) ***Absolute encoders :***

— These encoders provide *actual position relative to a fixed reference (zero) position.*

— Their output is a digitally coded signal with distinct digital code indicative of each particular least significant increment of resolution.

— The encoder is able to give linear or rotational position even if power has been applied to the electromechanical system using the sensor. Thus a robot joint equipped with an absolute encoder will *not require any calibration cycle* since the controller will immediately, upon power-up, know the actual joint position.

(ii) ***Incremental encoders :***

— *These encoders* sense the position from previous position. Their output is a pulse for each increment of resolution but these make no distinction between increments.

— Such a sensor only provides positional information *relative to some reference point.* A robot utilising an incremental encoder *must,* therefore, *first execute a calibration sequence before "true" positional information can be obtained.*

Although either linear or rotary encoders for both of the foregoing classes are available, the *rotary device is almost exclusively used in robotic applications.* One of the main reasons for this is that revolute joints far outnumber prismatic ones in robots currently being used. Even for joints that move in a linear fashion, as in the case of a spherical coordinate manipulator, the linear encoder is normally much more costly, and so rotary encoders are still employed.

Construction of an optical encoder : Refer to Fig. 3.4 the optical encoder consists of the following three basic comfortable.

1. *A light source.* It can be an incandescent lamp or infrared light-emitting diode (LED).

2. *A rotary or translatory disc.* The disc has alternate opaque and transparent sectors, which are etched by means a photographic process on a plastic disc or slots are cut on a metal disc.

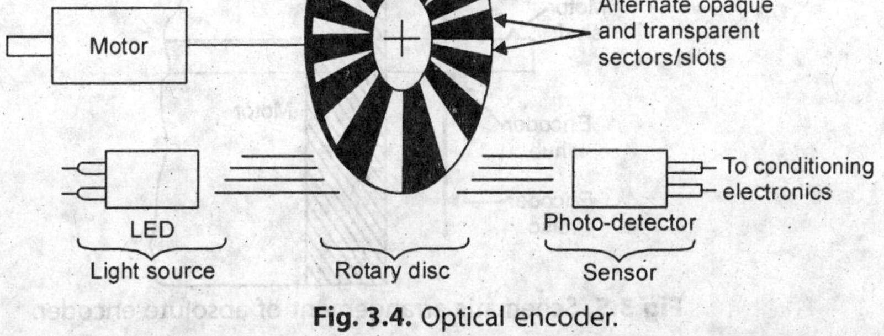

Fig. 3.4. Optical encoder.

3. *A sensor.* It is a photodiode or phototransistor, which senses the light (or no light) passing through sectors or slots of the disc.

Finally, *a conditioning electronics detects the light and dark signals and converts these signals into usable form of pulses or digital code.*

3.5.2.1. Incremental optical encoder

Refer to Fig. 3.4 :

— The disc of the encoder is connected to the motor shaft and rotates with motor.

— The disc rotation permits the light to reach the sensor, whenever the transparent slot comes in front of LED, thereby generating a *series of pulses* as output of sensor.

— These pulses are fed to a *counter,* which counts the number of pulses; the count being a measure of angle (or distance) through which the shaft has moved.

— *The speed of shaft is obtained by sampling the counter at regular intervals by means of clock pulses.*

- In incremental optical encoder,

$$\text{Basic resolution} = \frac{360°}{N_s}$$

where, N_s = Number of sectors on disc, each sector is half transparent and half opaque.

The basic resolution of a typical incremental recorder having disc with 2500 sectors

$$= \frac{360°}{2500} = 0.144°$$

— The accuracy and resolution of the increment optical encoder can be *increased* by using another disc as a mask, using two or four optoelectronic channels or using multiplier circuits.

3.5.2.2. Absolute optical encoder

The absolute encoder, as earlier mentioned, is capable of giving the correct rotary position at all times even after power-up has occured.

The device produces a separate and *unique* coded word for each shaft position and unlike the incremental encoder, *every reading is independent of the proceding one.*

Fig. 3.5 shows an absolute encoder mounted on a motor :

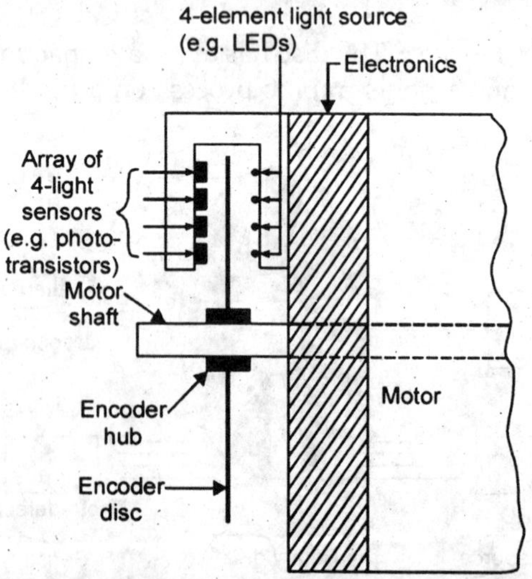

Fig 3.5. Schematic arrangement of absolute encoder.

- Absolute encoders usually consist of the following *three components* :
  (*i*) A multiple-track (or channel) light source.
  (*ii*) A multiple-channel light receiver.
  (*iii*) A multiple-track rotary disc.
- Normally, light emanating from a linear, N-element source (*e.g.* LEDs) is made to pass at right angles through the disc and is received for collected by a corresponding linear array of N light sensors (*e.g.* phototransistors) mounted on opposite side of the disc. The disc is divided into circumferential *tracks* and radial *sectors.*
- Absolute rotational information is obtained by utilising one of several possible code formats. For example Fig. 3.6 shows a 4-track 16-sector pure binary code formats. Other coding schemes that can be used include *binary-coded decimal (BCD) and Gray code.*

Refer to Fig. 3.6 :

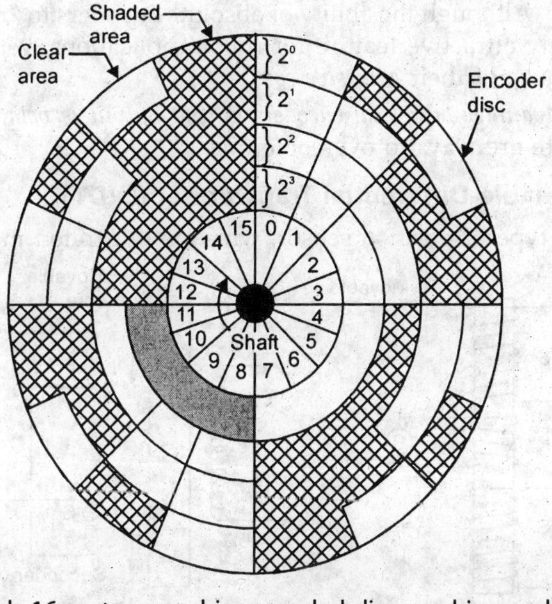

Fig. 3.6. A 4-track, 16-sector pure binary-coded disc used in an obsolute encoder.

— It can be observed that the resolution of the disc is equal $\frac{360°}{16}$ or 22.5° (since one complete disc revolution is 360° and there are 16 sectors)

— If the shaded areas are assumed to represent a binary "1" and the clear area as a binary "O", the outputs of each of the four light sensors, will represent a *4-bit sequence of ones and zeros.* For the binary code used in Fig. 3.6, the decimal equivalent of this number is the sector number. For example, if sector 11 is in the region of LED, the output of the phototransistors will be 1011 or decimal 11. Thus, the *absolute disc position is known simply by reading the outputs of the photodetector.*

— The resolution of an absolute encoder depends on the number of tracks (N) and is given by,

$$\text{Resolution} = \frac{360°}{2^N} \text{ degree ...... for a single complete rotation of the disc.}$$

In practice, it is possible to produce absolute encoders with upto 13 separate channels (*i.e.* 13 bits) which means that resolutions of upto $\frac{360}{(2)^{13}} = 0.044°$ are possible for a *single complete rotation of the disc.* However, oftenly it is necessary for the device being monitored by the encoder to undergo many rotations. In such a case it is possible to use a *second disc* placed on the same shaft as the first but geared down so that a complete revolution of the first moves the second only a distance of *one sector.* The *first one* is used for absolute positional information for any single shaft rotation, whereas the *second disc gives the actual rotation number.*

Advantage. An absolute encoder is a digital transducer in the true sense as its output does not require any conditioning but *can be directly read to get the absolute position information.*

Disadvantage. Although the ability of absolute encoders to "remember" positions is an extremely attractive feature for robot applications, yet these are generally *not* used because of their *excessive cost.*

The chief disadvantage of optical encoders is that *output is volatile when power is off;* however, there are ways to overcome this.

3.5.3. Linear Variable Differential Transformer (LVDT)

LVDT is another type of position sensor, which is illustrated in Fig. 3.7.

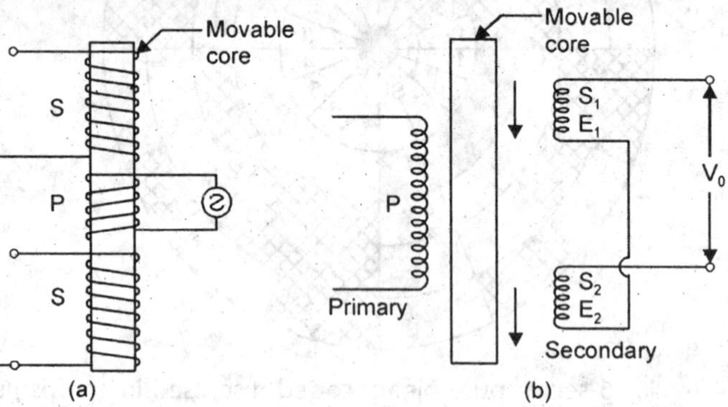

Fig. 3.7. Linear-variable-differential transducer (LVDT).

Construction. Refer to Fig. 3.7(*a*) :

— It consists of one primary winding (P) and two secondary winding (S_1 and S_2) which are placed on either side of the primary mounted on the same magnetic core. The magnetic core is free to move axially inside the coil assembly and the motion being measured is mechanically coupled to it.

— The two secondaries S_1 and S_2 have equal number of turns but are connected in series opposition so that e.m.fs, (E_1 and E_2) induced in them are 180° out of phase with each other and hence, cancel each other out. [See Fig. 3.7 (*b*)]

— The primary is energised from a suitable A.C. source.

Working :

— When the core is in the centre (*called reference position)* the induced voltages E_1 *and* E_2 *are equal and opposite.* Hence they cancel out and the output voltage V_0 is zero.

— When the external applied force moves the core towards coil S_2, E_2 is increased but E_1 is decreased in magnitude though they are still antiphase with each other. The net voltage available V_0 is $(E_2 - E_1)$ and is *in phase with* E_2.

Similarly, when the movable core moves towards coil S_1, $E_1 > E_2$ and $V_0 = E_1 - E_2$ and is in *phase with* E_1.

Thus, from above discussion, we find that the magnitude of V_0 is a *function of the distance moved by the core and its polarity or phase* indicates as to in which direction it has moved.

If core is attached to moving object, the *magnitude of* V_0 *gives the position of that object.*

- LVDTs are often used to sense position where *pure linear motion is performed* (e.g. in variety of machine tools, such as lathes). Rotary motion can be sensed with an RVDT (where 'R' stands for rotary), a device that works on the same principle as the LVDT but has a *rotational configuration*. When it comes to robots however neither device has been much used because of the following *problems/limitations:*
  - — It is necessary to use the device about its centre point.
  - — The centre or "null" position has a tendency to drift with time and temperature.
  - — Possibly the major difficulty with using an LVDT and/or RVDT as a *"position sensor"* is that the joints of most robots do not move in pure straight lines or circular arcs. Motion is normally a complex combination of these trajectories, consequently it is quite difficult to configure the LVDT/RVDT so that magnetic core is always collinear with the coils axis.
  - — RVDT has one additional limitation that makes its application to robots problematical. That is it can only sense rotary motions of approximately± 60°. Since most robotic axis are required to move more than 120°, this is most certainly a severe restriction.
  - — LVDT being an analog device cannot be conveniently utilised in a microprocessor-controlled sensor system. Further, the cost of sensor, signal-conditioning circuitry, and A/D converter would be significantly higher than that of an equivalent optical incremental encoder and its circuitry.

  In view of the above facts it seems more realistic to expect that such devices will be utilized in equipment that is used in *conjuction* with a robot.

3.6 FORCE/TORQUE SENSORS (FTS)

These sensors are primarily used for measuring the reaction forces developed at the interface between mechanical assemblies. The principal approaches to accomplish this are :

1. **Joint sensing :**
   - — *A "joint sensor" measures the cartesian components of force and torque acting on a robot joint and adds them vectorially.*
   - — For a joint driven by a D.C. motor, sensing is accomplished simply by measuring the armature current.
2. **Wrist sensing :**
   - — The wrist sensors are mounted on the tip of a robot arm and the end-effector.
   - — They consist of *"strain gauges" that measure the deflection of the mechanical structure due to external forces.*

- *"Pedestal sensing"* is another category, in which strain gauge transducers are installed between the base of a robot and its mounting surface in order to measure the components of force and torque acting on the base. In most applications, however, the base is firmly mounted on a solid surface and no provisions are made for pedestal sensing. The analysis of pedestal sensing is quite similar to that used for wrist sensing.

Wrist Force Torque Sensors (WFTS) :

— The major problem associated with force/torque FTS is mounting it on the joint. For this sensor to be sensitive, it must be flexible (not rigid).

— The joint torques and inertias can be estimated without the use of force/torque sensors; estimation algorithms based on actuator power input and inertia estimation are available. But these methods are complex and very sensitive to propagated errors.

— Besides the joint forces/torque, the interaction between robot (its end-effector) and *environment generates forces/torques that must be controlled to preserve the integrity of task being performed. In order to measure these interaction forces, sensors are mounted between the wrist and the end effector and are known as wrist force torque sensors* (WFTS).

- Fig. 3.8 illustrates a strain-gauge based wrist force torque sensor (WETS) :
  — It consists of two annular rings, outer ring and inner ring, which are connected to wrist and the end-effector end respectively. They are connected to each other by a cross and on the arms of this cross, eight pairs of strain gauges are mounted as shown.
  — When the gauges (after suitably selected) are properly connected in the Wheatstone bridge the *sensor detects three force components along three principal axes and three moments along three axes.*

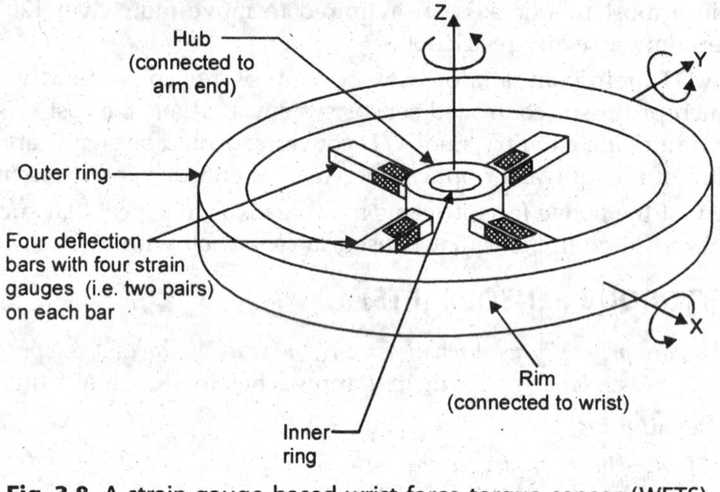

Fig. 3.8. A strain-gauge based wrist force torque sensor (WFTS).

Several other alternate designs of WFTS, based not only on strain gauges but also on *sonic* and *optic sensors*, are available.

Most WFTS function as transducers for transforming forces and moments exerted at the hand into measurable deflections or displacements at the wrist.

- A wrist force torque sensor (WFTS) must possess the following *"characteristics"* :
  (*i*) High stiffness (so that the positioning accuracy of the manipulator is not affected).
  (*ii*) Small in size.
  (*iii*) Light in weight.
  (*iv*) Compact in design.

(v) Sensitive.
(vi) Linear.
(vii) Low internal friction.

3.7 PROXIMITY SENSORS

Proximity sensors (sometimes referred to as external state sensors) generally have a binary output which indicates the presence of an object within a specified distance interval. "Range sensors," in contrast, yield an estimate of the distance between a sensor and a reflecting object.

These sensors, typically, are used in robotics for near-field work in connection with object grasping or avoidance.

3.7.1. Types of Proximity Sensors

1. Contact proximity sensor.
2. Non-contact proximity sensors.
   (i) Optical proximity sensors.
   (ii) Ultrasonic sensors.
   (iii) Eddy current sensors.
   (iv) Inductive sensors.
   (v) Hall-effect sensors.
   (vi) Capacitive sensors.
   (vii) Pneumatic sensors.
   (viii) Fibre optic scanning sensors.
   (ix) Scanning laser sensors.

3.7.2. Contact Proximity Sensors

Fig. 3.9. shows a simplest type of contact variety proximity sensor :

— It consists of a rod that protudes from one end and a *switch or other linear position-monitoring element located within the body of the sensor.*

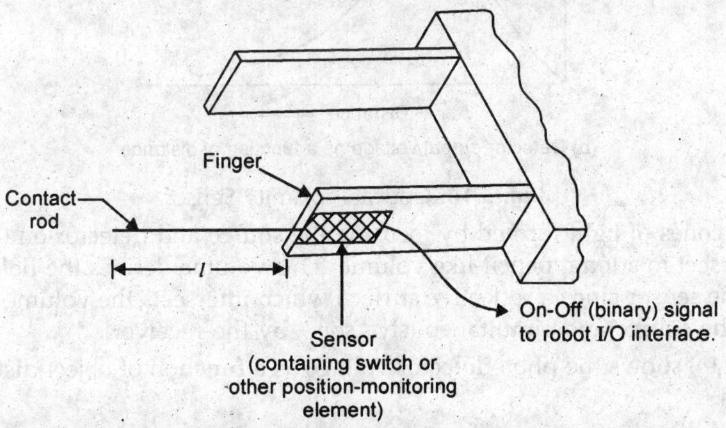

Fig. 3.9. Simple contact rod proximity sensor.

— When the robotic manipulator moves, the sensor becomes active as soon as the rod comes in contact with an object or obstruction. When this happens, the switch will close or open, whichever more convenient. The change of state of switch, monitored through the robot's I/O interface will cause an appropriate action to take place.

— It is possible to place such contact monitors anywhere on robot's arm and/or wrist and it is feasible to utilise more than one. As such, the *simultaneous obstacle and object sensing is possible.*

- A single contact proximity device can not provide any information about the shape or nature of the object or obstacle; however, by utilising *arrays* of such devices, some degree of object recognition can be obtained.

3.7.3. Non-contact Proximity Sensors

The various non-contact proximity sensors are discussed in the following sub-articles.

3.7.3.1. Optical proximity sensor

Fig. 3.10 (*a*) shows one of the simplest types of proximity sensors that uses light reflected from an object.

— It consists of a source of light (LED) and a detector (Photodiode) separated by about 8 mm and tilted symmetrically toward one another.

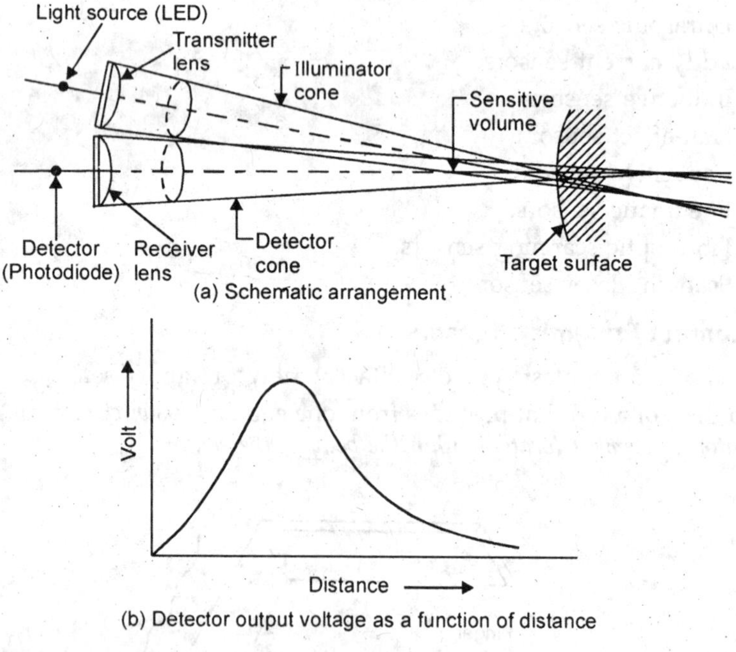

Fig. 3.10. Optical Proximity Sensor.

— The cones of light formed by focusing the source and detector on the same plane intersect in a long, pencil-like volume. This volume defines the field of operation of the sensor since a reflective surface which intersects the volume is illuminated by the source and simultaneously *"seen"* by the receiver.

Fig. 3.10 (*b*) shows the photodetector voltage as a function of object distance from the receiver lens.

— By placing a number of these sensors on a robotic gripper, proximity in several directions can be monitored simultaneously *e.g.*, ahead and below the robot's hand.

3.7.3.2. Ultrasonic proximity sensors

Ultrasonic has been used to provide *ranging and imaging information* for several years. Ultrasonic ranging systems like the one used on the automatic focusing Polaroid camera have been *widely used to give environmental awareness to a mobile robot.*

An ultrasonic sensor determines the range by measuring the elapsed time between the transmission of certain frequencies and their detected echoes. Different discrete frequencies are used because surface characteristics could cancel a single waveform, preventing detection.

- **Fig. 3.11.** Shows a typical ultrasonic sensor *used for proximity sensing :*

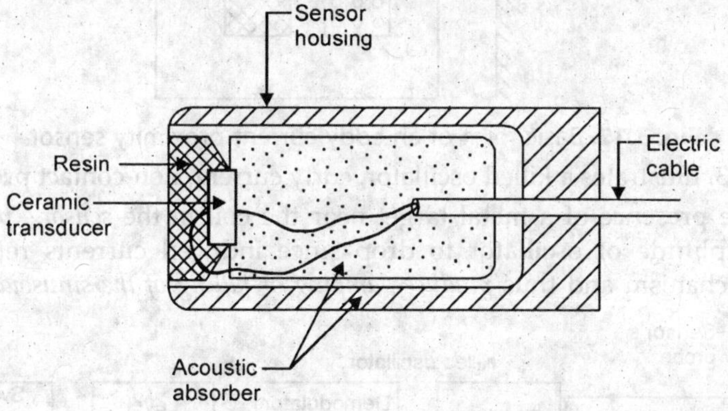

Fig. 3.11. Ultrasimic proximity sensor.

— It consists of an *electroacoustic transducer* (the basic element), often of piezoelectric ceramic type.

— The *resin layer protects the transducer* against humidity, dust and other environmental factors; it also *acts as an acoustical impedance matcher.*

— As the same transducer is generally used for both transmitting and receiving, *fast dumping of the acoustic energy is necessary to detect objects at close range.* This is *accomplished by providing acoustic absorbers, and by decoupling the transducer from its housing.*

— The housing is designed in such a way *as to produce a narrow accoustic beam for efficient energy transfer and signal directionality.*

The operational details of this tipe of sensor can be best understood by analyzing the waveforms used for both transmission and detection of the acoustic energy signals.

3.7.3.3. Eddy current proximity sensors

An "eddy current sensor" is used as a proximity switch and for determining the accuracy and repeatability of commercial robotic manipulators.

Working principle :

When a coil is supplied with an alternating current an alternating magnetic field is produced. If there is a metal object in close proximity to this alternating magnetic field, then eddy currents are induced in it. The eddy currents themselves produce a magnetic field which *distorts* the magnetic field responsible for their production. Consequently, the *impedance of the coil changes and so the amplitude of the alternating current. This change, at some preset level, can be used to trigger a switch.*

Fig. 3.12. shows the basic form of an eddy current proximity sensor. It is *used for the detection of non-magnetic but conductive materials.*

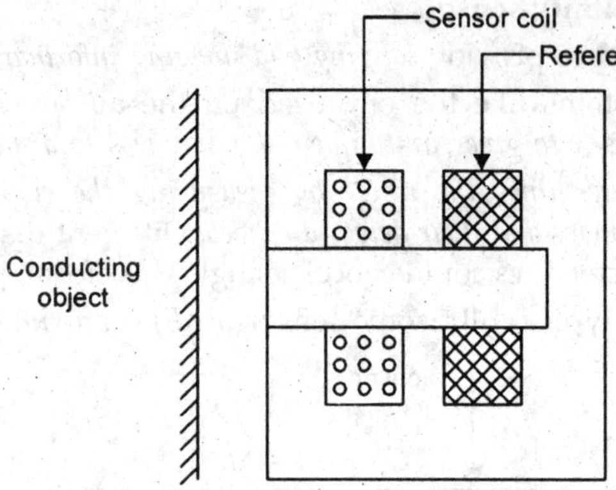

Fig. 3.12. Basic form of an eddy current proximity sensor.

- Fig. 3.13. Illustrates a killed oscillator, eddy current, non-contact proximity sensor :
  — The presence of a metal target near the coil in the *sensor's probe* causes the amplitude of oscillator to drop since induced currents represents a loss mechanism and thus *produce damping or killing of the sinusoidal waveform.*

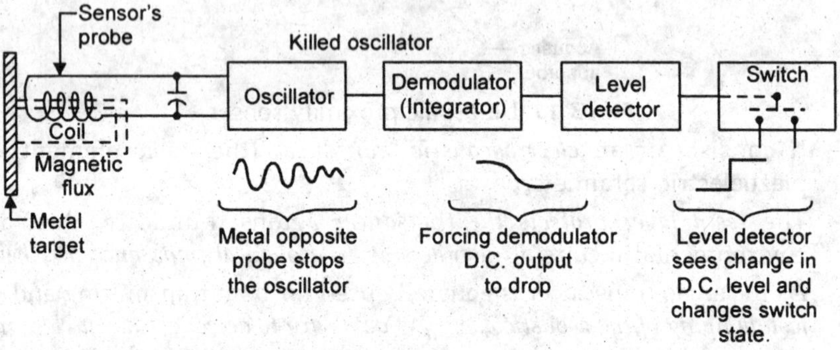

Fig. 3.13. Killed oscillator, eddy current, non-contact proximity sensor.

— A *demodulator* (essentially an integrator) responds to such a change by producing a smaller D.C. output.

— The *level detector* sees change in D.C. level and changes switch state (In proximity switch application, a *thresholding circuit* is used to detect when the level drops below some predetermined value, at which point the switch state is changed. *By adjusting the threshold value, a robot manipulator can be stopped at a desired distance from an object or a part*).

Advantages :

1. High flexibility.
2. High sensitivity to small displacements.
3. Small in size.
4. Relatively less expensive.

Disadvantages :

1. These devices must be calibrated for the type of the metal used for the target.
2. The size of the sensor depends on the effective linear measuring range – *larger probes are required for larger distances.*

Applications :

1. To determine both the repeatability and accuracy of the robots.
2. To involve the use of such a sensor and a robot to perform various types of inspections on manufactured parts such as :
   — Shaft and commutator runout;
   — Plate and disc flatness;
   — Thickness of nonconductive materials.
3. To measure static and dynamic displacements and hence are well suited for determining manipulator position overshoot under a variety of load conditions.
4. To measure discontinuation on moving metallic objects (since these sensors can sense frequency variations of upto about 50 kHz).

3.7.3.4. Inductive proximity sensors

This sensor is based on a *change of inductance due to the presence of a metallic object.* Inductive sensors are most widely use as industrial proximity sensors.

Fig. 3.14. shows a schematic diagram of an inductive sensor.

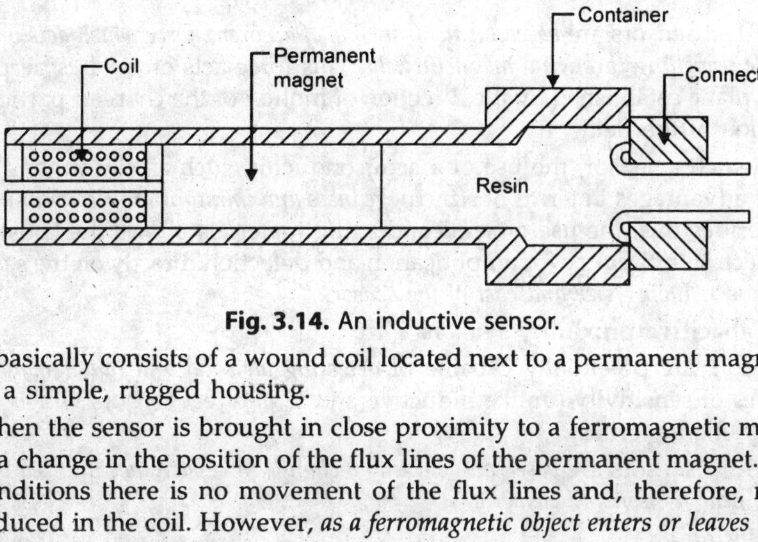

Fig. 3.14. An inductive sensor.

— It basically consists of a wound coil located next to a permanent magnet packaged in a simple, rugged housing.

— When the sensor is brought in close proximity to a ferromagnetic material, there is a change in the position of the flux lines of the permanent magnet. Under static conditions there is no movement of the flux lines and, therefore, no current is induced in the coil. However, *as a ferromagnetic object enters or leaves the field of the magnet the resulting change in the flux lines induces a current pulse whose amplitude and shape are proportional to the rate of change in the flux.*

— *The voltage waveform observed at the output of the coil provides an effective means for proximity sensing. The polarity of the voltage* out of the sensor depends on whether the object is *entering* or *leaving* the field.

— It has been observed that the sensitivity falls off rapidly with increasing distance, and that the sensor is *effective only for fractions of a millimetre (mm).*

3.7.3.5. Hall-effect sensors

In these sensors *the magnetic field intensity (e.g., produced by either permanent or electromagnets mounted on a object) determines the generated voltage output.*

"When any object carrying a current I is placed in the transverse magnetic field B, then an electric field E is induced in the object in direction perpendicular to both I and B the phenomenon is known as *Hall effect.*"

- These sensors *can only detect magnetised objects.* However, when used in conjunction with a permanent magnet in a configuration such as one shown in Fig. 3.15, they are capable of detecting *all feromagnetic materials.*

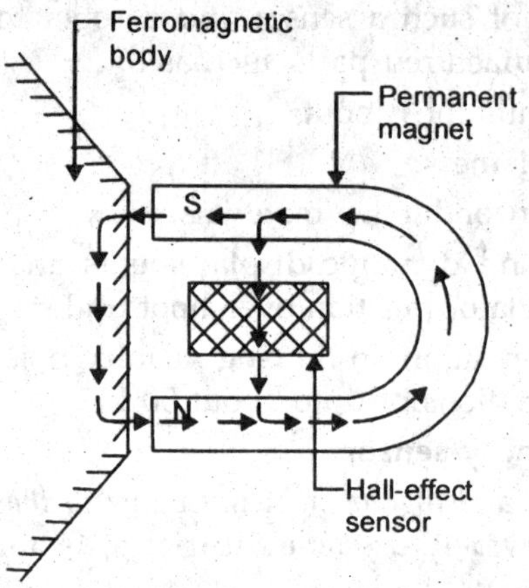

Fig. 3.15. Operation of a Hall-effect sensor in conjunction with a permanent magnet.

- Hall effect sensors are *based on the principle of a Lorenz force which acts on a charged particle travelling through a magnetic field.* This force acts on an axis perpendicular to the plane established by the direction of motion of the charged particle and the direction of the field.
  - — In such a sensor, the use of a semiconductor, such as silicon, has a number of advantages in terms of *size, ruggedness and immunity to electrical interference.* Besides this, the use of semiconducting materials allows the construction of electronic circuitry for amplification and detection directly on the sensor itself, thus *reducing size and cost of the sensor.*

3.7.3.6. Capacitive proximity sensors

These sensors are potentially capable of *detecting all solid and liquid materials,* with various degrees of sensitivity (unlike inductive and Hall-effect sensors which can detect only ferromagnetic materials).

The capacitive sensors are *based on detecting a change in capacitance induced by a surface that is brought near the sensing element.*

- ***Description :***

  The sensing element (see Fig. 3.16.) of a capacitive proximity sensor is a capacitor composed of a *sensitive electrode* (say metallic disc) and a *reference electrode* (a ring) separated *by a dielectric material.*
  - — A cavity of dry air is usually placed belind the capacitive element *to provide isolation.*
  - — The rest of the sensor consists of *electronic circuitry* which can be included as an integral part of the unit, in which case it is normally *embedded in a resin to provide sealing and mechanical support.*

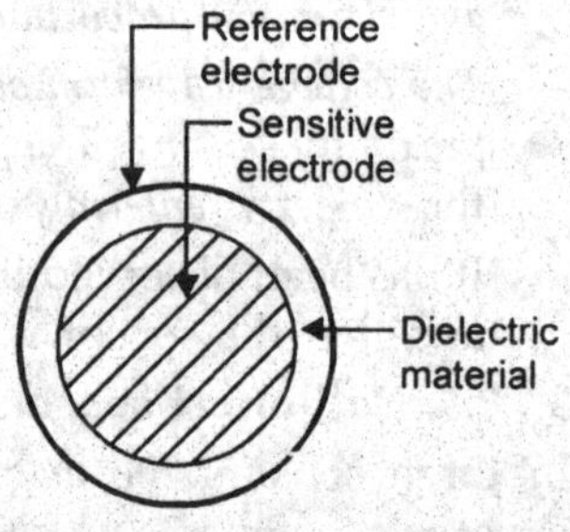

Fig. 3.16. Sensing element of a capacitive proximity sensor.

- Out of a number of electronic approaches *for detecting proximity based on a change in capacitance,* a couple of these are :

(i) The simplest one *includes the capacitor as a part of an oscillator circuit designed so that the oscillation starts only when the sensor's capacitance exceeds a predefined threshold value. The start of oscillation is then translated into an output voltage which indicates the presence of an object.* This method provides a *binary output whose triggering sensitivity depends on the threshold value.*

(ii) Another approach (more complicated) *utilises the capacitive element as a part of a circuit which is continuously driven by a reference sinusoidal waveform.* A change in capacitance produces a phase shift between the reference signal and a signal derived from the capacitive element. *The phase shift is proportional to the change in capacitance and can thus be used as a basic mechanism for proximity detection.*

3.7.3.7. Pneumatic proximity sensors

These sensors involve the use of compressed air, displacement or proximity of an object being transformed into a *change in air pressure.*

Fig. 3.17. shows the basic form of a pneumatic sensor.

— Low pressure air is allowed to escape through a port in front of the sensor.

— This escaping air, in the absence of any close by object, escapes and in doing so also *reduce the pressure* in the nearby sensor output port. However, if there is a close by object, the air cannot so readily escape and result is that the *pressure increases* in the sensor output port. The output pressure from the sensor thus depends on the proximity of objects.

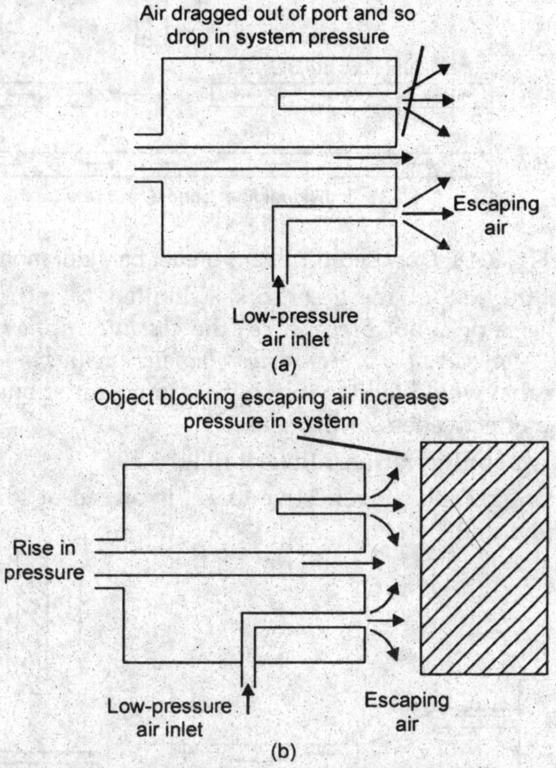

Fig. 3.17. Pneumatic proximity sensor.

- *"Peumatic proximity sensors" are useful in workpiece grasping* as they can provide accurate distance between fingers and workpiece, orientation of workpiece, finger closure signals and so on.

- Pressure-or vacuum-based sensors are *very sensitive and are suitable for handling delicate and fragile objects.* Counter pressure based sensors are used in detecting or identifying the features such as edges, holes, shapes etc. and detecting the movement.

3.7.3.8. Fiber optic scanning sensors

Following are the three systems which utilize the "fibre optics technology" in the robotics and/or the manufacturing field. :

1. Opposed configuration.
2. Retroreflective configuration.
3. Diffuse configuration.

1. Opposed configuration : Refer to Fig. 3.18.

— In the opposed (or beam break) configuration the object is detected when it *actually interrupts the beam of light.*

— Such an optical interrupter depends on the object being opaque and is obviously not useful where parts are made of transparent or translucent materials.

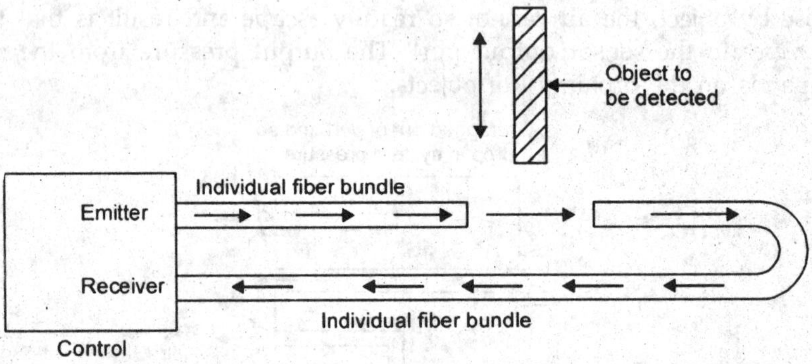

Fig. 3.18. Opposed (or beam break) configuration.

— Usually, the use of these sensors is limited to informing the robot that something is or is not present, *i.e.,* the *absolute position information cannot be obtained. Moreover, the receiver fiber bundle alignment is* fairly critical; thus anything that would tend to misalign it from emitter bundle would obviously affect the effectiveness of the sensor.

2. Retroreflective configuration : Refer to Fig. 3.19.

— In this configuration, a reflective target is placed at some distance from the body of the unit.

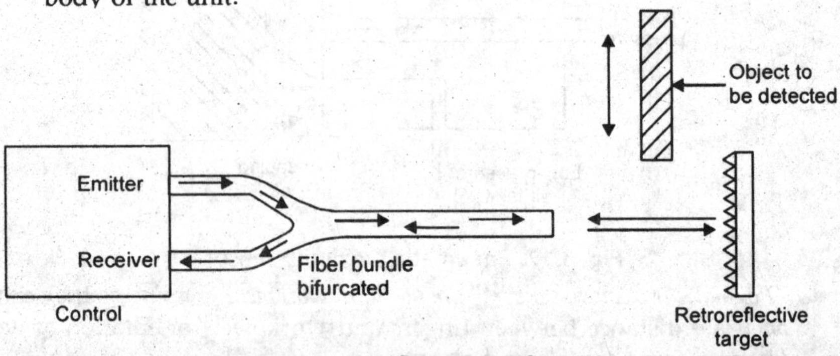

Fig. 3.19. Retroreflective configuration.

— An opaqe object entering the area between the end of the fiber bundle and the target is *sensed* since the reflected light and that reflected from the target are *both attenuated* when they pass through such an object.

— This arrangement utilizes a bifurcated fiber bundle so that incident and reflected light is carried by the same fiber (thus eliminating alignment difficulties associated with previous configuration).

Limitation. The need for a separate target somewhat restricts the use of such a sensor to *parts detection only*.

3. Diffuse configuration : Refer to Fig. 3.20.

— This configuration also uses a bifurcated fiber bundle but there is no retroreflective target.

— The sensor actually can measure the amount of light reflected from an objected upto a few centimetres away from the fiber bundle.

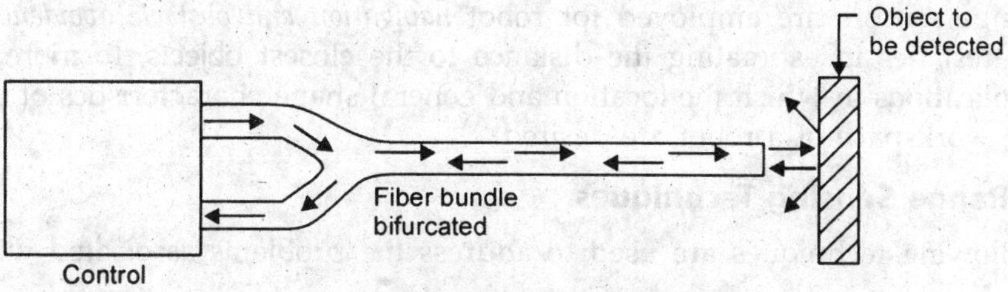

Fig. 3.20. Diffuse configuration.

— This device can be used to *detect transparent and translucent objects* since most materials reflect some light.

- This mode of fibre optic sensing is *most commonly used since it is self-contained* (*i.e.*, no target is required), *rugged, weighs very little, and is relatively inexpensive, making it ideally suited for a manufacturing environment*.

3.7.3.9. Scanning laser proximity sensors

Refer to Fig. 3.21.

This sensor (costly and considerably more involved) consists of :

(*i*) A laser light source;

(*ii*) Two mirrors, one of which is rotated by an A.C. motor;

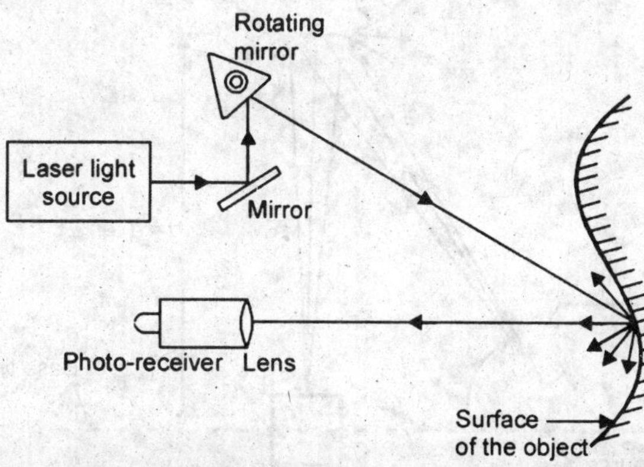

Fig. 3.21. Scanning laser proximity sensor.

(*iii*) A lens-photo-receiver assembly.

— The incident light beam from the laser (helium-neon) is swept across the surface of the object by the action of the motor-driven triangular miror; this occurs three times for each motor revolution.

— A lens mounted infront of a photodetector (*e.g.*, a phototransistor) *permits light reflected from one point on the object's surface to be acquired. The distance from the sensor to this point is determined by synchronising the A.C. motor voltage with a high-frequency clock. The number of clock pulses from the time this voltage is zero until the photodetector receives reflected light is a measure of the distance.*

3.8 RANGE SENSORS

3.8.1. Introduction

The function of a ***range sensor*** *is to measure the distance from a reference point (normally on the sensor itself) to objects in the field of operation of the sensor.*

- Range sensors are employed for robot *navigation and obstacle avoidence,* where interest lies in estimating the distance to the closest objects, to more detailed applications in which the location and general shape characteristics of objects in the workspace of a robot are desired.

3.8.2. Range Sensing Techniques

The following techniques are used to address the problems associated with range sensing :

1. Triangulation technique.
2. Structured lighting approach.
3. Time-of-flight range finders.

3.8.2.1. Triangulation technique

Fig. 3.22. shows the range sensing by triangulation technique :

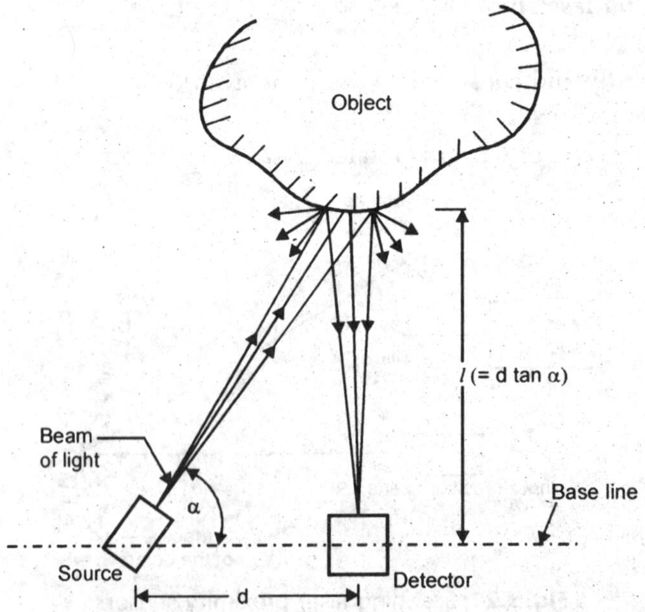

Fig. 3.22. Range sensing by triangulation technique.

— The object is illuminated by a narrow beam of light which is swept over the surface.

— The sweeping motion is in the plane defined by the line from the object to the detector and the line from the detector to the source.

— If the detector is focused on a *small portion* of the surface then, when the detector sees the spot of light, its distance l to the illuminated portion of the surface can be calculated from the geometry of Fig. 3.22 since the angle of the source (α) with the base line and the distance d between the source and detector are known (*e.g.*, $l = d \tan \alpha$)

— The above approach yields a *point measurement*. However, if the source-detector arrangement is moved in a fixed plane (up and down and sideways on a plane perpendicular to the paper and containing the base line in Fig. 3.22), then it is possible to *obtain a set of points whose distances from the detector are known*. These distances are easily transformed to three-dimensional coordinates by keeping track of the location and orientation of the detector as the objects are scanned.

3.8.2.2. Structured lighting approach :

- In this range sensing technique, a light pattern is projected on to a set of objects and range is calculated by using the distortion of the pattern.
- One of the most popular light patterns in use is a sheet of light generated through a cylindrical lens or a narrow slit :
  — The intersection of the sheet with objects in the workspace yields a light stripe which is viewed through a television camera displaced a distance d from the light source. The stripe pattern is easily analysed by a computer to obtain range information.
  — Specific range values are computed by first calibrating the system.

3.8.2.3. Time-of-flight range finders :

These methods involve the use of a *laser* and *ultrasonics*.

- In one approach of using a laser to determine range is to measure the time it takes an emitted pulse of light to return coaxially *i.e.*, along the same path from a reflecting surface. The distance d to the surface is given by :

$$l = \frac{ct_{pt}}{2} \qquad \text{...(3.1)}$$

c = Velocity of light, and

t_{pt} = Pulse transit time.

- In a *"pulsed-laser system"*, described by Jarvis, a two-demensional arrays with values proportional to the distance are produced. The two-dimeasional scan is accomplished by deflecting the laser light is a rotating mirror. The working range of this device is of the order of 1 to 4m, with an accuracy of ± 0.25 cm.
- In an *"ultrasonic range finder"* (the basic idea being same as that used with pulsed laser) an ultrasonic chirp is transmitted over a short time period and, since the speed of sound is known for a specified medium, a simple calculation involving the time internal between the outgoing pulse and return echo yields an estimate of the distance to the reflecting surface.

3.9 SLECTION OF A RIGHT SENSOR

The following points should be considered which selecting/choosing a sensor for a particular situation :

1. *The nature of the measurement required e.g.,*
   — The variable to be measured, its nomial value, the range of values;
   — The accura-y required;
   — The required speed of measurement;
   — The reliability required;
   — The environmental conditions under which the measurement is to be made.
2. *The nature of the output required from the sensor,* this determining the signal conditioning requirements in order to give suitable output signals from the measurement.

   Then, possible sensors can be identified taking into account such factors as their range, accuracy, linearity, speed of response, reliability, maintainability, life, power supply requirements, ruggedness, cost, availability.

ROBOTIC VISION

3.10 INTRODUCTION TO ROBOTIC VISION

Robotic vision *may be defined as the process of extracting, characterising and interpreting information from images of a 3-dimensional world.*

- Robotic vision (also termed as *"Computer vision"* or *"Machine Vision")* is an important sensor technology with potential applications in many industrial operations.
- Robotic vision is primarily targetted at *manipulation and interpretation of image and use of this information in robot operation control.* Robotic vision requires *two aspects to be addressed :*
  (*i*) Provision for visual input;
  (*ii*) Processing required to productively utilise the visual information in a computer-based system.
- The greatest virtues of a vision system are :
  (*i*) Repeatability;
  (*ii*) Accuracy;
  (*iii*) Ability to produce approximately the same results when given approximately the same inputs.

3.11 ROBOTIC VISION SYSTEMS

- *The basic perpose of robot vision system is to identify an object and determine its location (position and orientation).*
- The vision system must be capable of handling multiple views to deal with the multiple stable states. For this purpose, the system *has to be fast and work in parallel with the robot system.* Further, the system must be able to work in an industrial environment including factory illumination and be insensitive to normal light variation. A gray scale system is required so that numerous different shades of

gray can be assigned to each point of the image and the vision system's performance is not affected by the object's form, colour or surface texture.

3.11.1. Functions of Robot Vision System

The operation of robot vision consists of the following *three* functions :

1. **Sensing and digitizing :**

   This process yields a visual image of sufficient contrast that is typically digitized and stored in the computer memory.

2. **Image processing and analysis :**

   The digitized image is subjected to image processing and analysis *for data reduction and interpretation of the image.* This function may be further subdivided as follows :

   (*i*) *Preprocessing* .... It deals with techniques like noise reduction and enhancement details.

   (*ii*) *Segmentation* .... It partitions an image into objects of interest.

   (*iii*) *Description* .... It computes various features like size, shape, etc. suitable for differentiating one object from another.

   (*iv*) *Recognition* .... It identifies the object.

   (*v*) *Interpretation* .... It assigns meaning to an ensemble of recognized objects in the scene.

3. **Application :**

   The current applications of robot vision include the following :

   (*i*) Inspection.

   (*ii*) Part identification.

   (*iii*) Location and orientation.

- The various areas of vision processing may be grouped as follows, depending upon the sophistication involved in their implementation :

  (*i*) ***Low level vision*** .... *Sensing and processing.*

  (*ii*) ***Medium level vision*** .... *Segmentation, description and recognition of individual object.*

  (*iii*) ***High level vision*** .... *Interpretation.*

3.11.2. Components of Vision System

A complete vision system consists of hardware and software for performing the functions of sensing and processing the image (the scene) and utilising the results obtained to command the robot. Fig. 3.23. shows a block diagram of the components of a vision system, arranged schematically and their functions are explained below :

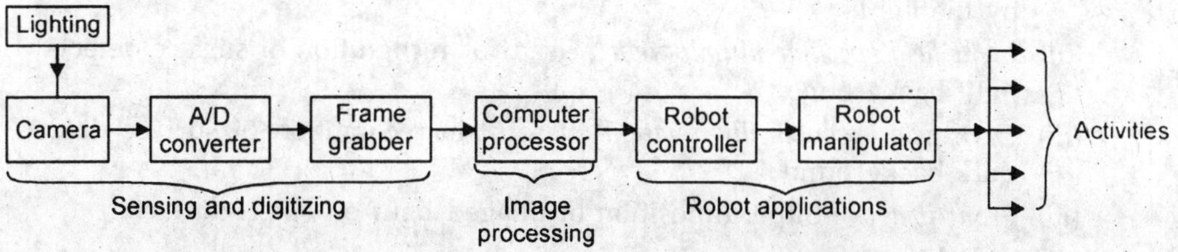

Fig. 3.23. Components of machine vision system.

3.11.2.1. Camera-Illumination

"Process of imaging" : In the basic process of imaging the light source illuminates the object and camera captures the reflected light. *The image formed in the camera is converted into analog signal (voltage) with the help of suitable transducers. Finally, the analog voltages are digitized and converted into an algebraic array.* This array is the *"image"* to be processed and interpreted by the computer according to predefined algorithms.

— The image presented to the camera is light reflected from the environment. This varies in wavelength and intensity throughout the image and is directly dependent on the illumination of the scene or lighting (*i.e.*, luminous power and placement of light sources).

— A poor lighting produces low-contrast images, shadows, and noise in two-dimensional (2-D) vision system, the desired contrast can often be accomplished by using a controlled lighting system. *A 3-D system may require more sophisticated lighting system.*

— *In a single camera system, triangulation is used to detect shape and depth.* Some systems use two cameras to get 2-D images to obtain a stereoscopic view of the scene.

- The robotic vision cameras are esentially *optoelectronic transducers,* which convert optical input signal to electrical output signal. They fall in the domain of TV cameras. There is a variety of camera technologies available for imaging. Some of these are :

  (*i*) Black-and-white vidicon tube;

  (*ii*) Solid-state cameras based on *charge-coupled devices (CCD),* and *charge injection devices* (CID);

  (*iii*) *Silicon bipolar sensor* cameras.

— **"Vidicons"** *are the most common tube-type cameras.* The image is focused on a photosensitive surface where a corresponding electrical signal is produced by an electron beam scanning the photosensitive surface. The electron beam passes easily through the photosensor at a highly constructive point caused by very intense light. Fewer electrons pass through the photosensor where lower levels have made it less conductive. *Scanning the electron beam carefully across the entire surface produces electrical information about the entire image.*

Illumination techniques :

It is of significance importance to have a well-designed illumination system since such a system *minimises the complexity of the resulting image, while the information required for object detection and extraction is enhanced.*

In robot vision, the following two basic illumination techniques are used.

1. ***Front light source :***

   (*i*) *Front illumination* : The feature of the image is defined by the surface flooded by the light.

   (*ii*) *Light field specular illumination* : Used for recognition of surface defects with light background.

   (*iii*) *Dark field specular illumination* : Useful in recognition of surface defects in dark background.

   (*iv*) *Front images* : Superimposition of imaged light on object surface.

2. ***Black light source :***

   (*i*) *Light-field rear illumination* : Used in simple measurement and inspection of parts.

(*ii*) *Condensed rear illumination* : High contrast of images produced in high magnification.

(*iii*) *Rear illumination collimator* : Parallel light ray source produced so that the same plane objects are featured.

(*iv*) *Offset rear illumination* : Highlights the object features in transparent medium.

Fig. 3.24 shows the basic illumination techniques.

- The basic types of *lighting devices* in robot vision may be grouped into the following categories :

(*i*) *Defuse surface devices;* (*ii*) *Condenser projectors;*
(*iii*) *Flood or spot projectors;* (*iv*) *Collimators;*
(*v*) *Imagers.*

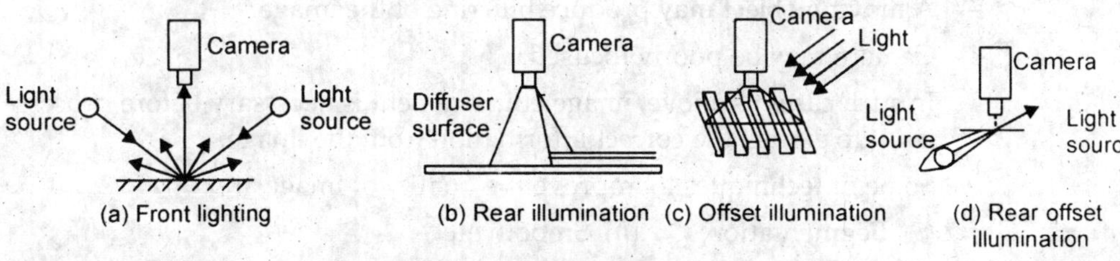

Fig. 3.24. Basic illumination techniques.

3.11.2.2 A/D (Analog-to-Digital) converter and frame grabber

A/D converter is required to convert analog picture signal from the camera into digital form that is *suitable for computer processing.*

— The analog voltage signal from the camera is sampled periodically at an appropriate sampling rate. Each sampled voltage is approximated to predefined voltage amplitude. The accuracy of this approximation depends on sampling rate of the A/D converter. The quantized voltage is encoded into a digital code represented by a binary number.

— Invariably, *A/D converter is a part either of the digital camera or the front end of frame grabber.*

- The *"frame grabber"* (a hardware device) is an image storage and computation device which stores a given pixel array. It can vary in capability from one which simply stores an image to a more powerful frame grabber in which thresholding, windowing and calculations for histogram modification can be carried out under computer control. The stored image is subsequently processed and analyzed by the combination of the frame grabber and the vision controller.

3.11.2.3. Image processing

— The processing of the visual information, which is voluminous, is slow. It is *preprocessed to filter the noise and to retain only the useful information* from the acquired image and enhance the details. The digitized image is subjected to *image enhancement;* a process that partitions the image into objects of interest.

— The digitized and preprocessed image matrix for each frame is stored in the memory and then subjected to processing as per the image processing algorithms for computation of features such as size, shape, etc., called *image description.*

— The preprocessed image frame is stored in computer memory for further processing.

(a) Image improvement :

- During image acquisition, the captured image may contain shadows, noise, distortions, and other imperfections as a result of sampling, transmission, improper lighting, or disturbance in the environment. The distorted poor-quality image is also produced by the faulty equipment or their incorrect use. *The image enhancement process is employed to obtain a second binary or gray-scale image of much improved quality.*
- Distorsions in the image may categorised as :
  (*i*) *Dimensional distortions* due to imperfections in the camera lens; or
  (*ii*) *Brightness distortions* due to improper illumination :
  — Overlapping frames or incomplete neutralisation of photodetectors in camera produces ghosting.
  — A moving object may produce blurring of the image.
  — Camera may be poorly focused.
  In such cases as above, image enhancement is necessary before processing it to extract the correct information from the image.
  Some of techniques to improve the quality of image are :
  (*i*) Segmentation; (*ii*) Smooth etc.

(b) Segmentation :

- *It is the process of identifying a group of related pixels for locating connected regions or areas of image having similar characteristics.* This process *divides the image into its constituent parts.*
- Segmentation algorithms are generally based on one of the two basic principles.
  (*a*) *"Similarity"* : Thr principal approaches in this category are based on :
  — Thresholding;
  — Region growing.
  (*b*) *"Discontinuity"* : This principle involves;
  — Edge detection.
  Thses concepts are applicable to both static and dynamic (or time varying) scenes.

Thresholding :

— *It is a binary conversion technique in which each pixel is converted into a binary value, either black or white.* This is accomplished by using a frequency histogram of the image and establishing what intensity (gray lend) is to be the border between black and white.

— This method is used by many commerical available robot vision systems these days.

— Global thresholds find applications in situations where there is a clear distinction between objects and the background and where illumination is relatively uniform.

Region growing :

— The region growing is a technique that *groups pixels having similar attributes into regions.* This is called *pixel aggregation.*

— Here, the gray scale raw image is scanned and the region is grown by appending or connecting together the neighbouring axel that have same property say gray level, texture, colour etc. Each region is labelled with a unique integer number.

— Region oriented segmentation process must meet the following conditions :

(*i*) Segmentaton *must be complete* (*i.e.*, every pixel must be in a region);

(*ii*) The points in a region *must be connected;*

(*iii*) The regions *must be disjoint.*

Edge detection :

This is another effective technique for segmentation.

— The outline boundary of an object within an image is equivalent to identifying the edges of the object that separate the object from its background. Algorithms to identify whether an image pixel lies on the edges of an object or not, are known as *"edge detection algorithms"*.

— A common method is *based on the intensity change or intensity discontinuity that occurs in adjacent pixels at the boundary or edge of an object.*

— *The idea underlying most of edge detection algorithms is the computation of local gradient of image intensity.*

— *The magnitude of the first derivative of intensity function can be used to detect edges in the image.*

(*c*) Smoothing (or noise reduction) :

— Smoothing operations are used to *improve the quality of the image* by reducing noise and other spurious effects that are introduced in a image as a result of sampling, quantization, transmission or disturbances in the environment during image acquisition and digitizing. These are *iterative in nature*.

— In this operation, the image intensity is modified using local techniques based on the assumption that the pixel value for a pixel in some sense is similar to that of its neighbours.

— *"Neighbourhood averaging"* is one of the several techniques of smoothing.

- The main *disadvantage* of smoothing is that *it blurs the images and other sharp details*. The blurring can be *reduced* by the use of so called *median filters* in which intensity of pixel is replaced by the median of the intensities in a predefined neighbourhood of that pixel, instead of by the average.

(*d*) Object descriptors :

— In order to identify a given object in different images, characteristics like shape, size, perimeter etc. are computed from the image. *The extracted geometric features of an image are known as "object descriptors."*

— Several descriptors are simultaneously obtained for a given object and then, the object may be identified with good statistical confidence level.

(*e*) Object recognition :

— The object recognition *deals with unique identification of each object in the image.*

— The algorithms for object recognition, for robotic application, should not only be powerful to uniquely, identify the object but should be fast enough to match the work cycle.

— For object identification, *image comparison technique* is a simple approach. The images of known objects are stored in the computer and the regions identified in the image are compared with these to recognise the parts in the image. This method is known as *"template matching."*

3.12.2.4. Image presentation

— The primary requirement of vision system is to *generate an appropriate method of representing image data. The method should provide for convenient and efficient storage and processing by computer and it should encapsulate all the information, which defines important characteristics of the scene in the image.*

— For computer-based processing, a digital form of the image is required. A suitable approximation of the intensity function I(x, y) is made for convenience of representation and processing. This *approximation is known as "digitization "* and is carried out in two stages :

(*i*) Spatial digitization;

(*ii*) Amplitude digitization.

3.12 ADVANTAGES OF MACHINE VISION

The advantages of machine vision are :

1. *Flexibility.* A wide variety of measurement can be made by the vision systems. When the applications change, the software can be easily modified or upgraded to accommodate new requirements.
2. *Cost effectiveness.* The machine vision systems are becoming increasingly cost effective, with the price of computer processing drooping rapidly.
3. *Precision :*
   — It is possible to measure dimensions to one part in a thousand or better by a well-designed vision system.
   — Since the measurements do not require contact, therefore, there is no wear or danger to delicate components.
4. *Consistency :*
   — Elimination of operational variability (since vision systems are not prone to the fatigue suffered by human operators);
   — Multiple systems can be configured to produce identical results.

3.13 APPLICATIONS OF MACHINE VISION

The applications of "machine vision" include the following :

1. Industrial manufacture on large scale.
2. Manufacture of short-run unique objects.
3. Retail automation.
4. Quality control and refinement of food products.
5. Monitoring of agriculture production.
6. Medical remote examination and procedures.
7. Medical imaging processes. .
8. Safety systems in industrial environments.
9. Automated monitoring of sites for security and safety.
10. Consumer equipment control.

11. Inspection of pre-manufactured objects.
12. Control of Automated Guided Vehicles (AGVs).

3.14 INDUSTRIAL APPLICATIONS OF VISION-CONTROLLED ROBOTIC SYSTEMS

The effective use of robotic system makes assembly, quality control parts handling, and classification tasks more robust. By using a single camera, it is possible to track multiple objects in visually cluttered environments. A visions – controlled robotic system can be deployed for a number of different applications mentioned below :

1. **Presence :**
   — Several sensors such as proximity or touch sensors can be used to find the presence or absence of a part at a specific location (say on a conveyor belt or in a bin).
   However, the visual detection of presence combined with other applications gives much more accurate and versatile information.
2. **Object location.** For tracking motion of manipulator, end-effector, objects, or ostacles, *accurate coordinate assessment is used.* For this purpose sometimes special identification markings are made on the part.
3. **Object identification.** The object identification, which may be a part of sorting process, involves determination of location and pick-and-place tasks to effect physical movement of parts.
4. **Pick and place :**
   — It is possible to guide the manipulator to pick a part from a specific location after its presence has been detected or from any imaged location in the workcell and place it at the desired location.
   — The gripper can be oriented according to orientation of the part to hold the part properly.
5. **Visual inspection :**
   — The visual inspection is based on *extracting specific quantitative measurements of desired parameters from an image.*
   — The robot employed in visual inspection :
   (*i*) can do sorting of 'good' and 'bad' parts, or
   (*ii*) can control the manufacturing process based on measured values of parameters.
6. **Visual guidance :**
   — The image of the scene can be used for accurate specification of relative positions of the manipulator and the part in the scene as well as their relative movements.
   Examples : (*i*) Assembly operation, which requires accurate position and orientation control of two parts to be fitted together.
   (*ii*) Guiding the motion of the maipulator through stationary or mobile obstacles in the environment of the robot.

QUESTIONS WITH ANSWERS

Q. 3.1. Explain briefly the architecture of a computer-based intelligent robotic manipulator.

Ans. Fig. 3.25 shows the multilevel control architecture of a computer- based intelligent robotic manipulator :

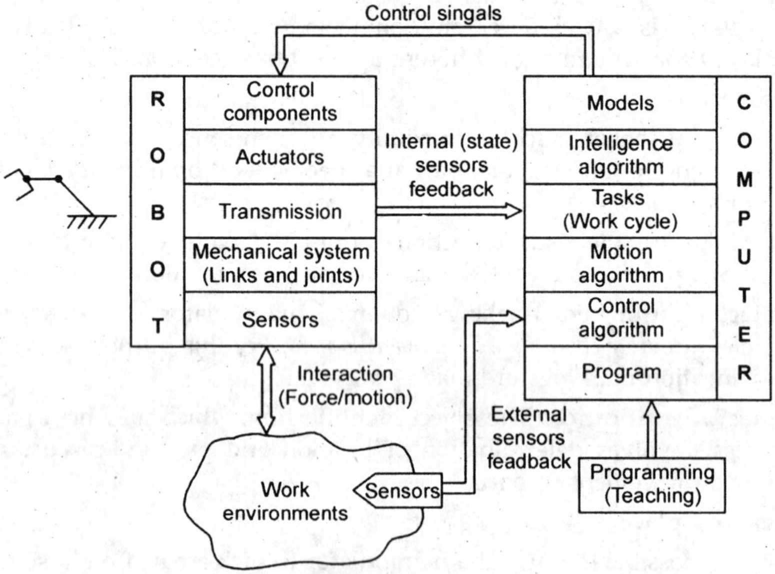

Fig. 3.25. A computer-based intelligent robotic manipulator's architecture.

— The robot having a fusion of the available sensory data with task planning is characterised as an *intelligent robot;* it has the capability of perception of action. The control algorithms can guarantee the coordinated motion of the mechanical structure in correspondence to the task planning.

— The signals from all the sensors on the manipulator and in the environment are fed to the computer-controller which analyses them as per system models task programs, control algorithms, motion algorithms, and intelligence algorithms and generates the control signals.

— These control signals are applied to the actuators to command and control the manipulator. The interaction of the manipulator with the work environment is sensed by the sensors and again fed back and the cycle is continued.

Q. 3.2. With the help of neat diagram, illustrate the scheme of robot sensors.

Ans. Fig. 3.26 shows a scheme of robot sensors:

— *External sensors* include CCD camera, force/torque sensors, proximity sensors and various tactile sensors to acquire knowledge or get information from the environment.

— *Internal sensors* may include encoders, tachometers, force/torque transducers, electrical and optical devices that transfer information about the position, speed and articulations of the manipulator to interact with the environment.

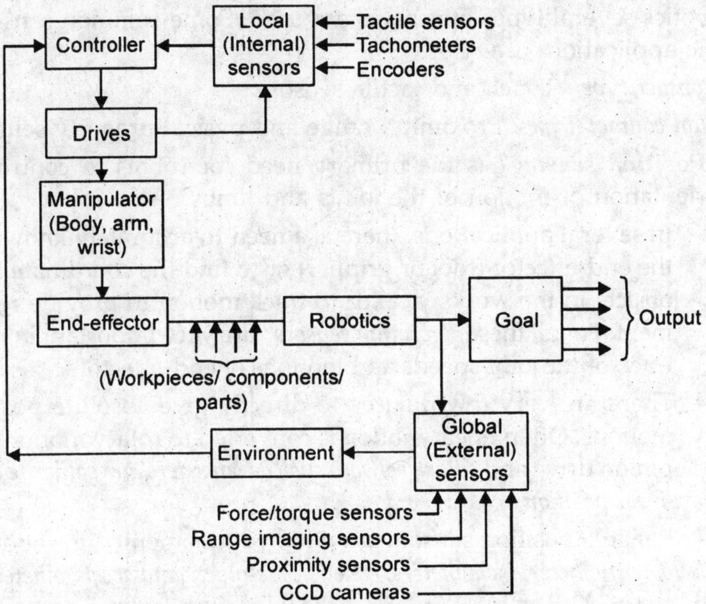

Fig 3.26. Scheme of robot sensors.

Q. 3.3. State the desirable features for sensors and transducers.

Ans. Sensors and transducers should possess the following desirable *features* :

1. ***Precision :***
   — Should be *as high as possible.*
   — *Deviation in* the measurment of reading should *minimum.*
2. ***Accuracy :***
   — Should be *as high as possible.*
   — Error between sensed value and actual value should *approach zero.*
3. ***Speed of response :***
   — Time taken to respond to variation should be *minimum.*
   — Response should be *instantaneous.*
4. ***Operating range :***
   — Should be *wide.*
   — Accuracy over the range to be *acceptable.*
5. ***Reliability :***
   — The life to be *high.*
   — Frequent failures are *not* acceptable.
6. ***Cost and ease :***
   — The cost of purchase should be *low.*
   — The installation and operation should be *easy* and *less costly.*
7. ***Calibration :***
   — Should be easy to *calibrate.*
   — Drift to be *minimum.*
   — Should take *less time to calibrate* without much trouble.

Q. 3.4. Discuss briefly the various kinds of sensors used in robotics.

Ans. In robotics several types of sensors are used. Some common sensors suitable for robotic applications are :

(i) *Contact type :* Touch and tactile sensors.

(ii) *Non-contact Type :* Proximity, range and pysical property sensors.

- ***"Position sensing"*** is the primary need for robots to control position and orientation or motion of the joints and limbs of the robot.
  - — In several applications, there is a need to accurately know the position of the end-effector (tool or gripper) or to find the coordinate locations of the objects in the workspace, or to track motion to provide real-time adjustment. For all these, *"rotation sensors"* are very important in robotics to keep track of the joint speeds and motions of end-effectors.
  - — There are very few devices to directly give absolute position for linear motions. Often linear motion is converted to rotary motion using rack and pinion drive and *rotary sensors such as 'encoders' are calibrated to give absolute linear position.*
  - — Reliable position sensors are widely implemented in robots *in the form of potentiometer, tachometers, encoders, resolvers and accelerometers.*
- ***"Proximity and touch sensors"*** *are generally used in intelligent grasping of workpiece,* while ***"force-torque sensors"*** *provide feedback for object manipulation and control of its interaction with the environment.* For example these sensors are useful to prevent slipping or damage of the object once grasped or to apply the right torque to tighten a nut. In contrast the *gross guidance information for a manipulator is provided by range and vision sensing. "Vision sensing"* can also provide other useful information in addition to gross guidance information.

Q. 3.5. State the functions of "external or environment sensors".

Ans External or environment sensors are required to perform one or more of the following *functions :*

1. To detect presence of workpiece.
2. To determine the position and/or orientation of workpiece and of objects to be articulated, or other objects present in the workcell.
3. To identify the workpiece.
4. To determine the properties (such as size, shape etc.) of the workpiece.
5. To detect, identify obstacles in the environment, and provide information about their size, shape, location, speed and so on.
6. To provide information about the manipulator-environment interaction forces and torques.
7. To provide information about environmental variables such as temperature, humidity etc.

Q. 3.6. State the characteristics of a "touch sensor".

Ans. A touch sensor's characteristics are :

1. It should ideally be a *single-point contact,* though the sensory area can be of any size. In practice 1-2 mm^2 is considered a satisfactory compromise between the difficulty of fabricating a sub-miniature sensing element and the coarseness of a large sensing element.

2. The minimum bandwidth of sensor should be *100 Hz.*
3. The touch sensor's *sensitivity* is dependent on a number of variables determined by the sensor's basic physical characteristics.

   For most of the industrial applications a sensitivity *within the range 0.4 to 10N together with an allowance for accidental mechanical overload is considered satisfactory.*
4. It should be robust and protected from environmental damage.
5. The characteristics of the sensor must be stable and repeatable with low hysteresis.

   – A linear response is not absolutely necessary as information processing technique can be used to compensate for any moderate non-linearities.

Q. 3.7. (*a*) What do you mean by 'Touch sensing,' 'Tactile sensensing' and 'Slip'?

(*b*) Name a few important tactile sensors.

Ans. ***Touch sensing :***

- This is the *detection and measurement of a contact force at a defined point.*
- A touch sensor can also be restricted to binary information, namely touch, and no touch.

Tactile sensing :

- This is the *detection and measurment of the spatial distribution of forces perpendicular to a predetermined sensory area, and the subsequent interpretation of the spatial information.*
- A tactile-sensing array can be considered to be a coordinated group of touch sensors.

Slip :

- This is the *measurement and detection of the movement of an object relative to the sensor.*
- This can be achieved either by a specially designed slip sensor or by the interpretation of the data from a touch sensor or a tactile array.

(*b*) Some important *tactile sensors* are :

(*i*) Proximity rod tactile sensor.
(*ii*) Photodetector tensile sensors.
(*iii*) Conductive elastomer sensors.
(*iv*) Pneumatic switch sensors.
(*v*) Polymer tactile sensors.
(*vi*) A hybrid tactile sensor.
(*vii*) VLSI-conductive elastomer sensor.
(*viii*) Optical tactile sensor.

Q. 3.8. Describe briefly "Joint sensors".

Ans. ***Joint sensors :***

- Several robots are powered by D.C. servomotors, for which the output torque is linearly related to the armature current in the motor. Thus, it is possible to measure the torque of each robot joint indirectly by inserting a suitable series resistor in one lead of each servomotor and measuring the voltage across it which is *proportional to the motor current* and hence, is *related to torque of the motor.*

- Although motor current measurment is simple and inexpensive it has the following *limitations/drawbacks* :
  - (*i*) The friction in motor bearings, associated gears and joint bearing affect the accuracy.
  - (*ii*) The measurements not only reflect the forces being applied at the tool but also the forces and torques required to accelerate the links of the arm and to overcome the friction and transmission losses of the joints.

Q. 3.9. Explain briefly the following sensors :

(*i*) Acoustic sensors; (*ii*) Optic sensors.

Ans. (*i*) *Acoustic sensors :*

- These sensors are commonly used for a wide variety of *non-contact presence, proximity, distance measuring, or navigation applications.* They can be simple or very complex.

— Ultrasonic sound is vibrations at a frequency above the range of human hearing, usually *greater than 20kHz.*

— *The acoustic sensors typically transmit a short burst of ultrasonic sound towards the target, which reflects the sound back to the sensor. The system then counts the time for the echo to return to the sensor and calculates the distance of the target using the speed of sound in the medium.* The wide variety of sensors corrently available differ from one another acoustically. They operate at different frequencies and have different radiation patterns.

— Most ultrasonic sensors use a single transducer to both transmit the sound pusle and receive the reflected echo, typically operating at frequencies between 40 kHz to 250 kHz.

- *Silicon-based ultrasonic sensors and acoustic wave sensors are extremely versatile microelectromechanical system (MEMS) sensors.* They have the following characteristics:
  — Competitively priced;
  — Inherently rugged;
  — Very sensitive;
  — Intrinsically reliable;
  — Consistent in performance.

These can be employed for a large variety of applications such as *presence, proximity, velocity, acceleration, force/torque, humidity, temperature etc.*

- "*Acoustic wave sensors*" send out accoustic waves, as the acoustic wave propagates through or on the surface of the material, changes in the characteristics of propagation path affect the velocity and/or amplitude of the wave. These changes can be monitored by measuring the frequency or phase characteristics of the reflected wave. These changes can be corelated to the corresponding physical quantity being measured.

— *Array of ultrasonic sensors or accoustic wave sensors are useful for imaging and sense analysis.*

(*ii*) ***Optical sensors :***

- These sensors are non-contact sensors and wide range of optic sensors based on different techniques are available for use in robotics.
- Simple light sensors work with visible light and work on the *main principle of detecting the change in the intensity of light sent and received.*

The signal can be binary 0 (black or no light) or 1 (white or bright), or magnitude of light, say, between 0 and 100.

- Optic sensors in novisible light include:
  — Infrared sensor;
  — Ultraviolet sensor;
  — Laser beam sensor.
  — An infrared sensor based tracking sensor system allows a full 3-D tracking within a space of few meters, with resolution of a fraction of a millimeter.
  — Another eye safe laser based scanning sensor is used for navigation on automated guided vehicles, robots and other material handling equipment.

Q.3.10. Explain briefly the following sensors :
1. Synchro; 2. Resolver; 3. Inductosyn.

Ans. ***Synchro :***

- *A "synchro" is a rotary transducer that converts angular displacement into an A.C. voltage or an A.C. voltage into an angular displacment* (This sensing device avoids the physical contact in order to produce an output, a significant problem with the potentiometers)
- Normally, a synchro system is made up of a number of separate 3-phase components such as:
  (*i*) Control transmitter (CX)
  (*ii*) Control transformer (CT)
  (*iii*) Control differential transmitter (CDX)

  All these elements work on essentially the *principle of the rotating transformer.*

  Typical, two or three of the devices are used to *measure angular position or the difference between this and a command position (i.e.,* the position error).
- In the recent past, a number of advances in digital and hybrid technologies have produced a varicty of devices that permit *synchro systems to be easily interfaced with digital system.*

2. ***Resolver :***

- The resolver is actually a form of synchro and for that reason it is often called a "synchro resolver", one of the major differences between the two devices is that the stator and rotor windings of the resolver are *displaced mechanically 90° to each other instead of 120° as in the case of synchro.*
- The most common form of resolver has a single rotor two stator windings as shown in Fig. 3.26. *With an A.C. carrier voltage input to the rotor the output*

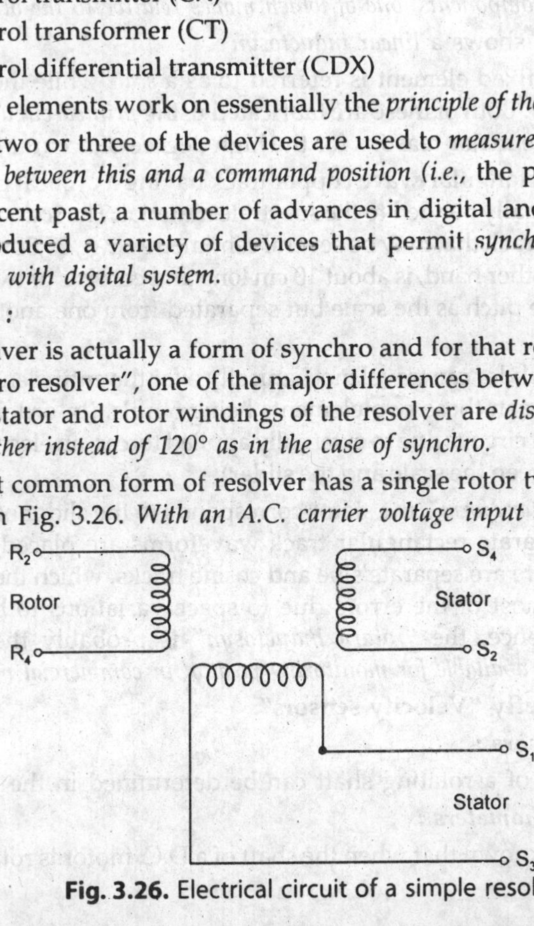

Fig. 3.26. Electrical circuit of a simple resolver.

voltage amplitude of the two stator windings will be dependent on the sine or cosine of resolver shaft angle α.

3. ***Inductosyn :***

- This device is used extensively in *NC (numerically controlled)* machine tools. Being one of the most accurate means of measuring position, it is capable of accuracies of 0.1 mil linear or 0.00042° rotary.

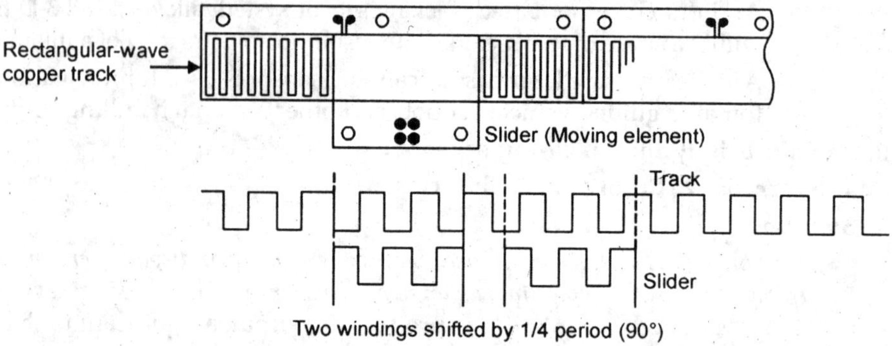

Fig. 3.27. Linear inductosyn.

- Inductosyn, in actual operation, is quite similar to the resolver. Regardless of whether the configuration is linear or rotary, there are always *two magnetically coupled components, one of which moves relative to the other.*

  Fig. 3.27 shows a *linear inductosyn* :

  — The fixed element is referred to as a *scale* while the moving element as a *slider*. Both of these are fabricated using *printed-circuit technology* (this is one of the major reasons for the high degree of accuracy that can be obtained)

  — A rectangular-wave copper track having a cyclical pitch of 0.1, 0.2, 2 mm is normally bonded to the substrate material. The *scale* usually has one continuous track that may be several centimetres (*e.g.*, 25,50 or longer). The *slider*, on the other hand, is about 10 cm long and consists of two separate tracks of the same pitch as the scale but separated from one another by 1/4 of a period (90°).

  — The slider can travel mechanically over the entire length of the scale, the gap between these two elements being 5 mils. (In order to prevent accidental short circuits due to externally applied forces, an electrostatic screen is placed between the scale and the slider).

- In its *"rotary form"*, the stator corresponds to the slider of the linear inductosyn. Two separate rectangular track waveforms are placed radially on a circular disc. There are separate sine and cosine tracks, which they alternate physically, permit most of the errors due to space variations to be averaged out. As a consequence, the *"rotary inductosyn"* is probably the *most accurate means currently available for monitoring positon in commercial machine tools.*

Q.3.11. Discribe briefly "Velocity sensors".

Ans. ***Velocity sensors :***

The velocity of a rotating shaft can be determined in the following ways:

1. ***D.C. tachometers :***

— It is known that when the shaft of a D.C. motor is rotated an analog voltage is produced which increases (or decreases) with increasing (or decreasing) shaft angular speed. In effect, the motor becomes a D.C. generator and can therefore be utilised to measure the shaft speed.

— Although it is possible to use almost any D.C. motor to measure the shaft speed, D.C. tachometers are usually *specially designed devices.*

— The underlying *principle* of the tachometer is as follow : "A wire moving in a magnetic field will induce a voltage across the wire which is proportional to its velocity and the sine of the angle between the magnetic field direction and the coil's plane. This angle is 90° when the wire's plane and the field are perpendicular to each other and results in the maximum voltage being developed."

— It is found that a permanent-magnet iron–copper armature tachometer (analog) is quite suitable for the purpose.

- The most common class of industrial robot that makes use of an analog tachometer is the SCARA. The main reason is that the configuartion of such a robot does not require the actuator to be lifted against gravity.

2. ***Using an optical encoder :***

In robotic applications, where an actuator must be moved with the particular manipulator link against gravity, a permanent-magnet tachometer is often unacceptable. In such applications, the optical encoder can be used for monitoring shaft velocity. To do this, the following two techniques exist :

(*i*) Encoder and frequency - to - voltage converter.

(*ii*) Encoder and software.

Q.3.12. Explain briefly an "Accelerometer."

Ans. ***Accelerometer :***

It is possible to monitor position, velocity and acceleration of a physical system. Normally, "linear acceleration" is *measured,* whereas "angular acceleration" is *derived,* from angular velocity by differentiation.

The device that can be used to obtain linear acceleration is called an *"accelerometer"*.

Fig. 3.28 shows the basic elements of a *linear accelerometer"*.

It consists of three elements:

(*i*) A mass '*m*';

(*ii*) A linear displacement sensor [(e.g. a linear variable differential transformer (LVDT)];

(*iii*) A set of springs having an equivalent spring constant, *k*.

Assume that the entire accelerometer begins to move (*i.e.* is accelerating) in a downward direction. Then, the force required to do this (i.e. *ma*) will be opposed by the springs supporting '*m*'. As

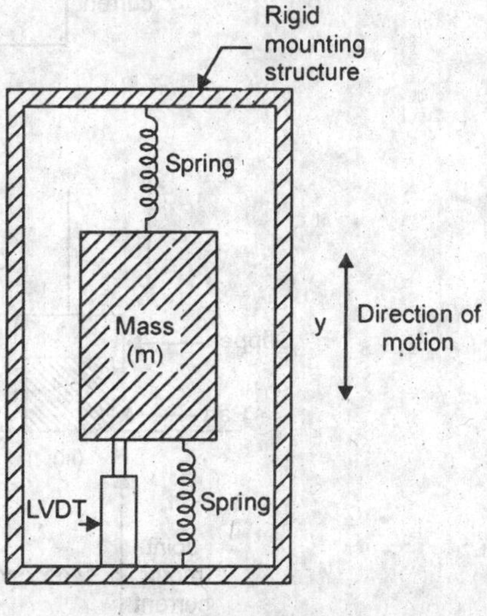

Fig 3.28. Linear accelerometer.

they bend upward a distance 'y' this force will be equal to ky, according to Newton's second law of motion. Thus,

$$ma = k.y \qquad ...(i)$$

or, Linear acceleration, $$a = \frac{k.y}{m} \qquad ...(ii)$$

From the above equation, it is abvious that $a \propto y$ *i.e.* the acceleration of the mass is *proportional* to the distance.

If an LVDT is employed to determine linear position as is often the case in commercial accelerometers the *output of that sensor will be proportional to the actual acceleration* (A signal-processing device used in an LVDT can be made to read acceleration directly).

Q.3.13. (*a*) Explain briefly 'Lift and try technique' for slip detection.

(*b*) Name some slip sensor developed in past.

Ans. (*a*) The *lift-and-try technique* (see Fig. 3.29) perhaps is the simplest technique to determine if a part is slipping. It entails using motor current of a particular joint or set of joints on a robot as a measure of whether or not part is slipping. In this respect, current monitoring can be performed either digitally or in an analog manner.

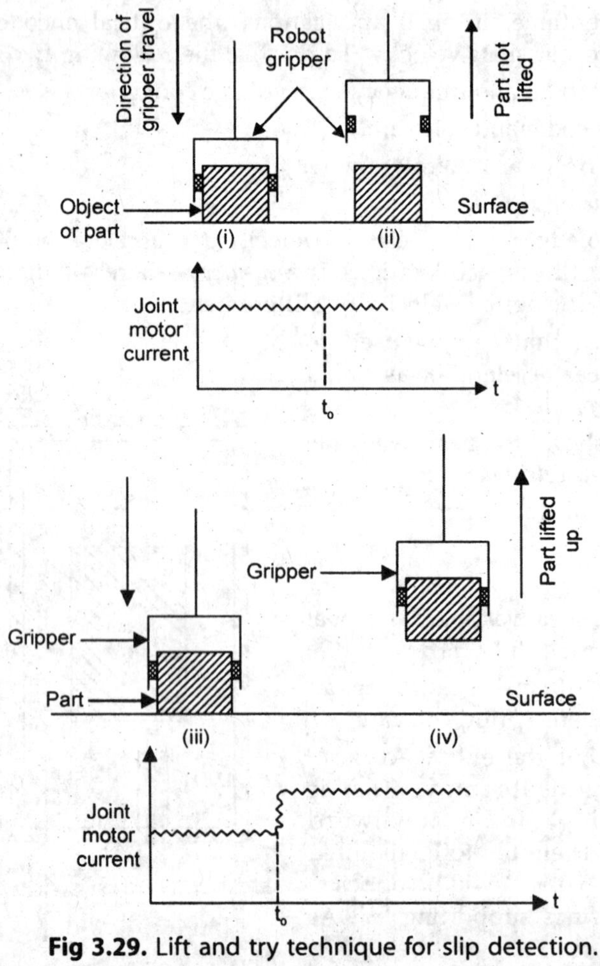

Fig 3.29. Lift and try technique for slip detection.

Referring to Fig. 3.29 the details of this technique are given below:

(*i*) Minimum gripper force applied at $t = t_0$.

(*ii*) Part is not lifted since force in *not* large enough. Joint motor current is *not changed* as gripper is raised above surface.

(*iii*) Gripper descends down and *force is incremented.*

(*iv*) Part is *successfully lifted (force is high enough).* Joint motor current is seen to *increase at* $t = t_0$ indicating that *gripper is loaded.*

(*b*) Some *slip sensors* developed in the past are :

1. Forced oscillation slip sensors.
2. Interrupter-type slip sensors.
3. Slip sensing finger.
4. Belgrade hand slip sensors.

Q.3.14. Explain briefly the use of strain gauges in tactile sensors.

Ans. • One of the simplest methods of sensing a force (or pressure) exerted on an object is to detect the deflection of the fingers of the robot gripper in response to such an applied force. The *strain gauge* provides a convenient and accurate means of doing this.

— The *principle* underlying the operation of a strain gauge is that a *mechanical deformation produces a change in resistance of the gauge, which can then be related to the applied force.*

• The strain gauge is manufactured from either resistive elements (foil, wire, or resistive link) or from semiconducting material.

— A typical *resistive strain gauge* consist of the resistive grid being bonded to an epoxy backing film. If the strain gauge is pre-stressed prior to the application of the backing medium, it is possible to measure both tensile and compressive stresses.

— The *semiconducting strain gauge* is fabricated from a suitably doped piece of silicon, *in this case the mechanism used for resistance change is "piezoresistive effect".*

• When applied to *robotic touch applications,* the strain gauge is normally used in *two configurations :*

(*i*) As a *load cell,* where the stress is measured directly at the point of contact; or

(*ii*) The strain gauge is positioned within the *structure of the end-effector*.

Q.3.15. Explain briefly the following sensors:

1. Quality control sensors.

2. Safety sensors.

3. Workcell control sensors.

Ans. 1. ***Quality control sensors :***

• For robotic sensors, inspection and quality control is an important function. Traditionally, quality control is performed on statistical sampling basis, using manual inspection techniques.

• Because sensors can be used to determine a variety of part quality characteristics, the *use of sensor permits 100 percent inspection.*

• The *external sensors* used for environmental feedback can also be used for detecting faults and failures in the finished product.

- The inspection process can be made a part of the programmed work cycle and sequence of operations performed by the manipulator may be linked to the result of inspection process.
- Computer vision, ultrasonic, sonic, and other sensors can be used for inspection. Vision system can provide a large variety of information about the workpiece in addtion to its position and orientation.

2. ***Safety sensors :***
   - An important function of the sensors in robotics is *safety and hazard monitoring*. The safety of the workers and other equipment in the work environment, and that of the manipulator itself, are important concerns.

     Example : If there is a *power failure,* all the links of the manipulator may fall to zero-gravity position instantaneously. This may injure the human beings in the vicinity or may damge the equipment or the manipulator itself. A *solution is to sense the power failure and apply brakes to prevent the uncontrolled falling of links due to gravity.*

3. ***Workcell control sensors :***
   - In robotics, sensor technology is also used to *implement interlocks in workcell. An interlock in the workcell is a situation that requires sensing of tasks such that completion of a task must be ensured before proceeding to the next task.* For example, a part must arrive on the conveyor before the gripper of the manipulator can pick it. This kind of verification of interlocked tasks can be done by incorporating the signals from a variety of sensors in the robot program.
   - Sensors may also be employed to *detect and resolve the interlocks in the workcell.*

Q.3.16. Emunerate improtant robotic arc welding sensors.

Ans. There exists a number of special purpose robots that are employed specifically for arc welding. However it is also possible to utilise the continuous-path programming option that is available on other robots to perform such a job.

The various sensors used for *arc welding* are :

(*i*) Simple active optical seam tracker.

(*ii*) Passive seam tracking sensors.

(*iii*) Active nonoptical seam tracking sensors.

Q.3.17. What do you mean by robot vision? Explain briefly.

Ans.
- The *process of deriving, featuring and analysing the details from 3-dimensional object in the form of a picture is known as "Robot vision"* or "machine vision" or computer vision".
- The areas of processing and analysis of the *images* are categorised as follows:

  (*i*) *"Sensing"* : This process describes and *gives out visual image.*

  (*ii*) *"Preprocessing"* : It deals with the *process of disturbance reduction and image development.*

  (*iii*) *"Segmentation"* : It is associated with the technique of dividing the image into parts of need.

  (*iv*) *"Description"* : It involves *distinguishing the the parts of focus from other object for computation of the image feature.*

  (*v*) *"Recognition"* : It is the *identifying stage* where in parts like spanner, bolts and nuts are recognised.

(*vi*) *"Interpretation"* In this process the *recognised parts are given useful meaning for task operation.*

Q.3.18. Enumerate the components of digital image processing.

Ans. The various component of digital image processing are :

- Capture (Digitization of image);
↓
- Noise clean up (Media filtration);
↓
- Determining the content of the image;
↓
- Adjust brightness, contrast and tonal range;
↓
- Image sharpening and blurring (convolution and unsharp masking)
↓
- *Output (Print and display image).*

Q.3.19. Explain briefly "Vision hardware".

Ans. The *"vision hardware"* consist of :

(*i*) A *camera* (usually charge-coupled device–CCD camera) placed above the scene.

(*ii*) A *TV monitor.*

(*iii*) An *image preprocessor.*

(*iv*) A *conventional microcomputer* housed in a control cabinet.

- CCD camera has no geometric distortion, a high signal-to-noise ratio is obtained because of absence of deflection coil, thus insensitive to disturbance from electromagnetic fields.

  The video signal from the camera enters the 'preprocessor', is digitized and loaded into an image memory. The image is then filtered and translated into a gradient image. In gradient processing, the algorithm looks for differences in the level of the pixel and is thus insensitive to light variation because the gradient level remains same irrespective of lighting across the scene. The gradients are then classified in order to create contours. The whole process is concluded with a *structuring of the resulting contour image.* The preprocessor thus produces a host of all the features in the image which is then fed to a general purpose *microcomputers* where the object (s) are *identified and the position and orientation determined.*

The micro-computer is also used :

— to supervise the preprocessor;

— to handle the robot vision system communication;

— to perform error supervision;

— to handle the self-diagnostic routines.

- The *programming of an object is performed by placing the object beneath the camera and processing the image.* The vision system will then display the object's contour (inner and outer) on the *"TV monitor"*. The operator then points out the object's significant features (characteristic forms) A cursor, which is moved by using the joystick on the programming unit is utilised for this purpose. The acquired information is stored in the computer memory and is available for use whenever the same object is to be identified. The object is then moved a couple of times in the scene, and image is processed in between each move.

Thus the system collects statistical data about the object and forms decision areas. The statistical data aids in the certainty factor with respect to part identification. In this way a number of different objects, each having different views of all possible stable positions, are programmed and stored on the floppy disc unit.

These days such robotic vision systems are available which allow robots to pick randomly oriented parts out of a bin, orient them for further processing, thus enabling handling of delicate parts that can not be fed using *conventional means such as belt feeders or vibrating bowls.*

Q.3.20. Describe briefly "Automated visual inspection."

Ans. • Automated visual inspection systems are *fast, provide high resolution and high repeatability.* Such systems not only operate from a computer (CAD/CAM) database, but also *generate a database, which can be used to evaluate, understand, and control the process.*

— A *CCD camera* is used to acquire and generate a bit map of the object under test.

— *Computer* used with such a system can provide a powerful tool to evaluate several different flaw detection algorithms.

• In such systems, the *setting of inspection parameters is very important.* The statistical variation around the design value inherent in the manufacturing process need to be clearly understood and defined, otherwise the machine will reject parts unnecessarily.

Q.3.21. Enlist the "Machine vision technologies".

Ans. The various machine vision technologies are :

1. Image processing.
2. Scene analysis.
3. Pattern recognition.
4. Model-based vision.
5. Motion analysis.
6. Infrared image analysis.
7. Dimensional inspection.
8. Industrial inspection and process control.

Q.3.22. What are the causes of robot accidents ? How can they be prevented ?

Ans. Causes of robot accidents :

I. *Unsafe acts :*

(*i*) Making errors in programming, interfacing peripheral equipment, and connecting input/output sensors.

(*ii*) Entering the envelope inadvertently, because of unfamiliarity with the safeguards in place or not if they are activated.

(*iii*) Placing oneself in hazardous positions while programming or performing maintenance within the work envelope of the robot.

II. *Unsafe conditions :*

(*i*) Safeguards deactivated.

(*ii*) Mechanical failure.

(*iii*) Injury to the personel may be caused due to electric shock and release of stored energy from the accumulating devices.

(*iv*) Hazards from robot's power system.

Prevention of accidents :

(*i*) Adequate room must be provided for a robot's movement as well as for the workers.

(*ii*) System components must be so designed, installed and secured that the hazards associated with stored energy are minimized.

(*iii*) A detailed risk assessment should be performed to ensure the safety of workers who operate, service and maintain the robotics system.

- *Safeguarding devices :*
  — Fixed barriers (to prevent contact with moving parts);
  — Interlocked barrier guards;
  — Mechanical and non-mechanical limiting devices;
  — Presence-sensing safeguarding devices.

HIGHLIGHTS

1. A *sensor* is a transducer used to make a measurement of a physical variable.
2. Robots capable to decision making are called *intelligent robot.*
3. *"Internal sensors"* are used to measure position,velocity or acceleration of robot joints and/or the end-effector. *"External sensors"* are used to monitor the robot's geometric and/or dynamic relation to its task, environment, or the objects that it is handling.
4. *"Tactile sensor"* are contact sensors that must be brought in contact with the object to obtain signals to measure the necessary quantities. *"Non-tactile sensors"* are contactless sensors which *sense the signals remotely, but only within the specified range of distance from the object.*
5. The output of an analog sensor is proportional to a total force.
6. *Potentiometers* are analog devices whose output voltage is proportional to the position (angular or linear) of a wiper.
7. An *optical encoder* converts linear or angular displacement into digital code or pulse signals.
8. *Force/torque sensors* are primarily used for measuring the reaction forces developed at the interface between mechanical assemblies.
9. *Proximity sensors* generally have a binary output which indicates the presence of an object within a specified distance interval.
10. Range sensing techniques are :
    (*i*) Triangulation technique;
    (*ii*) Structured lighting approach;
    (*iii*) Time-of-flight range finders.
11. *Robotic vision* may be defined as the process of extracting, characterising and interpreting information from images of a 3-dimensional world.
12. Function of robot vision system are :
    (*i*) Sensing and digitizing; (*ii*) Image processing and analysis;
    (*iii*) Application.

OBJECTIVE TYPE QUESTIONS

1. .... refers to the minimum and maximum changes in input signal to which the sensor can respond.

2. Sensitivity should be as low as possible.
3. Potentiometer is an external state sensor.
4. Tachometer is an .... state sensor.
5. Optical encoders are internal state sensors.
6. Force/torque sensors are typical non-contact type robotic sensors.
7. LVDT is a tactile sensor.
8. Microswitch is a status sensor.
9. Contact or touch sensors are rarely used in robotics.
10. A binary sensory is a .... device.
11. The output of an .... sensor is proportional to a total force.
12. A potentiometer is the simplest device that can be used to measure ....
13. Potentiometer are .... devices.
14. An optical encoder converts linear or angular displacement into digital code or pulse signals.
15. .... encoders provide actual position relative to a fixed reference (zero) position.
16. .... encoders sense the position from previous position.
17. LVDT is an .... device.
18. Proximity sensors generally have .... output which indicates the presence of an object within a specified distance interval.
19. An .... sensor determines the range by measuring the elapsed time between the transmission of certain frequencies and their detected echoes.
20. An eddy current sensor is used as proximity switch and for determining the accuracy and repeatability of commercial robotic manipulators.
21. .... is the process of extracting, characterising and interpreting information from images of a 3-dimensional world.
22. Sensing and processing are areas of .... level vision.
23. In a single camera system, triangulation is used to detect shape and depth.
24. .... is the process of identifying a group of related pixels for locating connected regions or areas of image having similar characteristics.
25. .... is a binary convestion technique in which each pixel is converted into a binary value, either black or white.

ANSWERS

1. Range	2. No	3. No	4. internal	5. Yes
6. No	7. Yes	8. Yes	9. No	10. contact
11. analog	12. position	13. analog	14. Yes	15. absolute
16. incremental	17. analog	18. binary	19. ultrasonic	20. Yes
21. Robotic vision	22. low	23. Yes	24. Segmentation	25. Thresholding.

THEORETICAL QUESTIONS

1. What is a sensor ?
2. What is the basic function of an electronic sensor ?
3. What are the characteristics of a sensing device ?
4. How are sensors classified ?
5. List the various sensors used in robotic systems.

6. With neat sketches explain the tactile sensors and range sensors.
7. Discuss the desirable engineering features of sensors and transducers.
8. Enumerate six commonly used internal sate sensors.
9. Explain briefly any two of the following internal state sensors :
   (*i*) Potentiometers;
   (*ii*) Optical encoders;
   (*iii*) Tachometer.
10. What is the difference between internal and external state sensors ?
11. Describe briefly the following external state sensors :
   (*i*) Strain gauges; (*ii*) Electromagnetic sensors.
12. What is the difference between tactile and non-tactile sensors ?
13. Name four non-tactile sensors.
14. What is the difference between the status and environment sensors ? Give four examples of each of these sensors.
15. Explain briefly the following touch/contact sensors :
   (*i*) Binary sensors ; (*ii*) Analog sensors.
16. What do you mean by "Artificial skins" ? Explain briefly.
17. With the help of a neat sketch, explain the construction and working of an optical encoder.
18. Explain briefly 'Absolute' and 'Incremental' optical encoders.
19. Describe briefly linear variable differential transformer (LVDT)–a position sensor.
20. Explain briefly a wrist force torque sensor (WFTS). What characteristics must this sensor possess ?
21. Discuss briefly contact proximity sensor.
22. Explain briefly any two of the following non-contact proximity sensors :
   (*i*) Optical proximity sensor.
   (*i*) Ultrasonic proximity sensor.
   (*iii*) Capacitive proximity sensor.
   (*iv*) Hall-effect sensor.
23. Describe briefly fiber optic scanning sensors.
24. What is the function of a range sensor ?
25. Explain briefly the following range sensing techniques :
   (*i*) Triangulation;
   (*i*) Structured lighting opproach;
   (*iii*) Time-of-flight range finders.
26. List the points which should be considered while selecting a right sensor.
27. What do you understand by the term "Robot vision" ?
28. What is a "Robot vision system" ? Explain briefly.
29. What are the functions of a 'Robot vision system' ? Explain briefly.
30. What is the difference between low level, medium level and high level vision ?
31. Explain briefly the components of a vision system.
32. With the help of neat sketches, explain briefly the various illumination techniques.
33. Discuss briefly the following operations/processes covered under image processing :
   (*i*) Image improvement; (*ii*) Segmentation;
   (*iii*) Smoothing; (*iv*) Object recognition.
34. State the industrial applications of vision-controlled robotic systems.

CHAPTER

4 Robot Programming and Languages

4.1 INTRODUCTION

Robot manipulators differentiate themselves from fixed automation by being *flexible* which means *"programmable"*. Not only are the movements of manipulators programmable, but, through the use of sensors and communication with other factory automation, manipulators can *adopt* the variations as the task proceeds.

- *A* **robot program** *may be defined as a path in space to be followed by the manipulator, combined with peripheral actions that support the work cycle.*

Peripheral actions include :

— Opening and closing the gripper ;

— Performing logical decision making ;

— Communicating with other pieces of equipment in the robot cell.

- **Robot languages** *have been developed for the ease of control of motions of robots having different structures and geometrical capabilities.* Some of the programming languages are :

  WAVE, AL, VAL,AML, MCL, RAIL, HELP, RPL, PAL and ADA.

— *Robot languages provide flexibility to the user in defining the task to be performed.* Not only do they permit the motion of the task to be defined but they also provide the user with the ability to imbue intelligence in the control program. In its simplest forms, this intelligence may check binary sensors and change a location, or make a simple decision based on sensory information to handle an exception. *As the capability of the language increases, the intelligence of the algorithm controlling the robot in a specific application can also increase. Thus corrections based on sensory inputs (such as vision or tactile sensors) are possible along with communication with other computer and data bases.*

— High level programming languages provide a more general approach to solving the human-robot communication problems.

4.2 ROBOT PROGRAMMING

The use of robots to perform assembly tasks requires high-level programming techniques because robot assembly usually relies on sensory feedback, and this type of unstructured interaction can only be handled by conditionally programmed methods.

Robot programming is substantially different from traditional programming. The following are the *considerations which must be handled by any robot programming method* :

(*i*) The objects to be manipulated by a robot are three-dimensional objects which have a variety of physical properties;

(*ii*) Robots operate in spatially complex environment;

(*iii*) The description and representation of three-dimensional objects in a computer are imprecise; and

(*iv*) Sensory information has to be monitored, manipulated, and properly utilised.

Current approaches to programming are classified into two major categories :

1. ***Robot-oriented programming :***
   — In this type of programming, an assembly task is *explicitly described as a sequence of robot motions;*
   — The robot is guided and controlled by the program through the entire task with each statement of programme roughly corresponding to one action of the robot.
2. ***Object oriented or task-level programming*** :
   — It describes the assembly task as sequence of *positional goals of the objects* rather than the motion of the robot needed to achieve the goals;
   — *No explicit robot motion is specified.*

- The various considerations in robot programming are :

  (*i*) The three-dimensional objects with different physical properties are to be manipulated.

  (*ii*) The environments of robot operations can be complex.

  (*iii*) The visualisation of the object can be discrete.

  (*iv*) The processing and the analysis of the digital data from sensors and vision system has to be done in real time.

4.3 METHODS TO PROGRAMME THE ROBOT'S WORK CYCLE

In order to programme the work cycle of a robot the following *methods* are used :

1. Manual programming method.
2. Walk through programming method.
3. Teach pendant or lead through programming method.
4. Off-line programming method.

4.3.1. Manual Programming Method

- Various controlling components of a robot, like cams, stops, switches, control relays etc. are set through this method.
- This method can be employed only for such robots (with *relatively short work cycles*) which have to perform simple tasks like *pick and place, loading and unloading* etc.

4.3.2. Walk through Pragramming Method

- In this method, the arm and hand of the robot are initially moved manually and these movements are stored in computer memory for being followed during further operations.
  — In this method, the person doing the teaching is in a potentially hazardous position because the operational safeguarding devices are deactivated or inoperative.
- This method suites well for robots used in arc *welding and spray painting operations.*

4.3.3. Teach Pendant or Lead through Programming Method

Lead through teaching/programming may be categorised as follows :

1. **Manual lead through teaching/programming :**
   — An operator grasps physically and moves the wrist end in the path of operation, which is recorded in the memory.
   — He uses a teach button while in teach mode.
   — During run mode the wrist end repeats the taught motion.
   — When robot structure is complicated a geometrically similar model is used to ease the robot handling.
   — This is more useful in *continuous operations.*

   Applications : (*i*) Spot welding;
   (*ii*) Point-to-point movements;
   (*iii*) Machine loading and unloading, etc.

2. **Powered lead through teaching/programming:**
   — In this method, a control box with buttons known as ***teach pendant*** is used to control joint motors.
   — The points of motion are recorded in the memory, and converted to motion programs.
   — Subsequently the program is utilised to play back the motion during the cycle of operation.
   — This is more useful in point-to-point movements.

   Applications : (*i*) Spray painting;
   (*ii*) Full welding, etc.

Teach pendants : Refer to Fig 4.1.

"Teach pendants" are hand held boxes that allow control of each manipulator joint or of each cartesian degree of freedom.

— Some such controllers allow testing and branching, so that simple programs involving logic can be entered.

— Some teach pendants have alphanumeric displays and are approaching hand-held terminals in complexity.

- *The primary functions of a teach pendant are :*

(*i*) To guide the robot or motion device, while teaching locations.
(*ii*) To serve as the primary point of control for initiating and monitoring operations.
(*iii*) To support application programs.

- Teach pendants of two types :

(*a*) *Programmer's pendant* — Designed for use while an application program is being written and debugged.

(*b*) *Operation's pendant* — Designed for use during normal system operation.

The major areas of teach pendant are : (*i*) Liquid Crystal Display (LCD); (*ii*) Data entry points; (*iii*) Emergency stop switch; (*iv*)User LED; (*v*) Mode control buttons; (*vi*) Manual control buttons; (*vii*) Manual state LEDs; (*viii*) Speed bars; (*ix*) Slow button; (*x*) Pre-defined function buttons; (*xi*) Programmable function buttons; (*xii*) Soft buttons.

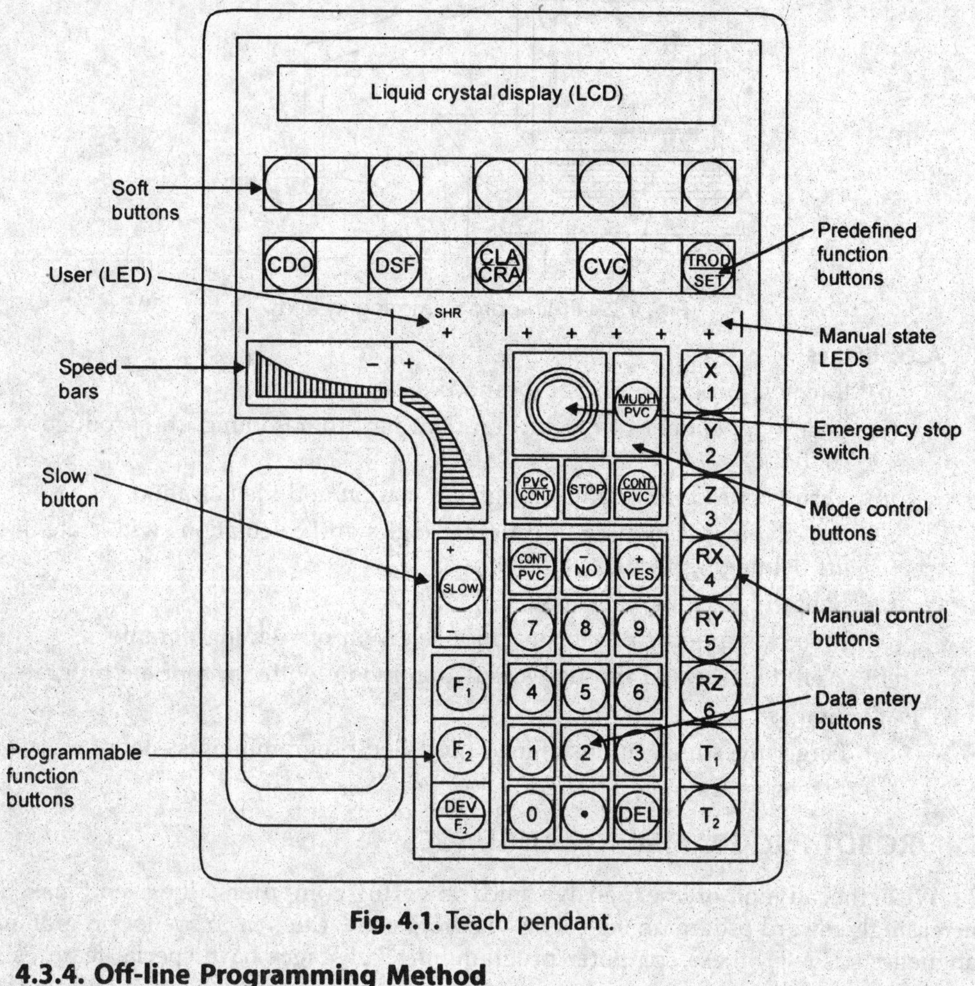

Fig. 4.1. Teach pendant.

4.3.4. Off-line Programming Method

- Such a program is separately prepared and fed into the computer memory. Then it is readily available for use whenever the operation is to be performed. Since it is separately prepared, it can be *prepared simultaneously* while the robot is operating on some other task and, therefore, a lot of time can be *saved* and *robot can be utilised more.*
- This type of robot can be connected to a central CAD/CAM data base system. Fig. 4.2 shows an off-line programming system.

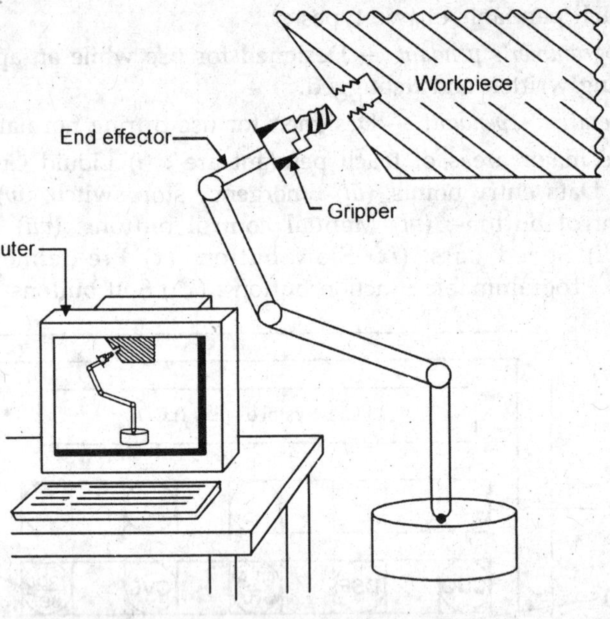

Fig. 4.2. Off-line programming system.

Advantages :

(*i*) Effective programming of locations.

(*ii*) Cost independent of production (During programming, the production can continue).

(*iii*) Verification of program through simulation and visualisation.

(*iv*) Effective programming of program logics and calculations with state-of-the-art debugging facilities.

(*v*) Reuse of exciting CAD data.

(*vi*) Process support tools for instance selection of welding parameters.

(*vii*) Well documented through simulation model with appropriate programs.

Disadvantages :

(*i*) Large investment in installing an off-line programming system.

(*ii*) Extensive training required.

4.4 ROBOT PROGRAMMING LANGUAGES

With the advent of inexpensive and powerful computers, the trend has been increasingly toward programming robots via programs written in computer programming languages. Usually, these computer programming languages have special features that apply to the problems of programming manipulators and so are called *"robot programming languages" (RPLs). Most of the systems that come equipped with a robot programme language have nonetheless retained a teach-pendant-style interface also.*

Robot programming languages can be categorised as follows :

1. **Specialised manipulation languages :**

— These languages have been built by developing a completely new language that although addressing robot-specific areas, might well be considered a general computer programming language :

Example : (*i*) VAL language developed to control the industrial robots by Unimation, Inc.

(*ii*) AL – a specialised manipulation language developed at Stanford University.

2. **Robot library for an existing computer language** :

— These languages have been developed by starting with a popular computer language (*e.g.* Pascal) and adding a library of robot-specific subroutines;

— The user then writes a Pascal program making use of frequent calls to predefined subroutine package for robot-specific needs.

Example :

(*i*) AR – BASIC, essentially a subroutine library for a standard BASIC implementation.

(*ii*) JARS – based on Pascal

3. **Robot library for a new general-purpose language** :

— These languages have been developed by first creating a new general-purpose language as a programming base and then supplying a library of pre-defined robot-specific subroutines :

Examples :

(*i*) RAPID;

(*ii*) AML;

(*iii*) KAREL.

- ***Task-level programming languages*** :

— These languages relate to the third level of robot programming methodology.

— *They allow the user to command desired subgoals of the task directly, rather than specify the details of every action the robot is to take.* In such a system, the user is able to include instructions in the application program at a significantly *higher level* than in an explicit robot programming language.

— A task-level robot programming systems must have the ability to perform several planning tasks *automatically*.

There is quite a *distinct border* between explicit robot programming languages and task-level programming languages. Incremental advances are being made to explicit robot programming languages to help to ease programming, but these enhancement *cannot* be counted as components of a task-level programming system. Till now, true task-level programming of manipulators does not exist; however, research is going on in this direction.

4.5 REQUIREMENTS OF A ROBOT PROGRAMMING LANGUAGE

The various requirements of a robot programming language are :

1. **World modelling** :

— Any robot programming language needs a means for moving objects in three-dimensional space. The most common element of robot programming languages is the existence of special *"geometric types"*; *types* are introduced to represent joint-angle sets, Cartesian positions, orientations, and frames.

— All motions are described as tool frame relative to station frame, with goal frames being constructed from arbitrary expressions involving geometric types.

— *Given a robot programming environment that supports geometric types, the robot and other machine parts, and fixture can be modelled by defining named variables associated with each object of interest.*

2. **Motion specification :**

— The basic function of a robot programming language is *to allow the description of desired motions of the robot*.

— Motion statements permit the user to specify via points, the goal point, and whether to use joint-interpolated motion or Cartesian straight-line motion. Additionally, the user might have control over the speed or duration of motion.

3. **Flow of execution :**

— A robot programming system permits the user to specify the flow of execution, that is concepts such as *testing and branching, looping, calls to subroutines, and even interrupts* are generally found in robot-programming languages.

— Further, in robot programming languages, *signal* and *wait* primitives are often found and occasionally more sophisticated parallel-execution constructs are also provided.

4. **Sensory control :**

— In order to take care of the uncertainty in location and dimension of the objects in the work-envelope, sensing is to be performed.

— The information gathered by the sensors acts as a feedback from the environment.

5. **Programming environment :**

— A good programming environment fosters programmer activity.

— Most of the robot programming languages are *interpreted*, so that individual language statements can be run at a time during program development and debugging (manipulator programming is difficult and tends to be very iteractive, with a lot of trial and error). In this case, typical programming support, such as text editors, debuggers, and a file system are also needed.

4.6 PROBLEMS ASSOCIATED WITH ROBOT PROGRAMMING LANGUAGES

Inspite of several advances in recent years, it is still difficult to programme robots. Robot-programming shares all the problems of conventional computer programming, plus some additional difficulties *caused by the effects of the physical world*. The problems peculiar to robot programming languages are discussed below.

1. **Internal world model versus external reality :**

- Discrepencies between internal world model and external reality result in poor or failed grasping of objects, collisions, and a host of more subtle problems.
- Besides the uncertainity inherent in each object's position, the manipulation itself is limited to a certain degree of accuracy. The manipulator's accuracy usually varies over its workspace.
  — During debugging of manipulation programs, it is very useful to be able modify the program and then back up and try procedure again

2. **Context sensitivity :**

- For writing a large computer program in which one develops small, low level pieces of a program and then puts them together into large pieces, eventually attaining a completed program, the standard approach is "*Botton-up programming*". This approach requires that the small pieces be relatively insensitive to the language statements that precede them and there be no assumptions concerning the context in which these program pieces execute. For manipulator programming this is often not the case.

- Manipulator programs can be highly insensitive to initial conditions. Due to insufficient manipulator accuracy, a program segment written to perform an operation at one location is likely to *need to be tuned* to make it work at different location. Changes in location within the workcell results in changes in the manipulator's configuration in reaching goal locations. Such attempts at relocating manipulator motions within the workcell test the accuracy of the manipulator kinematics and servosystem, and problems frequently arise.
- *Limitations in most manipulator control systems cause greater servo errors, which are to be expected if the quicker trajectory is followed.* The manipulator's configuration also affects the delicacy and accuracy of the forces that can be applied with it.

3. **Error recovery** :

- The first problem that arises for error recovery is identifying that an error has indeed occurred. *Error detection* is often difficult, since robots generally have quite limited sensing and reasoning capabilities.
- In order to detect an error, a robot program must contain some type of explicit test. This test might involve checking the manipulator's position to see that it lies in the proper range. Every motion statement in the program might fail, so these explicit checks can be quite cumbersome and can take up more space than the rest of the program. The use of parallelism in manipulator's program can further complicate recovery from errors.

4.7 DESCRIPTION OF ROBOT LANGUAGES

Following are the three major classes into which robot languages can be broadly grouped :

1. **First generation language** :

— This type of language provides an *off-line programming in combination with the programming through the robot pendant teaching.*

— Its capability is limited to be handling of sensory data (except ON/OFF binary signals) and communication with other components. However, branching, input/output interfacing and commands leading to a sequence of movements of arm and body, and opening and closing of the end-effectors are possible.

Example : **VAL**

2. **Second generation languages** :

— These are structured programming languages performing complete tasks.

— They can generate complex motions, can handle both analog and digital signals besides the binary signals.

— These languages have the added advantage of *better interfacing facilities with other computers.* Data processing, file management and keeping all the records of events happening in the work-cell can be done more efficiently.

Example : AML, RAIL, RCL, VALII etc.

3. **World modelling and task-oriented object level language** :

— A more advanced future language is *world modelling*. Here, a task is defined through a commond (say 'TIGHTEN THE NUT'). In such a case *intelligence is required and the robot should be capable of making decision.*

— Future generation robot languages *involve technology of artificial intelligence and hierarchial control systems.*

- Some the the high-level computer languages now used to program robots are : WAVE; AL; VAL; AML; MCL; RAIL; HELP; JARS; RPL; LADDER LOGIC etc. Some of these languages are briefly discussed below :

1. **AL :**

The AL (for Am Language) was developed at the robotics research centre of Stanford University. Currently AL can be executed on a VAX computer and real-time control of the arms is performed on a stand alone PDP-11. Its characteristics are :

- High-level language with features of ALGOL and PASCAL.
- Supports both robot-level language and interpreted on a real time control machine.
- Compiled into low-level language and interpreted on a real time control machine.
- Has real-time programming language constructs like synchronisation, concurrent execution, and on-conditions.
- ALGOL like data and control structure.
- Support for world modelling.

2. **AML :**

It was developed by IBM. It is the control language for the IBM RS-1 robot. It runs on a series-1 computer (or IBM personal computer) which also controls the robot. RS-1 robot is a cartesian manipulator with 6 DOFs. Its first three joints are prismatic and the last three joints are rotary. Its *characteristics* are :

- Provides an environment where different user-interface can be built.
- Supports features of LISP-like and APL-like constructs.
- Supports data aggregation.
- Supports joint-space trajectory planning subject to position and velocity constraints.
- Provides absolute and relative motions.
- Provides sensor monitoring that can interrupt motion.

3. **RAIL :**

- This language was developed by Automatix for use with robots and vision systems.
- It is an interpreter, loosely based on Pascal.
- Several constructs have been incorporated into RAIL to support inspection and arc-welding systems, which are a major product of Automatix.
- The central processor of RAIL system is a Motorola 68000.
- Peripherals include a terminal and a teach box.
- RAIL is being supplied with *three* different systems :
  (*i*) Vision only, no arm;
  (*ii*) A custom-designed Cartesian arm for assembly tasks;
  (*iii*) A Hitachi process robot for arc welding.

4. **RPL :**

- This high level programming languages was developed at SRI international to facilitate development, testing, and debugging of control algorithms for modest automatic manufacturing systems that consist of a few manipulators, sensors, and pieces of auxiliary equipment.

- The SRI Robot Programming System (RPS) consists of a compiler that translates RPL programs into interpretable code and an interpreter for that code.
- PUMA 500 can be controlled with the help of RPL.
- It is modular and flexible.

5. **VAL :**

- It is a robot programming language and control system originally designed for use with Unimation robots.
- Its stated purpose is to provide the ability to define robot tasks easily.
- The intended user of VAL will typically be the manufacturing engineer responsible for implementing the robot in a desired application.
- It has the structure of BASIC, with many new command words added for robot programming. It also has its own operating system, called the VAL monitor, which contains the user interface, editor, and file manager.
- It has been released for use with all PUMA robots and with the Unimate 2000 and 4000 series.

6. **HELP :**

- It was developed by General electric company.
- It acts more or less like RAIL.
- It has capability to control two robot arms at the same time.
- The structure of the language is like PASCAL.

7. **MCL :**

- USA Airforce ICAM project led to the development of another manufacturing control language known as MCL by McDonnel-Douglas.
- This is a modification of the popular APT (Automatically Programmed Tooling) language used in CNC machine tools as many similar commands are used to control machine tool in CAM applications.
- Lines, circles, planes, cylinders and many other complex geometrical features can be defined in MCL.

8. **JARS :**

- It was developed by NASA's JPL.
- The base of the language is PASCAL.
- JARS can be interfaced with PUMA 6000 robot for running robotic programs.

9. **RIPL** (*Robot Independent Programming Language*) :

- It is based on object-oriented Robot Independent Programming Environment (RIPE).
- RIPE computing architecture consists of a hierarchial multiprocessor approach, which employs distributed general and special-purpose processors. This architecture enables the control of diverse complex subsystems in real time while co-ordinating reliable communications between them.

10. **LADDER LOGIC :**

- This programming language is designed to be used by electricians.
- It closely resembles the relay logic that appears on the inside lids or doors of dishwashers and washing machines.
- The only robots that use ladder programming are those that come without a controller or with a programmable logic controller.

4.8 COMPUTER CONTROL AND ROBOT SOFTWARE

Although a robot can be taught and programmed through teach pendant but it is difficult to exercise close control through pendant teaching. Therefore, textual programming is attempted and the computer instructions are given following the syntax of a certain robot language. *The program and control methods are actuated through software running on an operating system in which manipulation of data takes place.* The control functions are activated through monitors.

The *three basic modes of operations* in a robot are :

1. **Monitor mode :**

- In this mode the programmer can carry out the following functions/activities :
  - — Define locations;
  - — Load a particular piece of information in a particular register;
  - — Store information in the memory;
  - — Save, transfer programs from storage into computer control memory;
  - — Enable or disable, and move back and forth into its edit and run mode.

2. **Edit mode :**

- In this mode, following operations can be carried out :
  - — The programmer can edit or change a set of instructions of existing programs or introduce a new set of information.
  - — The user can erase some instructions and can replace them by new lines.
  - — Any error, if shown on the monitor, can be corrected.

In order to come out of the mode an *end command should be given.*

3. **Run or executive mode :**

- During this mode, the sequential steps as written by the programmer are followed.
  - — Sometimes *"dry run"* can be tested by making the switch disable. After dry run, the switch may be made operational by the instruction *enable.*
  - — A program can be tested in *"run mode" and by debugging, the errors in the program can be rectified.*
- For implementing robot language program, the operating system uses either an 'interpreter' or 'compiler'.
  - — An *'interpreter'* takes the source program one line at a time and generates equivalent code that is understood by the *compiler*. Any mistake in the source program will be indicated to the user, who then can correct the source program and the line is interpreted.
  - — *Compiler* is a software in the operating system that *converts source code into the object code (machine code) after compliation of the whole program*. The robot controller can then read and process the machine codes.

    Executive time for a compiled program is fast while editing of an interpreted program is very fast.

4.9 COMPARISION OF VARIOUS EXISTING ROBOT CONTROL LANGUAGES

The comparision of various existing robot control languages is given in Table 4.1.

Table 4.1 : Comparision of various existing Robot Control Languages

S.No.	Languages	ASPECTS										
		Institute	Robot controlled	Robot or object level	Language basis	Compiler or interpreter	Geometric data type	Motion specified by	Control structure	Sensing command	Parallel processing	Multiple robot
1.	*AL*	Stanford	PUMA stanford arm	Mix	Concurrent Pascal	Both	Frame	Frame	Pascal	Position force semaphores	COBEGIN	Yes
2.	*AML*	IBM	IBM arm	Robot	Lisp, APL Pascal	Interpreter	Aggregate	Joints	Pascal	Position	None	No
3.	*AUTOPASS*	IBM	IBM	Object	PL/I	Both	Model	Implicit	PL/I	Force, tactile	IN PARALLEL	No
4.	*HELP*	GE	Allegro	Robot	Pascal	Interpreter	None	Joints	Pascal	Force, vision	Semaphores	Yes
5.	*JARS*	JPL	PUMA Stanford arm	Robot	Pascal	Compiler	Frame	Joints, frame	Pascal	Proximity vision	None	No
6.	*MAPLE*	IBM	IBM	Robot	PL/I	Interpreter	None	Translation, rotation	PL/I	Force, proximity	IN PARALLEL	Yes
7.	*MCL*	McDonnell Douglas	Cincinnatic Milacrosn T-3	Robot	APT	Compiler	Frame	Translation, rotation	If-then-else-while-do	Position	INPAR	Yes
8.	*PAL*	Pardue	Stanford arm	Robot	Transform base	Interpreter	Frame	Frame	If-then-else	Force	None	No
9.	*RAIL*	Automix	Custom-designed Cartesian arm	Robot	Pascal	Interpreter	Frame	Joints, frame	Pascal	Force, vision	None	No
10.	*RFL*	SRI	PUMA	Robot	Fortran, Lisp	Both	None	Joints	Fortran	Position, vision	None	No
11.	*VAL*	Unimate	PUMA	Robot	Basic	Interpreter	Frame	Joints, frame	If-then	Position, force	Semaphores	No

QUESTIONS WITH ANSWERS

Q. 4.1. How is sensing in robot programming classified ? Explain briefly.

Ans. Sensing in robot programming can be classified into the following *three* types :

1. **Position sensing** :

— This type of sensing is used to *identify the current position of the robot.*

— It is usually done by *encoders* that measure the joint angles and compute the corresponding hand position in the workspace.

2. **Force and tactile sensing** :

— It can be used to *detect the presence of objects in the workspace.*

— It is used in compliant motion to provide feedback for force-controlled motions.

— *Tactile sensing can be used to detect slippage while grasping an object.*

3. **Vision** :

— It is used to identify objects and provide a rough estimate of their position.

Q. 4.2. State the role played by "programming support"to a robot language.

Ans. A language with programming support (editor, debugger etc.) is useless to the user.

A sophisticated language must provide a programming environment that allows the user to support it.

Complex robot programs are difficult to develop and can be difficult to debug. Morever, robot programming imposes additional requirements on the development and debugging facilities :

(*i*) *On-line modifications and immediate restart* : Because robot task requires complex motions and long execution time, *robot programming system must have the ability to allow programs to be modified on-line and restart at any time.*

(*ii*) *Sensor outputs and program traces* : Since the real-time interactions between the robot and environment are not always repeatable, *the debugger should be able to record sensor values along with program traces.*

(*iii*) *Simulation* : With this feature, the program can be tested *without actually setting up robot and workspace.*

Q. 4.3. Give a comparison between robot-oriented languages and task-level programming languages.

Ans. In "*robot-oriented languages*", an assembly task is *explicitly described as a sequence of robot motions.* The robot is guided and controlled by the program throughout the entire task with each statement of the program roughly corresponding to one action of the robot.

On the other hand, "task-level programming languages" describe the assembly task *as sequence of positional goals* of the objects rather than the motion of the robot needed to achieve these goals, and hence no *explicit robot motion is specified.*

- A *"robot-oriented language" is difficult to use* because it requires the user to *program each detailed robot motion in completing a task.*

  "The task-level languages," in contrast, are *much easier to use.* However, the following problems must be solved before they these can be effectively used :

— Task planning;

— Object modelling;

— Obstacle avoidance;
— Trajectory planning;
— Sensory information utilisation;
— Grasping configurations etc.

Q. 4.4. Explain briefly "Motion programming".

Ans. • With today's robot languages, *"motion programming" requires a combination of textual statements and lead-through techniques.* Accordingly, this method of programming is sometimes referred to as *On-line/off line programming.*

• The *'textual statement'* are used to describe the motion and the *'lead-through methods'* are used to define the position and orientation of the robot during and/or at the end of the motion.

— The lead-through methods provide a very natural way of programming motion commands into the robot controller.

— In *manual lead-through* the operator simply moves the arm through the required path to create the program. In *power lead-through* the operator uses a *'teach pendent' to drive the manipulator.*

• Since coordinating the individual joints with teach pendant is sometimes an awkward way to enter motion commands to the robot, therefore, it is difficult to coordinate the individual joints of a jointed-arm robot to drive the end-of-arm in a straight line motion. Hence, many of the robots using power lead-through provide *two alternative methods for controlling movement of the manipulator during programming, in addition to individual joint controls; these methods are :*

(*i*) *World coordinate system*, and (*ii*) *Tool coordinate system*. Both systems use 'Cartesian coordinate system'.

— In the *"world coordinate system"*, the origin and frame of reference are defined w.r.t. some fixed position and alignment relative to the robot base.

— In the *"tool coordinate system"*, the alignment of the axis system is defined relative to the orientation of the wrist faceplate, to which the end-effector is attached.

The robot's speed is controlled by means of a dial or other input device located on the teach pendant and/or the main control panel.

Q. 4.5. Discuss briefly the "Intermediate Language for Mobile Robots (ILMR).

Ans. ILMR *is a structure which would link all subsystems together in a way that would enable the developer of one subsystem to control and utilise other subsystems easily and in a desired way and would give the user or the high level planner the means to design and control a new task or subtask for a robot.*

The various *requirements* of ILMR are as follows :

(*i*) Easy control and access to all subsystems in a robot;
(*ii*) Monitoring of the status of subsystems;
(*iii*) Emphasis on movement commands;
(*iv*) Usability with service and field robots;
(*v*) Separation of robot dependent and independent parts;
(*vi*) Inhibition of harmful/damaging commands;
(*vii*) Possibility to use behaviors;

(viii) Easy way to use concúrrent and subsequential subtask/operation execution and monitoring etc.

Q. 4.6. Explain briefly with a neat sketch a simple robotic insertion task.

Ans. Fig. 4.3 shows the robotic task of inserting a bolt into a hole. This task involves the following *operations* :

1. Moving the robot to the feeder;
2. Picking up the bolt;
3. Moving the bolt to the beam;
4. Inserting the bolt into one of the holes.

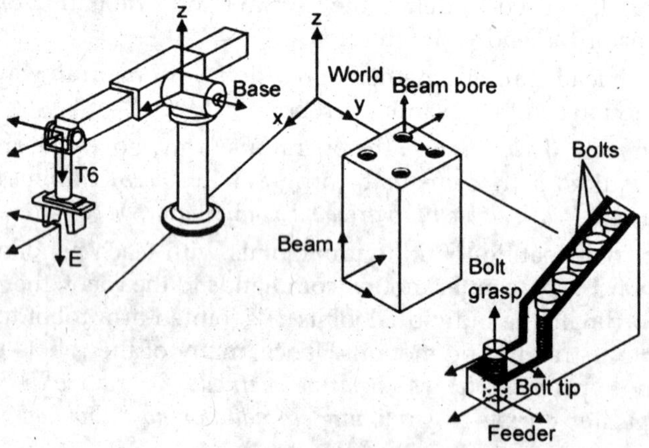

Fig. 4.3. A Simple robotic task of inserting a bolt into a hole.

The development of the program, typically, includes the following *"steps"* :

(i) To set-up the workspace and fix the parts by the use of fixtures and feeders.

(ii) To define the location (orientation and position) of the parts (feeder, beam, etc.) and their features (beam-bore, bolt-grasp etc.)

(iii) To partition the assembly task into a sequence of actions such as moving the robot, grasping objects, and performing an insertion.

(iv) To add sensory commands to detect abnormal situations (such as inability to locate the bolt while grasping) and monitor the progress of the assembly task.

(v) To debug the program and refine it by repeating steps (ii) to (iv).

Q. 4.7. Write a program for PNP (Pick and Place) activity.

Ans. A program for PNP activity is given below :

VAL Statement		Statement description
• BRANCH PICK	....	The branch of program indicating part pick.
• MOVE INTER	....	Move to an intermediate position above chute.
• WAIT 12	....	Wait for an incomping part to the chute.
• SIGNAL 5	....	Open gripper fingers (sensor control).
• MOVE PICK-UP	....	Move gripper and pick-up the object.
• SIGNA 6	....	Close the gripper to grab the object.

- MOVE INTER .... Depart to intermediate position above chute.
- END BRANCH .... End of pick-up activity.
- BRANCH PLACE .... Start of placing activity.
- MOVE Z (-50) .... Position part and gripper above the pallet.
- SIGNAL 5 .... Open gripper to release the part.
- MOVE Z (50) .... Depart from the place (pallet) point.
- END BRANCH .... End of place activity.

Q. 4.8. Enlist the components/layout of "Robot task planning".

Ans. *Robot task planning* includes the following :

Phase

↓

1. ***World modelling*:**

(*i*) *Geometric and physical models :*

↓

Representation of primitives :

- Set of edges and points
- Set of surfaces
- Generalized cylinders
- Cell decomposition.

(*ii*) *World state representation :*

↓

Relations :

- Attachment between objects
- Constraints-translational or rotational
- Assembly components.

2. **Task specification and program synthesis :**

↓

Decomposition of tasks :

- Initial state
- Grasping position
- Operand
- Specification
- Attachment
- Final state.

↓

Knowledge ↔ *Subtask planner* ↔ Models

↓

Program synthesis :

- Choosing candidate grasping configuration
- Pruning of the set up

- Selection of finál configuration.

↓

Robot program

HIGHLIGHTS

1. A *robot program* may be defined as a path in space to be followed by the manipulator, combined with peripheral actions that support the work cycle.
2. Robot languages have been developed for the ease of control of motion of robots having different structures and geometrical capabilities. Some of the programming languages are : WAVE, AL, VAL, AML, MCL, RAIL, HELP, RPL, PAL, etc.
3. Methods to programme the robot's work cycle are :
   (*i*) Manual programming method.
   (*ii*) Walk through programming method.
   (*iii*) Lead through programming method.
   (*iv*) Off-line programming method.

OBJECTIVE TYPE QUESTIONS

Fill in the Blanks or Say "Yes" or "No" :

1. Robot languages provide flexibility to the user in defining the task.
2. In .... oriental programming, an assembly task is explicitly described as a sequence of robot motion.
3. In task-level programming no explicit robot motion is specified.
4. .... through programming method suites well for robots used in arc welding and spray painting operations.
5. In powered lead-through programming method, a control box with buttons known as .... is used to control joint motors.
6. Programmer's pendant is designed for use while an application program is being written and debugged.
7. AL is a specialised manipulation language developed at .... University.
8. JARS is based on Pascal.
9. VAL II is a first generation language.
10. RAIL language was developed by ....

ANSWERS

1. Yes	2. robot	3. Yes	4. Walk	5. teach pendant
6. Yes	7. Stanford	8. Yes	9. No	10. Automix.

THEORETICAL QUESTIONS

1. What is a 'Robot program' ?
2. Enlist the considerations which must be handled by any robot programming method.
3. Explain briefly the following two current approaches to programming :
   (*i*) Robot-oriented programming.

(*ii*) Object-oriented or task-level programming.

4. Name the methods which are used to programme the work cycle of a robot.
5. Describe briefly any two of the following programming methods :
   (*i*) Walk through programming method.
   (*ii*) Lead-through (or teach pendant) programming method.
   (*iii*) Off-line programming method.
6. Explain briefly the off-line programming method and state its advantages and disadvantages also.
7. Explain the different ways by which robot teaching can be performed.
8. Explain the following :
   (*i*) Manual programming.
   (*ii*) Lead through teaching.
9. What are the critical information required for task programming of robot ?
10. Discuss the programming methods used in robots mentioning their specific field of application.
11. Explain briefly the following categories of robot programming languages :
    (*i*) Specialized manipulation languages.
    (*ii*) Robot library for an existing computer language.
    (*iii*) Robot library for a new general-purpose language.
12. Discuss briefly the requirements of a robot programming language.
13. Explain briefly the following classifications of robot languages :
    (*i*) First generation languages;
    (*ii*) Second generation languages;
    (*iii*) World modelling and task-oriented object level language.
14. Discuss briefly the following robot languages :
    AML, RAIL, RPL, HELP
15. Explain briefly the following three basic modes of operation in a robot :
    (*i*) Monitor mode;
    (*ii*) Edit mode;
    (*iii*) Run or excutive mode.
16. Discuss briefly the problems associated with robot programming languages.

CHAPTER

5 Robot Drives, Actuators and Control

5.1 Introduction; 5.2 Fluid power-general aspects – Introduction to fluid power – Advantages and disadvantages of fluid power systems – Fluid and its properties; **5.3 Hydraulic actuators** – General aspects – Hydraulic power sypply – Pumps – Pressure regulator – Hydraulic valves – Linear actuators – Rotary actuators; **5.4 Pneumatic actuators** – Introduction – Components of a pneumatic system – Pneumatic valves – Linear and rotary actuators – Special features of pneumatic actuators – Example of fluid control system; **5.5 Electrical actuators** – General aspects – Switching devices – Drive systems – Electric motors – D.C. motors – Single phase motors – Three phase induction motors – Electronic control of A.C. (Induction) motors – Synchronous motor – Digital control of electric motors; **5.6 Comparison of robot drive systems; *Questions with Answers*** – Highlights – Objective Type Questions – Theoretical Questions.

5.1 INTRODUCTION

The drives are the main motive power of the robots.

The various types of transmitting elements are employed to drive the mechanical links to a desired position and orientation in the envelope of the robot-manipulator.

Now-a-days, direct drives to the arms of robots are preferred though power is transmitted or applied to various joints of the robots through gears, belts, cable chains and several other means.

The various methods of drives are :

(*i*) *Hydraulic* ; (*ii*) *Pneumatic* ; (*iii*) *Electrical* A.C. and D.C. Motors; stepper motors.

The *choice of the drive systems* is based on the following *factors* :

(*i*) Power consumption ;
(*ii*) Positional accuracy ;
(*iii*) Repeatability ;
(*iv*) Operational speed ;
(*v*) Stability ;
(*vi*) Reliability ;
(*vii*) Cost and several related factors.

The *selection* of the drive methods is also made *on the basis of using open loop or closed loop controls.*

Based on the drive methods, the following *"actuators"* are used :

(*i*) ***Hydraulic cylinders*** : They handle *'oil'* under pressure and use electrohydraulic valve.

(*ii*) ***Pneumatic cylinders*** : They use *'air'* as the fluid medium with solenoid controlled valve.

(*iii*) ***D.C. electric motors*** with electrical amplifiers and controllers and *electrical stepper motors* with suitable circuitry (ramping circuit) to control the pulses.

— **Actuators** *produce physical changes such as linear and angular displacements. They also modulate the rate and power associated with these changes.*

- *"Hydraulic actuators" can produce large force/torque to drive the manipulator joints without the use of reduction gearing and are easily applied for robotic positional control.*
  — However, the hydraulic systems are *cumbersome* and *messy* and *require a great deal of equipment* such as pumps, actuators, hoses and servo valves.
  — In applications where position and/or torque must be accurately controlled, hydraulic actuators prove disadvantageous due to friction of seals, leakage, viscosity of oil, and its complex temperature dependence.
- *"Pneumatic actuators"* possess all the disadvantages of hydraulic actuators except that these are *relatively cleaner*. These are *difficult to control* accurately due to high friction of seals and compressibility of air (or other pneumatic medium).

☞ The *"electric motor" is the most popular choice for actuators for robotic systems.* Although these do not match the power-to-weight ratio of hydraulic or pneumatic actuators, these are *easy to control and interface.* This is an added advantage of the electric actuators but at the cost of increased joint friction, elasticity and backlash.

5.2 FLUID POWER

5.2.1. Introduction to Fluid Power

The term **fluid power** *refers to the energy that is transmitted via a fluid under pressure.* With *"hydraulics,"* that fluid is a liquid such as *oil* or *water;* while in *"pneumatics"*, the fluid is typically compressed air or nitrogen gas (inert gas).

"Fluid power technology" is a means to convert, transmit, control and apply fluid energy to perform useful work.

Applications *of fluid power* include the following areas :

(*i*) Agriculture ; (*ii*) Construction ;
(*iii*) Ships ; (*iv*) Aviation ;
(*v*) Defence ; (*vi*) Transportation ;
(*vii*) Fabrication ; (*viii*) Material handling ;
(*ix*) Automation.

5.2.2. Advantages and Disadvantages of Fluid Power Systems

The *advantages and disadvantages of fluid power systems* are :

Advantages :

1. The fluid power drives are more compact than a mechanical drive because it eliminates the need for links like cams and gears.
2. As the medium of power transmission is a fluid, it is not subjected to any breakage of parts as in a mechanical transmission.
3. In fluid power system constant force is possible, regardless of special motion requirements, whether the work output moves a few millimetres or several metres per minute.
4. The parts of a hydraulic system are lubricated with the hydraulic liquid itself.
5. Multiplication of small forces to achieve greater forces for performing work.
6. Accuracy in controlling small or large forces with instant reversal is possible with hydraulic systems.

7. It easily provides infinite and stepless variable speed control which is difficult to obtain from other drives.

Disadvantage. The main disadvantage of a fluid power system is the *maintenance of the precision parts when they are exposed to bad climates and dirty atmospheres.*

5.2.3. Fluid and its Properties

Fluid :

A fluid may be defined as follows :

"A fluid is a substance which is capable of flowing."

Or

"A fluid is a substance which deforms continuously when subjected to external shearing force."

A fluid has the following *characteristics* :

1. It has no definite shape of its own, but conforms to the shape of the containing vessel.
2. Even a small amount of shear force exerted on a liquid/fluid will cause it to undergo a deformation which continues as long as the force continues to be applied.

A fluid may be *classified* as follows :

A. (*i*) Liquid, (*ii*) Gas, and (*iii*) Vapour. B. (*i*) Ideal fluids and (*ii*) Real fluids.

Liquid :

— A liquid is a fluid which possesses a *definite volume* (which varies only slightly with temperature and pressure).

— Liquids have bulk elastic modulus when under compression and will store up energy in the same manner as a solid. As the contraction of volume of a liquid under compression is extremely small, it is usually ignored and the *liquid is assumed to be incompressible.* A liquid will withstand a slight amount of tension due to molecular attraction between the particles which will cause an apparent shear resistance, between two adjacent layers. This phenomenon is known as **viscosity**.

— All known liquids vaporise at narrow pressures above zero, depending on the temperature.

Gas. It possesses *no definite volume* and is *compressible.*

Vapour. It is a gas whose temperature and pressure are such that it is very near the liquid state (*e.g.*, steam).

Ideal fluids. An ideal fluid is one which has *no viscosity* and *surface tension* and is incompressible. In true sense no such fluid exists in nature. However fluids which have low viscosities such as water and air can be treated as ideal fluids under certain conditions. The assumption of ideal fluids helps in simplifying the mathematical analysis.

Real fluids. A real practical fluid is one which has viscosity, surface tension and compressibility in addition to the density. The real fluids are actually available in nature.

Density :

- ***Mass density*** :

The density (also known as *mass density* or *specific mass*) of a liquid may be defined as the *mass per unit volume* $\left(\frac{m}{V}\right)$ at a standard temperature and pressure. It is usually denoted by ρ(rho). Its units are kg/m^3 *i.e.*, $\rho = \frac{m}{V}$ ...(5.1)

- ***Weight density* :**

The weight density (also known as specific weight) is defined as the weight per unit volume at the standard temperature and pressure. It is usually denoted by w.

$$w = \rho g \quad ...(5.2)$$

For the purposes of all calculations, relating to Hydraulics and hydraulic machines, the specific weight of water is taken as follows :

In S.I. Units : $w = 9.81 \ kN/m^2$ (or $9.81 \times 10^{-6} \ N/mm^3$)

In M.K.S. Units : $w = 1000 \ kg/m^3$.

- ***Specific volume* :**

It is defined as *volume per unit mass of fluid*. It is denoted by v. Mathematically,

$$v = \frac{V}{m} = \frac{1}{\rho} \quad ...(5.3)$$

- ***Specific gravity* :**

Specific gravity is the ratio of the specific weight of the liquid to the specific weight of a standard fluid. It is dimensionless and has no units. It is represented by S.

For liquids, the standard fluid is pure water at 4°C.

$$\therefore \quad \text{Specific gravity} = \frac{\text{Specific weight of liquid}}{\text{Specific weight of pure water}} = \frac{w_{liquid}}{w_{water}}.$$

Pressure of a fluid :

When a fluid is contained in a vessel, it exerts force at all points on the sides and bottom (and a top) of the container. The *force per unit area is called* ***pressure***.

If P = The force, and

A = Area on which the force acts, then intensity of pressure, $p = \frac{P}{A}$ ...(5.4)

The pressure of a fluid on a surface will always act normal to the surface.

The intensity of pressure of a liquid may be expressed in the following two ways :

1. As a force per unit area (*i.e.*, N/mm^2, N/m^2), and
2. As an equivalent static head (*i.e.*, metres, mm or cm of liquid).

- ***Absolute and gauge pressures* :**

Atmospheric pressure. The atmospheric air exerts a normal pressure upon all surfaces with which it is in contract, and it is known as atmospheric pressure. The *atmospheric pressure* is also known as *Barometric pressure*.

The atmospheric pressure at sea level (above absolute zero) is called standard atmospheric pressure.

Note. The local atmospheric pressure may be a little lower than these values if the place under question is higher than sea level, and higher values if the place is lower than sea level, due to the corresponding decrease or increase of the column of air standing, respectively.

- ***Gauge pressure* :**

It is pressure, measured with the help of pressure measuring instrument, in which the atmospheric pressure is taken as datum. The *atmospheric pressure* on the scale is marked as *zero*.

"*Gauges*" record pressure above or below the local atmospheric pressure, since they measure the difference in pressure of the liquid to which they are connected and that of surrounding air. If the pressure of the liquid is below the local atmospheric pressure, then the gauge is designated as *'vacuum gauge'* and the recorded value indicates the amouont by which the pressure of the liquid is *below* local atmospheric pressure *i.e., negative pressure.*

(*Vacuum pressure* is defined as the pressure *below the atmospheric pressure*).

- ***Absolute pressure* :**

  It is necessary to establish an absolute pressure scale which is independent of the changes in atmospheric pressure. *A pressure of absolute zero can exist only in complete vacuum.*

  Any pressure measured above the absolute zero of pressure is termed as an **'absolute pressure'**.

  A schematic diagram showing the gauge pressure, vacuum pressure and the absolute pressure is given in Fig. 5.1.

Mathematically :

1. *Absolute pressure* = Atmospheric pressure + gauge pressure

   i.e., $$p_{abs} = p_{atm} + p_{gauge}$$

2. *Vacuum pressure* = Atmospheric pressure – absolute pressure

- ***Units for pressure* :**

  The fundamental S.I. unit of pressure is newton per square metre (N/m^2). This is also known as *Pascal.*

  Low pressures are often expressed in terms of mm of water or mm of mercury. This is an abbreviated way of saying that the pressure is such that will support a liquid column of stated height.

  Note. When the local atmospheric pressure is not given in a problem, it is taken as 100 kN/m^2 or 10 m of water for simplicity of calculation.

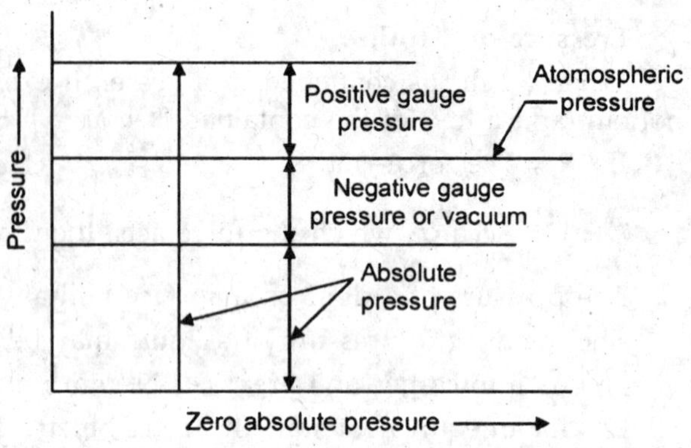

Fig. 5.1. Relationship between pressures.

  Standard atmospheric pressure has the following equivalent values :

  101.3 kN/m^2 or 101.3 kPa; 10.3 m of water, 760 mm of mercury, 1013 mb (millibar); $\simeq$ 1 bar $\simeq$ 100 kPa = 10^5 N/m^2.

Pascal's law

The Pascal's law states as follows :

"The intensity of pressure at any point in a liquid at rest, is the same in all directions".

Viscosity. It is the property of a fluid which determines its resistances to shearing stresses.

Compressibility. It is the property by virtue of which fluid undergoes a change in volume under the action of external pressure. It is the *reciprocal* of bulk modulus of elasticity (*K*).

$$K = \frac{dp \text{ (increase of pressure)}}{-\frac{dV}{V} \text{ (volumetric strain)}} \quad ...(5.5)$$

Bernoulli's equation. It states as follows :

"In an ideal, incompressible fluid when the flow is steady and continuous, then sum of pressure energy, potential (or datum) energy and kinetic energy is constant along a stream line."

Mathematically, $$\frac{p}{w} + \frac{V^2}{2g} + Z = \text{constant} \quad ...(5.6)$$

where, $\frac{p}{w}$ = Pressure energy or head,

$\frac{V^2}{2g}$ = Kinetic energy or head, and

Z = Datum (or elevation) energy or head.

Continuity equation. It is based on the principle of *conservation of mass*. It states as follows :

"If no fluid is added or removed from the pipe in any length then the mass passing across the different sections shall be same."

Gas laws :

(*i*) *Boyle's law* : It states that at constant temperature, the pressure (p) is inversely proportional to the volume (V) of a definite amount of gas (*i.e.*, $p \propto \frac{1}{V}$ or pV = constant)

Mathematically, $$p_1V_1 = p_2V_2 \quad ...(5.7)$$

where suffices 1 and 2 correspond to initial and final conditions of the gas.

(*ii*) *Charle's law* : It states that the volume (V) of a gas at constant pressure is directly proportional to the absolute temperature (T) [*i.e.* $V \propto T$ or $\frac{V}{T}$ = constant]

Mathematically, $$\frac{V_1}{T_1} = \frac{V_2}{T_2} \quad ...(5.8)$$

When it is difficult to keep either the temperature or pressure constant, the following combined gas equation is used :

$$\frac{p_1V_1}{T_1} = \frac{p_2V_2}{T_2} \quad ...(5.9)$$

5.3 HYDRAULIC ACTUATORS

5.3.1. General Aspects

An actuator wherein hydraulic energy is used to impart motion is called an ***hydraulic actuator.***

A system wherein energy is imparted to oil and this hydraulic energy so imparted to the oil is converted into mechanical energy is called an ***hydraulic system.***

Advantages and disadvantages of hydraulic system:

Following are the advantages and disadvantages of a hydraulic system :

Advantages:

1. Easy to produce, transmit, store, regulate and control, maintain and transform the hydraulic power.
2. Possible to generate high gain in force and power amplification.
3. Hydraulic systems are uniform and smooth, generate stepless motion and variable speed and force to a greater accuracy.
4. Limiting and balancing of hydraulic forces are easily performed.
5. Weight-to-power ratio of an hydraulic system is comparatively less than that for an electro-mechanical system.
6. Easy maintenance.
7. Hydraulic systems are cheaper if one considers the high efficiency of power transmission.
8. Hydraulics is mechanically safe, compact and adaptable to other forms of power and can be easily controlled.
9. Division and distribution of hydraulic power is simpler and easier than other forms of energy.
10. Frictional resistance is much less in an hydraulic system as compared to a mechanical movement.

Disadvantages:

1. The manufacturing cost of the system is high since the hydraulic elements have to be machined to a high degree of precision.
2. Hydraulic elements have to be specially treated to protect them against rust, corrosion, dirt, etc.
3. Petroleum based hydraulic oil may pose fire hazards thus limiting the upper level of working temperature.
4. Certain hydraulic systems are exposed to unfriendly climate and dirty atmosphere as in the case of mobile hydraulics like dumpers, loaders, etc.
5. Hydraulic power is not readily available compared to electric power.

5.3.2. Hydraulic Power Supply

Hydraulic systems are designed to move large loads by controlling a high-pressure fluid in distribution lines and pistons with mechanical or electromechanical valves.

5.3.2.1. Basic element of an oil hydraulic system

Fig. 5.2 shows the *element* of an hydraulic system.

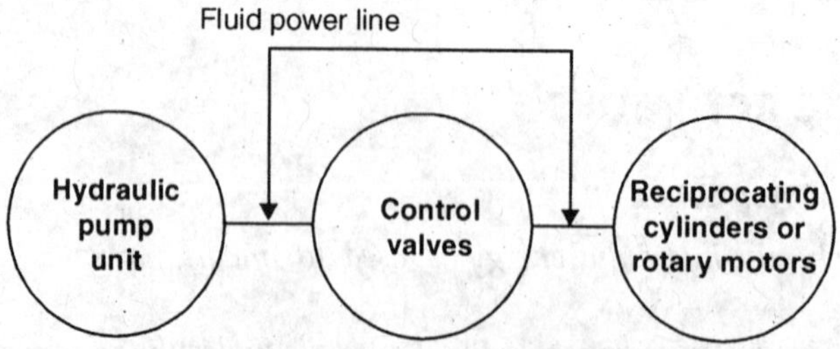

Fig. 5.2. Elements of an hydraulic system.

1. Hydraulic pump unit :

- In an actual hydraulic system a pump converts mechanical power into fluid power.

- The intake of the pump is connected to a liquid source usually called a tank or reservoir.

2. Control valves :

- The flow of pressurised liquid discharges by the pump is controlled by valves :
  — *"Pressure control valves"* control the liquid *pressure.*
  — *"Flow control valves"* control the liquid *flow rate.*
  — *Directional control valves* control the *direction* of flow of the liquid.

3. Hydraulic motor/cylinder :

- The liquid discharged by the pump is directed to hydraulic motors or cylinders by control valves.
- *Motors are used where rotary motion is desired and cylinders are used where linear motion is necessary.*

5.3.2.2. Components of an hydraulic system

Fig. 5.3; shows the various *components* of an hydraulic system; these are :

1. ***Pump:*** It delivers high pressure fluid.
2. ***Pressure regulator:*** It limits the pressure in the system.
3. ***Valves:*** These control flow rates and pressures.
4. ***Distribution system:*** It is composed of hoses or pipes.
5. ***Infrastructure:*** It consists of the elements contained in the dashed box in the figure and is typically used to power many hydraulic valve actuator subsystems.

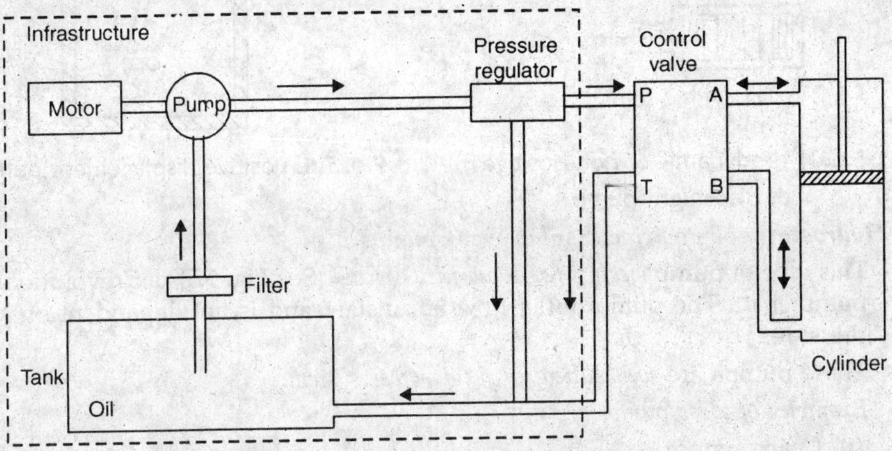

Fig. 5.3. Components of an hydraulic system.

5.3.3. Pumps

- An hydraulic pump is usually driven by an electric motor (*e.g.*, a large AC induction motor) or an internal combustion engine.
- Typical fluid pressures generated by pumps used in heavy equipment (*e.g.*, large industrial machines and construction equipment) are 6.9 MPa to 20.7 MPa range.
- The hydraulic fluid is selected to have the following *characteristics* :
  (*i*) Good lubrication to prevent wear in moving components.
  (*ii*) Corrosion resistance.
  (*iii*) Incompressibility to provide rapid response.

Classification of pumps :

There are two broad classifications of pumps as identified by the fluid power industry:

1. ***Hydrodynamic or Non positive displacement pumps :*** Refer to Fig. 5.4.

Examples of these pumps are : Centrifugal and propeller pumps.

— Although these pumps provide smooth continuous flow, their flow output is reduced when the circuit resistance is increased. Since there is a great deal of clearance between the rotating and stationary elements, when the resistance of the external system starts to increase, some of the fluid slips back into the clearance spaces causing a reduction in the discharge flow rate. This slippage is due to the fact that the fluid follows the least resistance path. Thus the pump flow rate depends not only on the rotational speed but also on the resistance of the external system.

— These types of pumps are used for *low pressure, high volume flow applications.*

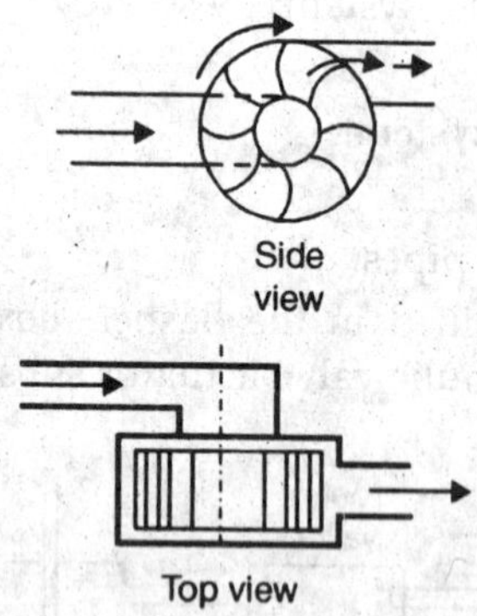

Fig. 5.4. Hydrodynamic or non-positive displacement pump.

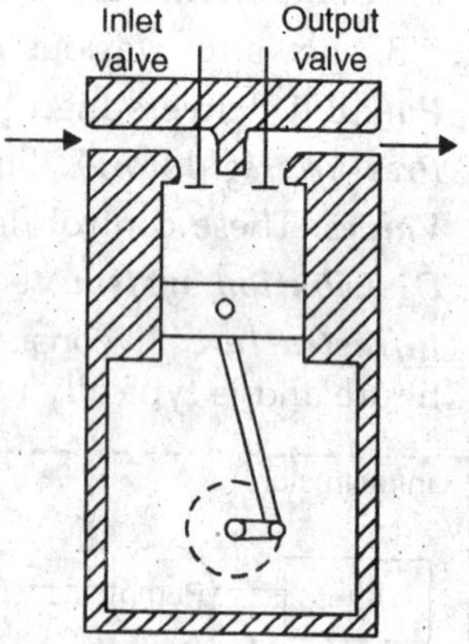

Fig. 5.5. Positive displacement pump.

2. ***Hydrostatic or positive displacement pumps :***

— This type of pump *ejects a fixed quantity of fluid* (See Fig. 5.5) per revolution of the pump shaft. The pump outlet flow is constant and is not dependent on system pressure.

— These pumps are *well suited for fluid power system.*

Examples of these pumps are :

(*i*) Gear pumps.

(*ii*) Vane pumps.

(*iii*) Piston pumps.

(i) Gear pump : The gear pump unit (Fig. 5.6) consists of two identical intermeshing spur gears with involute teeth. One of the gears is keyed to the driving shaft of the motor and the other gear revolves *idly.* These gears rotate in opposite directions in a closely fitting stationary *housing.*

Although the gear pump, which consists of two gears, is a rotating machine, yet its action on the liquid to be pumped is *not dynamic* and it *merely displaces the liquid to be pumped, is continuous and uniform and there is no change of velocity and acceleration under normal stable conditions.*

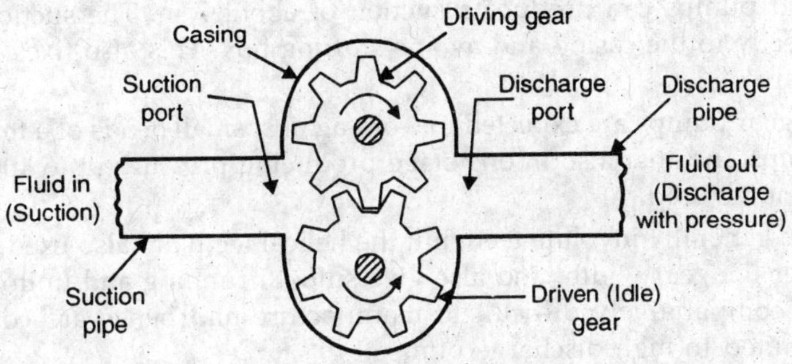

Fig. 5.6. Spur gear pump.

Working Principle. Referring to Fig. 5.6, the oil coming in at the suction port fills the space between the teeth, is carried around the periphery of the revolving gears and is finally pushed out to the discharge port. The teeth of the gears have a perfect meshing and that serves both to transmit the drive and to maintain a seal between the suction and discharge side. Care is taken to ensure that the oil trapped between the successive lines of contact does not built up pressure.

— For good working it is necessary to have teeth made precise and further they should have hard surface. Above all the whole gear and casing should have good surface and accurate dimenisions. The casing should have, in addition, a packing to prevent leakage. For highly viscous liquids the casing is provided with a heating jacket to reduce viscosity.

The direction of flow can be changed by reversing the direction of gear assembly. But the spur gear pump delivers hydraulic fluid always at right angles to the axis of rotation.

Let, A = Area enclosed between two adjacent teeth and casing,

L = Axial length of teeth,

T_g = Number of teeth in each gear wheel, and

N = Speed in r.p.m.

Then, volume of liquid discharged in one revolution,

$$q = 2LAT_g$$

∴ Theoretical volume displaced by the pump per second,

$$Q_{\text{th}} = (2LAT_g) \times \frac{N}{60} \qquad \text{...(5.10)}$$

Normally the area between the adjacent teeth is larger than the cross-sectional area of the meshing tooth, and that causes some liquid to flow back to the low pressure side. Also, there is some leakage because of clearance between the casing and gear wheel. Thus if η_v is the volumetric efficiency of the pump, then

Actual discharge, $$Q_{\text{actual}} = Q_{th} \times \eta_v$$

or, $$Q_{\text{actual}} = (2LAT_g) \times \frac{N}{60} \times \eta_v \qquad \text{...(5.11)}$$

If it is not possible to determine easily the area enclosed between the adjacent teeth and casting, then volumetric displacement of the pump per *revolution* is calculated by using the following emperical relation.

$$q = 0.95\, \pi c\, (D - c)t \qquad \text{...(5.12)}$$

where, D = Outside diameter of gears, and

c = Centre-to-centre distance between the axes of gears.

High speed pumps can produce a suction of about 7 m. The suction pipe is often connected directly to the casing and avoids stuffing box. This also fixes the direction of rotation of gears.

Normally gear pumps are expected to work against small heads of a few atmospheres. However, pumps and that also in one stage producing pressure upto 100 atm. absolute have been manufactured.

Gears have generally involute teeth but the helical teeth are also used, which provide better grip over the gear width, and also give smooth running and uniform flows. They are, however, comparatively difficult to manufacture and hence are costly. Therefore, their use in limited to high discharge pumps.

Applications. This type of pump is widely used for cooling water and pressure oil to be supplied for *lubrication to motors, turbines, machine tools etc.*

(*ii*) *Vane pump:* Fig. 5.7, shows the schematics of a vane pump.

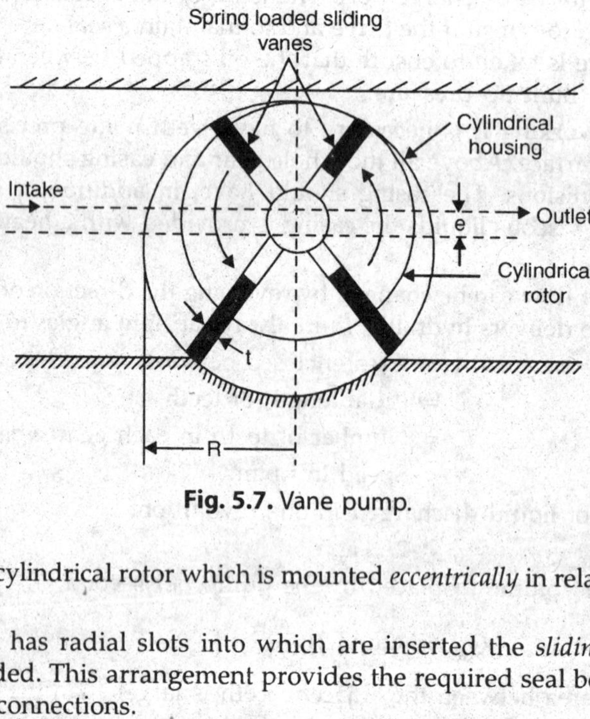

Fig. 5.7. Vane pump.

Construction :

It consists of a cylindrical rotor which is mounted *eccentrically* in relation to a cylindrical housing.

— The motor has radial slots into which are inserted the *sliding vanes*, which are spring loaded. This arrangement provides the required seal between suction and discharge connections.

Working/Operation :

— As the rotor rotates, the vanes undergo free to-and-fro sliding movement in the slots. On suction side, the pocket (space) between the vanes tend to increase in volume and the space gets filled with liquid.

— After the point maximum distance between the rotor and casing has passed, the space opens to delivery and the liquid is discharged. The quantity of liquid pumped and the flow direction can be controlled by affecting a change in the degree of eccentricity.

Let, b = Width of vane,

t = Thickness of vane,

n = Number of vanes,

R = Inner radius of casing,

e = Eccentricity between the rotor and casing, and

N = Speed in r.p.m.

Then, the volume of liquid discharged in one revolution,

$$q = 2eb\,[2\pi\,(R - e) - nt]$$

∴ Theoretical volume of liquid displaced by the pump per second,

$$Q_{th} = 2eb\,[2\pi\,(R-e) - nt] \times \frac{N}{60} \qquad ...(5.13)$$

A *single stage* vane pump can develop about 15 to 65 atm. In order to get higher pressures, more than one stage may be used.

(iii) Piston pump: Fig. 5.8 shows a *swash plate piston pump.*

— The cylinder block is rotated by the input shaft, and the piston ends are driven in and out as they slide in the fixed *swash plate* slot, which is angled with respect to the shaft axis.

— A piston draws fluid from an inlet manifold over half the swash plate and expels the fluid into the outlet manifold during the other half.

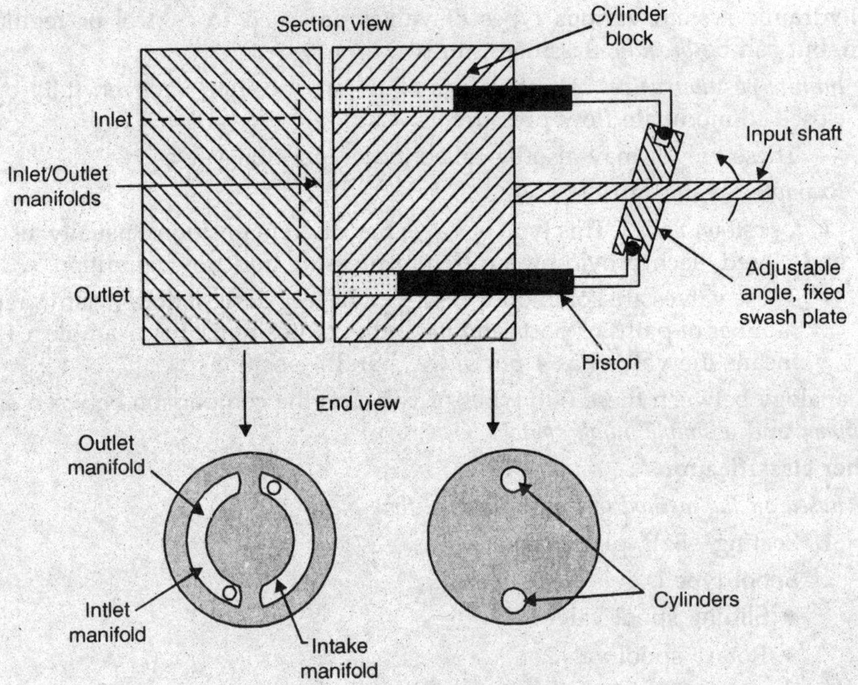

Fig. 5.8. Swash plate piston pump.

— The pump displacement can be altered simply by changing the angle of the fixed swash plate.

— Comparison of characteristics of various pumps are given below :

S. No.	Aspects	Gear pump	Vane pump	Piston pump
1.	*Displacement*	Fixed	Variable	Variable
2.	*Cost*	Low	Medium	High
3.	*Typical pressure*	140 MPa	270 MPa	420 MPa

5.3.4. Pressure Regulator

*A **pressure regulator** is a pressure relief valve which is provided in positive displacement pumps* (which provide a fixed volumetric flow) *to prevent the pressure from exceeding design limits.*

Fig. 5.9, shows the simplest pressure regulator—*spring ball arrangement* :

— When the pressure force exceeds the spring force, fluid is vented back to the tank, preventing a further increase in pressure.

— The *cranking pressure* (threshold pressure) is usually adjusted by altering the compressed length of the spring *i.e.*, its resisting force.

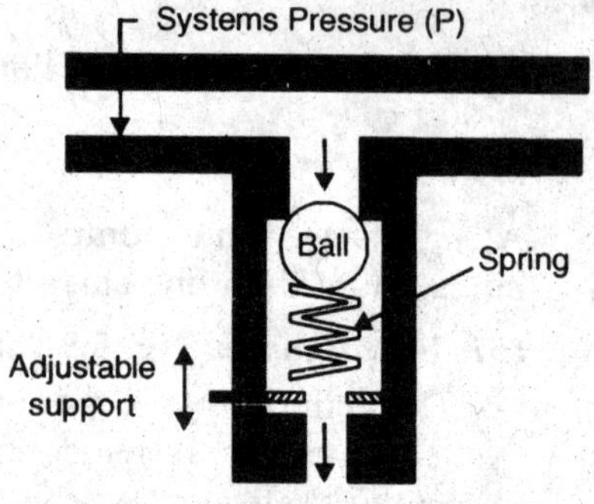

Fig. 5.9. Pressure regulator.

5.3.5. Hydraulic Valves

***Hydraulic valves** are mechanical devices used for controlling hydraulic parameters in hydraulic systems.*

5.3.5.1. Classification of valves

In hydraulic system various types of valves are used to control or regulate flow medium, but can be *broadly* classified into two types:

1. *Infinite position valves.* These valves allow any position between fully open and closed to modulate flow pressure.

   — These valves may also be called *analog* position valves.

   Examples: Water tap.

2. *Finite position valves.* This type of valve has discrete positions, usually just opened and closed, each providing a different pressure and flow condition.

   — These valves are commonly described by x/y designation, where x represents number of paths or ports and y number of positions; For example a 4/3 valve means the valve has 4 ports/ways and 3 positions.

An analogy between these two types of valves is the comparison between an *electric light dimmer* and a *simple on/off switch.*

Other classifications :

I. *Based on the method of controlling the flow medium :*

1. Seating—Ball, plate, cone.
2. Spool type :
   - Sliding spool valve
   - Rotary spool valve.

II. *Based on function and hydraulic quantity controlled :*

1. Pressure control valve—To control pressure of flow medium.
2. Flow control valve—To control quantity of flow medium.
3. Direction control valves—To control direction of flow medium.

- In mechatronics and hydro-mechanical systems control of pressure, flow and direction are of significant importance.

5.3.5.2. Graphic valve symbols

A valve is represented by a *square* for each of its switching positions.

Refer to Fig. 5.10.

— Fig. 5.10 (*i*), shows the symbol of a two position valve.

— Fig. 5.10 (*ii*), shows a three position valve. Valve position can be represented by letters *a*, *b*, *c* and so on, with *o* being used for a central neutral position.

— Fig. 5.10 (*iii*), shows two position valve with three ports. Parts of a valve are shown on the outside of the boxes in normal, *unoperated* or *initial position.*

— Fig. 5.10 (*iv*), shows two position valve in the four ports.

— Fig. 5.10(*v*), shows closing of port (S).

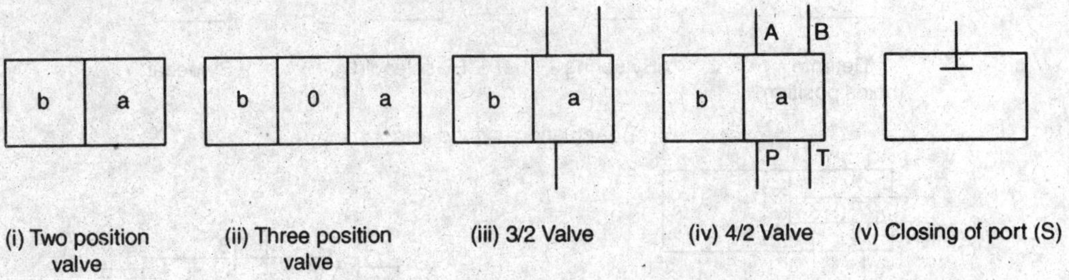

Fig. 5.10. Basis of graphic symbols.

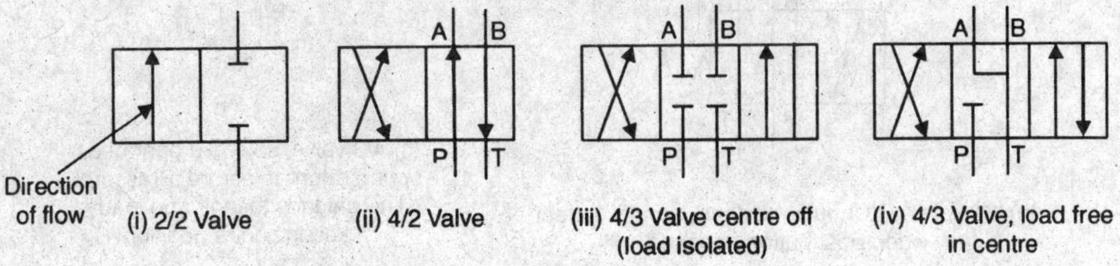

Fig. 5.11. Valve symbols.

- Fig. 5.11 shows valves symbols for 2/2 valve, 4/2 valve and 4/3 valve centre off (load isolated) and 4/3 valve, load free in centre respectively.

Following ***designations*** are normally given to the ports:

Working links = *A*, *B*, *C* and so on; *Pressure* (*power*) *supply* = *P*;

Exhaust return = *R*, *S*, *T* and so on (*T* normally used for hydraulic systems, *R* and *S* for pneumatic systems).

— Arrow-headed lines represent direction of flow. In Fig. 5.11(*ii*), for example, fluid is delivered form port *P* to port *A* and returned from port *B* to port *T*, when valve is in its normal state *a*. In state *b* the flow is reversed. Shut off position are represented by ⊥ as shown by the central position of the valve in Fig. 5.11(*iii*), internal flow paths can be represented as shown in Fig. 5.11(*iv*).

Refer to Fig. 5.12.

— Fig. 5.12(*a*) shows the various ways in which valves can be operated.

— Fig. 5.12(*b*) represents a 4/2 valve operated by a push button, with the push button depressed the ram extends. With the push button released, the spring pushes the valve to earlier position and the ram retracts.

— Fig. 5.12(*c*) represents a solenoid-operated 4/3 valve, with spring return to centre.

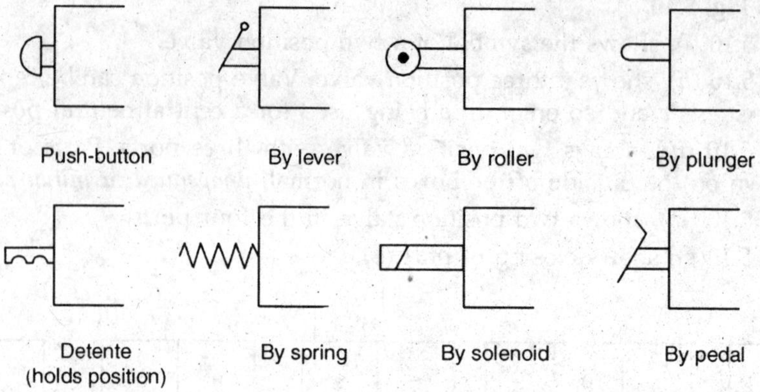

(a) Actuation symbols.

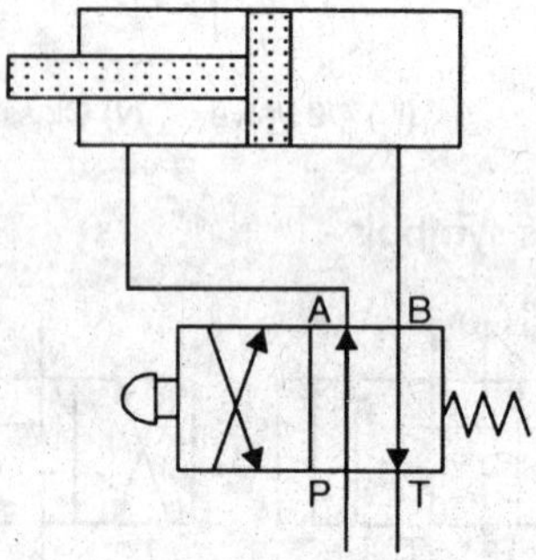

(b) 4/2 valve, Pushbutton extend, spring retract when pushbutton released

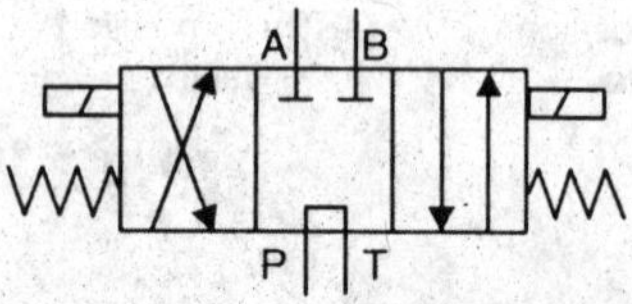

(c) 4/3 valve, solenoid operated, spring return to centre. Pressure line unloads to tank and load locked in centre position

Fig. 5.12. Complete valve symbols.

Refer to Fig. 5.13.

This figure shows the *"infinite position valve symbols"*.

— A basic valve is represented by a single square, with the valve being shown in normal, or non-operated, position.

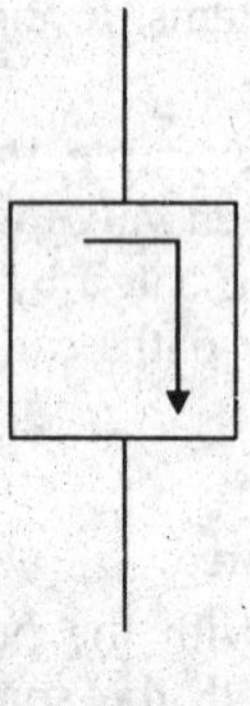

(a) Infinite position valve

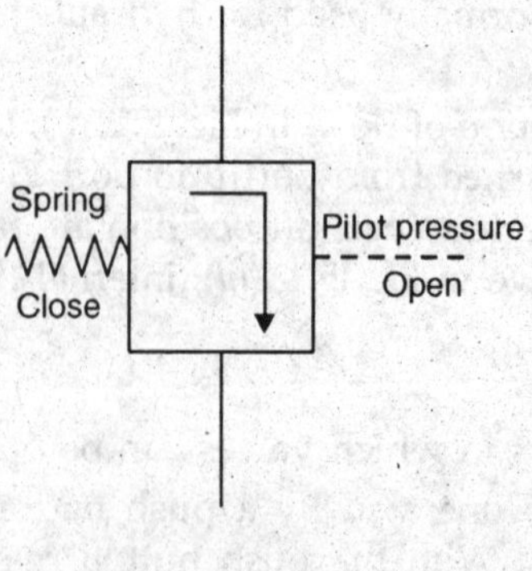

(b) With actuation symbols

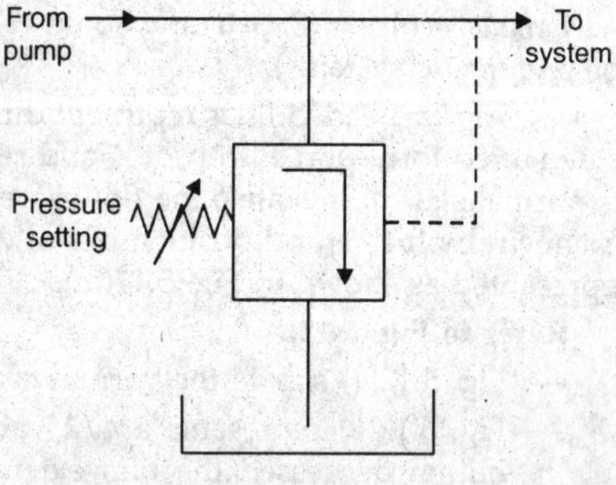

(c) Pressure relief valve

Fig. 5.13. Infinite position valve symbols.

5.3.5.3. Pressure control valves

These valves *control the pressure of flow medium required by the system.*

The pressure control valves perform the following *functions* :

(*i*) To regulate or reduce oil pressure in certain portions of the circuit.

(*ii*) To unload system pressure.

(*iii*) To limit maximum system pressure as a safety measure.

(*iv*) To assist sequential operation of actuators in a circuit by pressure control.

(*v*) To perform any other pressure related functions by virtue of pressure control.

These valves are classified based on (*i*) the primary functions they have to perform, (*ii*) type of connection; (*iii*) size and pressure operating range.

Based on their primary functions, the pressure control valves are ***classified*** as follows:

1. Pressure relief valve.
2. Pressure sequencing valve.
3. Pressure reducing or regulating valve.
4. Pressure unloading valve.
5. Pressure brake valve.

The operation of a pressure control valve is based mainly on balance between pressure and a mechanical load e.g. a spring force biased against the oil pressure. The valve can assume various positions between fully closed and fully open conditions depending on the flow and pressure differential.

1. Pressure relief valves:

These valves are found in every hydraulic system. It is a normally closed valve connected between the pressure line and the oil reservoir. Its main purpose is to *limit the pressure in a system to a prescribed maximum by diverting some or all of the pump output to the tanks, when the desired set pressure is reached.*

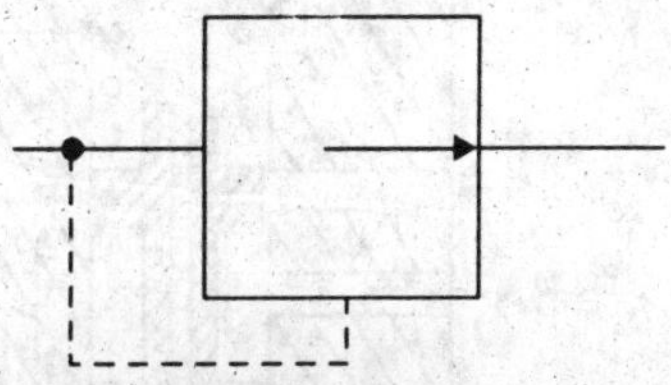

Fig. 5.14. Graphic symbol.

The pressure relief valves are of the following two types : (*i*) Direct acting relief valve; (*ii*) Pilot operated relief valve.

(*i*) ***Direct acting relief valve.*** For details refer to Art. 5.3.4. Fig. 5.14 shows the graphic symbol of this type of valve.

Fig. 5.15 shows the application of pressure relief valve in a hydraulic system containing accumulator as one of the elements.

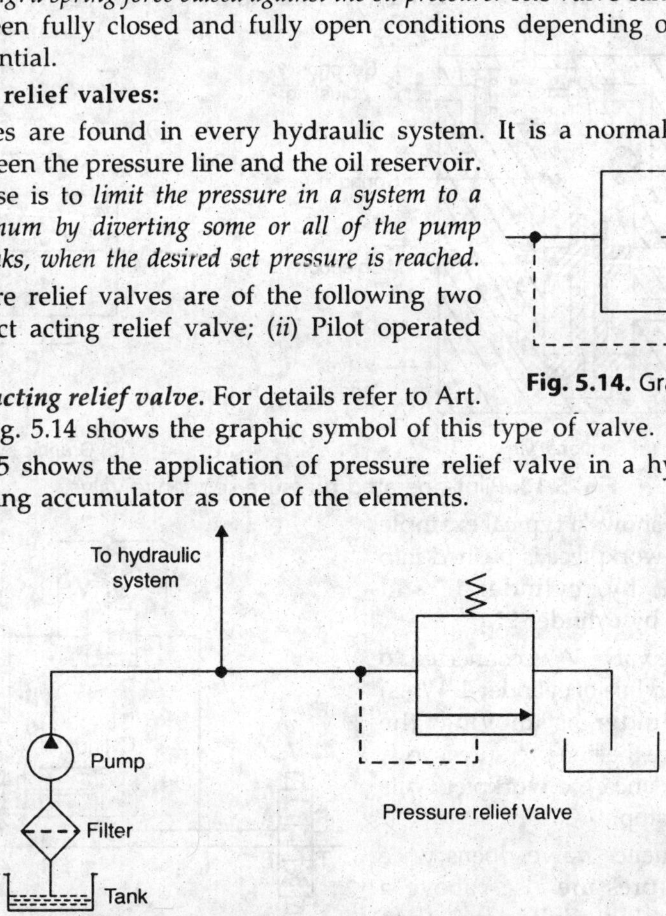

Fig. 5.15. Application of pressure relief valve.

(*ii*) ***Pilot operated relief valve.*** Direct controlled pressure relief valves are used where the flow rate and the system pressure are *reasonably smaller and there is not much variation in system pressure or flow rate.*

The use of pilot operated valve, however, is *most common for a larger flow rate and higher pressure.*

The main *advantage* of such a valve is that here the pilot valve can be kept spatially separated from the main valve such as on a control panel and can introduce suitable D.C. valves in between to set different pressures by pilot valves.

2. Pressure sequencing valve :

Sequencing valves extensively used in hydraulic systems are also pressure control valves.

A pressure sequence valve is used in hydraulic system *to cause various operations in a sequential order, i.e., one after the other.* For example a pressure sequence valve used in *clamping and machining circuit* may permit the clamping operation to take place first and when clamping cylinder is fully extended, the machining cylinder is actuated.

Fig. 5.16 shows the pilot operated sequence valve. Here the required sequential pressure can be adjusted manually.

— In this valve, the fluid flows freely through the primary passage to operate the first phase until the pressure setting of the sequence valve is reached.

— As the spool lifts, flow is diverted to the secondary port to operate the second phase.

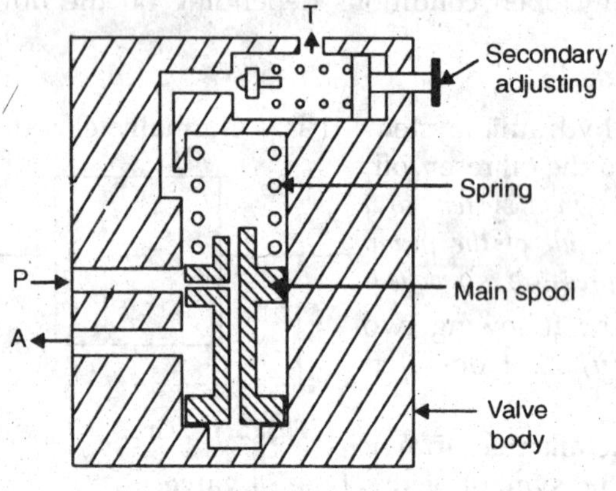

(a) Sectional view

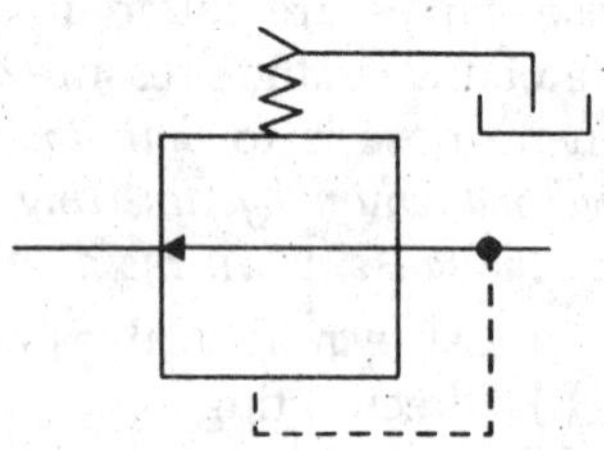

(b) Graphic symbol

Fig. 5.16. Pilot-operated pressure sequence valve.

- Fig. 5.17 shows a typical example where a workpiece is pushed into position by cylinder-1 and clamped by cylinder-2.

— Sequence valve V_2 is connected to the extend line of cylinder-1. When this cylinder is moving the workpiece the line pressure is low, but rises once the workpiece hits the end stop.

— The sequence valve opens once its inlet pressure rises above a preset value. Cylinder-2 then operates to clamp the workpiece.

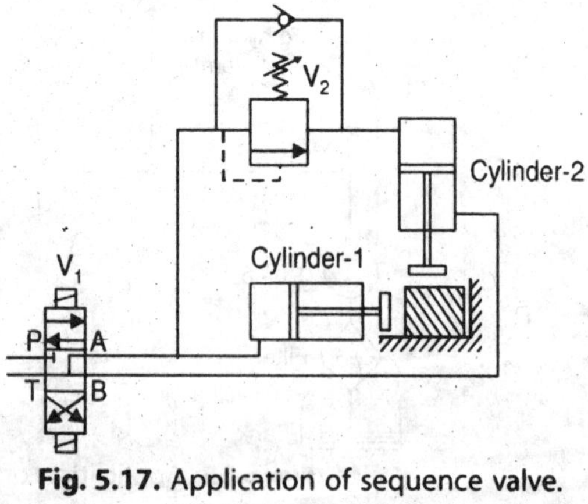

Fig. 5.17. Application of sequence valve.

A check valve across V_2 allows both cylinders to retract together.

3. Pressure reducing valve :

— The pressure reducing valves are normally open pressure control valves used *to maintain reduced pressures in certain portions of the hydraulic system.*

— These are actuated by the pressure sensed in the branch circuit and tend to close as it reaches the pressure of the valve setting preventing further build-up of pressure.

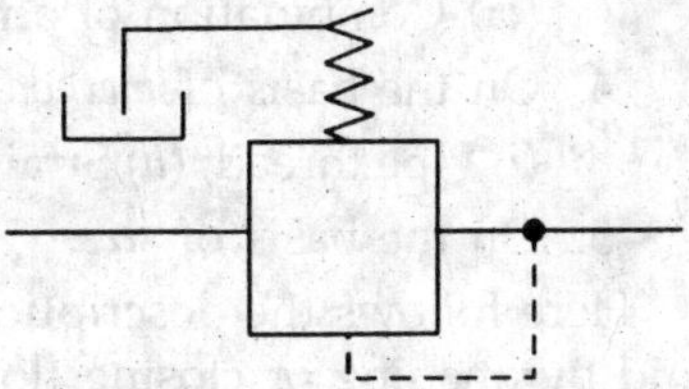

Fig. 5.18. Graphic symbol.

4. Pressure unloading valve :

This type of valve is used to *unload the energy in a system of a lower pressure* (whereas a pressure relief valve requires full system pressure to open thereby causing higher quantity of energy loss due to heat).

— These valves are employed in systems where two pumps provide a large volume of oil at a low pressure and one of them must be unloaded during a specific period requiring only a small volume of oil at a high pressure. This will save the system from undesired heat energy to a great extent.

5.3.5.4. Flow control valves (variable orifice)

— These valves are used to control the speed of hydraulic actuator by controlling the flow rate or discharge.

— The principle of flow valve is based on the flow rate or discharge and the velocity of the actuator.

Needle valves :

— It is the most common hydraulic flow-control device.

— It consists of a needle or pointed threaded stem that can be adjusted manually to control the flow or discharge through the valve. It is made of steel.

— This valve can also be used as a stop valve to prevent the flow of fluid from one part of the hydraulic circuit to another.

Glode valve :

— In this valve the flow area is larger than that of a needle valve.

— These valves are not suitable for throttling but can be used as a stop valves. Normally, they are not used as flow control valves.

5.3.5.5. Direction control valves

As the name implies, *the direction control valves start, stop and control the direction of flow for reversing the direction of motion of the actuator.*

- Direction control valves are employed in a hydraulic system to *determine the direction of the fluid* in the hydraulic circuit. Sometimes they are also used as a *selector switch.*

Direct control (DC) valves may be *classified* as follows:

1. On the basic of *internal valving element :*

   (*i*) Poppet (Ball or piston)

   (*ii*) Spool valve.

2. On the basis of *flow paths :*

   (*i*) Two way; (*ii*) Three way; (*iii*) Four way.

3. On the basis of actuation of *internal valving element* :

   (*i*) Manual; (*ii*) Mechanical; (*iii*) Electrical; (*iv*) Hydraulic; (*v*) Pneumatic; (*vi*) Combination of any of these.

4. On the basis of *method of connection:*

   (*i*) Pipe thread; (*ii*) Straight thread; (*iii*) Flanged or subplot; (*iv*) Manifold mounted.

5. On the basis of *size.*

Here follows the description of finite positioning directional valves which direct the fluid by opening or closing flow paths in definite valve position. Each finite position is represented graphically by a square and the arrows inside showing the path.

1. Check valve :

The check valves only allow flow in one direction and, as such, are similar in operation to *electronic diodes.*

(*i*) *Simple check valve.* It is simplest in construction and consists of a ball and seat arrangement of the valve as shown in Fig. 5.19(*a*). It is commonly used in pneumatic systems.

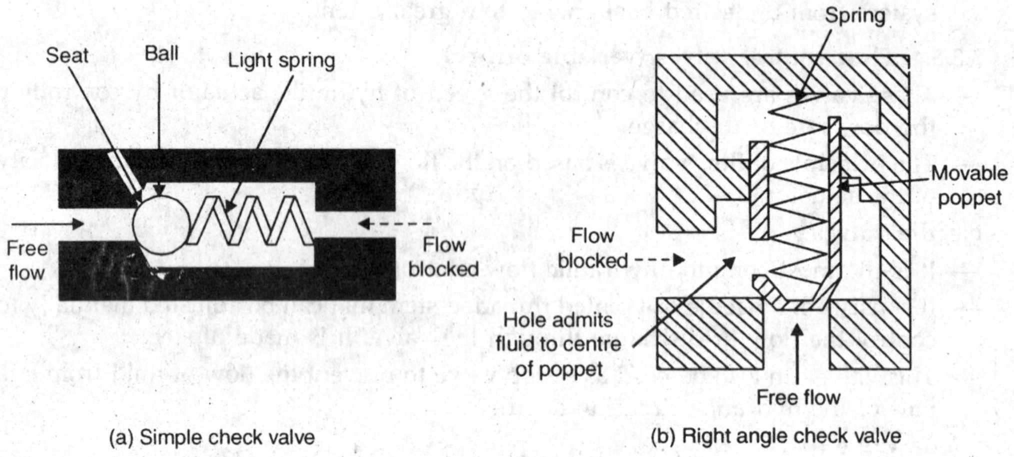

Fig. 5.19. Check valves.

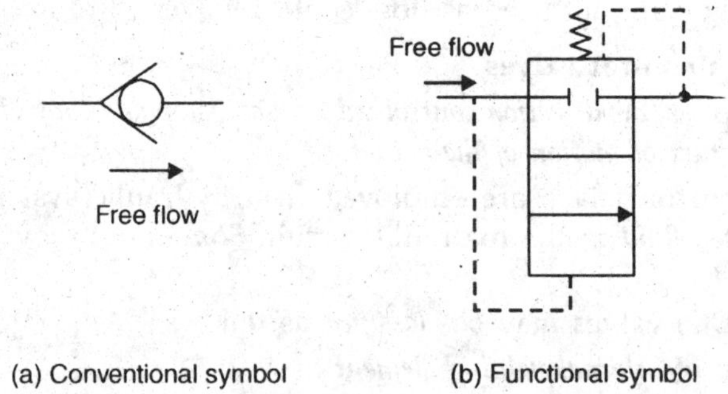

Fig. 5.20. Check valve symbols.

(*ii*) *Right angle check valve.* Its construction is shown in Fig. 5.19(*b*), it is better used to the higher pressures of a hydraulic system.

— A *pilot-operated check valve* is similar to a basic check valve but can be held open permanently by application of an external pilot pressure signal.

Fig. 5.21 shows the *application of a check valve.* In the figure is shown an hydraulic circuit with a pressure storage device called an *accumulator* (An accumulator, besides being a storage tank, supplies oil to the hydraulic actuator at constant pressure depending on the requirement). Here a check valve allows the pump to unload via the pressure regulating valve, while still maintaining system pressure.

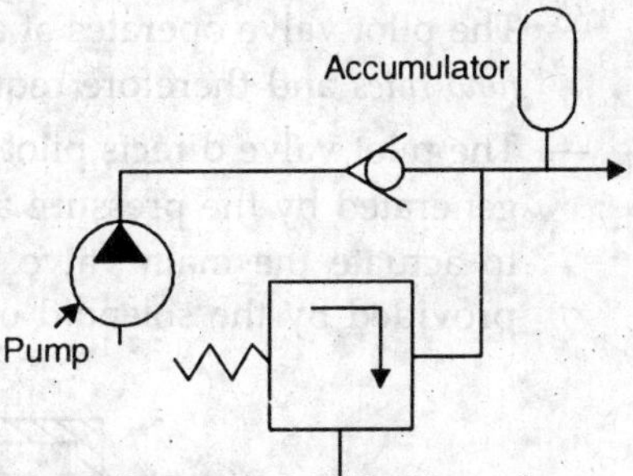

Fig. 5.21. Application of a check valve.

2. Poppet valve :

Fig. 5.22 shows a poppet valve. It is a *check valve that can be forced open to allow reverse flow.*

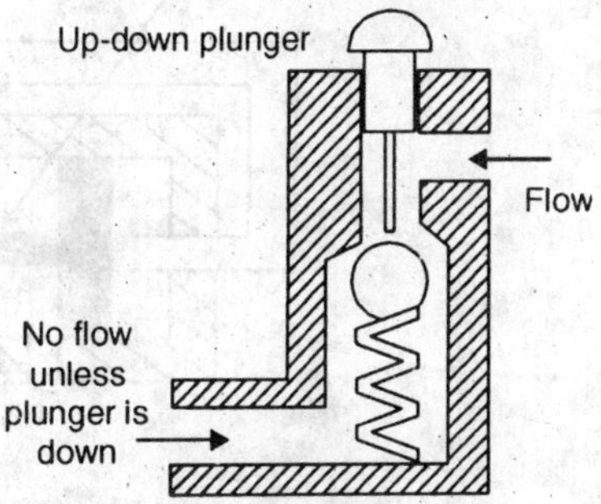

Fig. 5.22. Poppet valve.

3. Spool valves :

Fig. 5.23, shows the schematic of a spool valve:

— It consists of a cylindrical spool with multiple lobes moving within a cylindrical casing containing multiple ports.

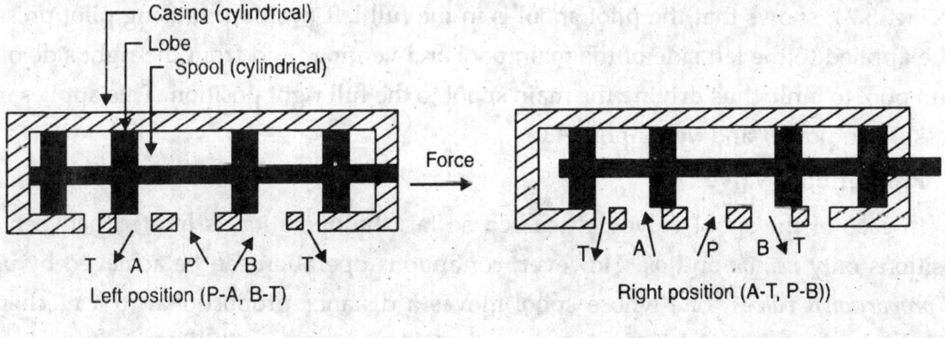

Fig. 5.23. Schematic of spool valve.

— The spool can be moved back and forth to align spaces between the spool lobes with input and output ports in the housing to direct high-pressure flow to different conduits in the system. The static pressure force on the spool is balanced since the pressure is always applied to opposing internal faces of the lobes.

— To move the spool, an axial force (from a solenoid or manual control lever) is required to overcome the hydrodynamic forces associated with changing the momentum of flow.

In the *left position,* port-*A* is pressurised and port-*B* is vented to the tank. A force is required to move the spool from this position to the *right position,* where port-*B* is pressurised and port-*A* is vented.

Pilot-operated spool valve :

Where large hydrodynamic forces occur, a *pilot valve* is added to the design of the spool valve, as shown in Fig. 5.24.

— The pilot valve operates at a *lower pressure,* called *"pilot pressure", and much lower flow rates* and therefore requires *less force to actuate.*

— The pilot valve directs pilot pressure to one side of the main spool, and the force generated by the pressure acting over the main spool lobe force is large enough to actuate the main valve. The effect of the pilot valve is to *amplify* the force provided by the solenoid or mechanical lever acting on the pilot spool.

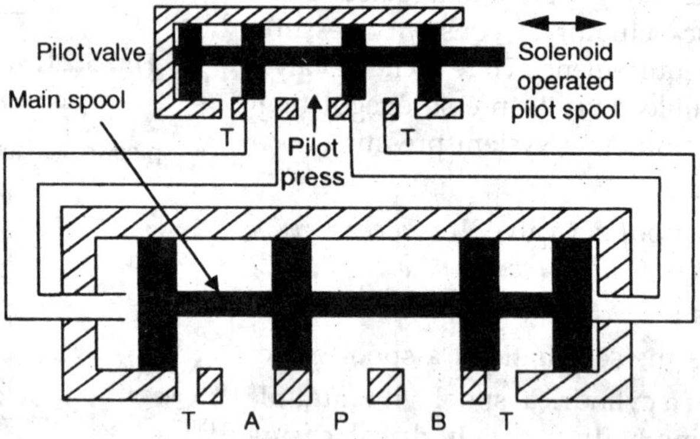

Fig. 5.24. Pilot-operated spool valve.

Fig. 5.24, shows that the pilot spool is in the full left position, causing pilot pressure to be applied to the left side of the main pool and venting fluid from the right side of the main pool to tank, thus driving the main spool to the full right position. This applies main pressure to port-*B* and vents port-*A*.

Proportional valve :

In case of the spool valves, discussed so far, the operation is limited between two positions only *i.e., on* and *off.* However, continuous operation can be achieved by using a *"proportional valve"*, one whose spool moves a distance proportional to a mechanical (say a lever) or electrical input (say an adjustable current solenoid), then changing the flow rate and varying the speed and force of the actuator.

- When the spool position is controlled by solenoids, the proportional valve is called an *"electrohydraulic valve"*. These valves may be used in open control situations with no feedback, but they often include sensors to monitor spool position or actuator output.
- *Proportional valves equipped with sensor and control circuitry* are often called *"servo-valves"*.
- Electrohydraulic valves are often pilot operated where the solenoids drive the pilot spool, which in turn controls the position of the primary spool.

Fig. 5.25 and 5.26, show the block diagrams of mechanical hydraulic servo-valve and electrohydraulic servo-valve.

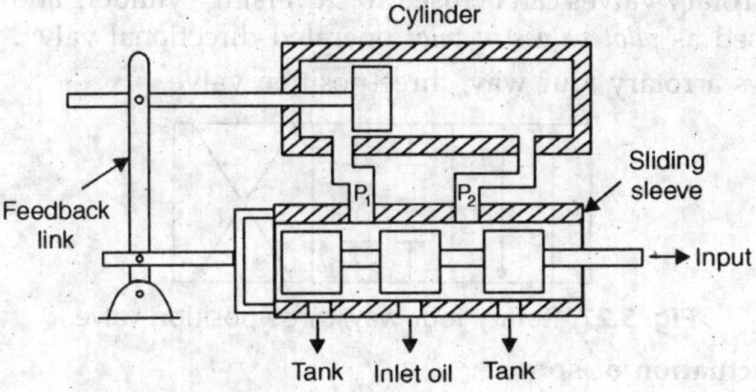

Fig. 5.25. Mechanical hydraulic servo-valve.

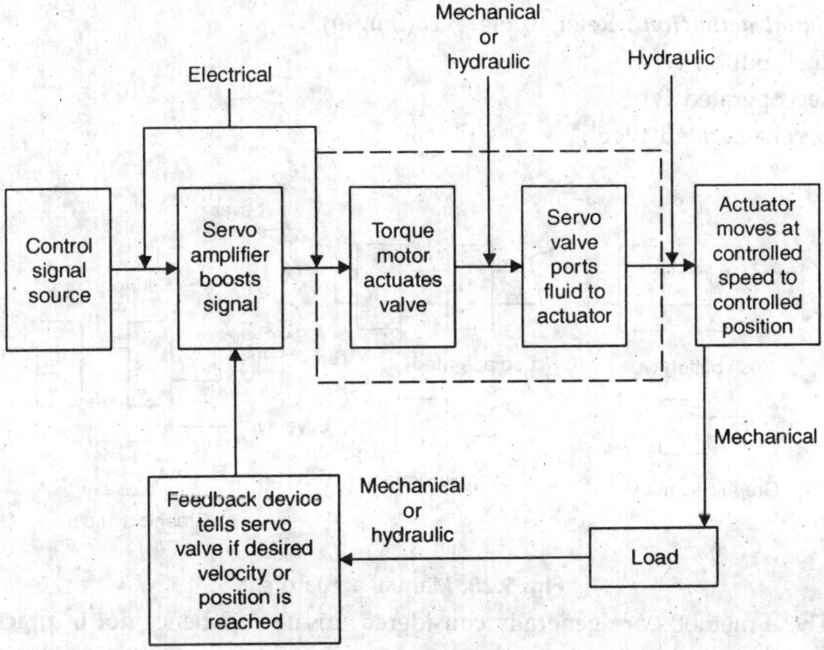

Fig. 5.26. Electrohydraulic servo-valve.

— *Electrohydraulic systems use low power electrical signals (1 W) for controlling the movements of large power hydraulic pistons (7640 W or more).* The typical applications are *aircraft controls and numerical control machines.*

4. Rotary valves :

With a rotary spool the hydraulic fluid is directed through longitudinal grooves machined into a rotatable piston. As a rotational movement is necessary to effect changeover, this type of spool is used predominantly in *manually operated valves.*

A *rotary valve* consists of a rotating spool that is fitted into a circular valve body with a close tolerance. The hydraulic fluid is directed through the longitudinal grooves or passages provided on the rotating spool. These passages connect or block the ports in the valve body and if necessary a centre position can be incorporated.

— Rotary valves can be line, panel or sub-plate mounted.

— These valves are generally *low flow valves.*

— Although rotary valves can be used for reversing cylinders and motors, these are usually used as *pilot valves in pilot* operated directional valve.

Fig. 5.27 shows a rotary four way, three-position valve.

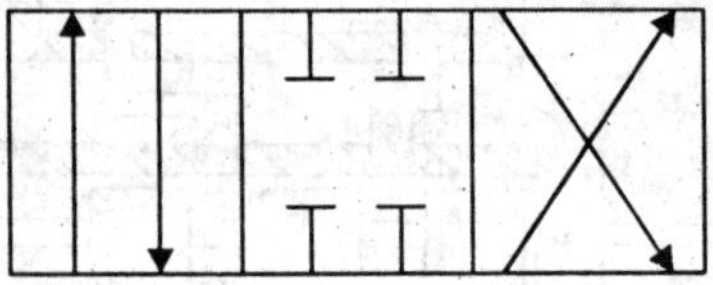

Fig. 5.27. Rotary four way, three-position valve.

Methods of actuation of spools :

The actuating force for providing axial motion to the sliding spool in a sliding spool direction control valve can be imparted by the following five methods :

1. *Manual actuation* : Refer to Fig. 5.28(*i, ii, iii*).

(*i*) Push button type;

(*ii*) Leg operated type;

(*iii*) Lever operated type .

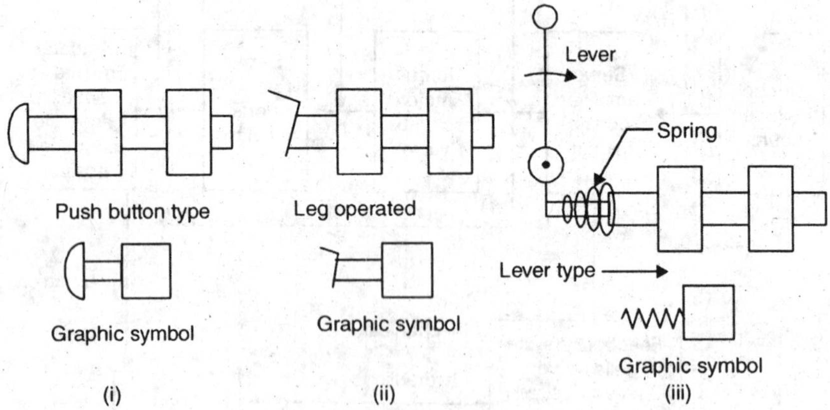

Fig. 5.28. Manual actuators.

- These methods are generally considered unsafe and hence not in practice.

2. *Mechanical actuation* : Refer to Fig. 5.29(*i, ii*).

(*i*) Plunger type;

(*ii*) Roller wheel type.

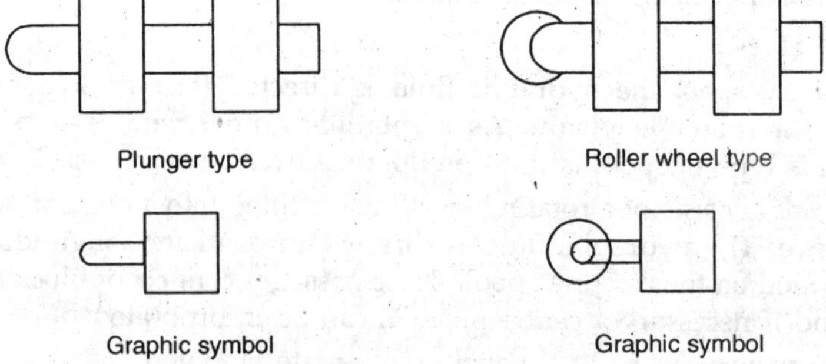

Fig. 5.29. Mechanical actuators.

3. ***Pneumatic actuation :*** Refer to Fig. 5.30.

The pneumatic actuator uses the force of air applied to a piston, which is connected to spool. As the piston moves due to air the spool actuates. In this case as the pressure is quite low the actuator piston must be relatively large to overcome the spring and flow forces.

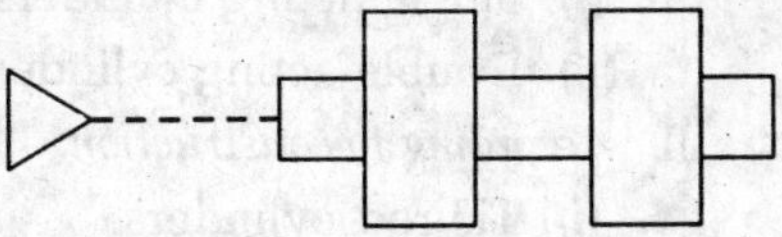

Fig. 5.30. Pneumatic actuation.

4. ***Electrical actuation :*** The most common type of electrical actuator is *"solenoid"*.

— A solenoid is made up of two parts : a coil and an armature as shown in Fig. 5.31. When coil is energised magnetic field is produced which attracts the armature, which is turn pushes the solenoid pin or the spool.

— Solenoid can be operated either on A.C. or D.C. signals. The designs may be different but the operating principle is same.

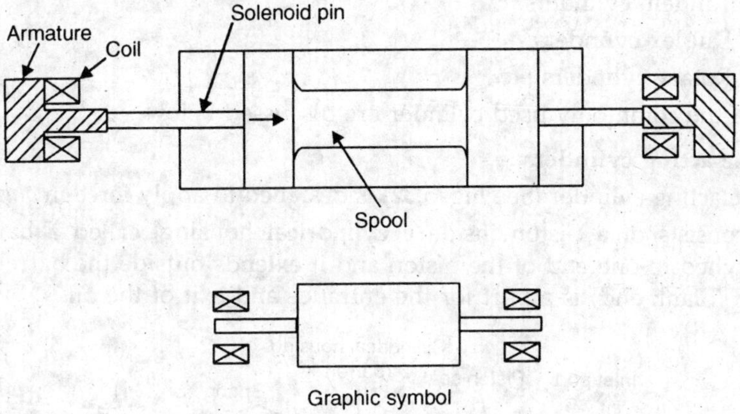

Fig. 5.31. Solenoid actuated valve.

5. ***Hydraulic actuation :*** The fluid discharged by the pump is directed to the 'hydraulic actuator'. The actuator converts the pressure energy of the fluid into mechanical energy. The hydraulic actuators are of the following two types :

(*i*) Linear actuators ;
- Hydraulic cylinder.

(*ii*) Rotary actuators ;
- Hydraulic motor.

5.3.6. Linear Actuators

A **fluid power hydraulic cylinder** *is a linear actuator which is most useful and effective in converting fluid energy to an output force in a linear direction for performing work such as pulling or pushing in a variety of engineering applications such as in machine tools and other industrial machinery, earth moving equipment, construction equipment and space applications.*

A hydraulic cylinder usually consists of a *movable element*, a *piston* and a *piston rod* operating within a cylindrical bore.

5.3.6.1. Types of cylinders

The cylinders may be *classified* as follows:

I. *According to function performed :*

(*i*) Single acting cylinders.
(*ii*) Double acting cylinders.

II. *According to construction:*
(*i*) Tie rod cylinders.
(*ii*) Mill type cylinders.
(*iii*) One-piece welded cylinders.
(*iv*) Threaded head cylinders.

III. *Special types :*
(*i*) Plunger or ram cylinders.
(*ii*) Telescoping cylinders.
(*iii*) Cable cylinders.
(*iv*) Diaphragm cylinders.
(*v*) Tandem cylinders.
(*vi*) Duplex cylinders.
(*vii*) Rotary cylinders etc.

Some of the commonly used cylinder are discussed below :

1. Single acting cylinders :

A single acting cylinder (See Fig. 5.32) is designed to apply force in *only one direction.*

— It consists of a piston inside a cylindrical housing, called a barrel. A rod is attached to one end of the piston and it extends outside the barrel. At the other end (blank end) is a port for the entrance and exit of the oil.

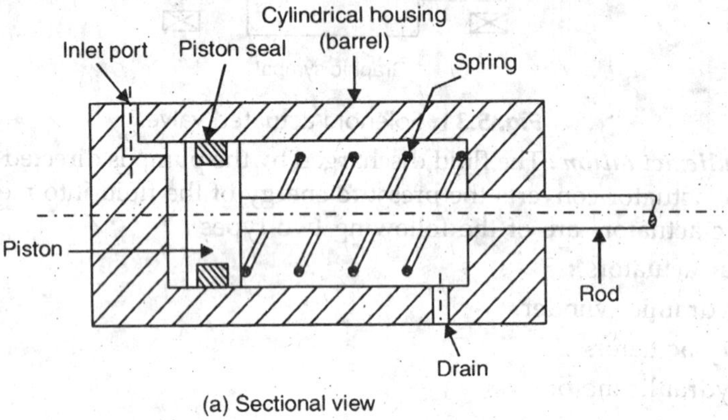

(a) Sectional view

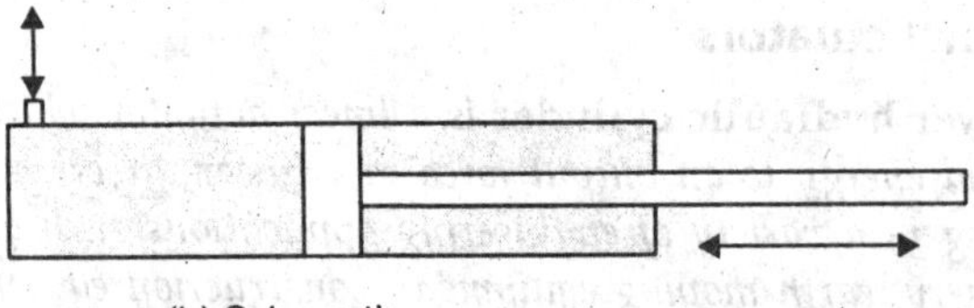

(b) Schematic

Fig. 5.32. Single acting cylinder.

— A single acting cylinder can exert a force only in the extending direction, as fluid from the pump enters through the blank end of the cylinder.

— These cylinders do not retract hydraulically. Retraction is accomplished by the inclusion of a compression spring or by using gravity.

2. Double acting cylinders :

A double acting cylinder (see Fig. 5.33) is capable of delivering forces in both directions and is most commonly used in industrial hydraulics.

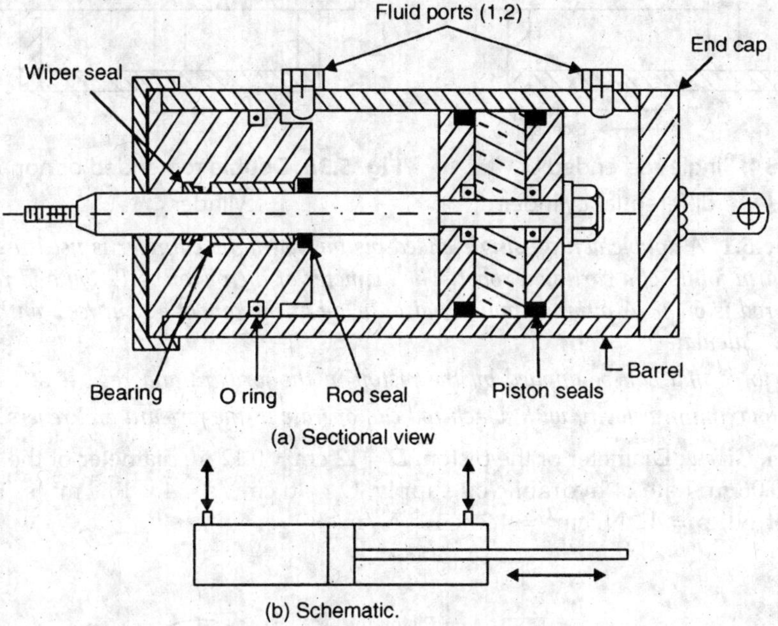

Fig. 5.33. Double acting cylinder.

— The *barrel* is made of seamless steel tubing, honed to a fine finish on the inside surface. The *piston* which is made of ductile iron contains *U* cup packings to seal the leakage between the piston and the barrel. The *ports* are located in the end caps which are secured to the barrel by tie rods. The load of the piston rod at the neck is taken by the bearing, which is generally made of brass or bronze. A rod *wiper* is provided at the end of the neck to prevent foreign particles and dust from entering into the cylinder along with the piston rod.

— As the fluid from pump enters the cylinder through port-1, the piston moves forward and the fluid returns to the reservoir from the cylinder through port-2. During the return stroke the fluid is allowed to enter the cylinder through port-2 and the fluid from the other side of the piston goes back to the reservoir through port-1.

— Double acting cylinders my be either single rod ended, (also called *differential cylinder*) or double rod ended, (also called *non-differential* cylinder) as shown in Figs. 5.34 and 5.35 respectively.

— The single rod cylinders have piston connected to a smaller diameter piston rod. For a given pressure, these cylinders exert greater force when extending than when retracting.

— The double rod ended cylinder is used when it is required to exert *equal forces in both direction*. However, the maximum force of the cylinder for a given tube size is smaller than the single rod end type.

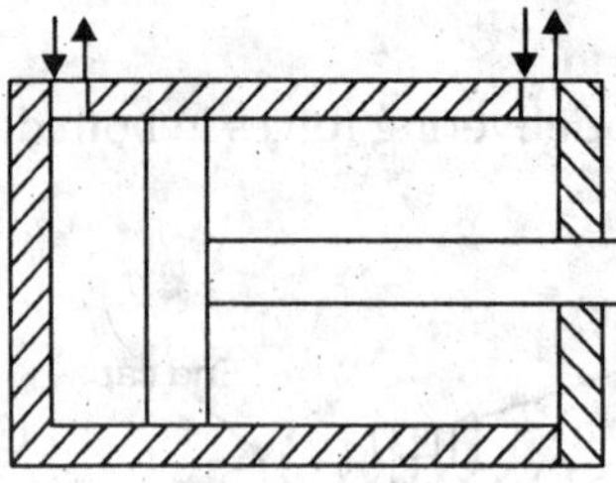

Fig. 5.34. Single rod ended or differential cylinder.

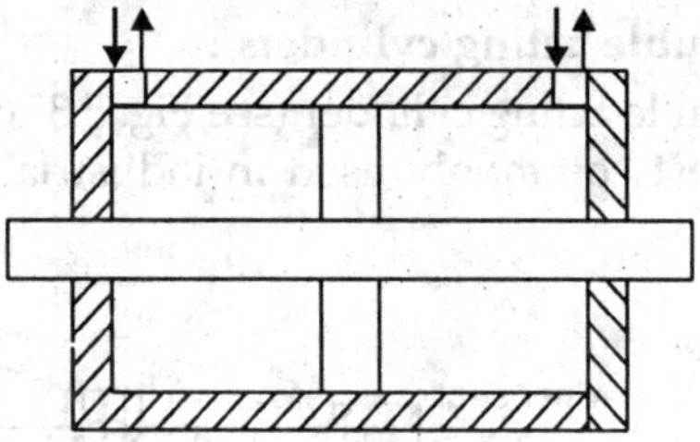

Fig. 5.35. Double rod ended or non-differential cylinder.

Example 5.1. *A double acting, single-ended piston hydraulic cylinder is used to actuate one of the linear arm joints of a cartesian robot. The diameter of the piston is 12 cm and the diameter of the piston rod is 6 cm. A pump supplies hydraulic oil at the rate of 40 cm^3/sec. with a pressure of 42 N/cm^2. Calculate :*

(*i*) *The force that can be applied by the piston in the forward and reverse strokes.*

(*ii*) *The maximum velocity with which rod can operate in the forward and reverse directions.*

Solution. *Given* : Diameter of the piston, D = 12 cm = 0.12 m; diameter of the piston rod, d = 6 cm = 0.06 m; Rate of hydraulic oil supply, Q = 40 cm^3/s = 4×10^{-5} m^3/s; The supply pressured of oil, p = 42 N/cm^2 = 4.2×10^5 N/m^2

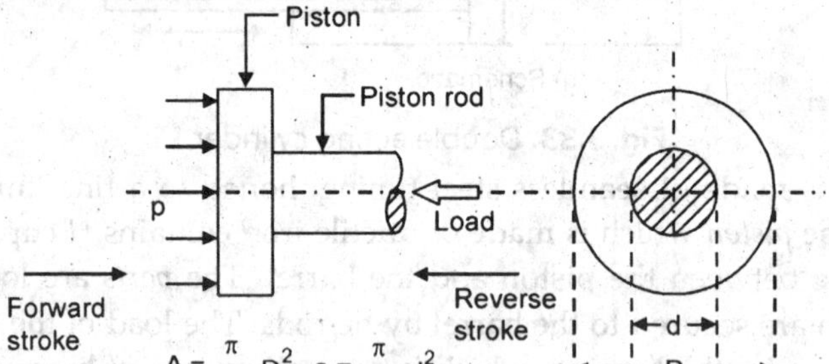

$$A = \frac{\pi}{4} D^2;\ a = \frac{\pi}{4} d^2$$

Fig. 5.36.

Refer to Fig. 5.36.

(*i*) **The force exerted by the piston :**

- The force exerted by the piston in the forward stroke,

$P_f = p \times A$, where A is the cross-sectional area of the piston

$$= p \times \frac{\pi}{4} \times D^2$$

$$= 4.2 \times 10^5 \times \frac{\pi}{4} \times (0.12)^2 = \mathbf{4750\ N} \quad \textbf{(Ans.)}$$

- The force exerted by the piston in the *reverse stroke*,

$P_r = p \times (A - a)$, where a is the cross-sectional area of the piston rod.

$$= p \times \left(\frac{\pi}{4}D^2 - \frac{\pi}{4}d^2\right) = p \times \frac{\pi}{4}(D^2 - d^2)$$

$$= 4.2 \times 10^5 \times \frac{\pi}{4}\left(0.12^2 - 0.06^2\right) = \mathbf{3562.6\ N} \quad \textbf{(Ans.)}$$

(*ii*) **The maximum velocity of the rod :**

- The maximum velocity of the rod in the forward stroke,

$$(V_{max})_f = \frac{Q}{A} = \frac{4 \times 10^{-5}}{\frac{\pi}{4} \times (0.12)^2}$$

$$= 3.54 \times 10^{-3} \text{ m/s} \quad \text{or} \quad \mathbf{3.54 \text{ mm/s}} \quad \textbf{(Ans.)}$$

- The maximum velocity of the rod in the reverse stroke,

$$(V_{max})_r = \frac{Q}{(A-a)} = \frac{4 \times 10^{-5}}{\frac{\pi}{4} \times (0.12^2 - 0.06^2)}$$

$$\simeq 4.72 \times 10^{-3} \text{ m/s} \quad \text{or} \quad \mathbf{4.72 \text{ mm/s}} \quad \textbf{(Ans.)}$$

3. Telescoping cylinders :

These cylinders are employed where *long work-strokes are required.*

A telescoping cylinder provides a relatively long working stroke for an overall reduced length by using several pistons which telescope into each other.

Figure 5.37, shows a two-stage double acting telescoping cylinder :

— Fluid for the retraction stroke is fed into port-*A* and passes through the hollow piston rod into the annulus behind the first stage piston. So the first stage piston is forced to the left until it uncovers the fluid ports connecting this with the second stage annulus, thereby moving the larger piston to the left until both the pistons are fully retracted into the body of the cylinder.

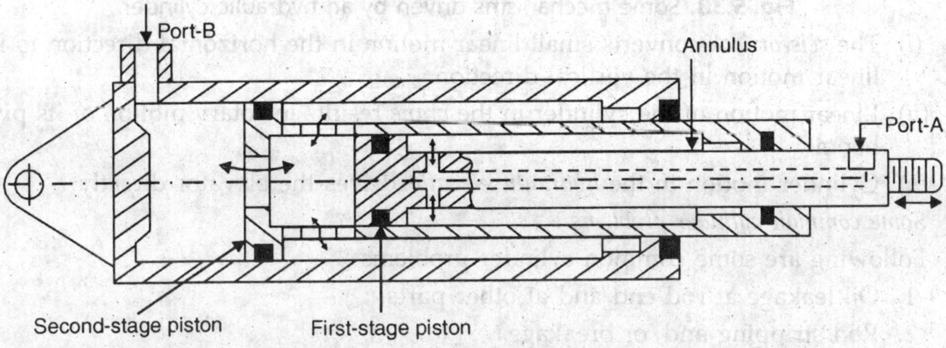

Fi.g 5.37. A two-stage double acting telescoping cylinder.

— Fluid for the extension stroke is then fed through port-*B*, forcing both pistons to the right until the cylinder is fully extended.

Cylinder ratings :

The ratings of cylinder are based on its size and pressure capability.

— The principal sizes features are :
(*i*) Piston diameter; (*ii*) Piston rod diameter; (*iii*) Stroke length.

— The pressure rating is established by the manufacturer and it is available in the manufacturer's catalogue.

Hydraulic cylinders applications :

The following are the *applications* of hydraulic cylinders:

1. Hydraulic jacks.

2. Aircraft landing system.
3. Hydraulic shapers and many machine tools.
4. Oceanography applications.
5. Tusk tabler for handling huge logs.
6. Power steering control for off-highway vehicles.
7. Earth excavators.
8. Automobile hoisting.

- As illustrated in Fig. 5.38, the *linear actuator* is very versatile in achieving a variety of motions.

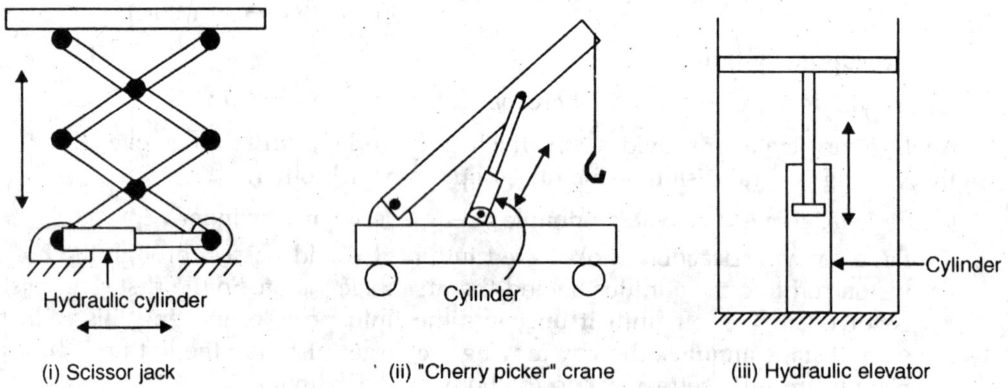

Fig. 5.38. Some mechanisms driven by an hydraulic cylinder.

(*i*) The *scissor jack* converts small linear motion in the horizontal direction to larger linear motion in the vertical direction.

(*ii*) Linear motion of the cylinder in the *crane* results in rotary motion of its pivoted boom.

(*iii*) Cylinder motion in the *hydraulic elevator* drives the elevator directly.

Some common cylinder problems :

Following are some common cylinder problems :

1. Oil leakage at rod end and at other parts.
2. Rod stripping and/or breakage.
3. Sticky, slow start-up cushion.
4. Tubing end leakage.
5. Problems associated with link mechanism and machine members.
6. Premature seal wear out.

5.3.7. Rotary Actuators

Rotary actuators *are the hydraulic pneumatic equivalents of electric motors.* For a given torque, or power, a rotary converter :

— is more compact than an equivalent motor;

— cannot be damaged by an indefinite stall;

— can be safely used in an explosive atmosphere.

For variable speed applications, the complexity and maintenance requirements of a rotary actuator are similar to a thyristor-controlled D.C. drive, but for fixed applications, the A.C. induction motor is simpler to install and maintain.

5.3.7.1. Hydraulic motors

An ***hydraulic motor*** *is a rotary actuator wherein hydraulic energy is converted into mechanical energy in the form of rotary motion and torque which may be used for doing work.* Thus, its function is just opposite to that of an hydraulic pump.

These motors very closely resemble pumps in construction. In an hydraulic motor high pressure oil pushes a freely moving element, thereby developing torque and continuous rotating motion. Since both inlet and outlet ports at times be pressurised, therefore, most hydraulic motors are drained externally.

- The maximum performance of a motor in terms of pressure, flow, torque, speed, efficiency, expected life and physical configuration is determined by the following factors :
  (*i*) Leakage characteristics;
  (*ii*) Efficiency of the means used to convert pressure surface and output shaft;
  (*iii*) Ability of the pressure surface to withstand hydraulic force.
- Hydraulic motors are *specified or rated* by :
  (*i*) Maximum operating pressure;
  (*ii*) Displacement (volume size);
  (*iii*) Speed;
  (*iv*) Torque capacity.

Rotary actuator symbols :

Fig. 5.39, shows the rotary actuator symbols :

— *Internal leakage always occurs in a 'hydraulic motor'* and a drain line, shown dotted, is used to return the leakage fluid to the tank. If this leakage return is inhibited the motor may pressure lock and cease to rotate or even suffer damage.

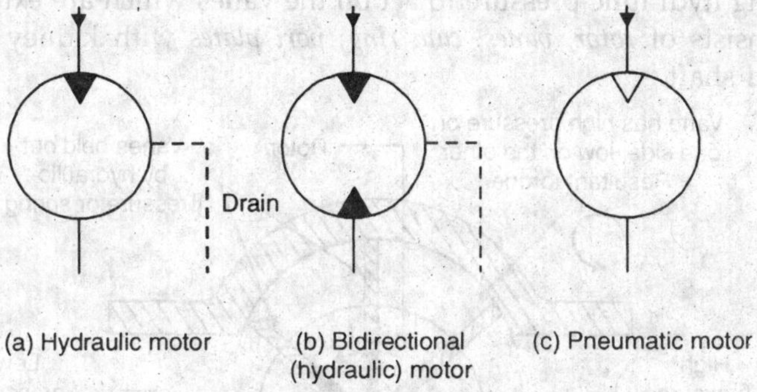

Fig. 5.39. Rotary actuator symbols.

Classification of hydraulic motors :

Hydraulic motors are *classified* as follows :

1. Gear motors.
2. Vane motors.
3. Piston motors :
   (*i*) Radial type;
   (*ii*) Axial type.

1. *Gear motors :*

Fig. 5.40, shows the schematic of a gear motor.

— Fluid enters at the top and pressurises the top chamber. Pressure is applied to two gear faces at A, and a single gear face at B. *There is, thus, an imbalance of forces on the gears resulting in rotation* as shown.

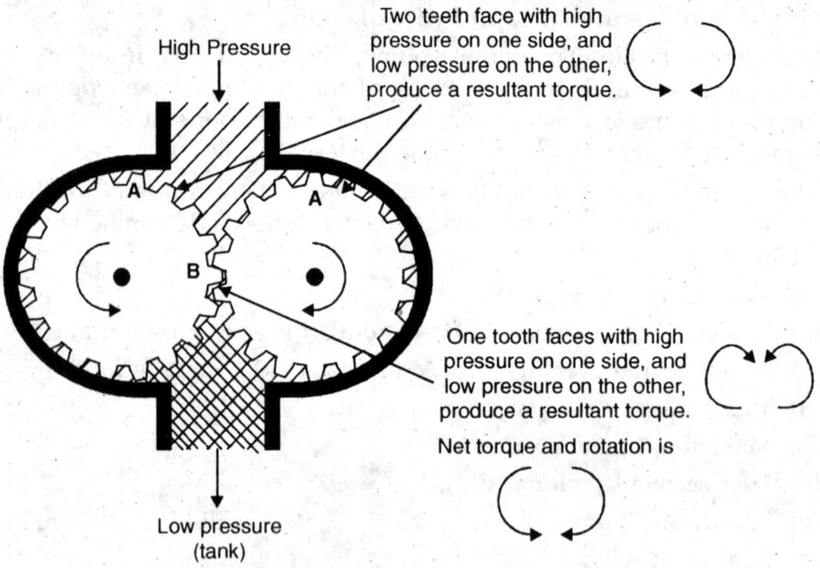

Fig. 5.40. A gear motor.

— Gear motors *suffer from leakage* which is more pronounced at low speed. Thus they *tend to be used in medium speed, low torque applications.*

2. *Vane motors :*

A vane motor is a *positive displacement motor* which develops an output torque at its shaft by allowing hydraulic pressure to act on the vanes which are extended. Basically a vane motor consists of *rotor, vanes, cam ring, port plates* with kidney shaded inlet and outlet ports and shaft.

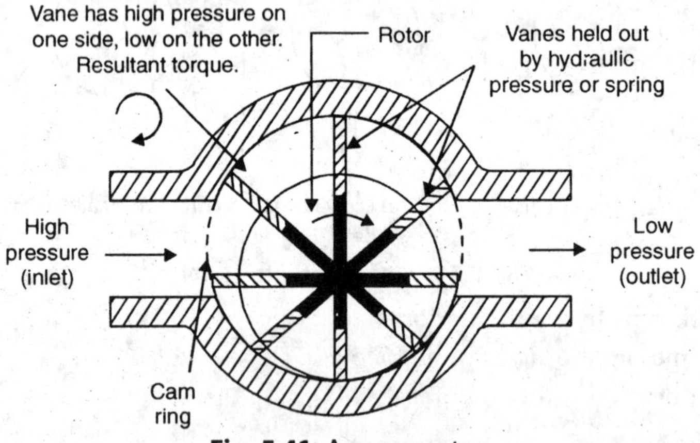

Fig. 5.41. A vane motor.

Fig. 5.41 shows the schematic of a vane motor:

— Vane motors develop torque by the hydraulic pressure that acts on the exposed surfaces of the vanes which slide in and out of the rotor connected to the driver shaft. Larger the exposed surface of the vane or higher the pressure of oil, more will be the torque developed.

— Since no centrifugal force exists until the motor begins to rotate, some means have to be provided to initially hold the vanes against the casting contour. Springs are often used for this purpose.

— In vane motor two different pressures, system pressure and outlet pressure are involved. System pressure will be greater than the outlet pressure resulting in *side loading* in the motor shaft. The side loading in the motor is avoided by using cam-shaped ring, instead of a circular ring. With this arrangement the two pressure quadrants oppose each other and the forces acting on the shaft are *balanced* such motors are balanced vane motors.

3. *Piston motors* :

- In *swash plate type*, the motor drive shaft and cylinder block are centered on the same axis. Pressure at the ends of the multiple pistons causes a sequential action against a tilted swash plate and rotates the cylinder block and the motor shaft.
- The operation of a *bent axis piston motor* is similar to the swash plate type, except that the thrust of the pistons is transferred against the drive shaft.
- In the *radial piston*, motors have a cylinder block with an attached output shaft to transmit the force imparted to the pistons. The cylinder block has an odd number of radial bores with precision fitted pistons.

  — When oil enters the cylinder bore, the piston is forced against the thrust ring, imparting a tangential force to the cylinder block and shaft, causing the assembly to rotate. Each piston is pushed inward by the thrust ring once it reaches the outlet port, thus pushing the fluid to the reservoir.

Semi-rotary/Limited motion rotary actuators:

These actuators are used to convert fluid pressure energy into torque which turns through an angle limited by the design of the actuator. With majority of the designs, the angle of rotation is within 360° although it is possible to considerably exceed this when using piston-operated actuators.

Some examples are illustrated in Figs. 5.42(*a*)(*b*).

— In Fig. 5.42(*a*), a double-acting piston is coupled to the output shaft by a rack in piston.

— The actuator shown in Fig. 5.42(*b*) is driven by a single vane coupled to the output shaft.

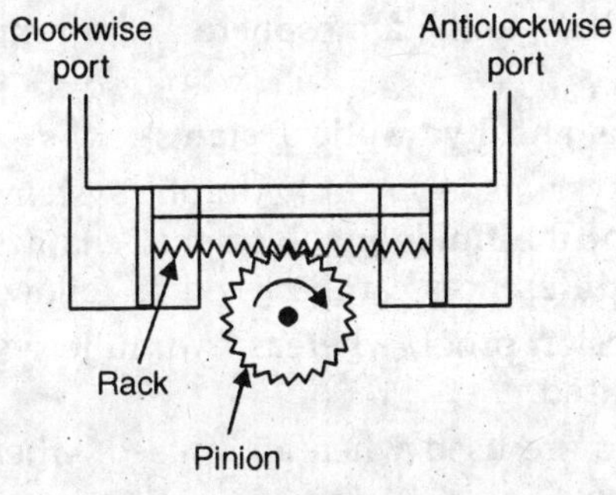

(a) Dual-acting piston type

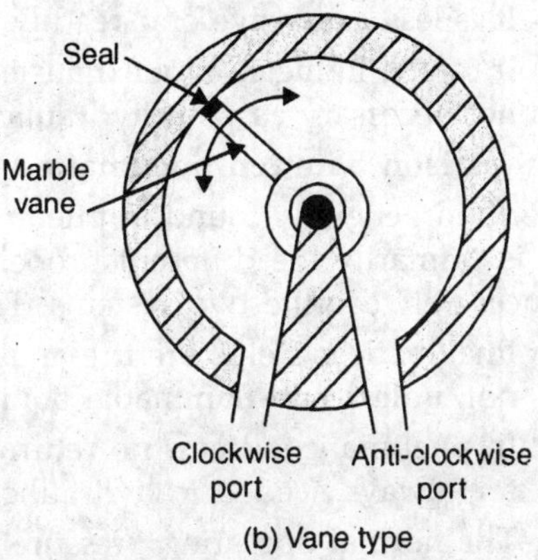

(b) Vane type

Fig. 5.42. Semi-rotary actuators.

In both the above cases the *shaft angle can be finelly controlled by fluid applied to the ports.*

5.3.7.2. Advantages and applications of hydraulic motors

Advantages:

1. These motors are explosion proof. An hydraulic motor requires no electrical connections except only two hydraulic lines.
2. It is easy to achieve frequent stopping, starting and reversing by a single four-way three position direction control valve.
3. Stepless variations in speeds are available from zero to that obtainable from the maximum output of pumping system.
4. The hydraulic motor can be suddenly stopped (by closing the control valve) without any harm to the mechanism.
5. The hydraulic motors can be employed for machines requiring controlled variable torque (*e.g.*, paper winding machines).

Applications:

The fields of applications of hydraulic motors include:

(*i*) Mining equipment;
(*ii*) Machine tools;
(*iii*) Pottery machines;
(*iv*) Drill rigs;
(*v*) Winding machines etc.

Note: • Hydraulic systems have the ***advantage*** of generating *extremely large forces from very compact actuators.* They also can provide *precise control of low speeds* and have built in travel limits defined by the cylinder stroke. • The ***drawbacks*** of hydraulic systems include the need for a *large infrastructure* (high-pressure pump, tank, and distribution lines); potential for *fluid leaks,* which are undesirable in a clean environment; *possible hazards,* associated with high pressures (*e.g.*, a ruptured line); *noisy operation, vibration, and maintenance requirements.* Because of *these disadvantages, electric motor drives are often the preferred choice.* However, in *large systems, which require extremely large forces, hydraulics often provide the only alternative.*

5.4 PNEUMATIC ACTUATORS

5.4.1. Introduction

Pneumatic systems *use pressurised air to transmit and control power.*

Air is used as the fluid because :

(*i*) It is safe.
(*ii*) It is less expensive and readily available.
(*iii*) It can be inducted and exhausted directly to the atmosphere and a return line is not necessary as with hydraulics.

Comparison between pneumatic systems and hydraulic systems :

The fluid generally found in pneumatic systems is *air,* in hydraulic systems it is *oil.* And it is primarily the different properties of the fluids involved that characterise the differences between the two systems. These differences can be listed as follows:

1. Pneumatic systems are fire-and-explosion proof, whereas hydraulic systems are not, unless non flammable liquid is used.
2. In pneumatic systems no return pipes are used when air is used, whereas they are always needed in hydraulic systems.
3. The normal operating pressure of pneumatic systems is very much lower than those of hydraulic systems.

4. Pneumatic systems are insensitive to temperature changes, in contrast to hydraulic systems, in which fluid friction due to viscosity depends greatly on temperature. Normally operating temperatures for pneumatic and hydraulic systems are 5°–60°C and 20°–70°C respectively.
5. The normal operating pressure of pneumatic systems is very much lower than those of hydraulic systems.
6. Accuracy of pneumatic actuators is poor at low velocities whereas accuracy of hydraulic actuators may be made satisfactory at all velocities.
7. Output powers of pneumatic systems are considerably less than those of hydraulic systems.
8. In pneumatic systems, external leakage is permissible to a certain extent, but internal leakage must be avoided because the effective pressure difference is rather small. In hydraulic systems internal leakage is permissible to a certain extent, but external leakage must be avoided.

☞ • *"Pneumatically actuated robots" are usually limited to small devices that are used to move light payloads from point to point without regard to the actual trajectory.* High-pressure compressed air permits fast, accurate, and generally simple moves (*e.g.*, non-coordinated) with mechanical limit stops often being utilised to stop the individual joints. However, *where large loads, requiring a controlled trajectory, must be handled either servomotors or hydraulic actuators are normally employed.*

5.4.2. Components of a Pneumatic System

Fig. 5.43, shows the various components of a pneumatic system; these are :

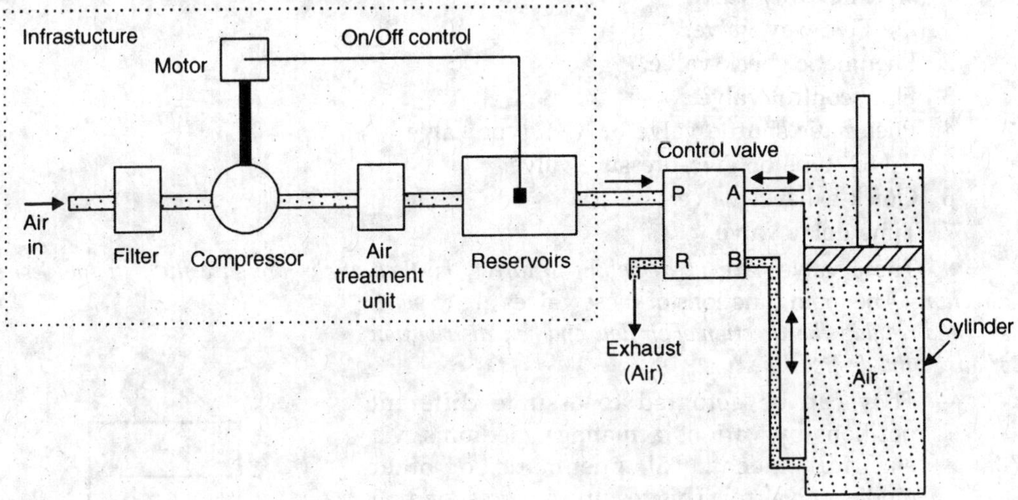

Fig. 5.43. Components of a pneumatic system.

1. **Compressor.** It is used to provide pressurised air, usually of the order of 500 kPa to 1.0 MPa, which is much lower than hydraulic system pressures.

   As a result of the lower operating pressures, pneumatic actuators *generate much lower forces* than hydraulic actuators.
2. **Air treatment unit.** After the inlet air is compressed, excess moisture and heat are removed from the air with an air treatment unit.

3. **Reservoir.** Unlike hydraulic pumps, which provide positive displacement of fluid at high pressure on demand, compressors cannot provide high volume of pressurised air responsively; therefore a large volume of compressed air is stored in a reservoir or tank.

   The reservoir is equipped with a *pressure sensitive switch* that activates the compressor when the pressure starts to fall below the desired level.

4. **Control valve and actuator.** *Control valves* and *actuators* act in much the same way as in hydraulic systems, but instead of returning fluid to tank, the air is simply returned (exhausted) to the atmosphere.

Following points are worth noting :

- Pneumatic systems are *open systems*, always processing new air (whereas hydraulic systems are closed systems, always returning the same oil).
- Since air is compressible, pneumatic cylinders are not typically used for applications requiring accurate motion between the end points, especially in the presence of a varying load.

5.5.3. Pneumatic Valves

In order to control pneumatic actuators, the air energy has to be regulated, controlled and reversed with a predetermined sequence. This is achieved with the help of pneumatic valves; these are enumerated and discussed as follows :

1. Direction control valves :
   (*i*) Two-way valve
   (*ii*) Three-way valve
   (*iii*) Four-way valve
   (*iv*) Five-way valve.
2. Pneumatic check valve.
3. Flow control valve.
4. Pneumatic shuttle valve or 'OR' type valve.
5. 'AND' type or Two-pressure valve.
6. Quick exhaust valve.
7. Time delay valve.

- These valves are *used mainly to direct the flow of the pressure fluid in the desired directions.* The main functions of these valves are to *start, stop and regulate the direction of air flow and help distribution of air in the desired line.*

— They can be actuated to assume different positions by various actuating mediums, viz electrical, mechanical, pneumatic, or other modes of control. This results in corresponding connection or disconnection of flow between various port openings. The various types of these valves are discussed below :

1. Directional control (D.C.) valves :

(*i*) *Two way valve* : Fig. 5.44 shows the symbolic representation of the valve. This is an *on-off* type of device.

— It is usually provided with two external ports, a supply port and an exhaust port.

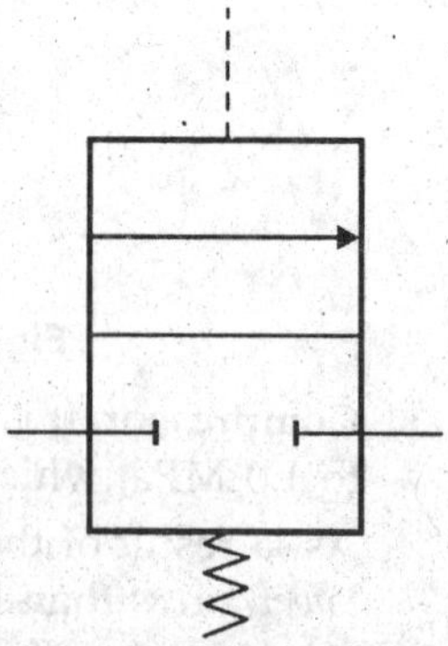

Fig. 5.44. Two-way direction control valve-poppet type.

— A normally open two-way valve permits the flow in its normal or in its rest position and blocks flow when actuated.

The normally closed valve blocks flow in its normal position and permits the flow when actuated.

(*ii*) *Three-way valve:* Refer to Fig. 5.45.

In this type of valve, one flow port is connected to either of the two ports. It may be used alternatively to pressurise one port and exhaust the other port.

These valves may be used :

- As a pilot relay to operate the other valves;
- To control single acting cylinder or in pairs to control double acting cylinders.

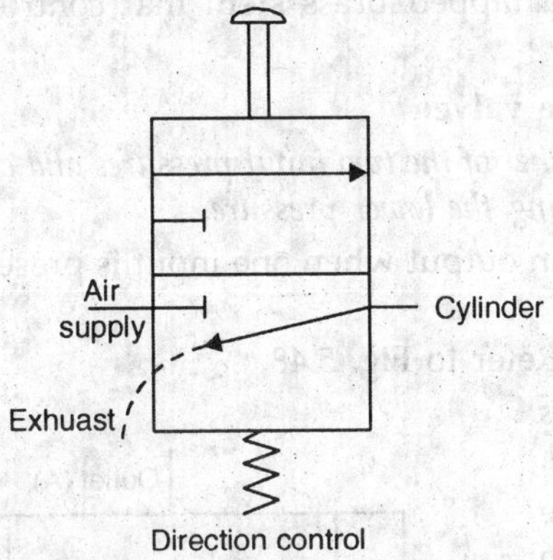

Fig. 5.45. Three-way valve-poppet type.

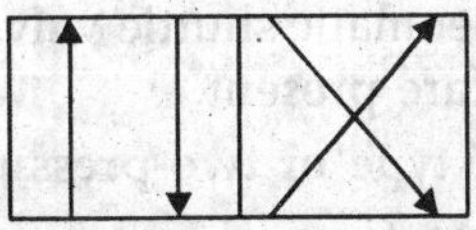

Fig. 5.46. Four-way valve-seat type

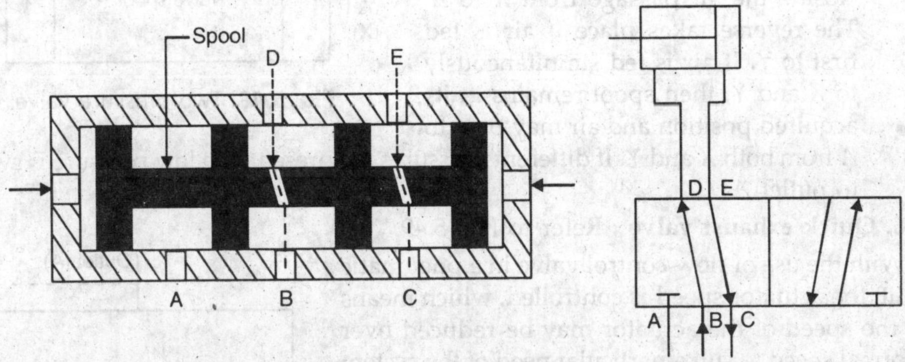

(a) Sectoral view

(b) Symbol

Fig. 5.47. Five-way valve–spool type.

(*iii*) *Four-way valve:* Refer to Fig. 5.46.

— A four-way valve has two working ports, a supply port and an exhaust port.

— In one position the valve allows air to flow from the supply source to one of the working ports. Simultaneously air is permitted to flow from the other working port to exhaust. The flow paths are reversed when the valve is shifted.

(*iv*) *Five-way valve:* Refer to Fig. 5.47.

This valve design permits the use of either dual supply or dual exhaust.

— Dual supply ports permit the use of different pressures for the cylinder movement.

— Dual exhaust enables easy exhaustion of the valve.

2. Pneumatic check valve :

The function of this valve is to *shut off* against reverse flow and *open* at a low working pressure in the forward direction.

Such valves having metal or light weight plastic body designs are available.

3. Flow control valve :

— This type of valve has a spring loaded disc which allows a *free flow in one direction and an adjustable or controlled flow in the opposite direction.*

— Flow adjustment is performed by a tapped brass stem that controls the flow through the cross hole in the disc.

4. Pneumatic shuttle valve or 'OR' type valve:

— This valve *automatically selects the higher of the two input pressures and connects that pressure to the output port while blocking the lower pressure.*

— A pneumatic shuttle valve delivers an output when one input is present or when both are present.

5. 'AND' type or two-pressure valve : Refer to Fig. 5.48.

In this type of valve, *an output is produced if both the input signals are fed.*

- This has two inlets (*X* and *Y*) and one outlet (*A*). When signal is fed first to *X*, the valve spool moves towards *Y*, closing the air passage from *X* to *A*. The reverse takes place if air is fed first to *Y*. If air is fed simultaneously to *X* and *Y*, then spool remains in its acquired position and air may pass to *A* from both *X* and *Y*. If different pressures are present, the low pressure is switched to outlet *A*.

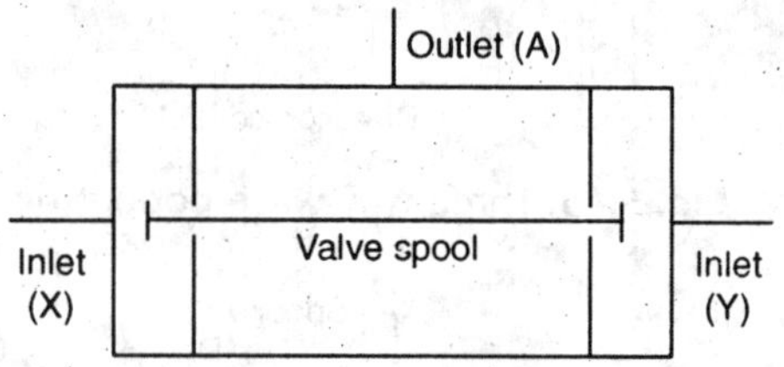

Fig. 5.48. Two-pressure valve.

6. Quick exhaust valve : Refer to Fig. 5.49.

With the use of flow control valve in a pneumatic circuit, the actuator speed is controlled, which means that the speed of the actuator may be reduced over its normal speed to suit a particular need of the system design. But this valve is *used to induce a higher speed in a cylinder by allowing the exhaust air to pass through the direction control valve from the cylinder, so that air energy can act quickly.*

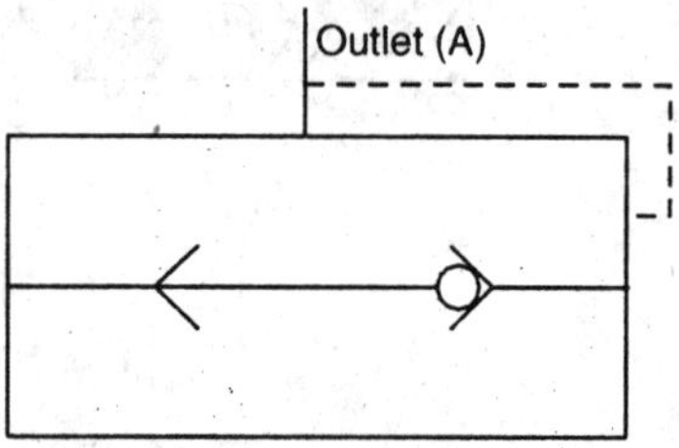

Fig. 5.49. Quich exhaust valve.

When air is fed to the piston side of the cylinder, air in the rod end of the cylinder exhausts to atmosphere quickly by using this valve. Here, the air flowing to the cylinder from the direct control valve will pass to the *port of the quick exhaust valve* and from here to the *port of the valve* and then to cylinder. But the return air from the cylinder will exhaust to the atmosphere without travelling through the exhaust valve port and thus avoids direct control valve. So the resistance to piston movement is eliminated to some extent and speed of the cylinder is *accelerated proportionally to the amount of reduced resistance.*

7. Time delay valve : This valve is used in the pneumatic system to *initiate a delayed signal.*

5.4.4. Linear and Rotary Actuators

5.4.4.1. Linear actuators—Pneumatic cylinders

The pneumatic cylinders *convert the pneumatic power into straight-line reciprocating motions.*

Pneumatic cylinder construction makes extensive use of aluminium and other non-ferrous alloy materials to reduce the weight and the corrosive effects of air and to improve heat transfer capabilities.

Air cylinders, according to the operating principle, are ***classified*** and described as follows : Refer to Fig. 5.50.

1. Single-acting cylinder.
2. Double-acting cylinder.
3. Tandem cylinder.
4. Three position cylinder.
5. Through rod cylinder.
6. Adjustable stroke cylinder.
7. Telescoping cylinder.

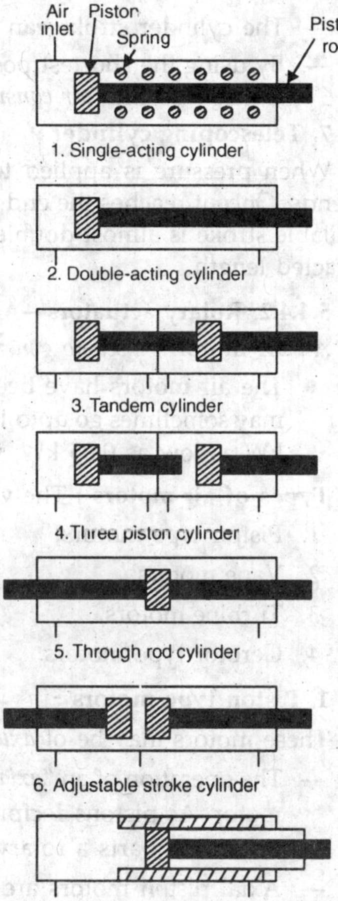

Fig. 5.50. Basic cylinder types.

1. Single-acting cylinder :

In a single-acting cylinder, the compressed air is fed only in one side. Hence, this cylinder can produce work only in one direction. The return movement of the piston is effected by a built in spring or by application of an external force. The spring is designed to return the piston to its initial position with a sufficiently high speed.

— The *advantage* of a single-acting cylinder lies in its reduced air consumption, since air is not wasted while retracting the piston.

2. Double-acting cylinder :

In this type of cylinder, the force exerted by the compressed air moves the piston in *two directions.* This cylinder produces less force during retraction, because the piston rod's cross-sectional area is subtracted from the piston area under pressure.

In principle, the stroke length is unlimited, although buckling and bending must be considered before we select a particular size of piston diameter, rod length and stroke length.

— These are used particularly when the piston is required to perform work not only on the advance movement but also on the return.

3. Tandem cylinder :

Here two cylinders are arranged in series so that the force obtained from the cylinder is almost double.

— Since the available force is doubled, this design is useful when *larger forces are required,* but a single cylinder with a larger diameter cannot be accommodated.

4. Three position cylinder :

A three position cylinder is quite similar to the tandem cylinder, except that the left-piston rod is *not connected* to the right piston and the left cylinder is *shorter* than the right one.

With the left piston extended, the retraction of the right piston is limited to an intermediate position which is determined by the ability of the right-piston to retract fully.

5. Through rod cylinder : Here the piston rod is extended on both ends of the piston. This will *ensure equal force and speed on both sides of the cylinder.*

6. Adjustable stroke cylinder :

— The cylinder stroke can be adjusted by screwing the left hand piston in or out.

— By using the shortest possible stroke needed for a given job, better rapid cycling is achieved and *air consumption is reduced.*

7. Telescoping cylinder :

When pressure is applied to the left side, the inner cylinder acts as a piston and extends. Once it reaches the end of its stroke, the inner most piston begins to extend. The available stroke is almost double when compared to a normal cylinder having the same retracted length.

5.4.4.2. Rotary actuators—Air motors

An **air motor** *is used to generate rotational motion in a pneumatic system.*

- The air motors have been found to provide very high rotational speeds, which may sometimes go upto 10,000 rpm. These motors are manufactured with fractional kW as low as 0.05 kW, while the higher range is upto 20 kW.

Types of air motors : The various types of air motors are :

1. Piston type motors.
2. Vane motors.
3. Turbine motors.
4. Gerotor type motors.

1. Piston type motors :

These motors may be of *axial* or *radial* type design :

— The operation of an *"axial"* piston air motors is similar to the piston type hydraulic motor. As pistons reciprocate in sequence, they actuate a wobble plate and this in term imparts a rotary motion to the output shaft through a gear train.

— Axial piston motors are low power (2.5 kW) motors while *radial* piston motors give upto 18 kW.

— *"Radial" piston motors are low speed motors.*

— Piston type motors may have 4, 5, or 6 cylinders. The power developed by these motors is dependent on the *inlet pressure, number of pistons, the area of the pistons, the stroke of the pistons* and *the speed.*

— The 5-cylinder design provides an *even torque* at any given operating speed due to the overlap of the five power impulses occurring in the stroke revolution.

2. Vane motor : Fig. 5.51 shows a vane motor.

— An eccentric rotor has slots in which vanes are forced outwards against the walls of the cylinder by rotation. The vanes divide the chamber into separate compartments which increase in size from the inlet port round to the exhaust port. The air entering such a compartment exerts a force on a vane and causes a rotor to rotate.

— The motor can be made to *reverse* its direction of rotation by using a *different inlet port.*

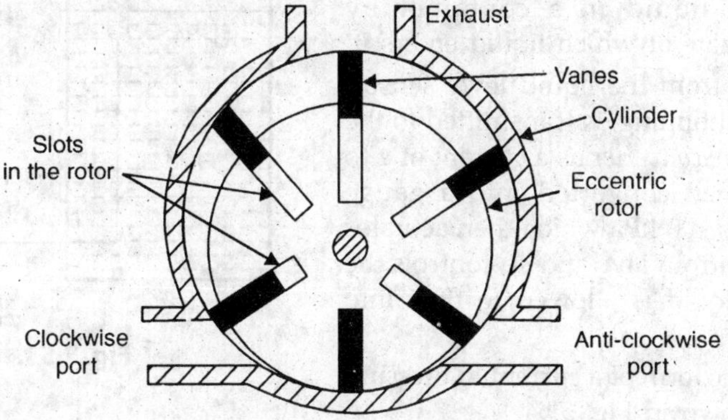

Fig. 5.51. Vane motor.

3. Turbine motors :

These motors *convert low velocity high pressure air to high velocity low pressure air by passing it through metering nozzles.* The advantage of this arrangement is that there is no rubbing or sliding contact between the rotating parts and the body cavity. This *reduces wear and lubricated air is not required to seal and lubricate parts.*

- These are *high speed low torque motors* for the same volume of air than piston vane type.

4. Gerotor type motor : A gerotor type motor is shown in Fig. 5.52.

- These air motors are mostly used for *low r.p.m.* (such as 20 to 30 r.p.m.) pressure, hence they *may not be suitable for high torque application.*

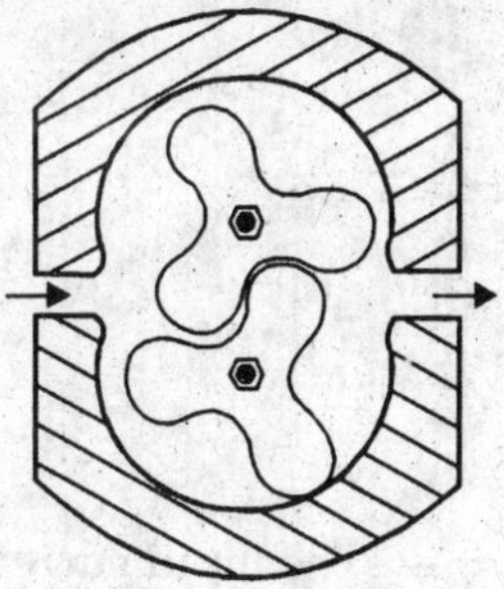

Fig. 5.52. Gerotor motor.

Applications of air motors :

The air motors may be used in conjuction with hydraulic power units in :

— Conveyor belts;
— Agitators and mixers;
— Pipe threaders;
— Tool devices;
— Bench grinders etc.

5.4.5. Special Features of Pneumatic Actuators

Pneumatic actuators should possess the following *special features* :

1. Better heat transfer capability.

2. Higher fatigue life.
3. Simple in construction.
4. High reliability against failure.
5. Should be made of anticorrosive materials.
6. Light weight (so that they are easier to manipulate).

5.4.6. Example of Fluid Control System

Fig. 5.53 shows the essential features of a system for the control of a variable such as the level of a liquid in a container by controlling the rate at which liquid enters it.

The output from the liquid level sensor, after signal conditioning, is transmitted to the *current to pressure converter* as a current of 4 to 20 mA. It is then converted into a gauge pressure of 20 to 100 kPa which then actuates a pneumatic control valve and so controls the rate at which liquid is allowed to flow into the container.

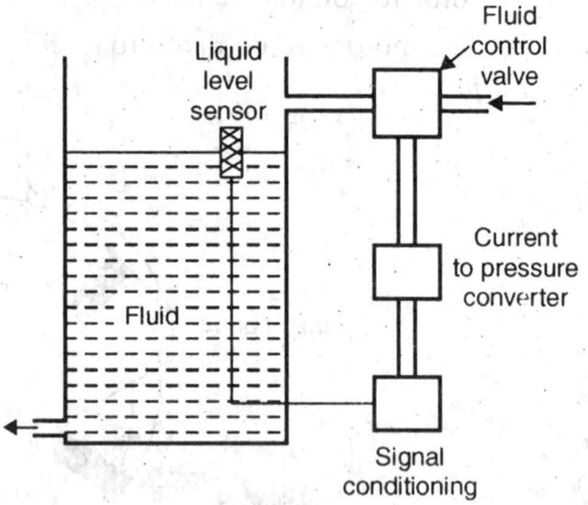

Fig. 5.53.

- The basic form of a current to pressure converter is shown in Fig. 5.54.

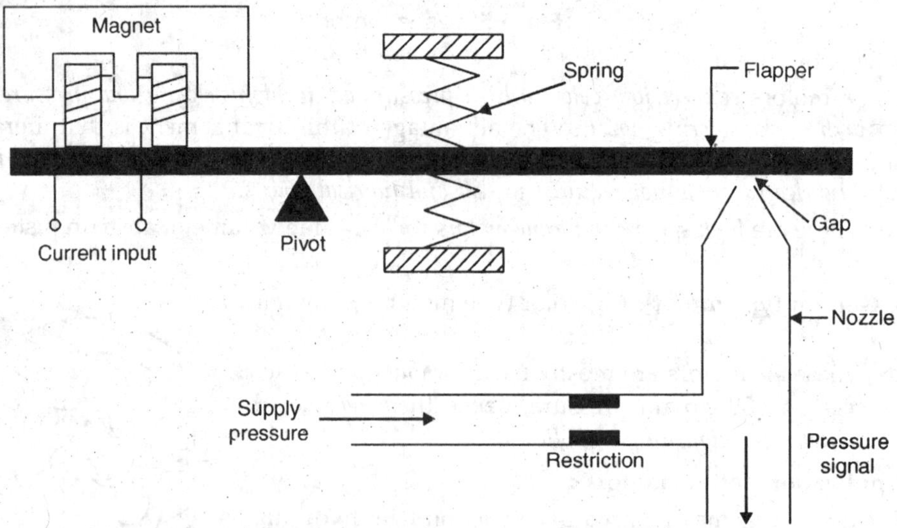

Fig. 5.54. Current to pressure converter.

— The input current passes through the coils mounted on the core which is attracted towards a magnet, the extent of the attraction depending on the size of the current.

— The movement of the core causes movement of a flapper above the nozzle. The position of the flapper in relation to the nozzle determines the rate at which air can escape from the system and hence the air pressure in the system.

Springs on the flapper are used to adjust the sensitivity of the converter so that currents of 4 to 20 mA produce gauge pressures of 20 to 100 kPa. These are the standard values that are generally used in such systems.

5.5 ELECTRICAL ACTUATORS

5.5.1. General Aspects

Actuator: *A mechanical device or a system which has motion or movement is called an actuator.*

Actuation system: *A group of elements which is responsible directly or indirectly for imparting motion to an actuator is called an* ***actuation system.***

Electrical actuator: *An actuator receiving electrical energy for motion is called an* ***electrical actuators.***

Electrical actuators systems include the following:

I. Switching devices:

1. Mechanical switches:
   (*i*) Solenoids.
   (*ii*) Relays.
2. Solid state switches:
   (*i*) Diodes.
   (*ii*) Tyristors.
   (*iii*) Transistors.

Here the *control signal switches on or off some electrical device, perhaps a heater or motor.*

II. Drive systems:

1. D.C. motors.
2. A.C. motors and stepper motors.

☞ • Amongst electric motors, *"D.C. motors" are most straight forward to interface and control and, hence, are most commonly used.*

- *"A.C. motors" and "stepper motors" are less utilised because A.C. motors are difficult to control and stepper motors have low torque capability. Moreover stepper motors cannot be used for cartesian space trajectory tracking operations (like plotting).*

5.5.2. Switching Devices

1. Mechanical switches :

Mechanical switches are elements which are often used as sensors to give input to systems *e.g., keyboards.* Here we are concerned with their use as actuators to perhaps switch on *electric motors* or *heating elements* or switch on the current to actuate *solenoid valves controlling hydraulic or pneumatic cylinders.*

Mechanical switches *are those where in switching action is by the application of force on the switch and during switching action mechanical elements move with the switch.* These switches consists of one or more pair of contacts which are mechanically closed or opened and in doing so make or break electrical circuit.

- Mechanical switches are specified in terms of *number of poles and throws.*
  — ***Poles (P)*** *are number of separate circuits that can be completed by the same switching action.*
  — ***Throws (T)*** *are number of individual contacts for each pole.*
- There are many designs for limit *"switches"* including push-button and levered microswitches. All switches are used to open or close connections within circuits.

As illustrated in Fig. 5.55, switches are characterised by the number of poles throws *and whether connections are "normally open (NO) or "normally closed (NC)".*

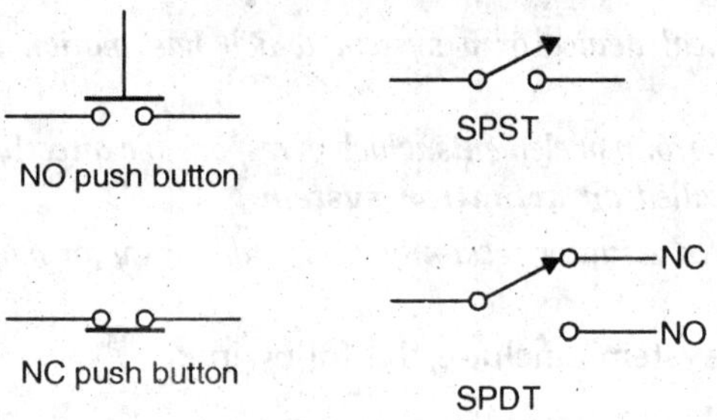

Fig. 5.55. Switches.

— The SPST switch is a single pole (SP), single throw (ST) device that opens or closes a single connection.

— The SPDT switch changes the pole between two different throw positions.

There are many variations on the pole and throw configurations of switches, but their function is easily understood from the basic terminology.

Bouncing and debouncing :

When mechanically switches are opened or closed, there are brief current oscillations due to mechanical bouncing or electric arcing; this phenomenon is called ***switch bounce.***

Fig. 5.56, illustrates that the mechanical contact associated with a switch closing results in multiple voltage transitions over a short period of time.

— Bouncing can occur when the switch is opened.

— Generally the bouncing time is about 20 ms.

The problem of bouncing can be solved by using the following *methods* :

(*a*) Specially designed switches.

(*b*) Software solution.

(*c*) Hardware solution.

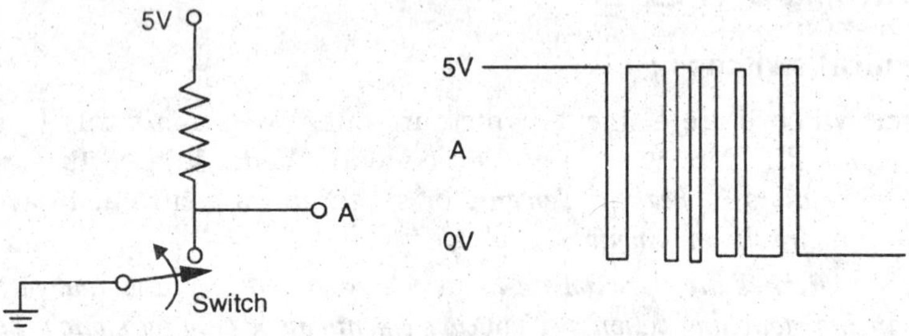

Fig. 5.56

(***a***) ***Specially designed switches:*** Specially designed switches include the following :

- Keys of a keyboard (toggle switch);
- Membrane switch;
- The keys used in calculators, mobile phones and telephones.

(***b***) ***Software solution:*** In this method, the microprocessor is programmed with a software to detect that the switch is closed and then wait for the bouncing period (say 20 ms). After checking that bouncing has ceased, the switch being in the closed position processing of next instruction can take place.

(*c*) ***Hardware solution:*** The hardware solutions to the bouncing problem are :

- *Set reset flip-flop circuit (also called latch circuit);*
- D flip-plop circuits
- Schmitt trigger.

Fig. 5.57, shows the sequential logic circuit which can provide an output that is free from multiple transitions associated with switch bounce.

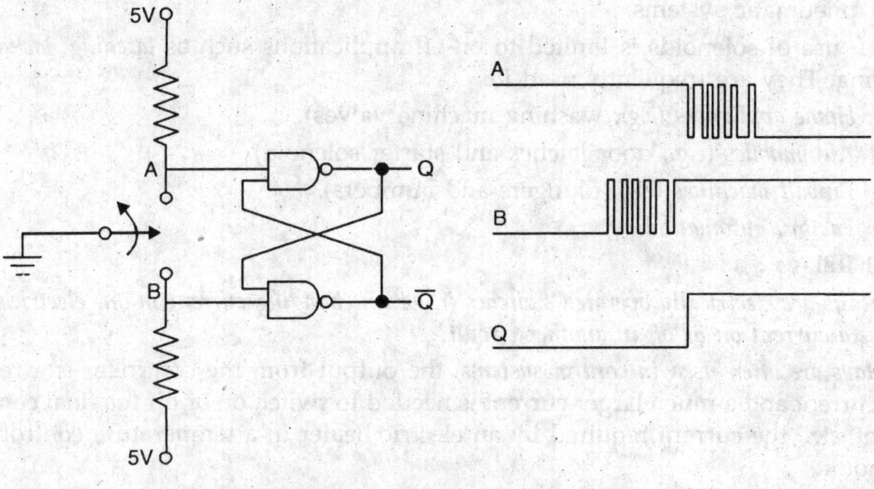

Fig. 5.57. Switch debounce circuits.

— As the switch breaks contact with *B*, single bounce occurs on the *B* line. There is a small delay as the switch moves from contact *B* to *A*, and then single bounce occurs on the *A* line as contact is established with *A*. The output of the debouncer *Q* is a single transition from 0 V to 5 V.

— The circuit functions very much like a flip-flop.

(*i*) **Solenoids** : Refer to Fig. 5.58.

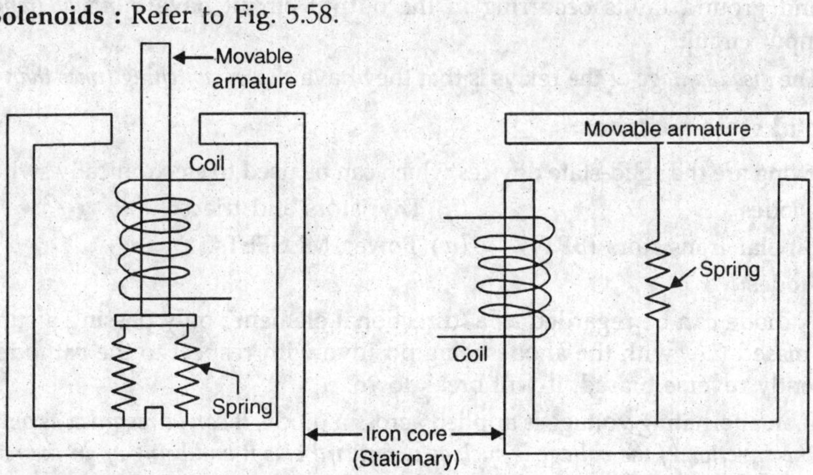

Fig. 5.58. Solenoids.

A *"solenoid"* consists of a coil and a movable iron core called the *armature*. When the current is passed through the coil it gets energized and consequently the core moves to increase the flux linkage by closing the air gap between the cores. The movable core is usually spring-loaded to allow the core to retract when the current is switched off. *The force generated is approximately proportional to the square of the current and inversely proportional to the square of the width of the air gap.*

- Solenoids are inexpensive.
- Solenoids can be used to provide electrically operated actuators. *Solenoid valves* are examples of such devices, being used to control fluid flow in hydraulic or pneumatic systems.

The use of solenoids is limited to on-off applications such as *latching, locking,* and triggering. They are frequently used in:

— *Home appliances* (*e.g.*, washing machine valves).

— *Automobiles* (*e.g.*, door latches and starter solenoid)

— *Pinball machines* (*e.g.*, plungers and bumpers).

— *Factory automation.*

(*ii*) Relays :

***Relays** are electrically operated switches in which changing current in one electrical circuit switches a current on or off in another circuit.*

Relays are often used in control systems; the output from the controller is a relatively small current and a much larger current is needed to switch on or off the final connection element, *e.g.*, the current required by an electric heater in a temperature control system or a motor.

Relays are used in '*power switches*' and '*electromechanical control elements*'.

- *A relay performs a function similar to a power transistor switch circuit but has the capability to switch much larger currents.*

  The input circuit of a relay is *electrically isolated* form the output circuit, unlike the common-emitter transistor circuit, where there is a common ground between the input and output. Since the relay is electrically isolated, noise, induced voltages, and ground faults occurring in the output circuit have *minimal impact* on the input circuit.
- The *disadvantage* of the relays is that they have *slower switching times than transistors.*

2. Solid state switches:

Following are the solid-state devices which can be used to electronically switch circuits:

(*i*) Diodes (*ii*) Thyristors and triacs

(*iii*) Bipolar transistors (BPT) (*iv*) Power, MOSFETs.

(*i*) Diodes:

— A diode can be regarded as a 'directional element', only passing a current when forward biased (*i.e.*, with the anode being positive with respect to the cathode). If diode is sufficiently reverse biased, it will breakdown.

— If an alternating voltage is applied across a diode, *it can be regarded as only switching on when the direction of the voltage is such as to forward bias it and being off in the reverse biased direction.*

- Generally diodes are not used as switches, but are used as *rectifiers.* It can also be used for full-wave rectification by forming a bridge using diodes.

***(ii)* Thyristors and triacs:**

Thyristors: The thyristor, or silicon-controlled rectifier (SCR) can be regarded as a diode which has a gate controlling the condition under which the diode can be switched on.

- The power-handling capability of a thyristor is high and thus it is widely used for *switching high power applications.*

  Thyristors are employed in:

  — D.C. controls;

  — Electric heaters;

  — Electric motors;

  — Lamp dimmers etc.

(*a*) *D.C. control :* Fig. 5.59 shows the thyristor D.C. control circuit and output :

— The thyristor is used as a switch to *on* and *off* the device.

— An intermittent voltage is generated by chopping of the supply voltage using an alternate signal to the gate. The average value of the D.C. voltage can thus be varied and hence controlled by the alternating signal at the gate.

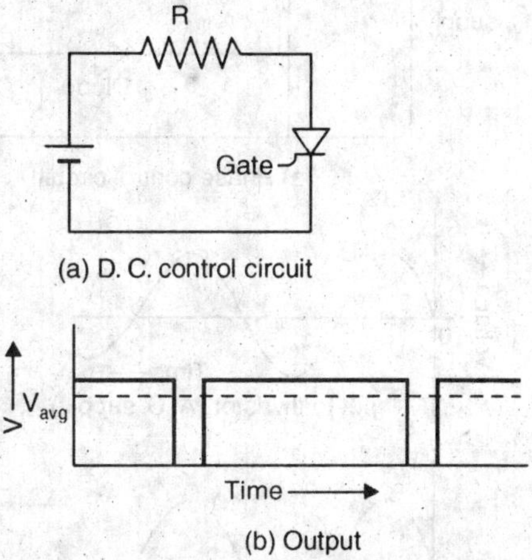

Fig. 5.59. Thyristor application.

(*b*) *Thyristor application in lamp dimmer (A.C. circuit) :*

Fig. 5.60, shows the circuit using thyristor for a lamp dimmer (also called phase control circuit) and output of the thyristor.

— A.C. supply is applied across R_L (may be a lamp or an electric heater) in series with a thyristor. R_P is the potentiometer (resistance) which sets the voltage at which the thyristor is triggered.

— Diode prevents the negative part of A.C. supplied to the gate.

— By adjusting the triggering voltage to the thyristor (using potentiometer), it can be made to trigger at any point between 0° and 90° in the +ve half cycle of A.C.

— When the thyristor is triggered at the beginning of the cycle *i.e.*, 0° full power supply is applied to the load and by varying the triggering voltage the supply to the load can be varied.

Thus *thyristor can be employed to control the A.C. supply to several devices.*

Triac :

— The triac is similar to the thyristor and is equivalent to a pair of thyristors connected in reverse parallel on the same chip.

— The triac can be turned on in either forward or reverse direction.

— Triacs are simple, relatively inexpensive, methods of controlling A.C. power.

***(iii)* Bipolar transistor (BPT):**

Bipolar transistors come in two forms, the NPN and the PNP. For the NPN transistor, the main current flows in at the collector and out at the emitter, a controlling signal being applied to the base. The PNP transistor has the main current flowing in at the emitter and out at the collector, a controlling signal being applied to the base.

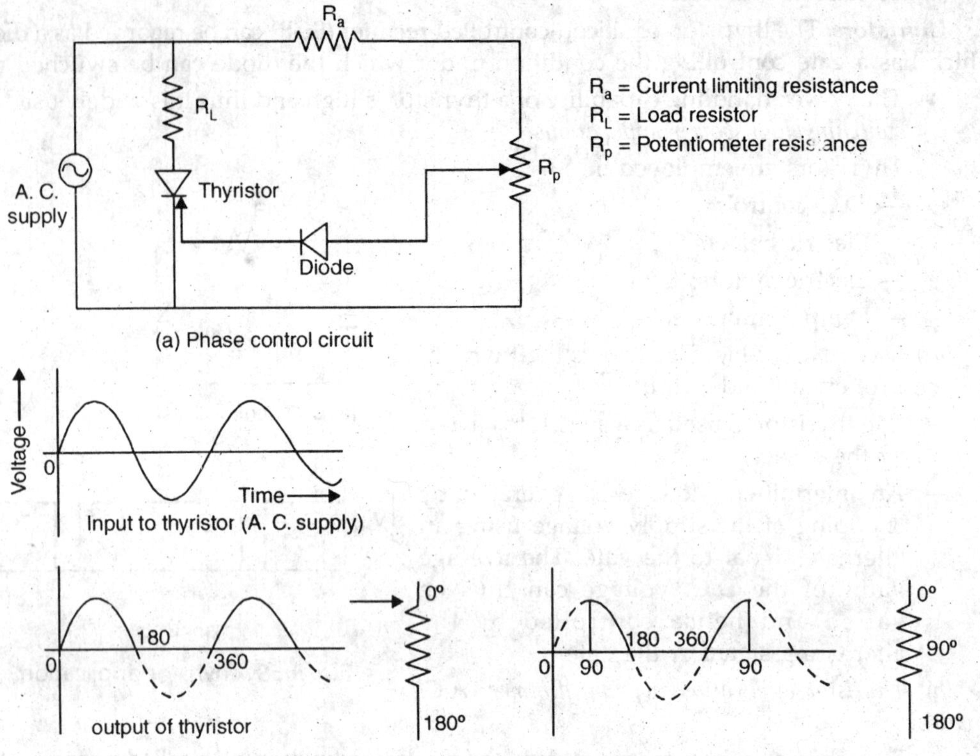

(a) Phase control circuit

(b) Input to and output of thyristor

Fig. 5.60. Thyristor application in lamp dimmer.

- In using transistor switched actuators with a microprocessor, attention has to given to the size of the base current required and its direction. The base current required can be too high and so a buffer might be used. The buffer increases the drive current to the required value. It might also be used to invert.
- *Bipolar transistor switching is implemented by base currents and higher frequencies of switching are possible than with thyristors. The power handling capacity is less than that of thyristors.*

(*iv*) MOSFETs:

MOSFETs (Metal-oxide field-effect transistors) are available in two types, the N-channel and the P-channel.

The main difference between the use of a MOSFET for switching and a bipolar transistor is that no current flows into the gate to exercise the control. The gate voltage is the controlling signal. Thus drive circuitry can be simplified in that there is no need to be concerned about the size of the current.

With MOSFETs, very high frequency switching is possible, up to 1 MHz, and interfacing with a microprocessor is simpler than with bipolar transistors.

Control of D.C. motor using MOSFET :

MOSFET can be employed as a control switch for a D.C. motor as *on-off* switch. Here, as compared to BJT for D.C. motor control, a level shifter buffer is used to raise the voltage level to that required for the MOSFET.

Figure 5.61 shows the circuit diagram for D.C. motor control using MOSFET.

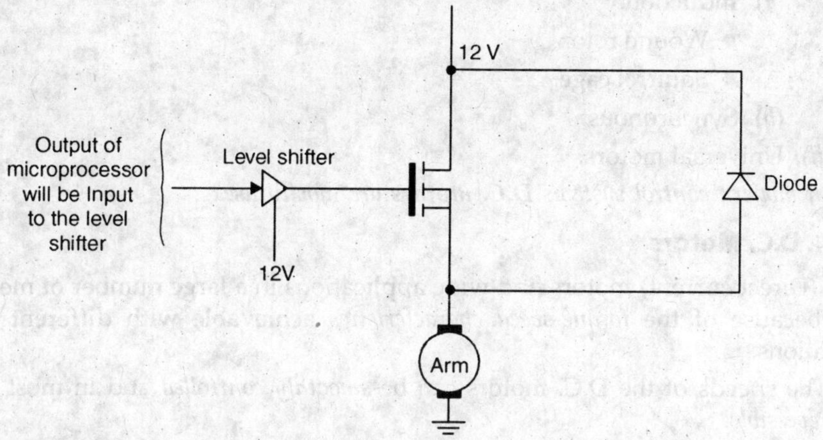

Fig. 5.61. MOSFET (N-channel type) application in D.C. motor control.

5.5.3. Drive Systems—Electric Motors

Electric motors are frequently used as the final control element in positional or speed control systems.

Electric motors for mechatronics/robotic applications, can be *classified* by electrical configuration as follows :

1. **D.C. Motors:**
   (*i*) *Permanent magnet.*
   (*ii*) *Series wound.*
   (*iii*) *Shunt wound.*
   (*iv*) *Compound wound.*
2. **A.C. Motors**
   (*i*) *Single phase:*
   (*a*) Induction:
   - Squirrel cage:
     — Split phase
     — Capacitor start
     — Permanent split capacitor
     — Shaded pole
     — Two-valve capacitor.
   - Wound rotor:
     — Repulsion
     — Repulsion start
     — Repulsion induction.

   (*b*) Synchronous:
   - Shaded pole
   - Hysteresis
   - Reluctance
   - Permanent magnet.

(*ii*) *Polyphase:*

(*a*) Induction:

- Wound rotor
- Squirrel cage.

(*b*) Synchronous.

(*iii*) Universal motors.

- *In modern control systems* ***D.C. motors*** *are mostly used.*

5.5.4. D.C. Motors

D.C. (Direct current) motors find wide applications in a large number of mechatronic designs because of the *torque-speed characteristics* achievable with different electrical configurations.

— The speeds of the D.C. motors can be *smoothly controlled* and in most cases are *reversible.*

— These motors *can respond quickly* since they have a *high ratio of torque to rotor-inertia.*

— *'Dynamic braking'* (where motor generated energy is fed to a resistor dissipater) and *'regenerative braking'* (where motor-generated energy is fed back to the D.C. power supply) can be implemented in applications where quick stops and high efficiency are desired.

5.5.4.1. Permanent magnet (PM) D.C. motors

In these motors (See Fig. 5.62) field excitation is obtained by suitably mounting permanent magnets (which require no power source and therefore produce no heating) on the stator. Magnets made froms ferrites or rare earth (cobalt samarium) are used.

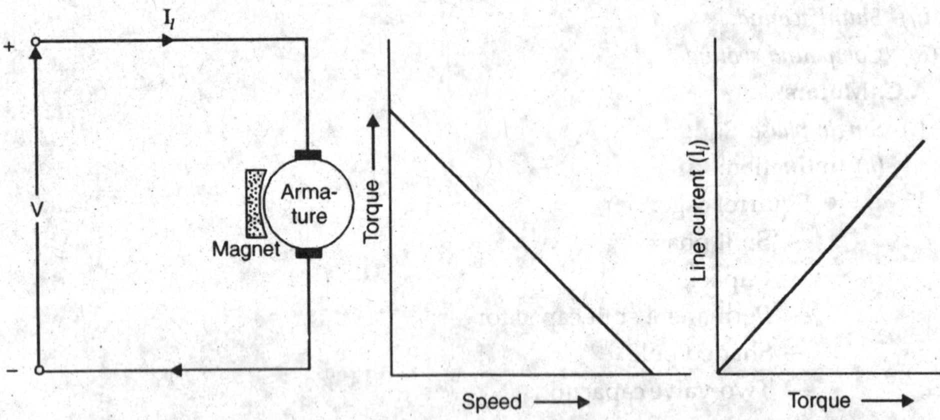

Fig. 5.62. Permanent magnet D.C. motor schematic and torque-speed and current-torque curves.

— A PM motor is *lighter and smaller* than others, equivalent D.C. motors because the field strength of permanent magnets is high.

— The radial width of a typical permanent magnet is roughly one-fourth that of an equivalent field winding.

— *PM motors are easily reversed by switching the direction of the applied voltage, because the current and field change direction only in the rotor.*

— PM motors can be brushed, brushless, or stepper motors.

Applications:

- The PM motor is *ideal in computer control applications* because of the *linearity* of its torque-speed relation.

— When a motor is used in a position or speed control application with sensor feedback to a controller, it is referred to as "*servomotor*".

- PM motors are used only in *low-power applications* since their rated power is limited to 5 H.P. or less, with fractional horsepower ratings being more common.

Advantages:

As compared to field wound motors, these motors possess the following *advantages*:

(*i*) More efficient.

(*ii*) More reliable.

(*iii*) More study and compact.

(*iv*) The field flux remains constant for all loads giving a more linear speed-torque characteristic.

(*v*) In a separately excited motor, failure of field can lead to runaway condition. This does not happen in PM motors.

Limitation. As the *flux is constant* in these motors, *speed cannot be controlled above base speed.*

5.5.4.2. D.C. Shunt motors:

Refer to Fig. 5.63.

In these motors armature and field windings are connected in *parallel* and powered by the same supply.

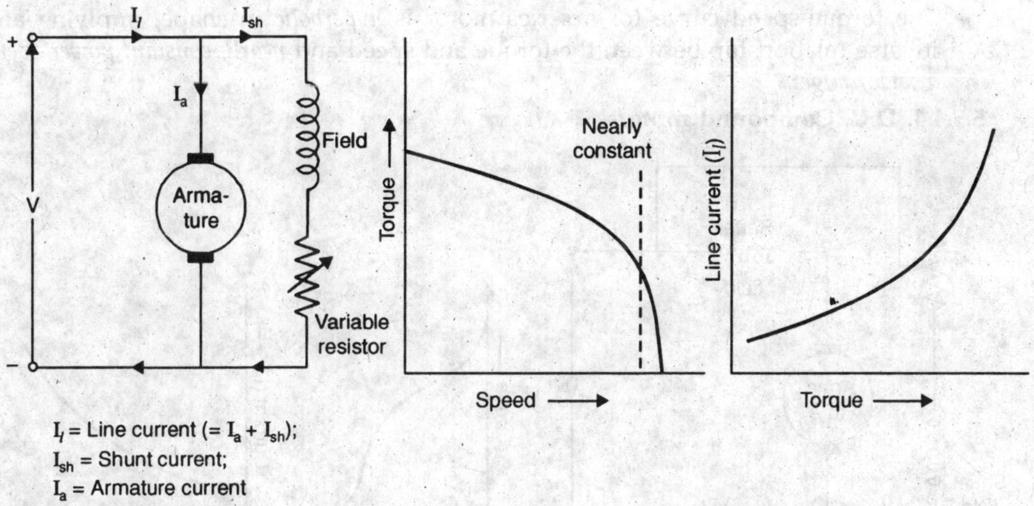

Fig. 5.63. D.C. shunt motor schematic and torque-speed and current-torque curves.

- These motors exhibit nearly *constant speed* over a long range of loading.
- They have starting torques about *1.5 times* the rated operating torque.
- They have *lowest starting torque* of any of D.C. motors.
- They can be economically converted to allow adjustable speed by placing a potentiometer in series with the field windings.

5.5.4.3. D.C. Series motors

In this type of motor (See Fig. 5.64) armature and field windings are connected in *series* so the armature and field currents are *equal.*

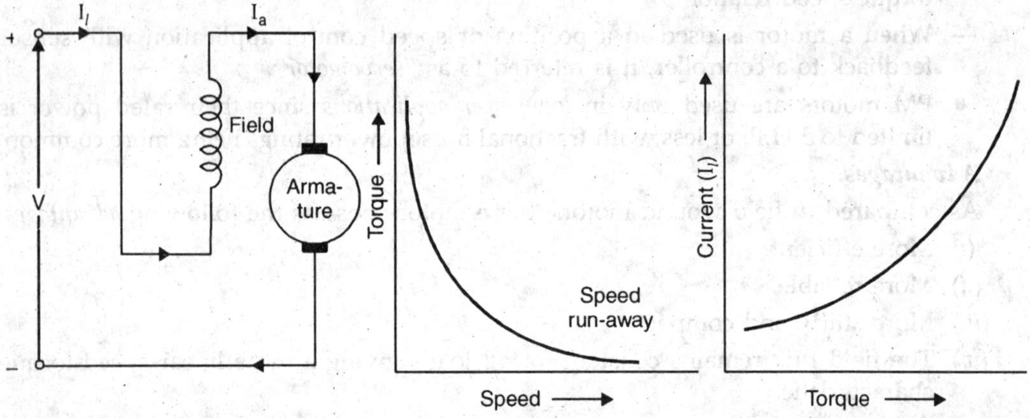

Fig. 5.64. D.C. series motor schematic and torque-speed and current-torque curves.

- *These motors exhibit very high starting torques, highly variable speed depending on load, and very high speed when the load is small.*
  - In fact, large series motors can fail catastrophically when they are suddenly unloaded (*e.g.*, in a belt drive application when the belt fails) due to dynamic forces at high speeds; this is called ***"run-away"***. As long as the motor remains loaded, this poses no problem.
- The torque-speed curve for a series motor is *hyperbolic* in shape, implying an inverse relationship between the torque and speed and *nearly constant power over a wide range.*

5.5.4.4. D.C. Compound motors

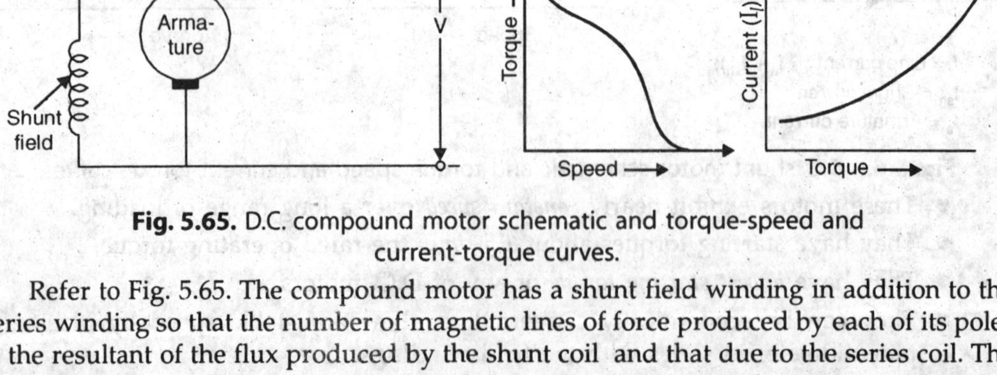

Fig. 5.65. D.C. compound motor schematic and torque-speed and current-torque curves.

Refer to Fig. 5.65. The compound motor has a shunt field winding in addition to the series winding so that the number of magnetic lines of force produced by each of its poles is the resultant of the flux produced by the shunt coil and that due to the series coil. The *flux so produced depends not only on the current and number of turns of each coil, but also on the winding direction of the shunt coil in relation to that of the series coil.* When the two fluxes *assist each other* the machine is a ***cumulative compound motor***, while if they *oppose each other*, it is said to be ***differential compound motor***.

Fig. 5.66 shows the field windings and interpole connections of a *differential compound wound motor*. The shunt coil is made up of *many turns of fine wire*, whilst the series coil comprises relatively *few turns of thickwire.*

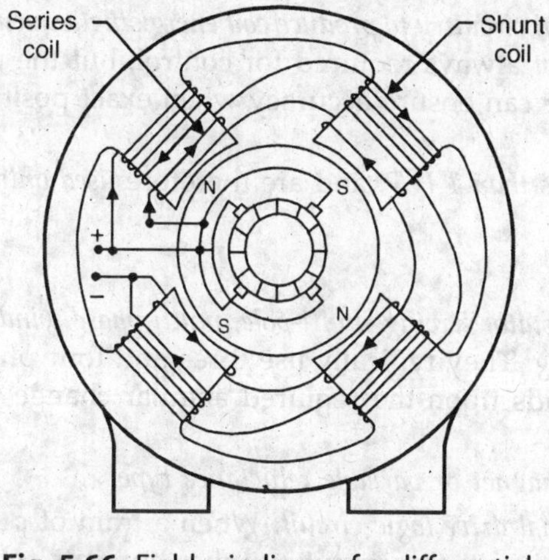

Fig. 5.66. Field windings of a differential compound motor.

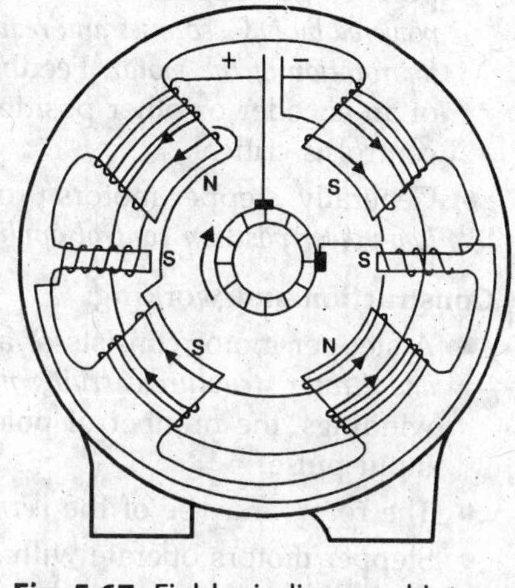

Fig. 5.67. Field windings and interpole connections of a cumulative compound wound motor.

Fig. 5.67, shows the field windings of a *cumulative compound motor*. The flow of currents in the shunt and series coil is worth noting in Fig. 5.66 and Fig. 5.67.

- *The maximum speed of compound motor is limited, unlike a series motor, but its speed regulation is not as good as with a shunt motor.*
- *The torque produced by compound motors is somewhat lower than that of series motors of similar size.*

Note: Unlike the permanent magnet motor, when voltage polarity for a shunt, series, or compound D.C. motor is changed, the direction of rotation does not change. The reason for this is that the polarity of both the stator and rotor changes, because the field and armature windings are excited by the same source.

5.5.4.5. Stepper motors

Introduction: A ***stepper motor***, *a special type of D.C. motor, is an incremental motion machine.* It is a *permanent magnet or variable reluctance D.C. motor* and has the following *characteristics:*

(*i*) It can rotate in both directions.

(*ii*) It can move in precise angular increments.

(*iii*) It can sustain a holding torque at zero speed.

(*iv*) It can be controlled with digital circuits.

— A stepper motor moves in accurate equal angular increments, known as ***steps***, in response to the application of digital pulses to an electric drive circuit. The number and rate of the pulses control the position and speed of the motor shaft.

— Generally, stepper motors are manufactured with steps per revolution of 12, 24,

72, 144, 180, and 200, resulting shaft increments of 30°, 15°, 5°, 2.5°, 2°, and 1.8° per step. Special ***micro-stepping*** circuitry can be designed to allow many more steps per revolution, often 10,000 steps/revolution or more.

— The stepper motor is used in digitally controlled position control system in open loop mode. The input command is in the form of a train of pulses to turn a shaft through a specified angle.

— Stepper motors are either ***bipolar,*** requiring two power sources or a switchable polarity power source, or ***unipolar,*** requiring only one power source. They are *powered by D.C. sources and require digital circuitry to produce coil energising sequences for rotation of the motor.* Feedback is *not* always required for control, but the use of an encoder or other position sensor can ensure accuracy when exact position control is critical.

- Generally, stepper motors produce *less than 1 H.P.* and are therefore *used only in low-power position control applications.*

Construction and working:

- A stepper motor consists of a *slotted stator having multi-pole, multi-phase winding and a rotor structure carrying no winding.* They typically use three and four phase windings, the number of poles depends upon the required angular change per input pulse.
- The rotors may be of the *permanent magnet or variable reluctance type.*
- Stepper motors operate with an *external drive logic circuit.* When a train of pulse is applied to the input of the drive circuit, the circuit supplies currents to the stator windings of the motor to make the axis of the air-gap field around in coincidence with the input pulses. The rotor follows the axis of the air-gap magnetic field by virtue of the permanent magnet torque and/or the reluctance torque, depending upon the pulse rate and load torque (including inertia effects).

1. Permanent magnet stepper motor :

In the case of a permanent magnet stepper motor, the stator consists of wound poles, the rotor poles are permanent magnets.

Fig. 5.68, shows the phases or stacks of a 2-phase, 4-pole permanent magnet stepper motor.

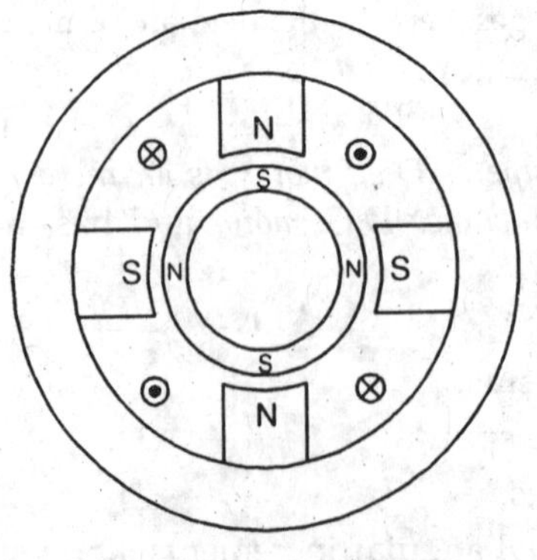

(i) Phase I

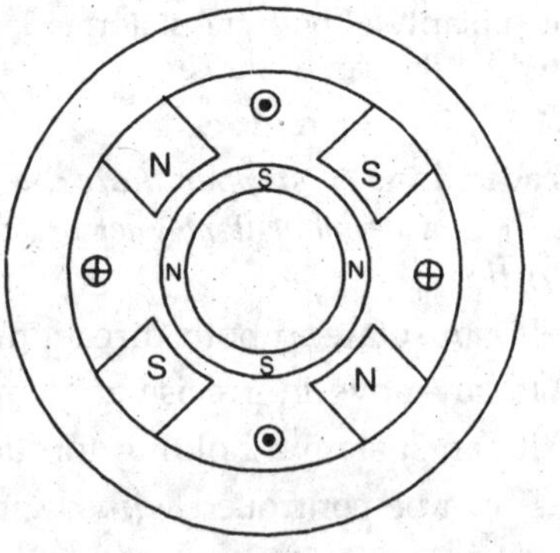

(ii) Phase II

Fig. 5.68. Permanent magnet stepper motor.

— The rotor is made of *ferrite or rare-earth material* which is *permanently magnetised.*
— The stator stack of phase II is staggered from that of phase I by an angle of 90°.
— When the phase 'I' is excited, the rotor is aligned as shown in Fig. 5.68(*i*), If now the phase 'II' is also excited, the effective stator poles shift anti-clockwise by 22.5° [Fig. 5.68(*ii*)] causing the rotor to move accordingly. Now, keeping the phase 'II' still energised, if the phase 'I' is now de-energised, the rotor will move another step of 22.5°. The reversal of phase 'I' winding current will produce a further forward movement of 22.5°, and so on.

It can be easily observed/visualised as to how the direction of movement can be reversed.

— Each phase is provided with double coils to simplify the switching arrangement (which is electronically accomplished).
— This type of motor has the *advantage of small residual holding torque, called* ***detent torque****, even when stator is not energized.*

2. Variable reluctance stepper motor :

— A variable-reluctance stepper motor has no permanent magnet on the rotor and the rotor employed is a ferro-magnetic multi-toothed one.
— *The large differences in magnetic reluctances that exist between the direct and quadrature axes develop the torque.* The stationary field developed by the direct current in some stator coils tends to develop a torque which causes the rotor to move to the position where the reluctance of the flux path is minimum.
Fig. 5.69 shows the basic form of the *variable reluctance stepper motor*.
— With this form the rotor is made of soft-steel and is cylindrical with four poles, *i.e.*, fewer poles than on the stator.

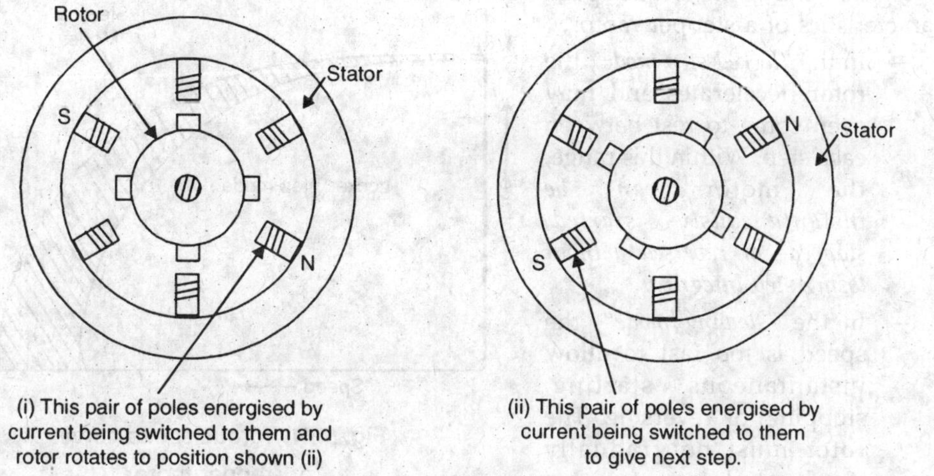

Fig. 5.69. Variable reluctance stepper motor.

— When an opposite pair of windings has current switched to them, a magnetic field is produced with lines of force which pass from the stator poles through the nearest set of poles on the rotor. Since lines of force can be considered to be rather like elastic thread and always trying to shorten themselves, the rotor will move until the rotor and stator poles line up. This is termed the position of minimum reluctance.

— This form of stepper generally gives step angles of 7.5° or 15°.

- *Stepping angle, irrespective of the type of stepper motor* is given as,

$$\alpha = \frac{360°}{\text{Number of phases} \times \text{number of poles}} = \frac{360}{np} \quad ...(1)$$

3. Hybrid stepper motor :

— This is *infact a permanent magnet stepper motor with constructional features of toothed and stacked rotor adopted from the variable-reluctance motor.*

— The stator has only one set of winding-excited poles which interact with the two rotor stacks.

— The permanent magnet is placed *axially along the rotor in the form of an annular cylinder over the motor shaft* (See Fig. 5.70).

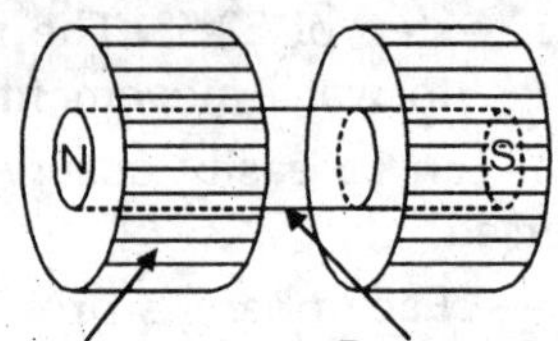

Fig. 5.70. Hybrid motor rotor.

— The stacks at each end of the rotor are toothed. So all the teeth on the stack at one end of the rotor acquire the *same polarity* while the teeth of the stack at the other end of the rotor acquire the *opposite polarity*. The two sets of the teeth are displaced from each other by *one half of the tooth pitch* (also called pole pitch).

— The primary ***advantage*** of the hybrid motor is that if *stator excitation is removed, the rotor continues to remain locked into the same position, as before removal of excitation.* This is due to the reason that the rotor is prevented to move in either direction by torque because of the permanent excitation.

Torque-speed characteristics of a stepper motor:

Fig. 5.71 shows the torque-speed characteristics of a stepper motor.

— In the "*locked step mode*', the rotor decelerates and may even come to rest between each step. Within this range, the motor *can be instantaneously started, stopped or reversed without losing step integrity.*

— In the "*slewing mode*", the speed is too fast to allow instantaneous starting, stopping, or reversing. The rotor must be gradually accelerated to enter this mode and gradually decelerated to leave the mode. While in slewing mode, the rotor is in synch with the stator field rotation and does not settle between steps.

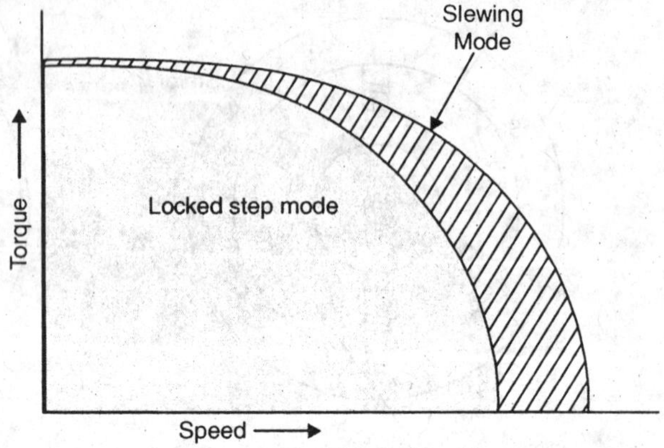

Fig. 5.71. Torque-speed curves of stepper motor.

The curve between the regions in the figure indicates the maximum torques that the stepper can provide at different speeds without slewing. The curve bordering the outside of the slewing mode region represents the absolute maximum torques the stepper motor can provide at different speeds.

Advantages and applications of stepper motor:

Advantages: The stepper motor (*a position control device*) entails the following advantages:

1. Compatibility with digital systems.
2. The angular displacement can be precisely controlled without any feedback arrangement.
3. No sensors are needed for position and speed sensing.
4. It can be readily interfaced with microprocessor (or computer based controller).

Applications: Stepper motors have a wide range of *applications,* mentioned below :

1. Paper feed motors in typewriters and printers.
2. Positioning of print heads.
3. Pens in XY-plotters.
4. Recording heads in computer disc drives.
5. Positioning of worktables and tools in numerically controlled machining equipment.
6. Also employed to perform many other functions such as *metering, mixing, cutting, blending, stirring* etc. in several commercial, military and medical applications.

5.5.4.6. Servomotors

Introduction: The term *servo* or *servo mechanism* refers to a *feedback control system* in which the controlled variable is:

— Mechanical position, or

— Time derivatives *e.g.,* velocity and acceleration.

Following *characteristics* are usually required for a feedback control system:

(*i*) High accuracy.

(*ii*) Remote operation.

(*iii*) Fast-response.

(*iv*) Unattended control.

Following are the *essentials* of a feedback control system :

1. ***An error detecting device.*** It determines when the regulated quantity is *different* from the reference quantity and sends out the error signal to the *amplifier.*
2. ***An amplifier.*** The amplifier receives the error signal and then supplies power to the error-correcting devices, which in turn changes the regulated quantity so that it matches the reference input.

A *servo-motor* should entail the following *characteristics* :

1. *The output torque of the motor should be proportional to the voltage applied* (*i.e.,* the control voltage which is developed by the amplifier in response to an error signal).
2. *The direction of the torque developed by the servo-motor should depend upon the instantaneous polarity of the control voltage.*

Types of servo-motors :

The servo-motors are of the following two types :

1. *D.C. servo-motors.*
2. *A.C. servo-motors.*

1. D.C. servo-motors :

These motors are preferred for very high power systems since they operate more efficiently (as compared to A.C. servo-motors).

These motors may be of the following types :

— Series motors;
— Split series motors ;
— Shunt control motors ;
— Permanent magnet (*fixed excitation*) shunt motor.

(i) Series motors :

- This motor has a high starting torque.
- It draws large current.
- The speed regulation is poor.
- Reversal can be obtained by reversing field voltage polarity with split series field winding.

(ii) Split series motor :

- The D.C. series motor with split field (small fraction kW) may be operated as a separately excited field-controlled motor (Fig. 5.72).

  The armature may be supplied from a constant current source.

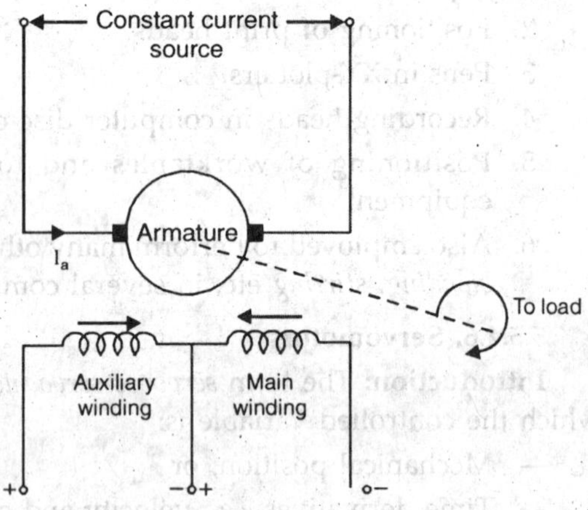

Fig. 5.72. From D.C. amplifier.

- A typical torque curve shows the following :

  — High stall torque ;

  — Rapid reduction in torque with increase in speed.

(iii) Shunt control motor :

- This type of motor has *two separate windings ; Field winding placed on the stator and the armature winding placed on the rotor of the machine.* Both the windings are connected to a D.C. supply source.
- Whereas in a conventional D.C. shunt motor, the two windings are connected in parallel across the D.C. supply mains, but in a *servo-application* the windings are driven by *separate D.C. supplies.*

(iv) Permanent magnet shunt motor :

- It is a fixed excitation shunt motor where the field is actually supplied by a permanent magnet.
- Its performance is similar to that of armature controlled fixed field motor.

2. A.C. servo-motors :

Applications :

- These motors are best suited for *low power applications.*

- Precision servo-motors are used in :
  — Instrument servos ;
  — Computers ;
  — Inertial guidance systems etc.
- The mechanical output power of A.C. servo-motor varies from 2 watts to a few hundred watts.
- An A.C. servo-motor is basically a two-phase induction motor except for certain special design features. The *main important difference* between a standard split-phase motor and an A.C. servo motor is that the *latter* has *thinner conducting bars in the squirrel cage motor, so that the motor resistance is higher*. The torque-speed characteristics should be *linear* as shown by the curve II in Fig. 5.73.

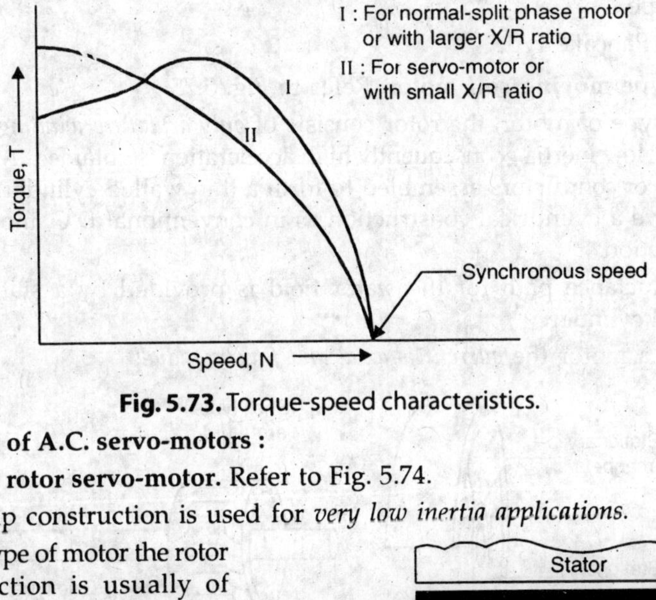

Fig. 5.73. Torque-speed characteristics.

Description of A.C. servo-motors :

1. Drag-cup rotor servo-motor. Refer to Fig. 5.74.

- Drag-cup construction is used for *very low inertia applications.*
- In this type of motor the rotor construction is usually of squirrel cage or drag-cup type; here only a *light cup rotates while the rotor core is stationary* (thus inertia is quite small).
- The servo-motors contains two windings namely, *main winding* (sometimes called fixed or reference winding) and *control winding*. The voltages applied to the windings are at *right angles* to one another. Usually one winding is excited with a fixed voltage while the other one is excited by the control voltage (which is the output from servo-amplifier).

Fig. 7.64

Fig. 5.74. Drag-cup rotor servo-motor.

- *While in operation, the output torque of the motor is roughly proportional to the applied control voltage, and the direction of torque is determined by the polarity of the control voltage.*

2. Shaded-pole type servo-motor :

- This type of motor employs a *phase-sensitive relay to actuate those contacts which produce a short-circuit of the shaded-pole winding to produce rotation in the desired direction.*
- *The main shortcoming of this motor is that it responds only when the amplifier error signal is of adequate magnitude to cause the relay to operate.*

5.5.4.7. Moving coil motors

There are certain applications which require acceleration much higher than what can be achieved in a conventional D.C. servo-motor. The armatures of moving coil D.C. motors have special constructions which allow a substantial reduction in armature inertia and inductance, permitting very high accelerations.

Moving coil motors are of the following *two* types :

1. Shell type.
2. Disc or Pancake type.

1. Shell type moving coil motor. Refer to Fig. 5.75.

- In this type of motor, the rotor consists of *only armature winding* due to which it has very low inertia ; consequently high acceleration is obtained. Armature winding consists of conductors assembled to form a thin walled cylinder. The *commutator* may have a cylindrical construction as in conventional D.C. motors or disc type construction.

  Low reluctance path for the stator field is provided by a stationary magnetic material cylinder.

  In such a motor the *current is axial and flux is radial.*

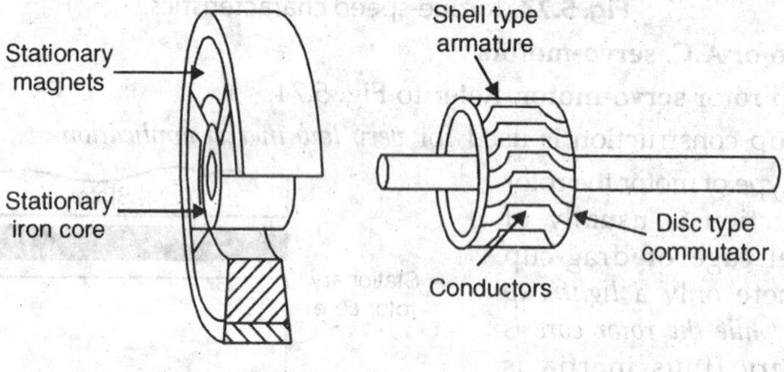

Fig. 5.75. Shell type moving coil motor.

- **Micromotors** (*Tiny motors with diameters around 1 cm*) have armature winding consisting of simply varnished wires arranged in cylindrical form and a disc type commutator.

Such motors find wide applications in *card readers, video systems, cameras* etc.

In *bigger size motors* the armature winding is made by bonding conductors together using polymer resins and fibre glass to provide adequate mechanical strength.

2. Disc or Pancake type moving coil motor. Refer to Fig. 5.76.

In this motor armature is made in disc or pancake form, and armature conductors resemble spokes on a wheel. The armature winding is formed by stamping conductors from a sheet of copper, welding them together and placing them on a light weight disc. Conductor segments are then joined with a *commutator* at the centre of the disc.

Here the direction of *flux is axial* and *armature current is radial* (just opposite to shell type conventional motors).

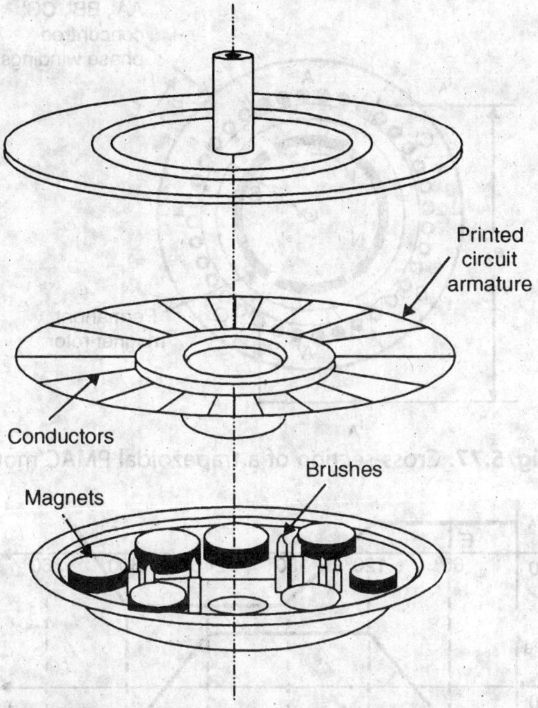

Fig. 5.76. Disc or Pancake type moving coil motor.

The *principle of operation is same as that of a conventional D.C. motor.*

- These motors are more robust and available in sizes *upto few kilowatts.*
- They find *applications* where *axial space is at a premium* such as *machine tools, disc drives* etc.

5.5.4.8. Torque motors

"Torque motors" are *the D.C. motors designed to run for long periods in a stalled or a low speed condition.* Some torque motors are designed to operate at low speeds intermittently.

The torque motor *applications* can be divided into the following three types :

(*i*) Motor is required to *operate in stalled condition.*

— The purpose of the motor is to develop the required tension or pressure on a material, similar to spring.

(*ii*) Motor is required to *move through only a few revolutions* or *degrees of revolution.'*

Examples. Opening of valves, switches, clamping devices etc.

(*iii*) This category involves *continuous movement of the motor at low speed.*

Example. Reel drive.

5.5.4.9. Brushless D.C. (or trapezoidal PMAC) motors

Fig. 5.77 shows the cross-section of a 3-phase 2-pole trapezoidal PMAC motor.

The stator has three concentrated phase windings (*AA'*, *BB'* and *CC'*) which are displaced by 120° and each phase winding spans 60° on each side. The voltages induced in three phases are shown in Fig. 5.78. The reason for getting trapezoidal waveforms is explain below :

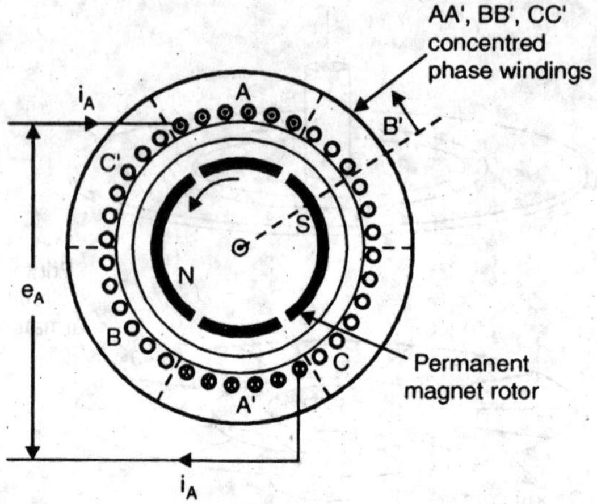

Fig. 5.77. Cross-section of a trapezoidal PMAC motor.

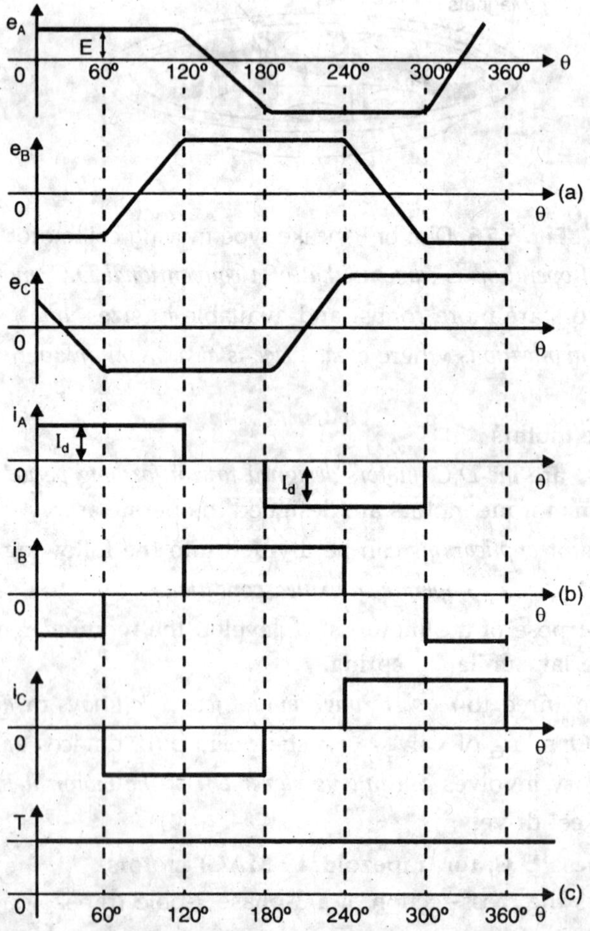

Fig. 5.78. Induced voltage, phase current and torque waveforms of a brushless D.C. motor.

When revolving in the counter-clockwise direction, upto 120° rotation from the position shown in Fig. 5.77, all top conductors of phase A will be linking the S-pole and all bottom conductors of phase A will be linking the N-pole. Hence the voltage induced in phase *A* will be the same during 120° rotation (Fig. 5.78). Beyond 120°, some conductors in the top link N-pole and others the S-pole. Same happens with bottom conductors. Hence, the voltage induced in phase *A* linearly reverses in next 60° rotation. Rest of the waveform of phase *A* and waveforms of *B* and *C* can be explained on the same lines.

Fig. 5.78 shows the induced voltage, phase current and torque waveforms of a brushless D.C. motor.

An *inverter fed trapezoidal PMAC motor drive operating in self-controlled mode is called a* **brushless D.C. motor.** This motor is also conceived as *electronically commutated D.C. motor,* because inverter here performs the same function as the brushes and commutator in a D.C. motor *i.e.,* to shift currents between armature conductors to keep the stator and rotor fields stationary, and in quadrature with respect to each other.

Advantages:

Owing to the absence of brushes and commutator, brushless D.C. motors claim the following *advantages over the conventional D.C. motors* :

(*i*) Long life.
(*ii*) Require practically no maintenance.
(*iii*) High reliability.
(*iv*) Low inertia and friction.
(*v*) Low radio frequency interference and noise.
(*vi*) Because armature windings are on the stator, cooling is much better, *i.e., high specific outputs* can be obtained.
(*vii*) They have a faster acceleration (due to low inertia and friction) and can be run at much higher speeds-upto 100000 r.p.m. and higher are common.
(*viii*) High efficiency, exceeding 75 percent (whereas wound field motors of low power ratings have much lower efficiency).

Disadvantages:

(*i*) High cost.
(*ii*) Low stalling torque.

The size of a brushless D.C. motor is nearly the same as that of the conventional D.C. motor.

Applications:

The brushless D.C. motor finds applications in :

(*i*) Tape drive for video recorders ;
(*ii*) Turn table drives in record players ;
(*iii*) Spindle drives in hard disk drives for computers ;
(*iv*) Low cost and low power drives in computer peripherals, instruments and control systems.
(*v*) Gyroscope motors ;
(*vi*) Cryogenic coolers ;
(*vii*) Artificial heart pumps ;
(*viii*) Cooling fans for electronic circuits and heat sinks.

5.5.4.10. Electronic control of D.C. motors

Introduction : Normally, it is essential to vary the speed of electrical drives in different fields of application. Usually, in all process industries, it is desired that the system be set at slow speed in the beginning and then gradually increased to meet the maximum production rate, *e.g., newspaper printing press.*

One of the major achievements of *thyristor technology* in the field of control is the control of D.C. and A.C. motor drives. *Thyristor controlled schemes* have totally dominated the field of control of D.C. as well as A.C. motors because of the following *advantages* :

(*i*) *Compactness.*

(*ii*) *Fast response.*

(*iii*) *More efficiency.*

(*iv*) *More control capabilities.*

(*v*) *More reliability.*

(*vi*) *Less cost etc.*

Advantages of electronic control systems :

The electronic control system claims the following *advantages over conventional methods* :

1. Very compact and small in size.
2. Consumes very less power.
3. Very fast in response.
4. Much more accurate and efficient than a conventional system.
5. Control ranges are much more than any other systems.
6. High reliability comparatively.
7. Economical, since maintenance cost in minimum.
8. Highly protective.
9. In electronic systems more automation, as required for highly sophisticated machines, is possible.

D.C. Motor speed control :

There are several methods by which the speed of a D.C. shunt motor can be controlled *using thyristors*, some of the commonly used methods are discussed below.

1. Armature voltage control method :

This is also called the *phase control method of speed control.* The complete diagram for this scheme is shown in Fig. 5.79.

- The field of motor is excited by a constant D.C. obtained from the *full-wave rectifier.*
- The armature voltage is *varied by varying the firing angle of the SCRs of the thyristor bridge.* Voltage across the armature terminals will be variable D.C. obtained from the full-wave half-controlled thyristor bridge. In the positive half-cycle SCR_1 and diode D_1 will conduct whereas in the negative half cycle SCR_2 and diode D_2 will conduct. Gates of *SCRs* will be given signal from the triggering circuit (not shown in the Fig. 5.79).

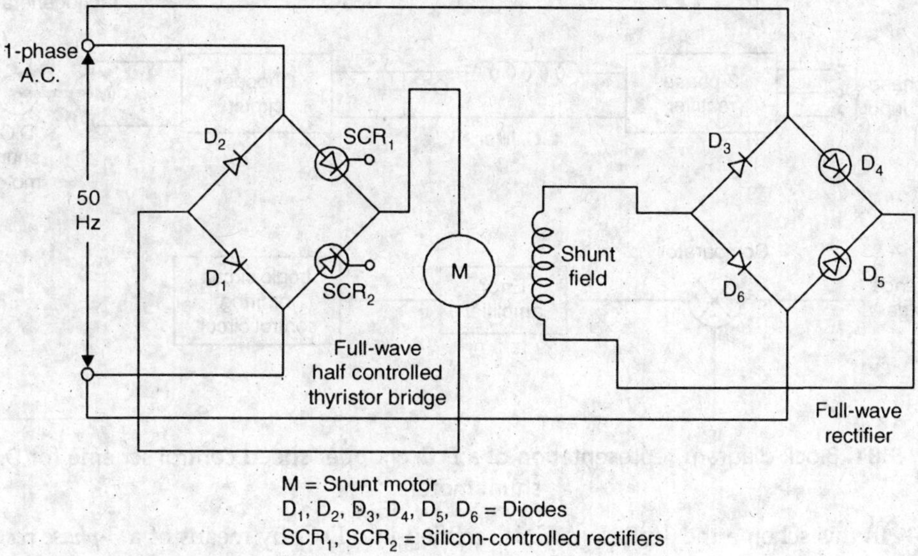

Fig. 5.79. Complete circuit diagram for the armature voltage control method for speed control of D.C. shunt motor.

- The wave shapes for the A.C. input voltage and controlled D.C. armature voltage are shown in Fig. 5.80.

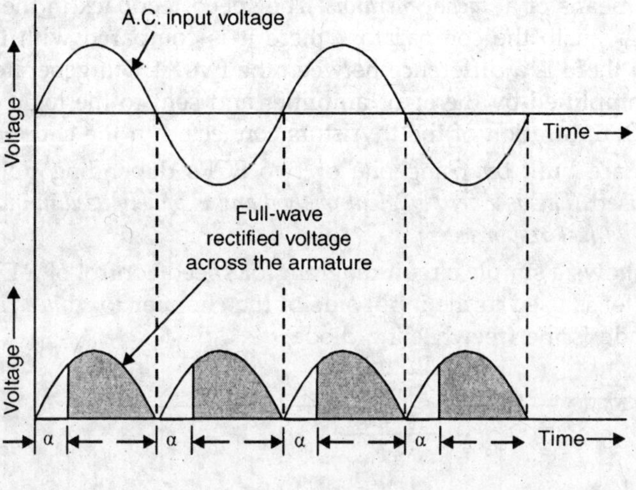

Fig. 5.80. Wave shapes for A.C. input voltage and controlled D.C. armature voltage.

2. ***D.C. chopper speed control :***

A *D.C. chopper can give variable D.C. at its output.* This variable can be utilised for the purpose of speed control of D.C. shunt motors. This method of speed control has gained popularity since the introduction of semiconductor devices.

Figure 5.81 shows the block diagram representation of a D.C. chopper speed control scheme for D.C. shunt motors.

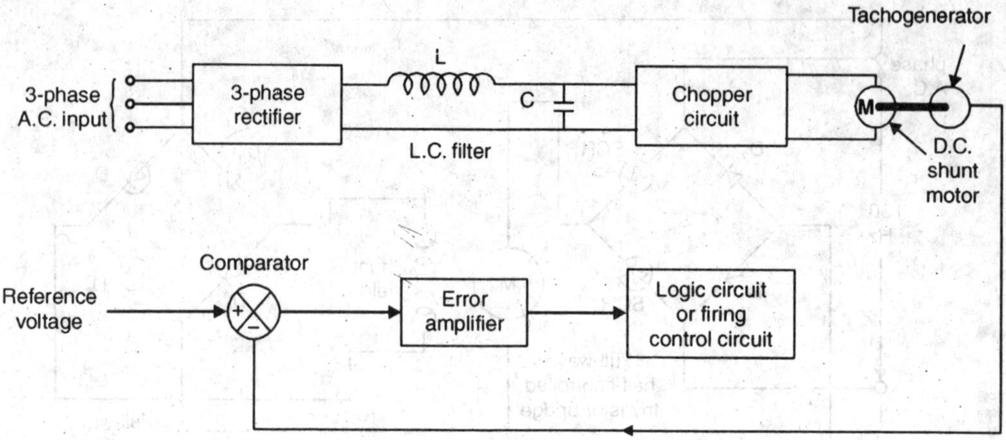

Fig. 5.81. Block diagram representation of a D.C. chopper speed control scheme for D.C. shunt motors.

— In this scheme the 3-phase A.C. is rectified into D.C. by means of a *3-phase rectifier*.

— The ripples are minimised with the help of a proper *'LC filter'*.

This filtered rectified D.C. serves as the *input* for the chopper circuit. There is a *'logic circuit'* which decides the firing of the thyristors used in the chopper. The ON, OFF durations for the thyristors used are decided by this unit. The input signal to this logic circuit is obtained from a *'comparator'* through an *'error amplifier'*.

— The speed feedback from the D.C. shunt motor is converted into equivalent voltage signal by means of a *'tachogenerator'*. The speed feedback in the form of voltage signal is given to the comparator where it is compared with the set reference voltage. If there is a difference between the two, it will generate an error signal which is amplified by the error amplifier and sent to the logic circuit to decide the ON, OFF duration of the thyristors connected in the chopper.

— Choppers are built by using one or two SCRs depending upon the type and circuitry used. *This is a very efficient method and is widely used in industries these days because of its fast response.*

— Fig. 5.82 shows a simple circuit diagram for speed control of a D.C. shunt motor. An *L-C* filter is used in the input side of the chopper to *reduce ripples* in the D.C. input. Diode is the freewheeling diode.

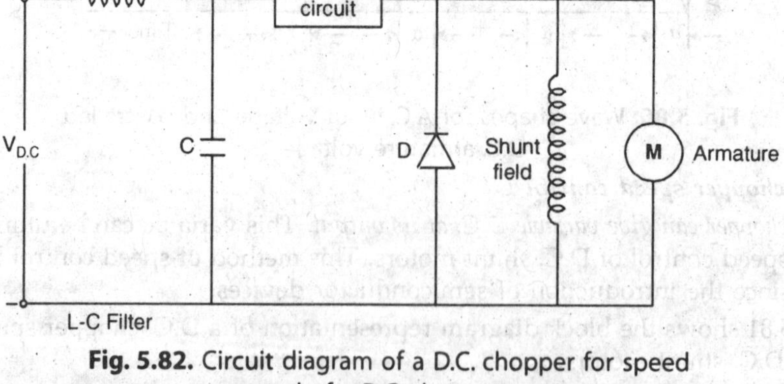

Fig. 5.82. Circuit diagram of a D.C. chopper for speed control of a D.C. shunt motor.

— Fig. 5.83 shows a simple chopper circuit which may be used for controlling the speed of a D.C. series motor.

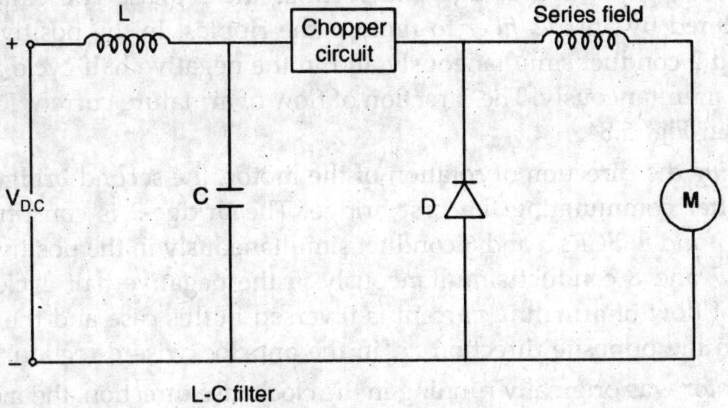

Fig. 5.83. D.C. chopper application for speed control of a D.C. series motor.

— Variation of T_{ON} and T_{OFF} will vary the load voltage at the output of the chopper which will change the speed of the motor accordingly.

— Diode *D* has been used as a *freewheeling diode to provide low resistance path for the current which will flow even at the OFF period of the thyristors.* This current flows for a little time due to the stored energy in the winding which is inductive in nature.

— An *L-C* filter has been used in the input side of the chopper to reduce the ripples in the D.C. input voltage.

3. Speed control by using a dual converter :

A dual converter, as the name indicates, uses two converters a *rectifier* and an *inverter.* Both the bridges are built by using *SCRs.* A dual converter may be used to obtain the following controls of a D.C. motor :

- Reversible speed control.
- Plugging.
- Regenerative braking.

The above controls are discussed below.

Fig. 5.84 shows the circuit diagram for speed control a D.C. shunt motor using a *dual converter.*

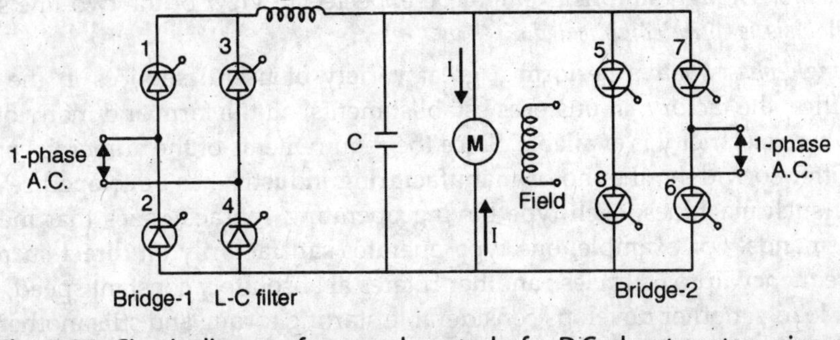

Fig. 5.84. Circuit diagram for speed control of a D.C. shunt motor using a dual converter.

1. *Reversible speed control and plugging :*

- Four *SCRs*, 1, 2, 3 and 4 from the first bridge (Bridge-1) which serves as a 1-phase full-wave fully-controlled bridge and rectifies the 1-phase A.C. into D.C. This D.C. is filtered by an *L.C. filter* to remove the ripples. In the positive half cycle *SCRs* 1 and 2 conduct simultaneously and in the negative half cycle *SCRs*, 3 and 4 conduct simultaneously. The direction of flow of armature current *I* is clockwise as shown in Fig. 5.84.
- For ***reversing*** the direction of rotation of the motor, the second bridge (Bridge-2) is gated after commutating the first bridge. The Bridge-2 is constituted by the *SCRs* 5, 6, 7 and 8. *SCRs* 5 and 6 conduct simultaneously in the positive half cycle and *SCRs* 7 and 8 conduct simultaneously in the negative half cycle. Thus, the direction of flow of armature current is reversed in this case and the motor tries to rotate in the opposite direction *i.e.* in the anticlockwise direction.

Because the motor was originally running in the clockwise direction, the inertia would oppose the torque developed in the anticlockwise direction. When the *two torques become equal, the motor becomes stationary provided bridge-2 is commutated.* This process of stopping the motor is called ***plugging***. If the bridge-2 further continues to conduct, the motor would start-running in the opposite direction resulting in speed reversal. In the opposite direction of rotation of the motor, the speed can be controlled by varying the *firing angle* of the second bridge.

2. *Regenerative braking :*

In this case, after bridge-1 is commutated and bridge-2 is triggered the counter e.m.f. generated in the armature of the motor acts as input for bridge-2 which is connected in the *inverter mode.* The output of bridge-2 which is 1-phase A.C. may be fed back to the mains supply. Thus, we see that bridge-1 acts as a rectifier and bridge-2 acts as inverter. Therefore, in regenerative braking the K.E. of the motor is converted into electrical energy and fed back to the supply system thereby saving energy.

5.5.5. Single-Phase Motors

5.5.5.1. General aspects

- The number of machines operating from single-phase supplies is greater than all other types taken in total. For the most part, however, they are only used in the smaller sizes, less than 5 kW and mostly in the fractional H.P. range. They operate at *lower power-factors* and are relatively *inefficient when compared with polyphase motors.* Though simplicity might be expected in view of the two-line supply, the *analysis is quite complicated.*
- *Single-phase motors* perform a great variety of useful services in the home, the office, the factory, in business establishments, on the farm and many other places where electricity is available. Since the requirements of the numerous applications differ so widely, the motor-manufacturing industry has developed several types of such machines, each type having operating characteristics that meet definite demands. For example, one type operates satisfactorily on direct current or any frequency upto 60 cycles ; another rotates at absolutely constant speed, regardless of load ; another develops considerable starting torque and still another, although not capable of developing much starting torque, is nevertheless extremely cheap to make and very rugged.

5.5.5.2. Applications and Disadvantages

Applications :

- Single-phase induction motors are in very wide use in industry especially in *fractional horse-power field.*

  They are extensively used for electric drive for *low power constant speed apparatus* such as *machine tools, domestic apparatus* and *agricultural machinery* in circumstances where a three-phase supply is *not* readily available.
- There is a large demand for single-phase induction motors in sizes ranging from a *fraction of horse-power upto about 5 H.P.*

Disadvantages:

Though these machines are useful for small outputs, they are not used for large powers as they suffer from many disadvantages and are never used in cases where three-phase machines can be adopted.

The main *disadvantages* of single-phase induction motors are :

1. Their output is only 50% of the three-phase motor, for a given frame size and temperature rise.
2. They have lower power factor.
3. Lower-efficiency.
4. These motors do not have inherent starting torque.
5. More expensive than three-phase motors of the same output.

5.5.5.3. Construction and working

Construction :

- A single phase induction motor is similar to a 3-ϕ squirrel-cage induction motor in physical appearance. Its rotor is essentially the same as that used in 3-ϕ induction motors. Except for shaded pole motors, the stator is also very similar. There is a uniform air-gap between the stator and rotor but no electrical connection between them. It can be wound for any even number of poles, two, four and six being most common. Adjacent poles have opposite magnetic property and synchronous speed equation, $N_s = \frac{120f}{p}$ also applies.
- The stator windings differ in the following two aspects :
  - — ***Firstly*** single phase motors are usually provided with concentric coils.
  - — ***Secondly,*** these motors normally have two stator windings. In motors that operate with both windings energised, the winding with the *heaviest wire* is known as the **main winding** and the other is called the **auxiliary winding.** If the motor runs with auxiliary winding open, these windings are usually referred as *running* and *starting.*
  - — In *most of motors the main winding is placed at the bottom of the slots and the starting winding on top but shifted 90° from the running winding.*

Working :

When the stator winding of a single phase induction motor is connected to single phase A.C. supply, a *magnetic field* is developed, *whose axis is always along the axis of stator coils.* The magnetic field produced by the stator coils is pulsating, varying sinusoidally with time. Currents are induced in the rotor conductors by transformer action, these currents being in such a direction as to oppose the stator m.m.f. Then the axis of the rotor m.m.f. wave coincides with that of the stator field, the *torque angle* is, therefore, *zero* and *no torque is developed on starting*. However, if the rotor is given a push by hand or by other means in either direction, it will pick-up the speed and continue to rotate in the same direction developing operating torque. Thus a single phase induction motor is *not inherently self starting and requires some special means for starting*.

The above mentioned behaviour of this type of motor can be explained by any one of the following theories :

1. Double revolving field theory 2. Cross-field theory.

The results given by both the theories are approximately same.

Double revolving field theory is described below :

The magnetic field produced by the stator coils is *pulsating*, varying sinusoidally with time. *Ferrari* pointed out that *such a field can be resolved into two equal fields but rotating in opposite directions with equal angular velocities. The maximum value of each component is equal to half the maximum of the pulsating field.*

If the initial time is such that the rotating vectors of the two component fields are along the *Y*-axis in the positive direction, the two component waves ϕ_1 and ϕ_2 coincide. The resultant of these two is ϕ_{max}. After a short interval of time the two vectors rotate, through an angle θ in their respective directions and the waves are shown to occupy the positions in Fig. 5.85. These waves intersect at *A* on the *Y*-axis and as the waves travel *A* moves along the *Y*-axis. Hence the resultant of these two component waves at any instant is equal to 2*OA*.

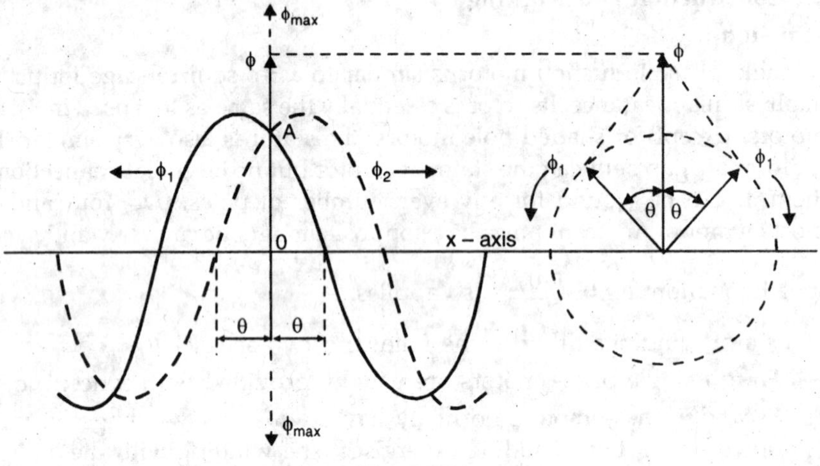

Fig. 5.85.

$$\phi_1 = OA = \phi_{1(max)} \cos(\omega t - \theta) \quad ...(i)$$

$$\phi_2 = OA = \phi_{2(max)} \cos(\omega t + \theta) \quad ...(ii)$$

and, $$\phi_{1(max)} = \phi_{2(max)}$$

By expanding and adding (*i*) and (*ii*),

$$\phi_1 + \phi_2 = 2\phi_{1(max)} \cos\theta \cos\omega t$$

$$2\,OA = \phi_{(max)} \cos\theta \cos\omega t \quad ...(iii)$$

which is the equation of the pulsating field and *proves "Ferrari's statement"*. Thus a single-phase induction motor is *not inherently self-starting*.

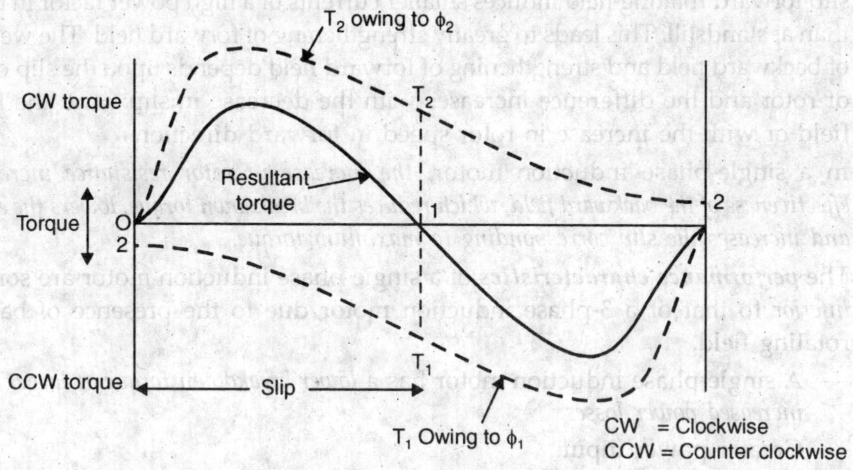

Fig. 5.86. Balanced torque at standstill in squirrel cage rotor excited by a single-phase winding.

The existence of these two fluxes (*forward and backward*) rotating in opposite directions can be verified by supplying a fractional horsepower single-phase induction motor with rated voltage. The motor does not start, but if the shaft is turned by hand, say in clockwise direction, the rotor picks up speed. This means that the rotor conductors are rotating in the direction of that field which rotates in clockwise direction. When the motor is braked and stopped without switching off the supply, the rotor remains at rest. If now the shaft is turned by hand in anti-clockwise direction, the motor picks up speed in that direction. This means that the rotor conductors are now rotating in the direction of the other field.

This behaviour of the motor is *due to the presence of two opposing torques due to the two fields.* When the rotor is at rest, (*i.e.*, slip = 1) the two torques are equal but opposite in direction. Hence the net torque is zero and therefore the rotor remains at rest. Fig. 5.86 shows the torque variations due to the two fields. If the rotor is made to speed up in one direction, say in that in which T_1 increases, T_1 exceeds the opposing torque T_2 and the motor begins to accelerate. T_2 goes on diminishing until at the working speed it is negligibly small. Hence the single-phase induction motor rotates in the direction in which it is made to run.

Thus, if the rotor is made to run at speed N by some external means in any direction, say in the direction of forward field, the two slips are now s and $(2 - s)$, as shown below :

The slip of the rotor w.r.t. the forward rotating field F_f,

$$s_f = \frac{N_s - N}{N_s} = s \qquad ...(1)$$

The slip of the rotor w.r.t. the backward rotating field F_b,

$$s_b = \frac{N_s - (-N)}{N_s} = \frac{2N_s - (N_s - N)}{N_s} = (2 - s) \qquad ...(2)$$

- Under normal running condition $(2 - s) >> s$ and as a consequence the backward field rotor currents are much larger than at standstill and have a low power factor. The corresponding opposing rotor m.m.f., owing to stator impedance, causes the backward field to be greatly reduced in strength. On the other hand, the low-slip forward rotating field induces smaller currents of a high power factor in the rotor than at standstill. This leads to greatly strengthening of forward field. The weakening of backward field and strengthening of forward field depends upon the slip or speed of rotor and the difference increases with the decrease in slip w.r.t. the forward field or with the increase in rotor speed in forward direction.

- In a single-phase induction motor, *the increase in rotor resistance increases the effectiveness of the backward field, which reduces the breakdown torque, lowers the efficiency and increases the slip corresponding to maximum torque.*
- The ***performance characteristics*** of a single phase induction motor are somewhat *inferior* to that of a 3-phase induction motor due to the presence of backward rotating field.
  - — A single-phase induction motor has a *lower breakdown torque at larger slip and increased power losses.*
  - — Greater power input.
  - — The *speed regulation* tends to be *poorer* than that for a polyphase motor.
  - — The *power factor tends to be lower* (since the normal slip of a single-phase induction motor under load conditions is rather greater than that of the corresponding 3-phase motor).
- In view of the above factors, a single phase induction motor has a *larger frame size* than that of 3-phase motor.
- Single-phase motors *tend to be somewhat noiser than 3-phase motors which have no such pulsating torque.*

5.5.6. Three Phase Induction Motors

5.5.6.1. Introduction

An induction motor is simply an *electric transformer* whose magnetic circuit is separated by an air gap into two relatively movable portions, one carrying the primary and the other the secondary winding. Alternating current supplied to the primary winding from an electric power system induces an opposing current in the secondary winding, when *latter is short-circuited or closed through an external impedance.* Relative motion between the primary and secondary structures is produced by the *electromagnetic forces corresponding to the power thus transferred across the air gap by induction.*

The *essential feature which distinguishes the induction machine from other types of electric motors is that the secondary currents are created solely by induction, as in a transformer instead of being supplied by a D.C. exciter or other external power source, as in synchronous and D.C. machines.*

Advantages : Three-phase induction motor is the *most commonly used motor* in industrial applications because of the *advantages* listed below :

1. Simple design.
2. Rugged construction.
3. Reliable operation.
3. Low initial cost.
5. Easy operation and simple maintenance.
6. High efficiency.
7. Simple control gear for starting and speed control.

Applications :

Induction motors are available with torque characteristics suitable for a *wide variety of loads :*

(*i*) The standard motor has a starting torque of about 120 to 150 per cent of full-load torque. Such motors are suitable for most applications.

(*ii*) For starting loads such as small refrigerating machines or plunger pumps operating against full pressure or belt conveyors, *high torque motors with a starting torque of twice normal full-load torque,* or more, are used.

(*iii*) For driving machines that use large flywheels to carry peak loads, such as punch presses and shears, a high-torque motor with a slip at full-load up to 10 per cent is available. *The high slip permits enough change in speed to make possible the proper functioning of the flywheel.*

(*iv*) By the use of a wound-rotor with suitable controller and external resistances connected in series with the rotor winding, it is possible to obtain any value of starting torque up to the maximum breakdown torque. *Such motors are well adapted as constant-speed drives for loads that have large friction loads to overcome at starting.*

5.5.6.2. Constructional details

The Stator :

- The stator frame consists of a symmetrical and substantial casting, having feet cast integral with it. The stator core, consisting of high grade, low loss electrical sheet-steel stampings, is assembled in the frame under hydraulic pressure. The thickness of stampings/laminations is usually from 0.35 to 0.6 mm. The stator laminations are punched in one piece for small induction motor (Fig. 5.87). In induction machines of large size the stator core is assembled from a large number of *segmental laminations.*

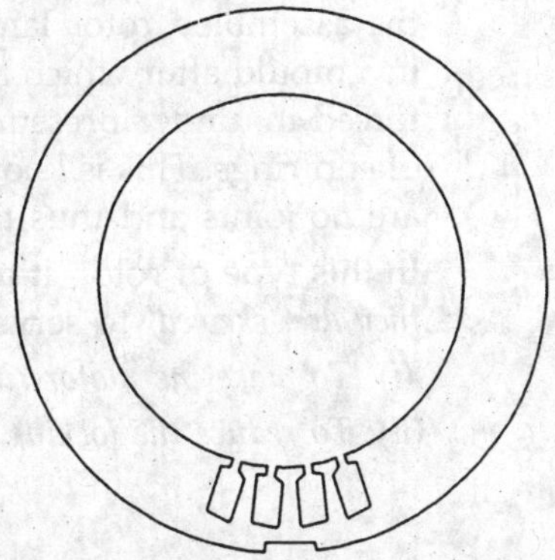

Fig. 5.87. Stator stamping.

The slots are *sometimes of the 'open type* (*i.e.,* having parallel walls) for the accommodation of former wound coils. But the usual practice is to have practically *'enclosed slots' in order to reduce the effective* length of air-gap.

- The *stator windings* are given the utmost care to make them mechanically and electrically sound, so as to ensure long life and high efficiency. After the winding is in position it is thoroughly dried out whilst still hot and is completely immersed in a high grade synthetic resin varnish. It is then *acid, alkali, moisture* and *oil proof.*

For *small motors* working at ordinary voltages, *single layer mush winding* is used. For *medium size machines double layer lap winding with diamond shaped coils* is used. *Single layer concentric windings are used for large motors working at high voltages.*

- *Frames* of electrical machines house the stator core. Frames of small and medium sizes of induction motors have hollow cylindrical form and that of large motors have the shape of a circular box. *In small induction motors,* having a frame diameter of up to about 150 cm, *the frame also supports the end shields.* The frame should be *strong and rigid* as rigidity is very important in the case of induction motors of large dimensions. *This is because the length of the air gap is very small and if the frame is not rigid, it would create an irregular air gap around the machine resulting in production of unbalanced magnetic pull.* Frames for small machines are made as a *single unit* and are usually cast. The frames of *medium and large sized* machines are *fabricated from rolled steel plates.*

The Rotor : The rotors are of two types :

1. Squirrel-cage ;

2. Wound rotor.

1. ***Squirrel-cage.*** The squirrel-cage rotor is made up of stampings (Fig. 5.88), which are keyed directly to the shaft. The slots are partially closed and the winding consists of embedded copper bars to which the short-circuited rings are brazed. The squirrel cage rotor is so robust that it is almost indestructible.

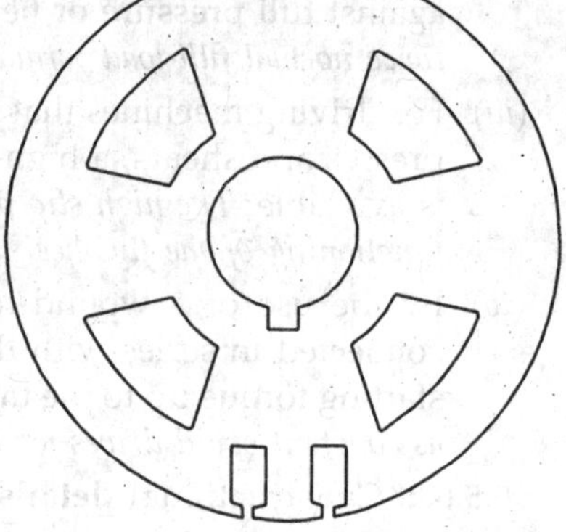

Fig. 5.88. Rotor stamping.

The great majority of present day induction motors are manufactured with squirrel-cage rotors, a *common practice being to employ winding of cast aluminium.* In this construction the assembled rotor laminations are placed in a mould after which molten aluminium is forced in, under pressure, to form bars, end rings and cooling fans as extension of end rings. This is known as *die cast rotor* and has become very popular as there are no joints and thus there is no possibility of high contact resistance.

In this type of rotor, it may be noted that *slots are not made parallel to the shaft but they are 'skewed'* to serve the following purposes :

(*i*) *To make the motor run quietly by reducing the magnetic hum.*

(*ii*) *To reduce the locking tendency of the rotor.*

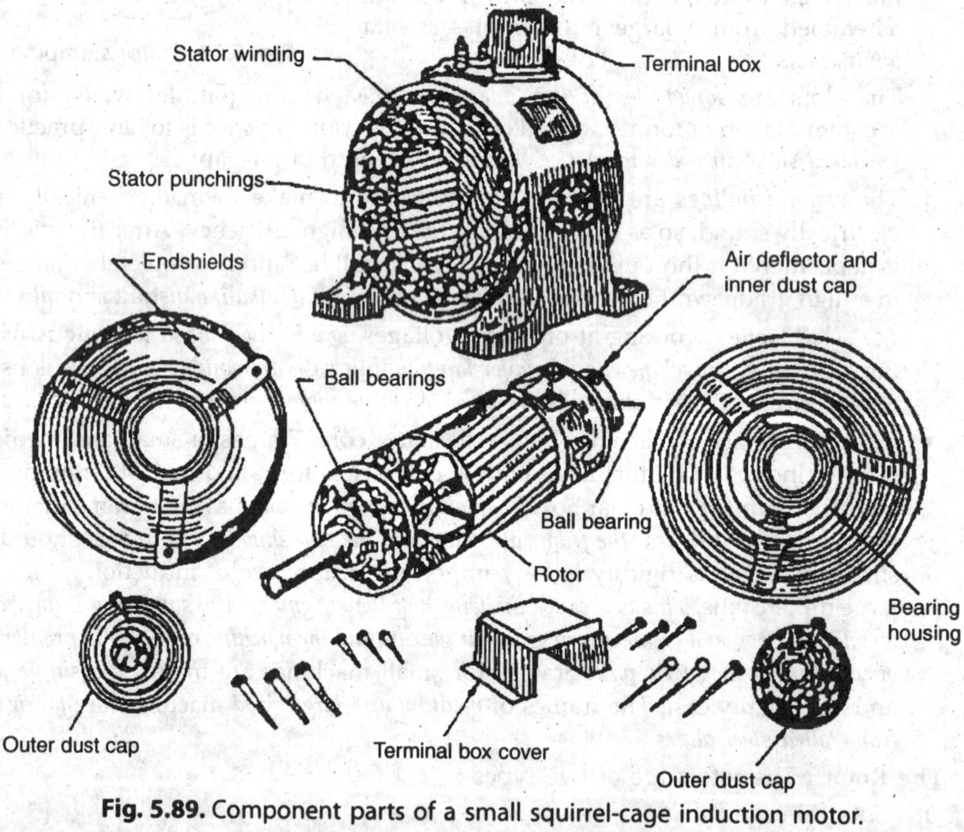

Fig. 5.89. Component parts of a small squirrel-cage induction motor.

2. ***Wound rotor.*** The wound rotor has also slotted stampings and the *windings are former wound*. The *wound rotor construction is employed for induction motors requiring speed control or extremely high values of starting torque.* The wound rotor has completely insulated copper windings very much like the stator windings. The winding can be connected in star or delta and the three ends are brought out at the three slip rings. The current is collected from these slip rings with carbon brushes from which it is led to the *resistances for starting purposes.* When the motor is *running, the slip rings are short-circuited by means of a collar, which is pushed along the shaft and connects all the slip rings together on the inside.* Usually the *brushes are provided with a device for lifting them from the slip rings when the motor has started up, thus reducing the wear and the frictional losses.*

The *number of slots in the rotor should never be equal to the number of slots in the stator.* If they are, there would be a *variation of reluctance of the magnetic path* from maximum, when teeth are opposite slots, to minimum when teeth are opposite teeth. The resulting flux pulsations would have a high frequency, since the periodic time would be the interval period for a tooth to occupy similar positions opposite two successive teeth. This will *not only cause extra iron loss but the rotor will tend to lock with the stator* if at the time of starting teeth are opposite teeth. The best plan is to make the number of the stator and the rotor teeth *prime to each other.*

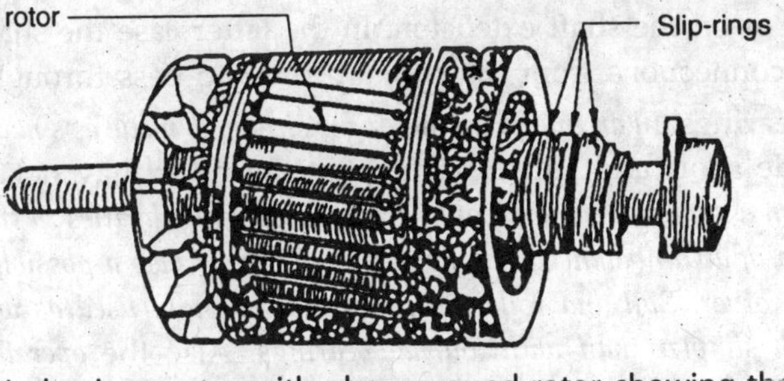

Fig. 5.90. Induction motor with phase-wound rotor, showing the three slip rings on the rotor shaft.

Figs. 5.91 and 5.92 show squirrel-cage and phase-wound induction motors respectively.

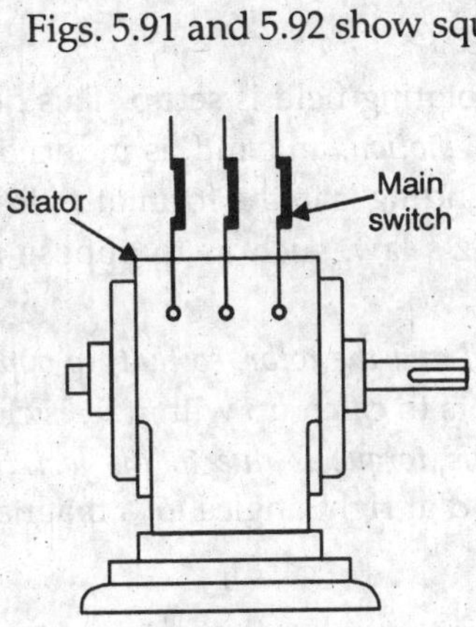

Fig. 5.91. Squirrel-cage motor.

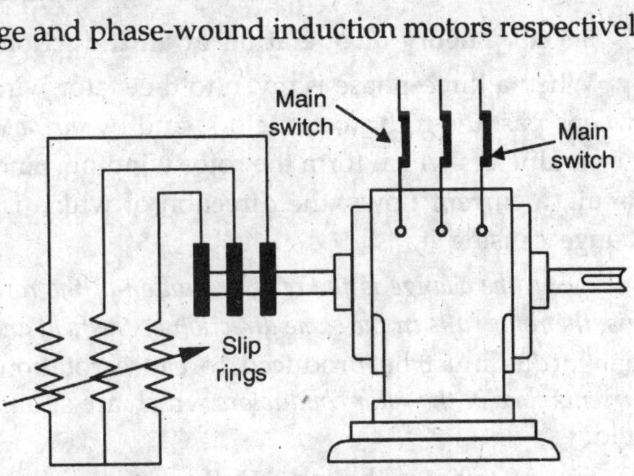

Fig. 5.92. Phase -wound motor connected to a three-phase star-connected starting resistance.

Advantages of a squirrel-cage motor over a phase-wound induction motor. As compared with a wound rotor a squirrel-cage induction motor entails the following *advantages* :

1. Slightly higher efficiency.
2. Cheaper and rugged in construction.
3. No slip rings, brush gear, short-circuiting devices, rotor terminals for starting rheostats are required. The star-delta starter is sufficient for starting.
4. It has better space factor for rotor slots, a shorter overhang and consequently a small copper loss.
5. It has a smaller rotor overhang leakage which gives a better power factor and a greater pull out torque and overload capacity.
6. It has bare end rings, a large space for fans and thus the cooling conditions are better.

The major *'disadvantage' of squirrel-cage motor* is that *it is not possible to insert resistance in the rotor circuit for the purpose of increasing the starting torque. The cage rotor has a smaller starting torque and large starting currents as compared with wound rotor.*

Slip rings. The slip rings for wound-rotor machines are made of *either brass or phosphor bronze.* They are shrunk on to a cast iron sleeve with moulded silica insulation. This assembly is passed on to the rotor shaft. The slip rings are rotated either between the core and the bearing or on the shaft extension. In the latter case the shaft is made hollow to allow the three connections from rotor to slip rings to pass through bearings.

Shaft and bearings. *In an induction motor the air gap is made as small as possible.* Therefore the shaft is made short and stiff in order that the rotor may not have any significant deflection, *as even a small deflection would create large irregularities in the air gap which would lead to production of an unbalanced magnetic pull. There is also a possibility of rotor and stator fouling with each other. Ball and roller bearings are generally used as with their use, accurate centering is much simpler than with journal bearings.* Also the *overall length of machine is reduced. For small motors,* a roller bearing may be used at the driving end and a ball bearing at the non-driving end. *For large and heavy rotors journal bearings are used.*

5.5.6.3. Theory of operation of an induction motor

When a three-phase is given to the stator winding a rotating field is setup. This field sweeps past the rotor (conductors) and by *virtue of relative motion,* an e.m.f. is induced in the conductors which form the rotor winding. Since this winding is in the form of a closed circuit, a current flows, the direction of which is, by Lenz's law, such as to oppose the change causing it.

Now, *the change is the relative motion of the rotating field and the rotor, so that, to oppose this, the rotor runs in the same direction as the field and* attempts to catch up with it. It is clear that torque must be produced to cause rotation, and *this torque is due to the fact that currents flow in the rotor conductors* which are situated in, and at right angles to, a magnetic field.

Fig. 5.93 shows the induction motor action.

- When the motor shaft is *not loaded,* the machine has only to rotate itself against the mechanical losses and the rotor speed is *very close to the synchronous speed.* However, the rotor speed cannot become equal to the synchronous speed because if it does so, *the e.m.f. induced in the rotor winding would become zero and there will be no torque.* Hence the speed remains slightly less than the synchronous speed. If the motor shaft is *loaded,* the rotor will slow down and the relative speed of the rotor with respect to the stator rotating field will increase. The e.m.f. induced in the rotor winding will increase and will produce more rotor current which will increase the electromagnetic torque produced by the motor. Conditions of equilibrium are attained when the rotor speed has adjusted to a new value so that the electromagnetic torque is sufficient to balance the mechanical or load torque applied to the shaft. The speed of the motor *when running under full load conditions is somewhat less than the no-load speed.*

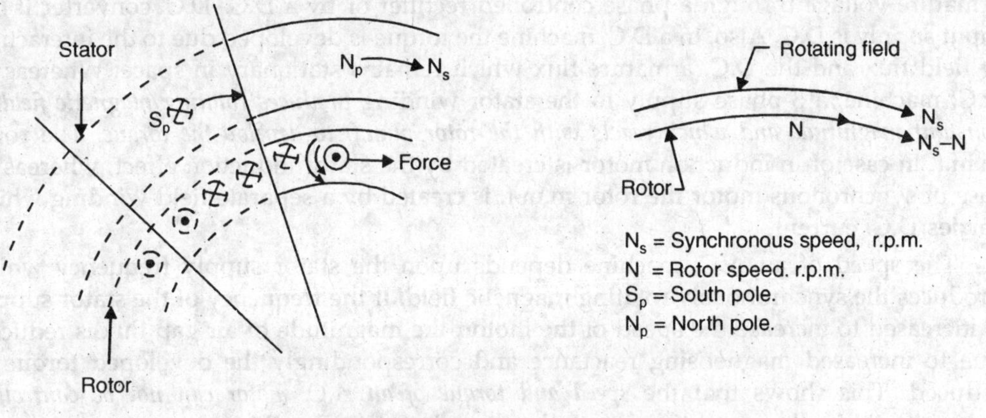

Fig. 5.93. Induction motor action.

5.5.6.4. Slip

- As earlier stated, *the rotor speed must always remain less than the synchronous speed. The difference between the synchronous speed and the rotor speed is known as* **'slip'**. It is usually expressed as a fraction of the synchronous speed. Thus slip s is

$$s = \frac{N_s - N}{N_s} \quad ...(5.6)$$

or, $$N = N_s(1 - s) \quad ...(5.6a)$$

where, N_s = Synchronous speed (r.p.m.), and
N = Motor speed (r.p.m.)

In practice the value of slip is very small. At no-load, slip is around 1% or so and at full-load it is around 3%. For large efficient machines the slip at full-load may be around 1% only. The induction motor, is therefore, a motor with substantially constant speed and fills the same role as *D.C. shunt motor.*

- When the rotor is stationary (standstill) its speed is zero and $s = 1$. The rotor cannot run at synchronous speed because then their will be no rotor e.m.f. and no rotor current and torque. If the rotor is to run at synchronous speed an external torque is necessary. *If the rotor is driven such that* $N > N_s$, *the slip becomes negative, the rotor torque opposes the external driving torque and the machine acts as* ***induction generator.***
- The induction motor derives its name from the fact that the *current in the rotor circuit is induced from the stator.* There is no external connection to the rotor except for some special purposes.

If the rotor reactance at standstill is X_2 its value at slip 's' becomes sX_2. *This is very desirable,* for at no-load the reactance becomes almost negligible and the rotor impedance is now all resistance. Further if the rotor resistance is small the rotor current is large, so that motor works with a large torque which brings the speed near to synchronous speed, *i.e.,* the slip is reduced.

5.5.7. Electronic Control of A.C. (Induction) Motors

5.5.7.1. Introduction

The speed of a D.C. motor can be controlled by varying the field current or the armature voltage through a phase controlled rectifier or by a D.C.–D.C. converter if the input supply is D.C. Also, in a D.C. machine the torque is developed due to the interaction of field flux and the D.C. armature flux which remains stationary in space. Whereas in A.C. machine, a 3-phase supply to the stator winding *produces rotating magnetic field of constant magnitude and which reacts with the rotor m.m.f. to develop the torque.* The rotor m.m.f. in case of in induction motor is created by the stator induction effect, whereas in case of synchronous motor the rotor m.m.f. is created by a separate field winding which carries D.C. current.

The speed of an A.C. machine depends upon the stator supply frequency which produces the synchronously rotating magnetic field. If the frequency of the stator supply is increased to increase the speed of the motor, the magnitude of air gap flux is reduced due to increased magnetising reactance and correspondingly the developed torque is reduced. This shows that the *speed and torque of an A.C. motor can not be controlled independently by the conventional methods of speed control.* For this reason, an *A.C. motor requires 'variable voltage variable frequency' power supply for its speed control* A *'D.C. link converter system'* consisting of a rectifier and an inverter or a *'cycloconverter'* can be used as a variable voltage-variable frequency source.

However, it may be noted that the *voltage and current waves obtained by solid state devices are rich in harmonics and cause problem of harmonic heating torque pulsation.*

5.5.7.2. Speed control of a single-phase induction motor

The most common method for speed control of a single-phase induction motor is the *stator voltage control method.*

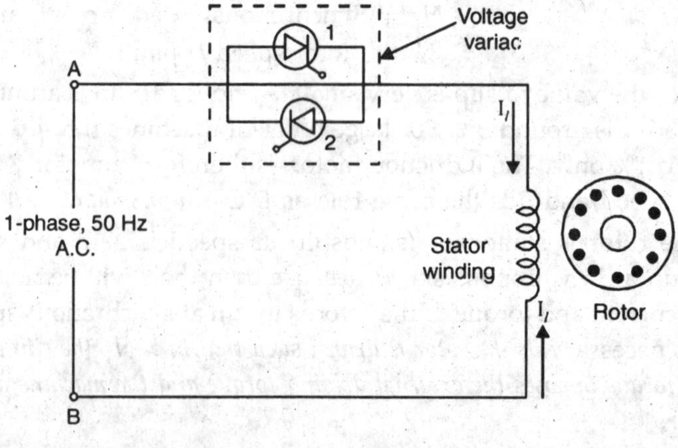

Fig. 5.94. Circuit diagram for speed control of a single-phase induction motor by stator voltage control method.

Figure 5.94, shows the circuit diagram for speed control of a single-phase induction motor by stator voltage control method. The circuit uses two *SCRs* connected in anti parallel. In the +ve half cycle when point *A* is positive and point *B* is negative SCR_1 is triggered. The direction of flow of current in the stator winding is from the top to bottom. In the –ve half-cycle point *A* becomes negative and *B* becomes positive. SCR_1 is turned OFF and SCR_2 is triggered. The direction of flow of current in the stator winding is reversed. In other words, the alternating current supply becomes available across the stator winding of the motor. By varying the firing angles of *SCRs* 1 and 2 the magnitude of the A.C. voltage across the stator winding of the motor can be *varied* ; this in turn will *vary the speed of the motor.*

Different schemes under this method are discussed below.

1. Speed control by using triac :

By using a triac, *very smooth speed control* of a single-phase induction motor can be obtained. A diac is used as a triggering agent for the triac in the circuit. Fig. 5.95 shows the circuit diagram for this arrangement. A diac-triac pair can provide the widest range of control.

There are two *R-C* networks. R_1-C_1 from the triggering circuit whereas R_2-C_2 along with C_1 form the π-network (filter) which would bypass any spike in the A.C. mains supply. The values of R_2 and C_2 are lower than the values of R_1 and C_1. R_2 also works as a current limiting resistance for the diac. The *R-C* triggering gate control process adopted in this circuit provides a very wide and smooth speed control for 1-phase induction motor.

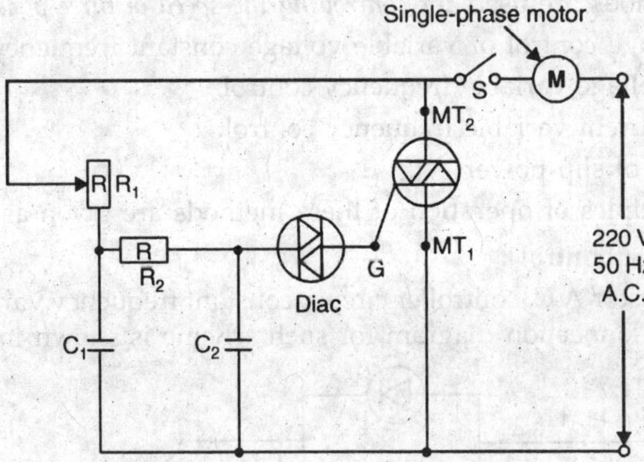

Fig. 5.95. Circuit diagram for the speed control of a single-phase induction motor using a triac.

- This circuit may be effectively employed for fabricating *fan regulators and illumination controllers.*

2. Speed control using single-phase inverter circuit :

With the help of an 'inverter circuit' we can obtain variable voltage fixed frequency A.C. supply which can be fed to the motor for speed control. Fig. 5.96, shows the block diagram representation for this scheme.

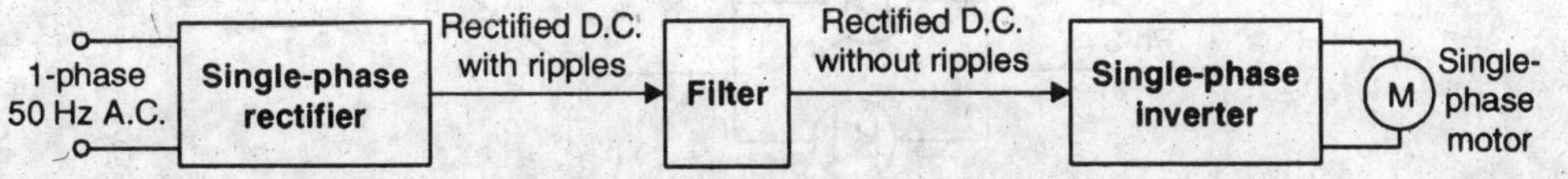

Fig. 5.96. Block diagram representation for the scheme of speed control of a single-phase induction motor using an inverter circuit.

- Single-phase A.C. is rectified with the help of a single-phase full-wave rectifier and then filtered to minimise the ripple content.
- The inverter output, which is a *fixed frequency variable A.C. voltage,* is fed to the motor whose speed is to be controlled. Inverter output (A.C. in nature) is *made variable by changing the firing time (angle) of SCRs.*

This process is known as *fixed frequency variable voltage control.*

3. Speed control by using cycloconverter circuits :

Fig. 5.97. Block diagram representation of the cycloconverter scheme for speed control of a single-phase induction motor.

Basically, this is a *variable frequency method for speed control.* By controlling the firing sequence of the *SCRs* connected in a cycloconverter the frequency of the A.C. input voltage can be changed. The variable frequency A.C. supply available at the point of the cycloconverter circuit may be fed to the motor for speed control. A block diagram representation of such a scheme is shown in Fig. 5.97.

- *"Cycloconverters" are mostly used for speed control of gearless drives.*

5.5.7.3. Speed control of three-phase induction motors

Following methods are used for *controlling the speed of three-phase induction motors :*

1. Stator voltage control or variable voltage constant frequency control.
2. Variable voltage variable frequency control.
3. Variable current variable frequency control.
4. Regulation of slip power.

The basic principles of operation of these methods are given as follows :

1. Stator voltage control :

By using a thyristor A.C. controller circuit, constant frequency variable voltage supply can be generated. Connection diagram for such scheme is shown in Fig. 5.98.

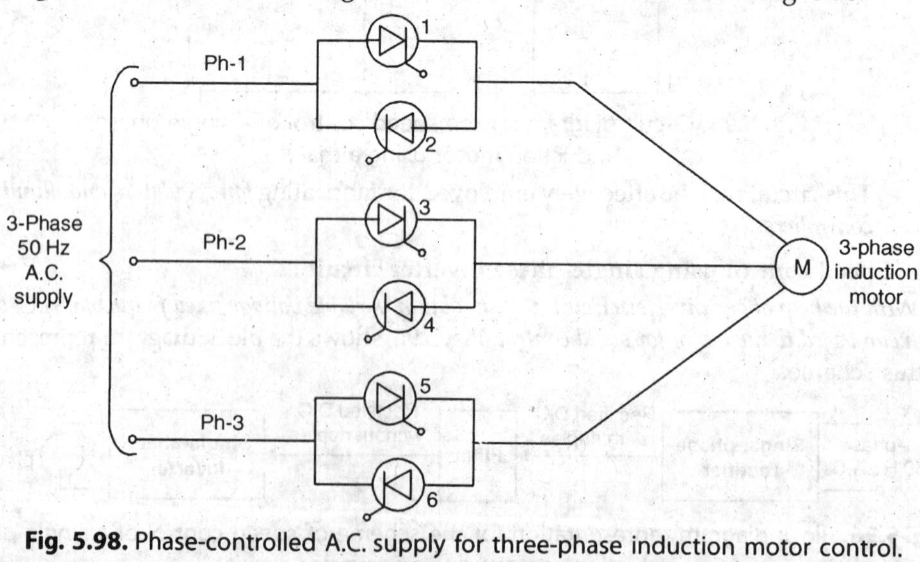

Fig. 5.98. Phase-controlled A.C. supply for three-phase induction motor control.

By changing the applied voltage, air gap flux can be changed so also the slip, and motor speed can be altered. To obtain a reasonable control a full thyristor controller is used. Two *SCRs* connected in antiparallel per phase are used to form three such bridges. *SCRs* 1, 2 form the bridge for phase-1, similarly, 3, 4 form the bridge for phase-2 and 5, 6 for phase-3. The controlled (variable) three-phase voltage, when fed to the 3-phase induction motor, will result in the desired speed control of the motor.

- This arrangement is *quite costly* and *its firing circuit will also be quite complicated.*

2. Variable-voltage variable-frequency control :

Fig. 5.99 shows the basic block diagram for a speed control scheme of a 3-phase induction motor. This is basically a *variable-voltage variable-frequency supply scheme* for the speed control.

- 3-phase A.C. is rectified into D.C. and then filtered to minimise the ripple content. *L-C.* filter is generally used for this purpose. This controlled D.C. is converted into controlled pulses by means of a voltage to frequency converter. These controlled pulses are fed to the inverter bridge for producing the variable-voltage variable-frequency output. This output is fed to the 3-phase induction motor for controlling its speed.

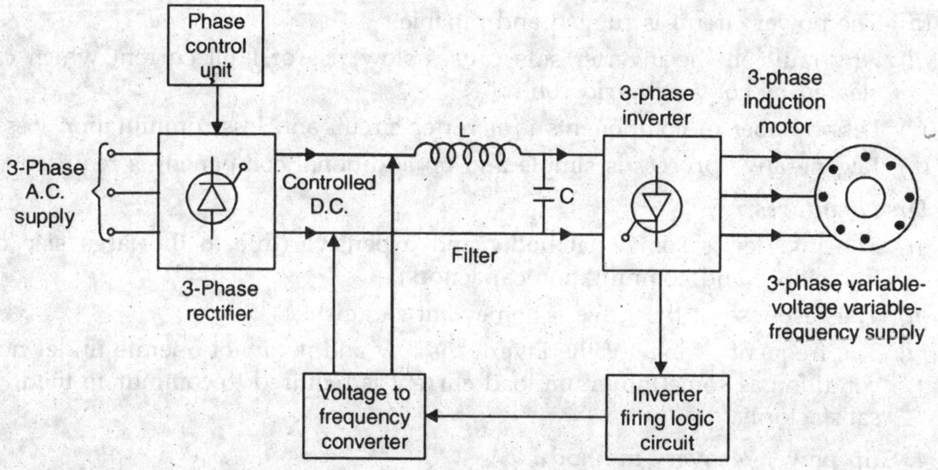

Fig. 5.99. Block diagram for basic scheme (variable-voltage variable-frequency control) for speed control of a 3-phase induction motor.

- The 'phase control circuit' is employed for triggering and logic sensing of 3-phase rectifier circuit. This circuit *controls the firing angle of the rectifier bridge.* The *'inverter firing logic circuit' controls the firing angle of the inverter bridge.*

3. Variable-current variable-frequency control :

Fig. 5.100, shows a variable-current variable-frequency control circuit for an induction motor.

A phase-controlled rectifier produces variable D.C. voltage which is converted to a current source by connecting a large inductor in series.

A diode rectifier followed by a D.C. chopper can also be used as a variable voltage D.C. source. It can be shown that the voltage at the terminals of 3-phase induction motor is almost sinusoidal with superimposed voltage spikes due to commutation.

The *'converter'* used is a *line commutated* whereas the *'inverter'* is *forced commutated* as the induction motor is a lagging p.f. load.

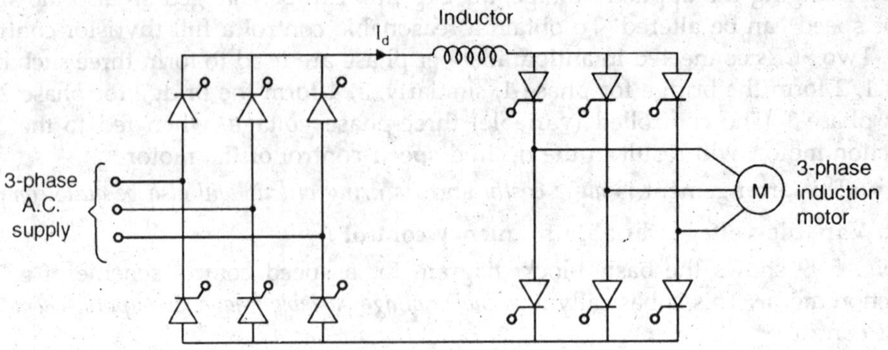

Fig. 5.100. Variable-current variable-frequency control circuit.

The following are the *advantages* and *disadvantages* of this circuit :

Advantages :

(*i*) The control circuit is simpler and more reliable since only six thyristors are to be controlled.

(*ii*) The power circuit is rugged and reliable.

(*iii*) Any fault on the inverter side causes slow rise of fault current which can be cleared by converter grid control.

(*iv*) Less number of components in inverter circuit and less commutation loss.

(*v*) Regenerative process is simple and no additional component is required.

Disadvantages :

(*i*) The inverter is somewhat bulky and expensive (due to the large size of the inductance and commutation capacitors).

(*ii*) The response of the drive is somewhat sluggish.

(*iii*) The frequency range of the inverter is low and it cannot operate under no load condition as some minimum load current is required to commutate the inverter satisfactorily.

4. Slip power recovery method :

When the supply frequency and the voltage are fixed, the speed of an induction motor can be *varied by injecting a counter e.m.f. into the rotor circuit of the motor.* This method is, therefore, *used for a wound-rotor induction motor.* The inefficiency of the drive system because of large slip power dissipation can be overcome by this method.

In this method the slip power of the motor is rectified by a diode rectifier and is them pumped back to the A.C. line through a line commutated inverter. Fig. 5.101 shows the schematic diagram for this method.

The required power handling capability of the converter corresponds to the slip power and hence if the range of speed control is small the rating of converter is also small. However, if speed control upto standstill is the required, the converters should be rated for the full machine rating. The initial in-rush current in the converters can be avoided by connecting separate starting resistances in the slip ring circuit.

The torque pulsation and additional heating must be considered while designing the drive system.

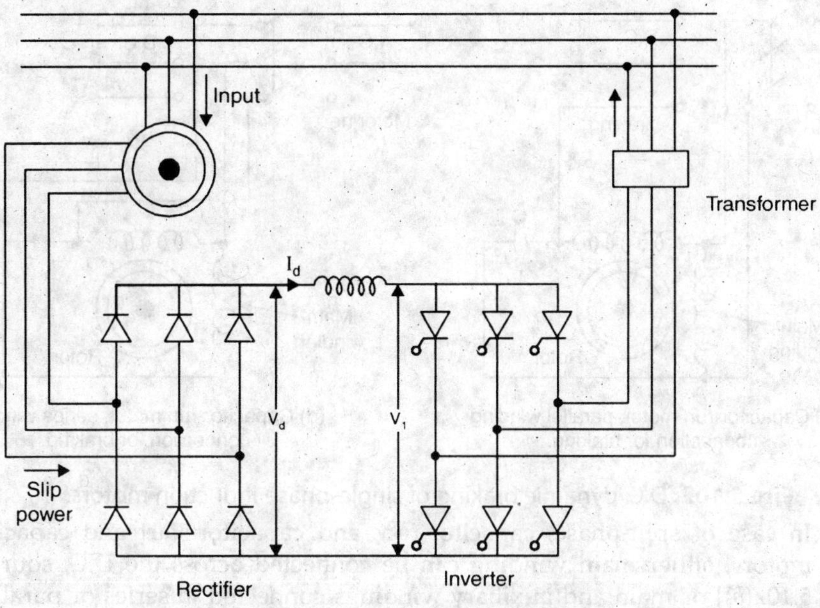

Fig. 5.101. Slip power recovery method.

The ***disadvantage*** of this system is that *regeneration and speed reversal are not possible in the drive system.*

- The drawback of this method is that it results in *very poor p.f.*
- This drive system is used in *large H.P. pump and blower type applications where limited range of speed control is required.*

5.5.7.4. Braking of single-phase motors

These motor can be braked by :

(*i*) D.C. dynamic braking (*ii*) Plugging.

(*i*) ***D.C. dynamic braking :*** This method is commonly used for braking of single-phase induction motors. With the help of a double-pole double-throw switch or triple-pole double-throw switch, motor connection is shifted from A.C. (motoring) to D.C. source for braking. These connections for various single-phase induction motors are shown in Fig. 5.102.

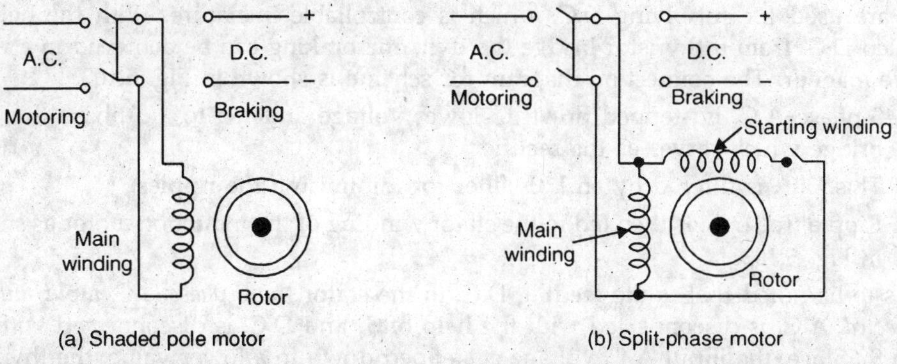

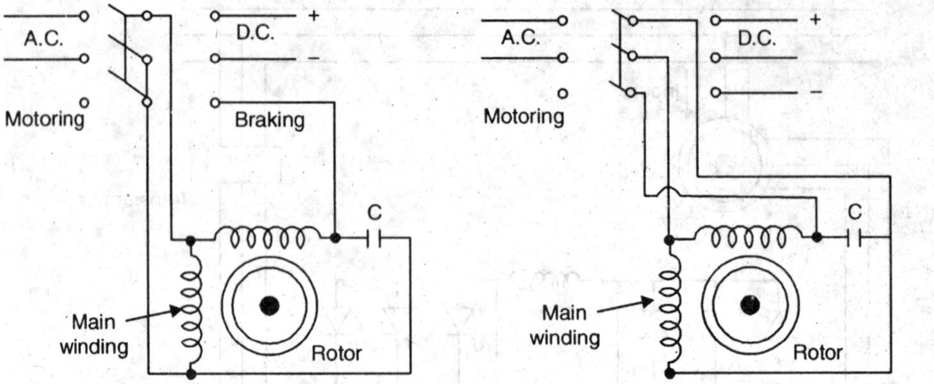

(c) Capacitor-run motor, parallel winding connection for braking.

(d) Capacitor-run motor, series winding connection for braking.

Fig. 5.102. D.C. dynamic braking of single-phase induction motors.

- In case of split-phase, capacitor-run, and capacitor-start and capacitor-run motors, either main winding can be connected across the D.C. source [Fig. 5.102(*b*)] or main and auxiliary windings connected in series or parallel [Fig. 5.102(*c*) and (*d*)].
- When in braking connection, D.C. current through the stator winding (or windings) produces a stationary field through which squirrel cage rotor moves. Current induced in rotor bars *interacts with D.C. field to produce braking torque,* as in 3-phase induction motor. Motor *decelerates and stops.* As induced currents are zero at zero speed, the braking torque is also zero. *For braking, supply is obtained by a diode rectifier connected to A.C. mains.* Motor winding can be connected directly across diode rectifier to obtain fast braking. Winding is disconnected from D.C. supply after the motor stops.

(*ii*) ***Plugging and reversal :*** Except in case of shaded pole motor, *plugging and reversal is obtained by changing phase sequence by reversing polarity of one of the windings.*

5.5.7.5. Dynamic braking of a 3-phase induction motor

The speed of an induction motor can be controlled by injecting D.C. voltage in its stator winding. A variable resistance may be used in the rotor (in case of a slip ring induction motor) for dissipating the required amount of power. Now-a-days *thyristor bridges* are used for supplying D.C. which is controllable in nature. With the help of controlled D.C. from a thyristor bridge the dynamic braking can be achieved in a more effective manner. The connection diagram for scheme is shown in Fig. 5.103.

- 3-phase A.C. is stepped down to lower voltage and fed to a 3-phase thyristor bridge which serves as the rectifier.
- This D.C. is filtered by an L.C. filter for minimising the ripples.
- Ripple free D.C. is then fed to the stator winding of the induction motor as shown in Fig. 5.103.

It is to be noted that while feeding D.C. to the stator the *3-phase A.C. input must be disconnected.* A.C. is disconnected with the help of S_1 and D.C. is disconnected with the help of S_2. Since, the input A.C. voltage is stepped down to a lower value, the thyristor converter may be of lower voltage rating.

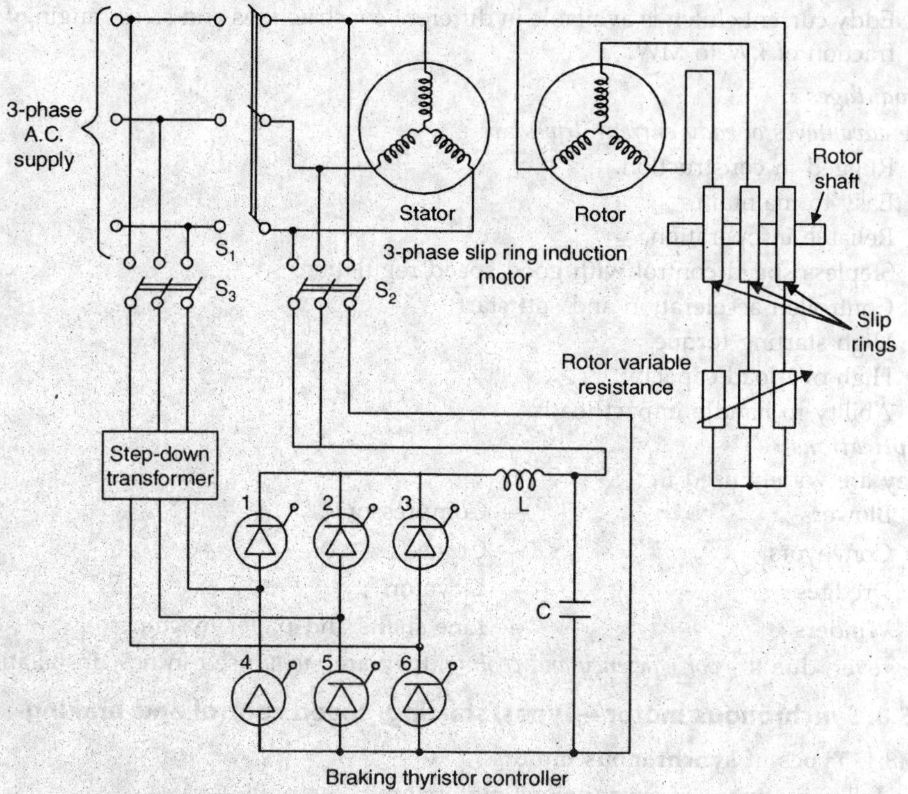

Fig. 5.103. Dynamic braking of a 3-phase slip ring induction motor.

5.5.7.6. Eddy current drives

- An eddy current drive consists of an *eddy current clutch placed between an induction motor running at a fixed speed and the variable speed load.* Speed is controlled by controlling D.C. excitation to magnetic circuit of the clutch. Since motor itself runs at a fixed speed it can be fed directly from A.C. mains.
- The principle of an eddy-current clutch is identical to an induction motor in which both stator and rotor are allowed to rotate. Stator, which is coupled to driving induction motor, has D.C. winding which produces magnetic field rotating at the speed of rotor. Rotor has a metal drum coupled to the load. Eddy currents are induced in rotor drum by stator magnetic field. Interaction between the stator field and eddy currents produces a torque which causes rotor to move with a slip. Slip, and therefore, the load speed, can be controlled by controlling D.C. current through rotor winding. Speed torque characteristics are identical to an induction motor.
  - — Speed reduction is obtained by wasting a power equal to sP_{in} in the rotor drum. Minimum speed is usually restricted to 30 percent below the synchronous speed, because efficiency becomes too low and cooling of the rotor drum becomes difficult below this speed.
  - — Load can be decoupled from induction motor by setting D.C. winding current to zero. Motor can now be started on no load. Load can be smoothly started by slowly increasing D.C. winding excitation.

- Eddy current clutch is available in different constructions and sizes ranging from fraction of kW to MW.

Advantages :

The *advantages of eddy current drives* are :

(*i*) Rugged in construction.
(*ii*) Easy to maintain.
(*iii*) Reliable in operation.
(*iv*) Stepless speed control with good speed regulation.
(*v*) Controlled acceleration and soft start.
(*vi*) High starting torque.
(*vii*) High overload capacity.
(*viii*) Ability to handle impact loads.

Applications :

They are widely used in :

— Blowers ;
— Compressors ;
— Conveyors ;
— Cranes ;
— Dredges ;
— Elevators ;
— Winders ;
— Line shafts and paper machines.

However, due to *poor efficiency and cooling*, they are *rarely* used in new installations.

5.5.8. Synchronous motor—Types, starting, speed control and braking

5.5.8.1. Types of synchronous motors

The following types of synchronous motors are commonly used :

1. Wound field synchronous motors.
2. Permanent magnet (PM) synchronous motors.
3. Synchronous reluctance motors.
4. Hysteresis synchronous motors.

— All these motors have a stator with a 3-phase winding, which is connected to an A.C. source.

— Fractional horse power synchronous reluctance and hysteresis motors employ a 1-phase stator.

1. Wound field motors :

Wound field synchronous motor rotor has a D.C. field winding, which is supplied from a D.C. source through slip-rings and brushes. The rotor can have cylindrical or salient pole construction.

— Cylindrical rotors have *higher mechanical strength* and are *employed in high power and high speed applications.*

— Salient pole motors, due to low cost, are preferred for other applications.

2. Permanent magnet (PM) synchronous motors :

In medium and small size motors, D.C. field can be produced by permanent magnets; thus dispensing with D.C. source, slip rings, brushes and field winding losses. Such motors are known as *permanent magnet (PM) synchronous motors.* Usually ferrite magnets are used. Rare earth (cobalt-samarium) magnets, although very expensive, are sometimes used to *reduce* the volume and weight of the motor.

PM synchronous motors are *classified* as follows :

(i) Surface mounted :

(*a*) *Projecting type.* In such motors magnets project from the surface of the rotor [Fig. 5.104 (*a*)].

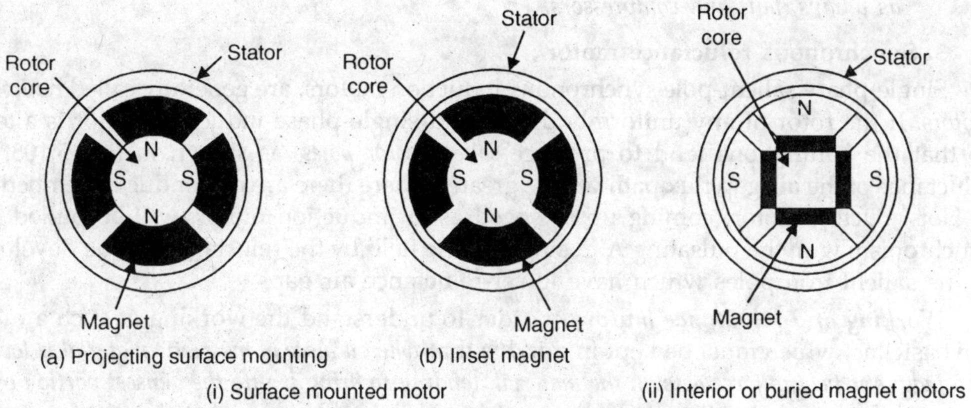

Fig. 5.104. Types of PM synchronous motors.

(*b*) *Inset type.* In this case, magnets are inserted into the rotor, providing a smooth rotor surface [Fig. 5.104(*i*) (*b*)]

- Epoxy glue is used to fix the magnets to the rotor surface in both.
- While these motors are *easy to construct and are less expensive, they are less robust compared to interior type rotors and are not suitable for high speed applications.*

(ii) Interior (or buried) : In these motors, magnets are imbedded in the interior of the rotor [Fig. 5.103(*ii*)].

Features of wound field and permanent magnet synchronous motors :

The wound field and PM synchronous motors have a *higher full load efficiency and power factor than an induction motor.*

Wound field motors can be designed for a higher power rating than induction motors. Since the air-gap flux is not produced solely by the magnetising current drawn from the armature, a larger air-gap suiting the mechanical design can be chosen. The ability to control power factor is an important advantage at higher power levels. *Operating at unity power factor minimizes the inverter rating.*

PM synchronous motor, apart from the *robust construction, has low losses and high efficiency.* Because of low losses, it is possible to make motors with very high power density and torque to inertia ratio. These make them *suitable for servo drives requiring fastest possible dynamic response.*

- One *significant difference* between the wound field and PM motors, which are designed to operate with a source of fixed frequency is discussed below :

When a wound field motor is started as induction motor, D.C. field is kept off. In case of a PM motor, the field *cannot* be 'turned off'. When at a speed below synchronous speed, the rotor field induces a voltage in the stator, which has a frequency different than the frequency of stator supply. The current produced by induced voltage interacts with the rotor field to produce a braking torque, which opposes the induction motor torque due to damper winding. The permanent magnet synchronous motor (PMSM) is designed so that the braking torque is very small compared to induction motor torque. Owing to the capability of starting direct on line these motors are called *line start PMSM.*

— PMSM are available in 3-phase and 1-phase construction.

— Although expensive to induction motors, they have advantages of *high efficiency, high power factor and low sensitivity to supply voltage variations.*

— These motors are preferred for industrial applications with *large duty cycles such as pumps, fans and compressors.*

3. Synchronous reluctance motor :

Single-phase salient-pole synchronous-induction motors, are generally called *reluctance motors.* If the rotor of any uniformly distributed single-phase induction motor is altered so that the laminations tend to produce *salient rotor poles,* as shown in Fig. 5.105, the reluctance of the air-gap flux path will be greater where there are no conductors embedded in slots. Such a motor, coming up to speed as an induction motor, will be pulled into synchronism with the pulsating A.C. single-phase field by the reluctance torque developed at the salient iron poles which have lower-reluctance air gaps.

Working of a reluctance motor. In order to understand the working of such a motor the basic fact which must be kept in mind is that *when a piece of magnetic material is located in a magnetic field, a force acts on the material, tending to bring it into the densest portion of the field. The force tends to align the specimen of material in such a way that the reluctance of the magnetic path that passes through the material will be minimum.*

When supply is given to the stator winding, the revolving magnetic field will exert reluctance torque on the unsymmetrical rotor tending to align the salient pole axis of the rotor with the axis of the revolving magnetic field (because in this position, the reluctance of the magnetic path would be minimum). If the reluctance torque is sufficient to start the motor and its load, the rotor will pull into step with the revolving field and continue to run at the speed of the revolving field. (Actually the motor starts as an induction motor and after it has reached its maximum speed as an induction motor, the reluctance torque pulls its rotor into step with the revolving field so that the motor now runs as synchronous motor by virtue of its saliency).

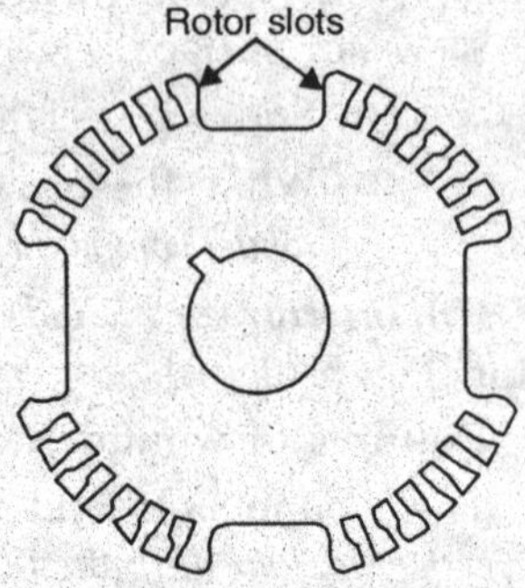

Fig. 5.105. Reluctance-motor lamination.

Reluctance motors have approximately *one-third* the horsepower rating they would have as induction motors with cylindrical rotors, although the ratio may be increased to one-half by proper design of the field windings. *Power factor and efficiency are poorer than for the equivalent induction motor.* Reluctance motors are subject to *'cogging'*, since, the locked-rotor torque varies with the rotor position, but the effect may be *minimized* by *skewing the rotor bars* and by *not* having the number of rotor slots exactly equal to an exact multiple of the number of poles.

Uses. Despite its shortcomings, the reluctance motor is widely used for many *constant speed* applications such as *recording instruments, time devices, control apparatus, regulators,* and *phonograph turntables.*

- *Reversing* is obtained as in any single-phase induction motor.

Speed-torque characteristics. Fig. 5.106 shows speed-torque characteristics of a typical single-phase reluctance motor.

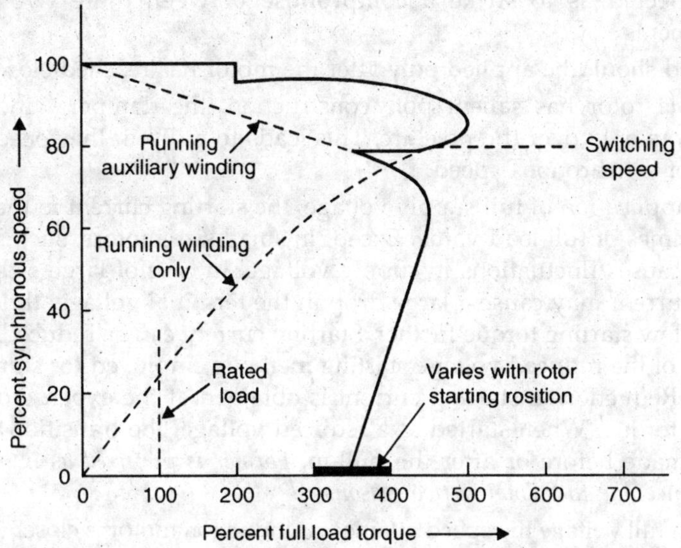

Fig. 5.106. Speed-torque characteristics of a single-phase reluctance motor.

- The motor starts at anywhere from 300 to 400 per cent of its full-load torque (depending on the rotor position of the unsymmetrical rotor with respect to the field windings) as a two-phase motor as a result of the magnetic rotating field created by a starting and running winding (displaced) 90° in both space and time.
- At about 3/4th of the synchronous speed, a centrifugal switch opens the starting winding, and the motor continues to develop a single-phase torque produced by its running winding only.

  As it approaches synchronous speed, the reluctance torque (developed as a synchronous motor) is sufficient to pull the rotor into synchronism with the pulsating single-phase field).
- *The motor operates at a constant speed up to a little over 200% of its full-load torque.* If it is loaded beyond the value of pull-out torque, it will continue to operate as a single-phase induction motor up to 500% of its rated output.

5.5.8.2. Starting of synchronous motor

The purpose of starting method is to bring rotor speed close to synchronous speed. Following methods are used to start synchronous motor :

1. Using damper windings as a squirrel-cage induction motor.
2. Using a low power auxiliary motor.

1. Using damper windings as a squirrel cage induction motor :

One widely used method is to start the synchronous motor as an induction motor with field unexcited and damper winding serving as a squirrel-cage rotor. Regarding this method, following points are worth noting :

(*i*) The currents and starting torque can be reduced and increased respectively, by increasing the damper winding resistance. The motor speed while running as an induction motor, for successful pull-in, must be close to synchronous speed. For this the damper winding resistance must be low. Further, for damping hunting oscillation damper winding resistance must be low. The damper winding resistance is so selected as to strike a compromise between these two contradictory requirements.

(*ii*) D.C. field should be applied only after the motor has reached close to full speed.

(*iii*) When the rotor has salient pole construction, the damper winding can have conductors only over the pole arc. This leads to a dip in the speed-torque curve at half of synchronous speed.

(*iv*) On the application of full supply voltage, the starting current in the motor can be 7 to 10 times of full load value. Except in small size motors, such a high starting current causes fluctuations in supply voltage. In case of large size motors, such a high current may cause a large drop in the terminal voltage, thus reducing the already low starting torque further. Starting current can be reduced by employing any one of the reduced voltage starting methods employed for starting induction motors. Reduction in starting current is obtained at the expense of reduction in starting torque. When started at a reduced voltage, the transition to full voltage can be made before or after the pull-in. *Former is preferred as it improves pull-in performance due to following two reasons :*

— With full voltage the speed attained as induction motor is closer to synchronous speed, and

— The pull-in torque increases in proportion to voltage squared, consequently pull-in can be achieved faster and with large motor loads.

2. Using a low power auxiliary motor :

(*i*) In this method a low power auxiliary motor is coupled to the synchronous motor shaft. With the help of auxiliary motor, the rotor speed is brought near synchronous speed and then D.C. field is switched-in.

(*ii*) This method has a very *low starting torque.*

Note: It is practically impossible to start a synchronous motor with its D.C. field energized. Even when left de-energized, the rapidly rotating magnetic field of the stator will induce extremely high voltages in the many turns of the field winding. It is customary, therefore, *to short-circuit the D.C. field winding during the starting period* ; whatever voltage and current are induced in it may then aid in producing induction motor action. In very large synchronous motors, field sectionalising or field-splitting switches are used which short-circuit individual field windings to prevent cumulative addition of induced voltages from pole to pole.

5.5.8.3. Braking of synchronous motors

- The motor can work in regenerative braking only at synchronous speed. Therefore, *"regenerative braking" cannot be used for stopping or decelerating a load.*
- *"Dynamic braking"* is obtained by *disconnecting stator from the source and connecting it to 3-phase resistor.* Machine works as a synchronous generator and dissipates generated energy in the braking resistor.

5.5.8.4. Speed control of synchronous motors

The *speed* of synchronous motors can be *controlled* as follows :

(*i*) By using current-fed D.C. link.

(*ii*) By using cycloconverter.

1. Speed control by current-fed D.C. link :

Fig. 5.107, shows the circuit diagram for speed control of synchronous motor by current-fed D.C. link.

The typical circuit consists of three converters two of which are connected between the 3-phase source and synchronous motor and the third converter (bridge rectifier) supplies D.C. field excitation for the rotor.

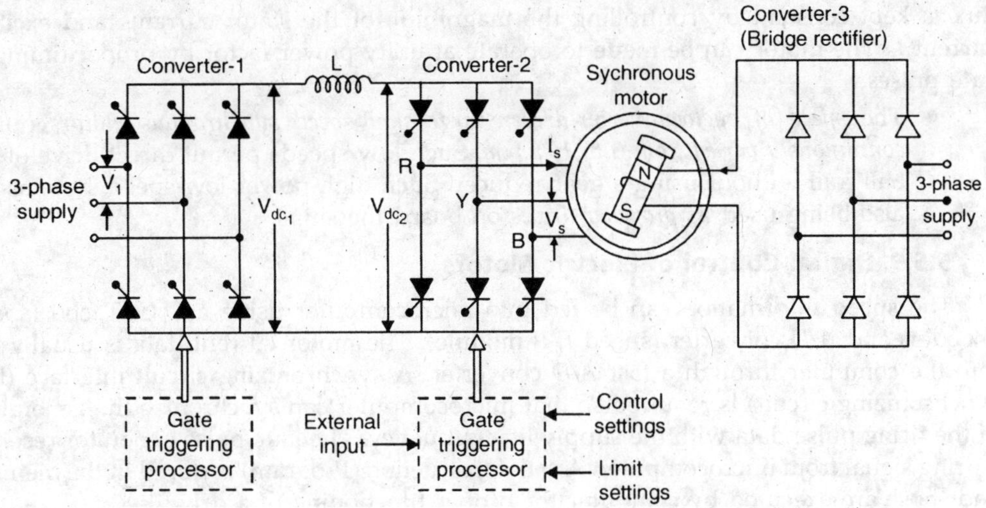

Fig. 5.107. Circuit diagram for speed control of synchronous motor by current-fed D.C. link.

— Converter-1 (*C*-1) acts as a *controlled rectifier* and feeds D.C. power to converter-2 (*C*-2). It acts as a current source and controls *I*.

— Converter-2 (*C*-2) behaves as a *naturally commutated inverter* whose A.C. voltage and frequency are established by the motor. The converter-2 is naturally commutated by voltage V_S induced across motor terminals by its revolving magnetic flux. The revolving flux which depends on the stator currents and the D.C. field exciting current is usually kept constant. Consequently V_S is proportional to motor speed.

— The function of the smoothing inductor *L* is to maintain a ripple-free current in the D.C. link between the two converters.

As regards various controls, information picked up from various points is processed in the gate-triggering processors which then send out appropriate gate firing pulses to converters 1 and 2. The processors receive information about the desired rotor speed, its actual speed, instantaneous rotor position, field current, stator voltage and current etc. The processors check whether these inputs represent normal or abnormal conditions and send appropriate gate firing pulses either to correct the situation or meet a specific demand.

Gate triggering of *C*-1 and *C*-2 is done at line frequency (50 Hz) and at motor frequency respectively. In fact, gate pulses of *C*-2 are controlled by rotor position which is pulsed by position transducers mounted at the end of the shaft. The speed of the motor can be increased by increasing either D.C. link *I* or exciting current I_f.

- This method of speed control is applied to *motors ranging from 1 kW to several MW.*
- *Permanent-magnet synchronous motors used in textile industry and brushless synchronous motors for nuclear reactor pumps are controlled by this method.*

2. Speed control and cycloconverter :

This arrangement consists of *three cycloconverters* connected to the three phases of the synchronous motor and *one controlled rectifier* for supplying field exciting current, I_f to the rotor. Each cycloconverter is composed of two 3-phase bridges and supplies a single-phase output. With a line frequency of 50 Hz, the cycloconverters output frequency can be varied from 0 to 10 Hz (It is well known that a cycloconverter can convert A.C. power at higher frequency to one at a lower frequency).

The cycloconverters and controlled rectifies function as *current sources.* The air-gap flux is kept constant by controlling the magnitude of the stator currents and exciting current I_f. The motor can be made to operate at unity power factor by proper timing of gate pulses.

- The *speed of cycloconverter-driven large slow-speed synchronous motors can be continuously varied from 0 to 15 r.p.m.* Such low speeds permit direct-drive of the ball mill without using a gear reducer. Such high-power low-speed systems are also being used as *propeller drives* on board the ships.

5.5.9. Digital Control of Electric Motors

The speed information can be fed into microcomputer using a D.C. Tacho (speed encoder) and *A/D* converter (speed *I/P* module). The motor current data is usually fed into the computer through a fast *A/D* converter. A synchronizing circuit interface (line synchronizing circuit) is required so that microcomputer can synchronize the generation of the firing pulse data with the supply line frequency. The gate pulse generator receives a firing signal from microcomputer. A set of instruction (Program) is stored in the memory and those are executed by computer for proper functioning of a drive.

Advantages of digital control :

1. High reliability.
2. Easy software control.
3. High precision and accuracy.
4. Better speed regulation.
5. Faster response.
6. Improved performance.
7. Economical.
8. Flexibility.
9. Better time response.
10. *The major advantage of the digital control is that by changing the program, desired control technique can be implemented without any change in the hardware.*

5.6 COMPARISON OF ROBOT DRIVE SYSTEMS

The comparison of robot drive systems is given in Table 5.1 below :

Table 5.1. Comparison of Robot Drive Systems

S. No.	Aspects	Robot drive systems		
		Hydraulic	*Pneumatic*	*Electric*
1	*Reliability of the components*	Low	Higher	High
2.	*Accuracy and response*	Good	Bad	Higher
3.	*Pressure, torque and inertia on the actuator*	High	Low to medium	Medium to high
4.	*Range of operational speeds*	Wide	Very little	Comparatively less
5.	*Power-to-weight ratio*	Highest	Lowest	Moderate
6.	*Leakage and cleanliness*	Worst	Better	Nil
7.	*Size and stiffness*	Very high	Very low	Low stiffness
8.	*Compliance of the system*	Low	Good	Better
9.	*Need for maintenance*	More	Less	Low

10.	*Controlling devices*	Hydraulic power pack required	Pneumatic power control devices required	Control system is required
11.	*Path generation application*	Continuous path	Only in pick and place type	Both continuous pick and place
12.	*Payload carried by the robot*	Highest	Lowest	Moderate

QUESTIONS WITH ANSWERS

Q. 5.1. Enlist some important pump selection parameters.

Ans. The various pump selection parameters are :

(*i*) Cost
(*ii*) Speed
(*iii*) Weight
(*iv*) Suction performance
(*v*) Fluid type
(*vi*) Pressure ripple and noise
(*vii*) Fixed or variable displacement
(*viii*) Maximum pressure and flow, or power
(*ix*) Contaminant sensitivity.

Q. 5.2. What are the features and applications of "hydraulic actuators"?

Ans. The *features* and *applications* of hydraulic actuators are listed below :

Features :

1. These actuators have longer life and reliability due to the lubricating properties of the oil.
2. Can operate safely and continuously under stall conditions.
3. Provide high power in small light components.
4. Can be easily built using readily available standard elements.
5. Have flat load-speed or torque-speed characteristics.
6. Provide stepless variation in speed.
7. Less deflection due to low compliance of the elements.
8. High inertia on the robot joints.
9. Operation is noisy.
10. Have contaminant sensitive elements.
11. Power loss and unclean work due to possibility of leakage of liquid.

Applications :

The hydraulic actuators are used :

— to drive the joints of assembly (heavy) robots ;
— to drive the spray coating robots ;
— in producing translatory motion in cartesian robots ;
— in heavy part loading robots ;
— in gripper mechanisms ;
— in material handling robot system ;
— in robots operating in hazardous, sparking environments.

Q. 5.3. State the advantages and disadvantages of using water as a fluid medium in the hydraulic circuit.

Ans. *Advantages* : Water has low viscosity, so it flows easily. This results in lesser loss of pressure in the pipes and fittings of the hydraulic system.

Disadvantages : (*i*) It does not have any lubriction property.

(*ii*) Its physical characteristics get changed with the variation in temperature.

(*iii*) It causes corrosion of metal components of the system.

Q. 5.4. Enlist the common contaminants in hydraulic fluids.

Ans. *Solid contaminants* :

Dirt ; dust ; scale ; small chips ; bits of seal material.

Liquid contaminants :

Moisture ; sludge ; varnish from oxides of oil.

Q. 5.5. Discuss why and where filters are fitted in a hydraulic circuit.

Ans.
- *Filtration* is the process of removal of solid contaminants from the fluid. The contamination of fluid may result in wear, corrosion and poor performance of the hydraulic system.
- Filter may be placed *before* the pump and it is known as *suction filter*. It may also be fitted *after* the pump in which case it must be capable of withstanding the full pressure of the system. In addition, it may also be incorporated in the *return line* to the tank.

Q. 5.6. State important characteristics of oil to be used in an hydraulic system.

Ans. The oil to be used in an hydraulic system should possess the following *characteristics* :

1. Good wear properties.
2. Good lubricating properties.
3. Reliable air separation.
4. Suitable viscosity.
5. Good corrosion inhibitor.
6. Good anti-aeration properties.

- The three most common types of hydraulic liquids are : *Petroleum-based; synthetic fire-resistant; and water-based fire-resistant.*

Q. 5.7. State the features of "pneumatic actuators".

Ans. The pneumatic actuators are characterised by the following *features* :

1. Highly compliant system.
2. Simple and low cost components.
3. Reliable and easily available components.
4. Lowest power-to-weight ratio.
5. Less leakage of air and not susceptible to sparks.
6. Drift under load constantly.
7. Use low pressure compressed air, hence less actuation force or torque.
8. The exact positons of the actuators can be controlled by servocontrol valves by differential movements.
9. Low, inaccurate response due to low stiffness.
10. Useful in on-off applications like pick and place robots.

Q. 5.8. Give the classification of air compressors.

Ans. The air compressors may be *classified* as follows :

Air compressors :

1. *Positive displacement compressors* :
   (*i*) Rotary :
   — Vane
   — Lobe
   — Screw
   (*ii*) Reciprocating :
   — Piston
   — Diaphragm
2. *Dynamic compressors* :
   (*i*) Centrifugal
   (*ii*) Axial.

Q. 5.9. How are cylinders specified ?

Ans. Cylinders are specified by the following parameters :

(*i*) Cylinder bore ;
(*ii*) Piston rod diameter ;
(*iii*) Stroke length ;
(*iv*) Pressure range ;
(*v*) Force output at maximum pressure ;
(*vi*) Mounting styles ;
(*vii*) Rod ends ;
(*viii*) Cushions (at one end or both ends)
(*ix*) Standard operating temperature.

Q.5.10. Enlist the various accessories used in a pneumatic systems.

Ans. The various accessories in a pneumatic systems are listed below :

1. Air receiver
2. Air intercooler
3. Air aftercooler
4. Air filter
5. Pressure regulator
6. Flow regulator
7. Air lubricators
8. F.R.L.—Filter regulator lubricator
9. Air dryers
10. Pipeline layout
11. Seals.

Q.5.11. Enumerate various accessories used in hydraulic systems.

Ans. Following accessories are used in hydraulic systems :

1. Hydraulic fluids
2. Hydraulic reservoir
3. Hydraulic filter
4. Pressure gauges and volume meters
5. Lines
6. Fittings and connectors
7. Hydraulic seals
8. Hydraulic accumulator
9. Intensifiers.

Q.5.12. Give a comparison between 'Air 'and 'Electric' motors.

Ans. Air motors differ from electric motors in the following respects :

1. Air motors are simple in design and construction is relatively inexpensive. They are *reliable* and *easy to maintain* and there is no heat build up even if it stalls for longer period.
2. They develop more kW/N weight and per cubic metre of displacement than most standard electric motor.

3. Speed can be varied over a wide range without complicated control arrangements in air mass, but in electric motors it is expensive.
4. The harder an air motor works, the cooler it runs because air expands as it passes through the motor, the air creates a cooling effect which permits air motors to operate in environments where ambient temperature may reach as high as 120°C which is simply the opposite case in case of electric motors.
5. They are inherently shock and explosion proof which is not in case of electric motors.
6. These are not damaged by overloading, rapid reversals or continuous stalling; when an air motor stalls it continues to produce high torque with no damage to itself, but electric motor can be severely damaged due to overloading.
7. Due to their low inertia, air motors accelerate and decelerate quickly from zero to fullspeed in milliseconds-ideal fast cycling, electric motors take more time to do this.
8. These are not affected by hot, wet or corrosive atmosphere. Beacuse they operate under internal pressure, dust, moisture and air cannot enter the motor case. Dust or explosion proof enclosures for electric motors involve extra cost.

Q.5.13. Explain briefly with neat diagrams hydraulic system having (*i*) constant delivery pump; (*ii*) variable delivery pump.

Ans. (*i*) **Hydraulic system having constant delivery pump :**

Fig. 5.108, shows the schematic arrangement of hydraulic system having *constant delivery pump*. The various components of the system are :

(*i*) Gear pump ; (*ii*) Relief valve ;
(*iii*) Stop and speed control valve ; (*iv*) A piston-cylinder arrangement.

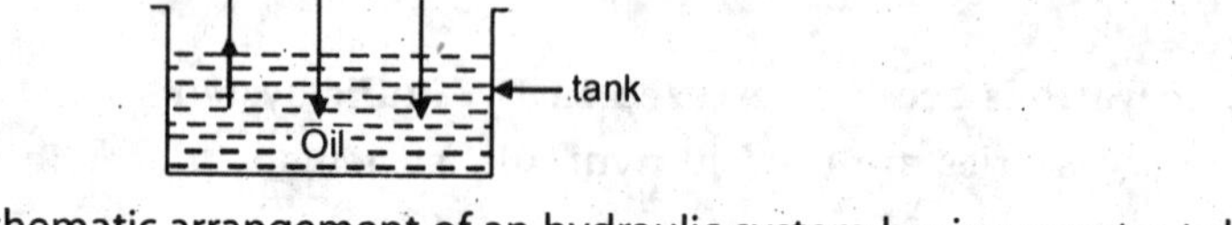

Fig. 5.108. Schematic arrangement of an hydraulic system having constant delivery pump.

— The pump provides fluid at *constant pressure* ;
— The relief valve controls the flow of oil into the system. Through this valve the requisite quantity of fluid may be allowed to pass through the stop cum speed control valve; rest of the fluid falls back into the oil tank.

- This system is applicable when the cost of the system is *low*.
- It is *not* suitable for use in cases where the variation in the resistance to the piston is to be registered and where both feed and rapid traverse motions are required. This system is *less* reliable.

(*ii*) **Hydraulic system having variable delivery pump :**

Fig. 5.109, shows the working of an hydraulic system having *variable delivery pump*.

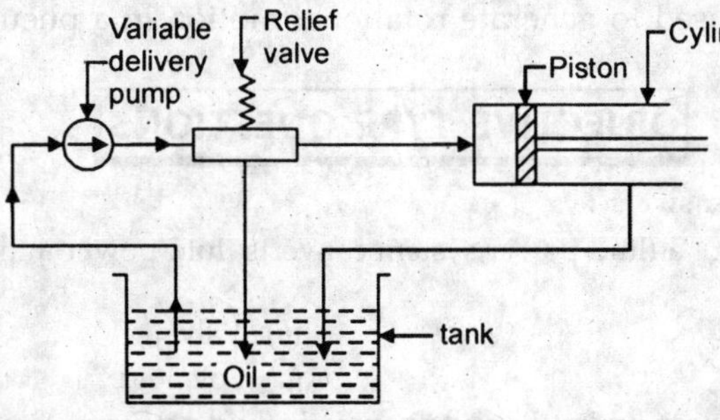

Fig. 5.109. Hydraulic system having variable delivery pump.

— The variable delivery pump gives better results in case of heavy duty, high speed machining systems. It also reduces wastage of power when the fluid is at full pressure for light duty work.

In order to avoid leakage of oil the mating parts of pumps, motors and valve gears must be accurately fitted.

— The increase or decrease in resistance to the movement of the piston will raise or lower the pressure in the cylinder and supply pipe, without materially changing speed of the piston.

- This system is *more reliable* than constant delivery pump hydraulic system.

Q.5.14. Discuss briefly future trends and developments in fluid power.

Ans. The future trends and developments in fluid power include the following :

1. Components in modular block form.
2. Noise control.
3. Electro-modulated components.
4. Computer control.
5. Electronic position sensors.

Note. The fluid power components are being developed in the form of modular block to reduce the size and price. Several of these blocks, or modules, when assembled together would perform certain desired control function. The main *advantages* of modular blocks are :

(*i*) Lesser chances of oil leakage ; (*ii*) Smaller pressure losses ; (*iii*) Very compact and space saving ; (*iv*) Easy replacement of components, etc.

HIGHLIGHTS

1. *Actuators* produce physical changes such as linear and angular displacements. They also modulate the rate and power associated with these changes.
2. When one of the links of a kinematic chain is fixed, the chain is known as *mechanism*.
3. *Electrical actuators* may be : Switching devices and Drive systems.
4. *A stepper motor*, a special type of D.C. motor, is an incremental motion machine.
5. An actuator wherein hydraulic energy is used to impart motion is called an *hydraulic actuator*.
6. *Pneumatic systems* use pressurised air to transmit and control power.
7. Pneumatic cylinders convert the pneumatic power into straight-line reciprocating motions.

8. An *air motor* is used to generate rotational motion in a pneumatic system.

OBJECTIVE TYPE QUESTIONS

A. Choose the Correct Answers :

1. What component of a fluid power system converts fluid power into mechanical force and motion?
   (*a*) Pump (*b*) Valve
   (*c*) Actuator (*d*) Solenoid.
2. The similarity between hydraulic and pneumatic fluid power systems is correctly indicated by which of the following statements?
   (*a*) The basic components of the systems are essentially the same.
   (*b*) Both systems depend upon internal lubrication by the system fluid.
   (*c*) Both a and b above correctly indicate the similarity.
   (*d*) The basic components of the systems are identical and interchangeable.
3. How can the specific gravity of a liquid or solid be expressed?
   (*a*) As a ratio between the weight of the substance and the density of a volume of water.
   (*b*) As a ratio between the weight of the substance and the weight of an equal volume of water.
   (*c*) As the number that shows the density of the substance in the metric system.
   (*d*) None of these.
4. Relief valves are used for which of the following functions ?
   (*a*) To maintain fluid flow below a predetermined rate.
   (*b*) To maintain pressures above a predetermined level.
   (*c*) To prevent pressure from rising above a predetermined level.
   (*d*) To prevent thermal expansion of the fluids.
5. Which of the following statements about specific gravity is false?
   (*a*) The density of a solid can be determined by multiplying its specific gravity times the density of water.
   (*b*) Specific gravity can also be described as specific weight or specific density.
   (*c*) Specific gravity will vary with the size of the sample being tested.
   (*d*) Specific gravity of a substance should be measured at a standardized temperature and pressure.
6. Hydraulic systems are equipped with one or more accumulators that serve to perform which of the following functions?
   (*a*) To provide pressure for emergency operation of the system in the event of system failure.
   (*b*) To act as a buffer and absorb surges and shock pressures that might damage pipes and other components of the system.
   (*c*) To equalize and readjust for any pressure losses in the system due to small leaks and thermal reaction of the fluid.
   (*d*) All of the above.
7. Which component of a hydraulic fluid power system performs the same function as the receiver in a pneumatic fluid power system?
   (*a*) Reservoir (*b*) Actuator
   (*c*) Compressor (*d*) Selector valve.
8. What type of directional control valve is normally used to control a single-acting, spring-loaded, piston-type actuating cylinder?
   (*a*) Shuttle (*b*) Two-way
   (*c*) Three-way (*d*) Four-way.

9. What type of valving element is most commonly used in directional control applications?
   (*a*) Sliding spool (*b*) Rotary spool
   (*c*) Poppet (*d*) Ball.
10. If a hydraulic pump is located below the reservoir, fluid is supplied to its inlet port by which of the following forces?
   (*a*) Fluid head (*b*) Atmospheric pressure
   (*c*) Gravity (*d*) A combination of all of the above.
11. Valves are used to control which of the following in fluid power systems?
   (*a*) Direction of fluid flow (*b*) Fluid pressure
   (*c*) Fluid flow (*d*) All of the above.
12. Hydraulic liquid must possess which of the following properties?
   (*a*) Chemical stability and freedom from acidity.
   (*b*) Lubricating ability and proper viscosity.
   (*c*) Minimum toxicity and high flashpoint.
   (*d*) All of the above.
13. A liquid that is satisfactory for use in a hydraulic system provides
   (*a*) a low viscosity index, good sealing quality, and lubricity
   (*b*) a high viscosity index, good sealing quality, and a low flashpoint
   (*c*) good lubrication and sealing qualities, and high viscosity
   (*d*) good lubrication and a viscosity that decreases as temperature increases.
14. A device used for measuring the specific gravity of a liquid is known as a
   (*a*) hydrography (*b*) hydrometer
   (*c*) hydroscope (*d*) hydrostat.
15. The reservoir serves the primary function of storing the hydraulic fluid required by the system, which of the following secondary functions does it also serve?
   (*a*) Separates air from the system (*b*) Traps foreign matter
   (*c*) Dissipates heat (*d*) All of the above.
16. All of the following contaminants are abrasive except
   (*a*) lint (*b*) rust
   (*c*) sand particles (*d*) sludge.
17. Which of the following is the function of a hydraulic pump?
   (*a*) To provide flow to the hydraulic system.
   (*b*) To create the pressure required in a hydraulic system.
   (*c*) To control the pressure required in a hydraulic system.
   (*d*) To compensate for atmospheric pressure at varying altitudes.

ANSWERS

1. (*c*)	2. (*c*)	3. (*b*)	4. (*a*)	5. (*c*)	6. (*d*)	7. (*b*)
8. (*b*)	9. (*a*)	10. (*d*)	11. (*d*)	12. (*d*)	13. (*a*)	14. (*b*)
15. (*d*)	16. (*d*)	17. (*b*).				

B. Fill in the blanks or Say "Yes" or "No" :

1. Rack-and-pinion can be used to convert rotational motion to linear motion.
2. .... is a device by means of which available energy can be converted into desired form of useful work.
3. A kinematic pair is a joint of two links that permits relative motion.
4. A mechanism with .... links is known as simple mechanism.

5. .... is the reverse of the diametral pitch.
6. A journal bearing is one in which the bearing pressure is parallel to the axis of the shaft.
7. A mechanical device or a system which has motion or movement is called an ....
8. Solenoids can be used to provide electrically operated actuators.
9. .... are electrically operated switches in which changing current in one electrical circuit switches a current on or off in another circuit.
10. MOSFET can be employed as a control switch for a D.C. motor as on-off switch.
11. A stepper motor is an incremental motion machine.
12. Torque motors are the D.C. motors designed to run for long periods in a stalled or low speed condition.
13. An inverter fed trapezoidal PMAC motor drive operating in self controlled is called a .... D.C. motor.
14. An actuator wherein hydraulic energy is used to impart motion is called an hydraulic actuator.
15. A .... delivers high pressure fluid.
16. A valve is represented by a .... for each of its switching positions.
17. The .... valves start, stop and control the direction of flow for reversing the direction of motion of the actuator.
18. Proportional valves equipped with sensor and control circuitry are often called .... valves.
19. The most common type of electrical actuator is ....
20. .... cylinders are used where long work-strokes are required.

ANSWERS

1. Yes	2. Machine	3. Yes	4. four
5. Module	6. No	7. actuator	8. Yes
9. Relays	10. Yes	11. Yes	12. Yes
13. brushless	14. Yes	15. pump	16. square
17. direction control	18. servo	19. solenoid	20. Telescoping.

THEORETICAL QUESTIONS

1. What is an 'actuator'? List the various types of actuators.
2. Define the following terms : Machine; Kinematic link; Kinematic pair; Kinematic chain; Inversion of mechanism.
3. What are the advantages and disadvantages of toothed gearing?
4. What is a 'bearing'? How are bearings classified?
5. What is an 'electrical actuator?
6. What is meant by 'electrical actuation system'? What are the devices used in such systems?
7. What are the mechanical switches? Explain.
8. Explain the terms 'bouncing' and 'debouncing' as applied to mechanical switches.
9. What are the various methods which can be used to tackle the problem of 'bouncing' in mechanical switches.
10. Explain briefly the working of following mechanical switching devices:
    (*i*) Solenoids; (*ii*) Relays.
11. Explain briefly any two of the following solid-state devices which can be used to electronically switch circuits:

(*i*) Diodes; (*ii*) Thyristors;
(*iii*) Bipolar transistors; (*iv*) Power MOSFETs.

12. Explain briefly D.C. motor control by using MOSFETs.
13. Give brief classification of electric motors.
14. Explain briefly the following electric motors:
    (*i*) Permanent magnet D.C. motors;
    (*ii*) Stepper motors.
15. State the advantages and applications of stepper motor.
16. What are servo-motors? Explain briefly.
17. Explain briefly the following :
    (*i*) Moving coil motor. (*ii*) Brushless D.C. motors.
18. What are the advantages of electronic control systems?
19. Explain briefly the various methods by which speed of D.C. motor can be controlled.
20. Discuss briefly the following electric motors:
    (*i*) Single phase motors.
    (*ii*) Three phase induction motor.
    (*iii*) Synchronous motor.
21. Discuss briefly the electronic control of A.C. (induction) motors.
22. Discuss briefly "Digital control of electric motors".
23. What factors/specifications should be considered while selecting a motor for mechatronic applications?
24. What is an 'hydraulic actuator'?
25. What are the advantages and disadvantages of hydraulic system?
26. What are the basic elements of an oil hydraulic system?
27. What are the components of an hydraulic system? Explain briefly.
28. What is the function of an hydraulic pump?
29. How are hydraulic pumps classified?
30. Explain briefly the following :
    (*i*) Centrifugal pump. (*ii*) Gear pump.
    (*iii*) Vane pump.
31. Explain briefly with the help of a sketch, the swash plate piston pump.
32. Describe briefly a pressure regulator.
33. What are hydraulic valves? How are these classified?
34. What are the functions of pressure control valves?
35. Give the classification of pressure control valves.
36. Explain briefly the following valves:
    (*i*) Pressure relief valves.
    (*ii*) Pressure sequencing valves.
    (*iii*) Pressure reducing valves.
37. Discuss briefly flow control valves.
38. What is the function of a 'direct control valve'?
39. How are direct control valves classified?
40. Explain briefly any two of the following valves:
    (*i*) Check valve. (*ii*) Spool valves.
    (*iii*) Proportional valve. (*iv*) Rotary valves.
41. What is fluid power hydraulic cylinder? Explain briefly.

42. How are hydraulic cylinders classified?
43. Explain briefly the following :
    (*i*) Double acting cylinders.
    (*ii*) Telescopic cylinders.
44. What are the applications of hydraulic cylinders?
45. List some common cylinder problems.
46. What is an 'hydraulic motor'?
47. How are hydraulic motors specified or rated?
48. Explain briefly the following hydraulic motors :
    (*i*) Gear motors; (*ii*) Vane motors.
49. State the advantages and applications of hydraulic motors.
50. Give the comparison between pneumatic systems and hydraulic systems.
51. Explain with the help of a neat diagram the components of a pneumatic system.
52. Enumerate various types of pneumatic valves and explain briefly and two of them.
53. Explain briefly the following :
    (*i*) Pneumatic check valve.
    (*ii*) Pneumatic shuttle valve.
    (*iii*) Quick exhaust valve.
54. What is the function of a pneumatic cylinder?
55. How are pneumatic cylinders classified?
56. Explain briefly any two of the pneumatic cylinders:
    (*i*) Double-acting cylinder (*ii*) Tandom cylinder
    (*iii*) Three position cylinder (*iv*) Telescoping cylinder.
57. What is the function of an air motor?
58. Name the various types of motors.
59. Explain briefly any two of the following air motors?
    (*i*) Piston type motors (*ii*) Vane motors
    (*iii*) Turbine motors (*iv*) Gerotor type motors.
60. What are the special features which pneumatic actuators should possess?
61. List the applications of air motors.

CHAPTER

6 Hydraulic and Pneumatic Circuits

6.1 Introduction; 6.2 Hydraulic and pneumatic symbols; 6.3 Comparison of hydraulic, pneumatic and hydropneumatic systems; 6.4 Hydraulic circuits – Introduction – Design of hydraulic circuits – Basic hydraulic circuits – Meter-in circuits – Meter-out circuits – Bleed-off circuits – Accumulator circuits – Sequencing circuits – Regenerative circuit – Pressure reduction circuit – Hydraulic circuit for a robot arm; **6.5 Pneumatic circuits** – Basic pneumatic circuit control of double acting cylinder – speed control circuit for a double acting cylinder – semi-automatic material handling circuit – Time delay circuit; **6.6 Pneumatic logic circuits** – Introduction – Advantages of pneumatic logic controls – Pneumatic sensors – Common methods of circuit design for automatic system – Common logic functions – Description of some pneumatic logic circuits. ***Questions with Answer*** – Highlights – Objective Type Questions – Theoretical Questions.

6.1 INTRODUCTION

- *A hydraulic or pneumatic circuit (designed to perform a particular job) consisting of a large number of equipments (like power unit, service unit, accessories, valves actuators etc.) combined in logical sequence to get desired output in the form of motion of actuators and shown diagrammatically by using certain standard symbols of components, constitutes a* **fluid circuit**.
- Circuit diagrams are prepared by arranging various components into a sequence according to their functions. In case of very complex fluid circuits, where it is difficult to understand the 'start' and 'end' of the circuit, the designation systems are classified as : (*i*) *Systems using digits*, and (*ii*) *Systems using alphabets*.
- ☞ *The major difference between hydraulic and pneumatic circuits is that hydraulic elements are designed for high pressures (20 bar to 400 bar) and pneumatic circuits are designed for low pressures (5 bar to 20 bar).*

6.2 HYDRAULIC AND PNEUMATIC SYMBOLS

A summary of selected hydraulic and pneumatic symbols is given below :

A. Hydrautic symbols :

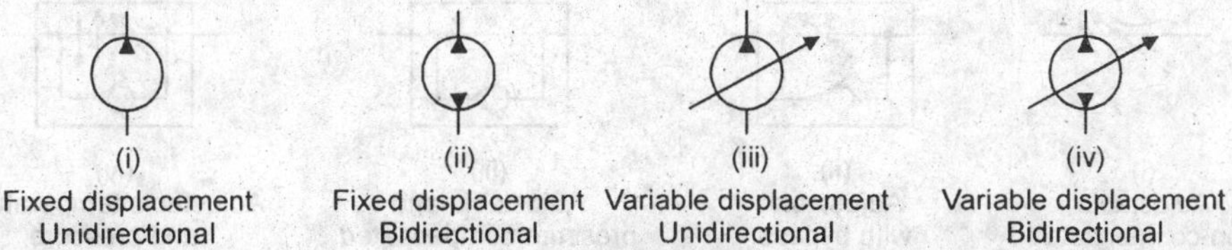

Fig. 6.1. Pumps.

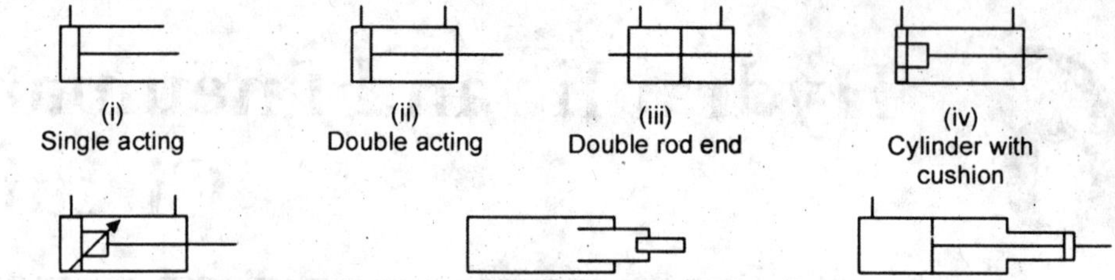

Fig. 6.2. Cylinders.

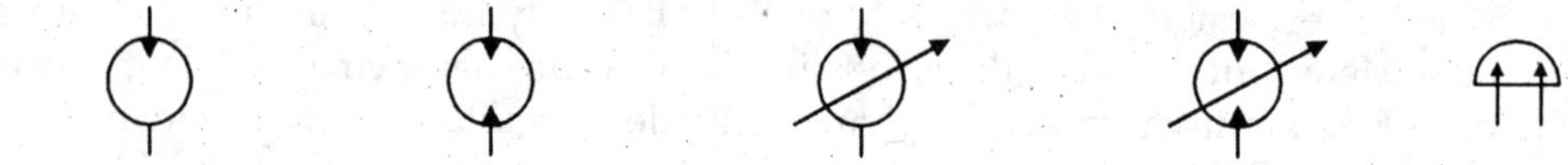

Fig. 6.3. Fluid motors.

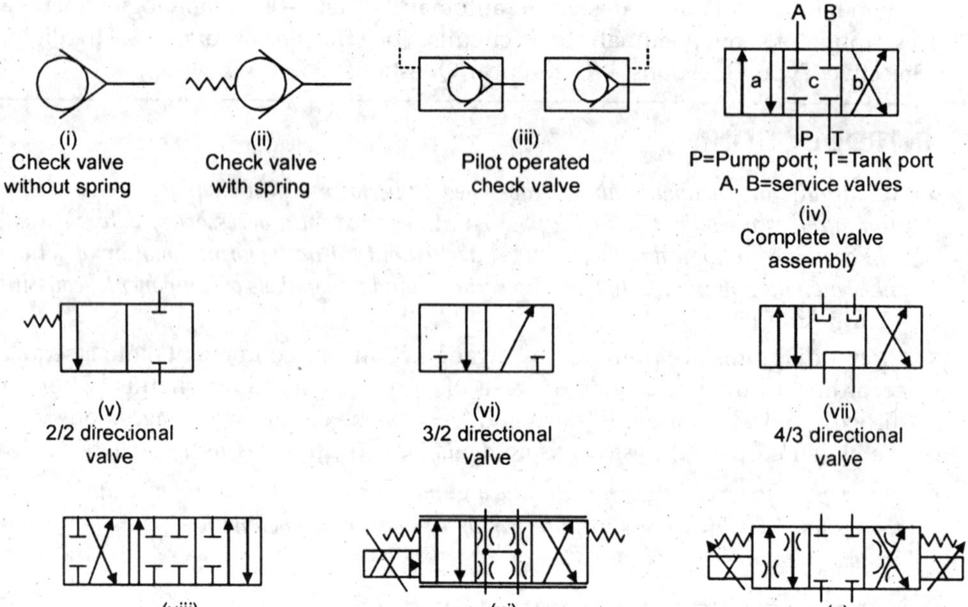

Fig. 6.4. Directional valves.

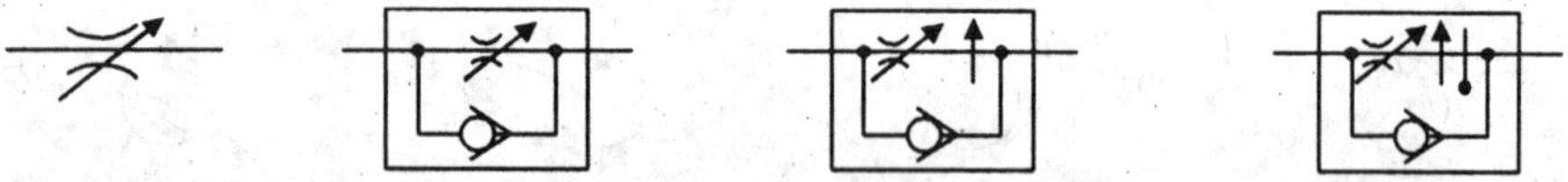

Fig. 6.5. Flow control valves.

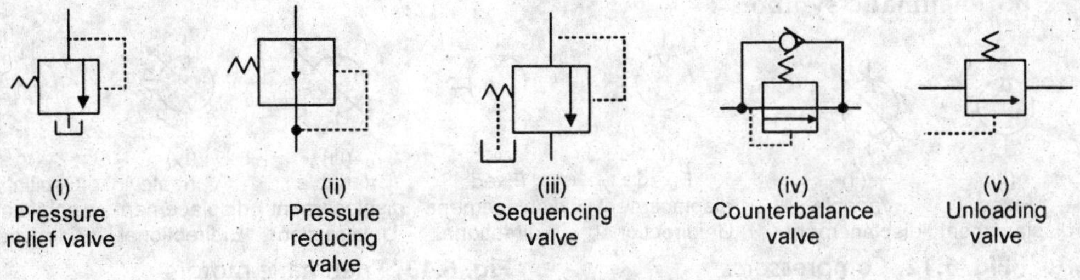

Fig. 6.6. Pressure valves.

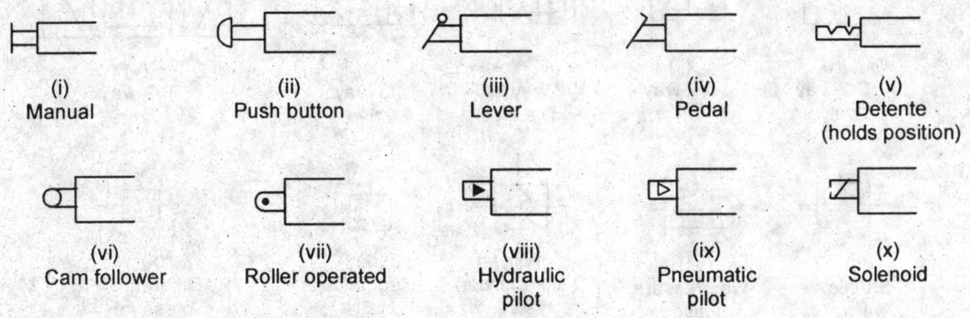

Fig. 6.7. Operating actuators.

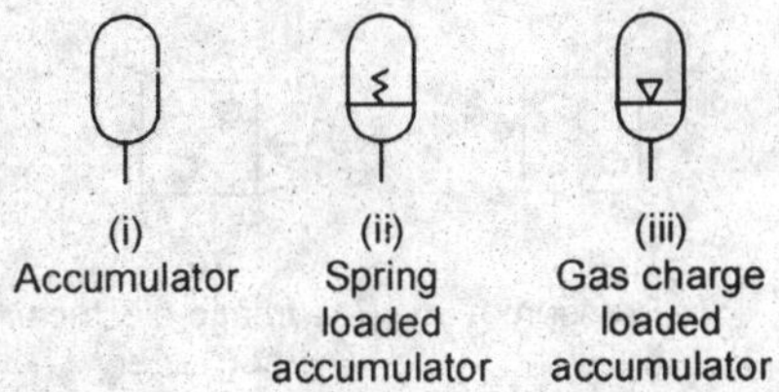

Fig. 6.8. Accumulators.

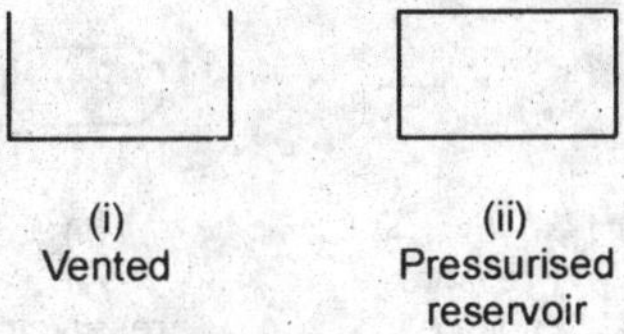

Fig. 6.9. Reservoirs

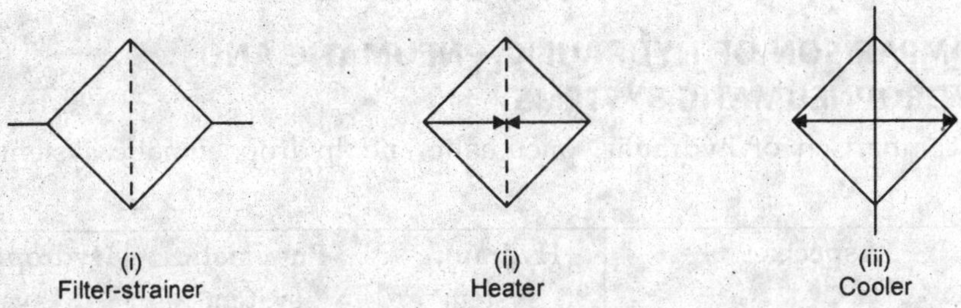

Fig. 6.10. Fluid conditioners.

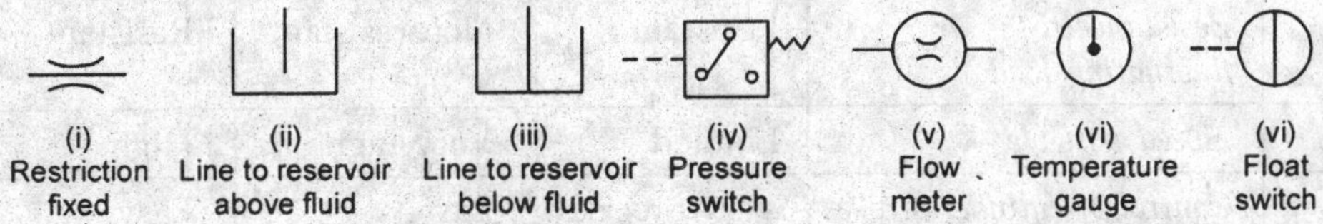

Fig. 6.11. Other hydraulic symbols.

B. Pneumatic symbols :

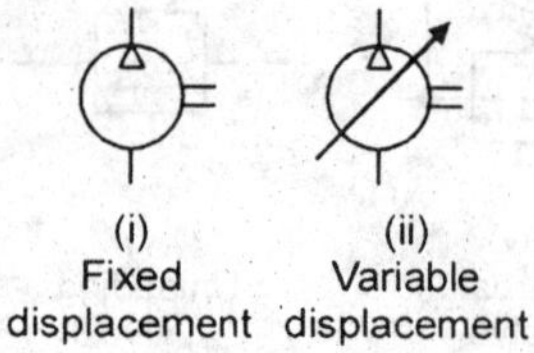

Fig. 6.12. Compressors.

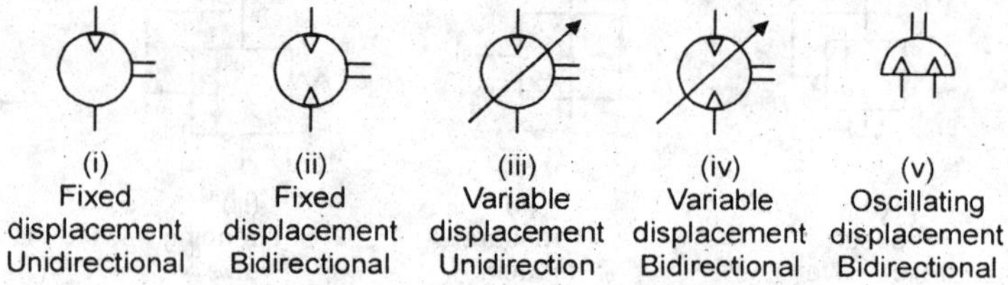

Fig. 6.13. Pneumatic motors.

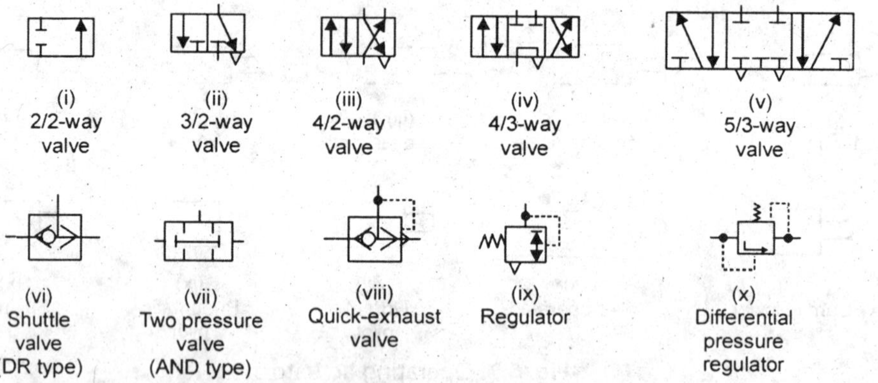

Fig. 6.14. Control valves.

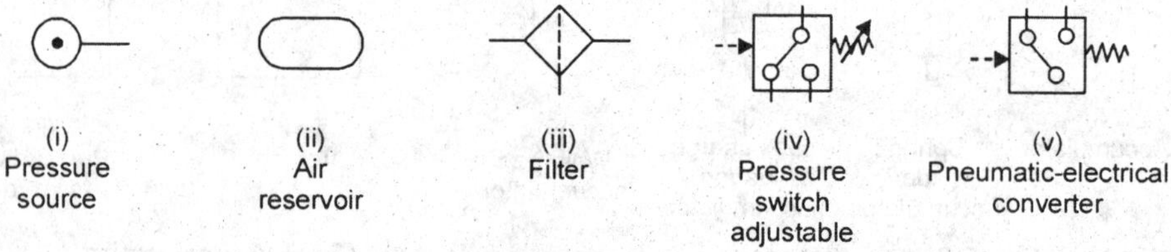

Fig. 6.15. Other pneumatic symbols.

6.3 COMPARISON OF HYDRAULIC, PNEUMATIC AND HYDROPNEUMATIC SYSTEMS

The comparison of hydraulic, pneumatic and hydropneumatic systems is given below :

S.No.	Aspects	Hydraulic system	Pneumatic system	Hydropneumatic system
1.	*Weight-pressure ratio*	Very small	Large	Large
2.	*Resistance to fluctuating load.*	Resistant	Non-resistant	Resistant
3.	*Speed possible*	Limited	Very high	High
4.	*Control of cylinder stroke*	Very precise	Easy	Easy
5.	*System rigidity*	Good	Poor	Good
6.	*Operating pressure*	From low to very high	Generally 6 bar	Low

7.	*Pump requirement*	Pump is necessary	Pump not required	No pump is necessary; small oil reservoir is adequate.
8.	*Suitability for feed movement in machine tools*	Yes	No	Yes
9.	*Cavitation problem*	Yes	No	No
10.	*Operating cost*	Moderate	Very low	Very low
11.	*Overall cost*	Maderate to high	Low	Low to moderate

6.4 HYDRAULIC CIRCUITS

6.4.1. Introduction

***Hydraulic circuits** can be defined as graphic representation of elements of an hydraulic system where these elements are represented by their corresponding symbols and these symbols are sequentially arranged in a specific manner depending upon the nature of the output required.*

In "*Hydraulic and pneumatic circuits*", all the elements can be divided into the following categories.

1. *Driving elements :*
   — Pumps/compressors;
   — Filters;
   — Regulators;
   — Pressure gauges;
   — Service units;
   — Pressure relief valves, etc.
2. *Controlling elements :*
   — Direction control valves;
   — Flow control valves;
   — Pressure control valves;
   — Signalling elements, etc.
3. *Energy consuming members :*
   — Hydraulic and pneumatic actuators.

6.4.2. Design of Hydraulic Circuits

An **hydraulic circuit** *is an arrangement of interconnected components, selected to achieve the desired work output.* The path taken by hydraulic fluid generates a *hydraulic circuit.*

From a design standpoint, hydraulic circuit consists of the following *four critical sections :* Refer to Fig. 6.1

Section - 1 : It represents the *energy input* that is available to perform work. Energy is transferred to the load which is across the mechanical interface, thus involving a cylinder in this circuit.

Section - 2 : It includes the energy control where fluid modulation and directional switching occur.

Section - 3 : In this section, the energy output portion of a hydraulic circuit is indicated. This section includes a *pump* which is *coupled to a prime mover.* Due to fictional and heat losses the energy at the output must exceed that of the required energy input to the load.

Section - 4 : This section represents *auxiliaries,* used to improve the performance and safety of the system.

All the above four sections are critical for the design of a functional, safe and efficient circuit.

- *While "drawing" a circuit, it must begin with pump and end with actuator. While "designing"* a hydraulic circuit, the *reverse pattern holds true* (the designer begin with the load or task and works backward through the actuator, controls, pump and ultimately the prime mover.

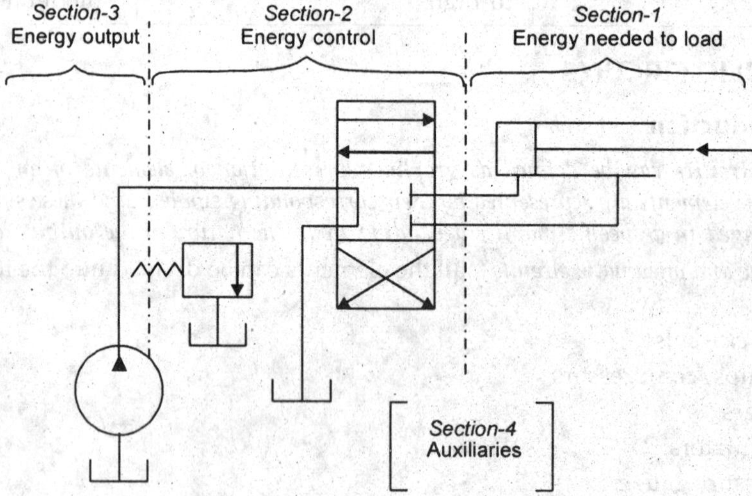

Fig. 6.1. Sections of an hydraulic circuit.

- The *main criteria in the design of any system are :*

  (*i*) Simplicity; (*ii*) Reliability;
  (*iii*) Efficiency; (*iv*) Cost effectiveness;
  (*v*) Maintainability; (*vi*) Safety.

  Among all the above criteria, *simplicity* of design will probably have the *greatest effect.*

  — A simple design will have the leasts number of components and therefore has fewer components to fail, thus improving the *reliability.*

  — As there are pressure drops through each component in a circuit, the design with the least components will have *minimum losses* and hence the *highest efficiency.*

  — *In general, lower the number of components, lower is the cost and higher is the cost effectiveness.*

  — Simplicity of design *can reduce maintenance cost* but need not necessarily make maintenance easier. Ease for maintenance cost has to be incorporated into circuit by the careful selection of component, layout of paperwork, component positioning and inclusion of test facilities in the circuit.

  — Above all, *safety* is of extreme importance and *all units must fail-safe.*

Design informations required :

The following design informations must be available with the designer :

1. *Sizing actuators from output objectives :*
   — The output objective may be described by the performance of a task against a load resistance.
2. *Sequence of operations :*
   — Type of sequence; event or time-based or a mixture.
   — Details of a sequence includes changes in pressure or flow rates.
3. *Method of control :*
   — Manual, mechanical, hydraulic, electrical, microprocessor, proportional or servo.
4. *Operational pressures :*
   — Is there a maximum pressure or can the designer decide this ?
5. *Operating conditions :*
   — The location of the system and the environmental conditions, skill of the operational personnel and the maintenance facilities.
6. *Special requirements :*
   — What are the typical hours of use per week ? Is this a key system.
7. *Component suppliers :*
   — Is the designer free to select the maufacturers of the components or is it based on the suggestions given by the customer ?

- *Selection of a hydraulic cylinder :*

  The following design considerations are required while selecting a hydraulic cylinder :

  (*i*) Type of cylinder (Double or single acting, telescoping).
  (*ii*) Cylinder stroke requirement.
  (*iii*) Thrust/stroke and speed/stroke profile.
  (*iv*) Construction (tie rod or non-tie rod design).
  (*v*) Cylinder control and special seal requirements.
  (*vi*) Counterbalance of loads.

- *Selection of hydraulic motors :*

  The selection of hydraulic motors include the following :

  (*i*) Hydrostatic drives.
  (*ii*) Hydrastatic device characteristics.
  (*iii*) Braking of hydrastatic drives.
  (*iv*) Hydraulic motor characteristics and performance curves.
  (*v*) Matching motor to load.

- *Selection of control valves :*

  It includes the following :

  (*i*) Selection of relief valves.
  (*ii*) Selection of flow control valve.
  (*iii*) Selection of direction control valve.

- *Selection of filters.*

- *Selection of hydraulic conduits :*

  It includes the following :

  (*i*) Tubing; (*ii*) Hoses.

- *Selection of pump :*

  The main parameters affecting the selection of a particular type of pump are :

  — Maximum operating pressure; — Maximum delivery;
  — Pump drive speed; — Type of fluid;
  — Fluid contamination; — Pump pulsation;
  — Size and weight of pump; — Efficiency;
  — Cost.

6.4.3. Basic Hydraulic Circuits

For simple applications following basic hydraulic circuits are available :

1. *Meter-in-circuit :*
   - To control incoming flow-rate.
2. *Meter-out-circuit :*
   - To control out-going flow-rate.
3. *Bleed-off flow control :*
   - To dump part of the pump flow back to tank.
4. *Open centre circuit :*
   - To drive an actuator through directional control valve.
5. *Closed centre circuit :*
   - To drive an hydraulic motor directly.
6. *Sequence circuit :*
   - For the sequence control when more than one actuator is involved.
7. *Two-pump unloading system :*
   - To have larger flow at low pressure and smaller flow at high pressure operation.
8. *Regenerative circuit :*
   - To have large flow during extension stroke.
9. *Counterbalance circuit :*
   - To hold a load in the neutral position of directional control valve without creeping movement.
10. *Brake circuit for an hydraulic motor :*
    - To brake an hydraulic motor without incurring high pressure system.
11. *Deceleration circuit :*
    - To advance the initial extension stroke rapidly, then slow it down at the end.
12. *Electro-hydraulic servo-control :*
    - To infinitely vary the position or speed of actuator incurring high system pressure.

6.4.4. Meter-in Circuits

Fig. 6.2. shows a meter-in circuit :

— In this circuit a flow-control valve is installed in a pressure line that leads to a work cylinder.

— All flow entering in a work cylinder is first metered through a flow-control valve. Excess fluid returns to a tank through a relief valve.

— The check valve allows free reverse flow when the cylinder retracts.

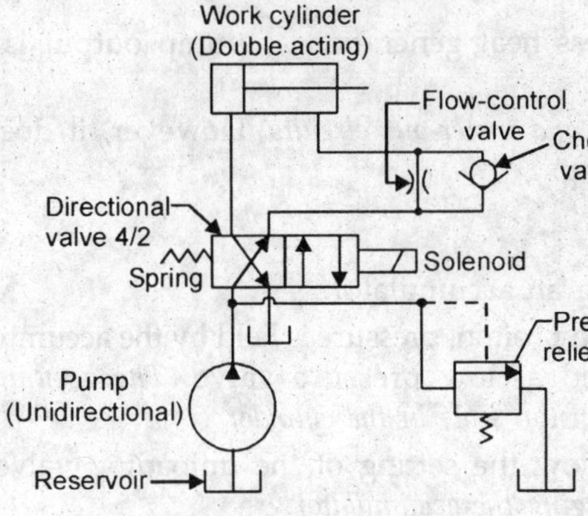

Fig. 6.2. Meter-in circuit

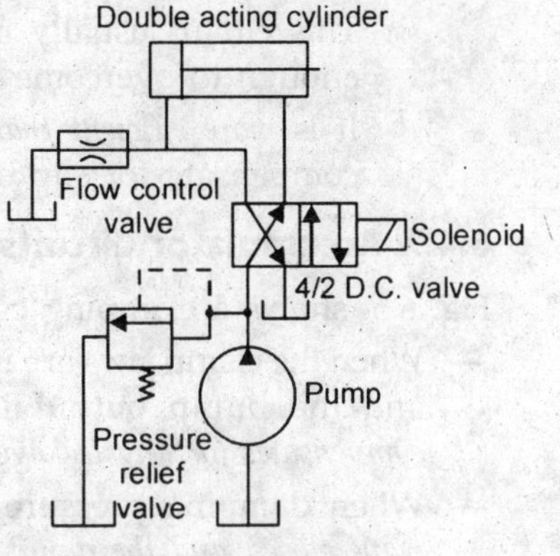

Fig. 6.3. Meter-out circuit

It *normally gives finer speed control than a meter-out circuit.*

- This type of circuit is *ideal in applications where a load always offers a positive resistance to flow during a controlled stroke.*

6.4.5. Meter-out Circuits

Fig. 6.3. shows a meter-out circuit.

— This circuit maintains a constant back-pressure during rod extension and prevents lunging if the load drops quickly or reverses.

— As shown in the figure, a flow-control valve is instilled on the return side of a cylinder so that it controls cylinder's actuation by metering its discharge flow.

6.4.6. Bleed-off Circuits

A bleed-off circuit is shown in Fig. 6.4.

— In this circuit, flow to the cylinder is regulated by metering part of the pump flow to tank.

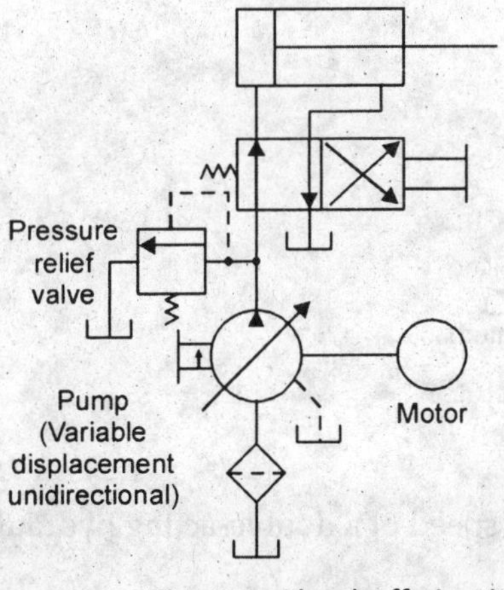

Fig. 6.4. Bleed-off circuit

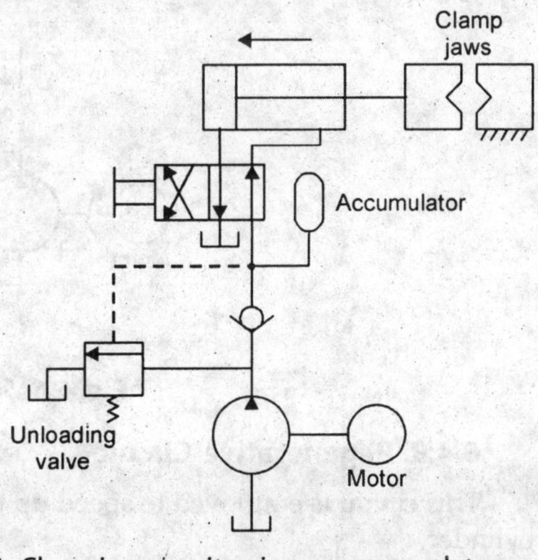

Fig. 6.5. Clamping circuit using an accumulator.

— A valve regulates flow to a cylinder by diverting an adjustable portion of a pump's flow to a tank. Since the fluid delivered to a work cylinder does not have to pass through a flow-control valve, therefore, excess fluid does not have to be damped through a relief valve.

- This circuit usually involves less heat generation as pump output is high enough to overcome resistance.
- It is *more efficient than meter-in and meter-out circuits.* However, it does not compensate for slip of the pump.

6.4.7. Accumulator Circuits

Fig. 6.5. shows a clamping circuit using an accumulator.

— When the clamp jaws are in the clamp positon, pressure is held by the accumulator, and the pump output is unloaded at low pressure valve. *The accumulator compensates for any leakage past the piston seals in the cylinder.*

— When damping pressure drops below the setting of the unloading valve, the *valve closes, and the pump starts charging the accumulator.*

6.4.8. Sequencing Circuits

In sequencing circuits, the *actuators may be sequenced by restricting flow to one and free flow to others.*

- One method of restricting flow is with *back pressure check valves.* They prevent flow until a set pressure is reached. Fig. 6.6. shows such a circuit in which cylinder-1 extends and retracts ahead of cylinder-2.

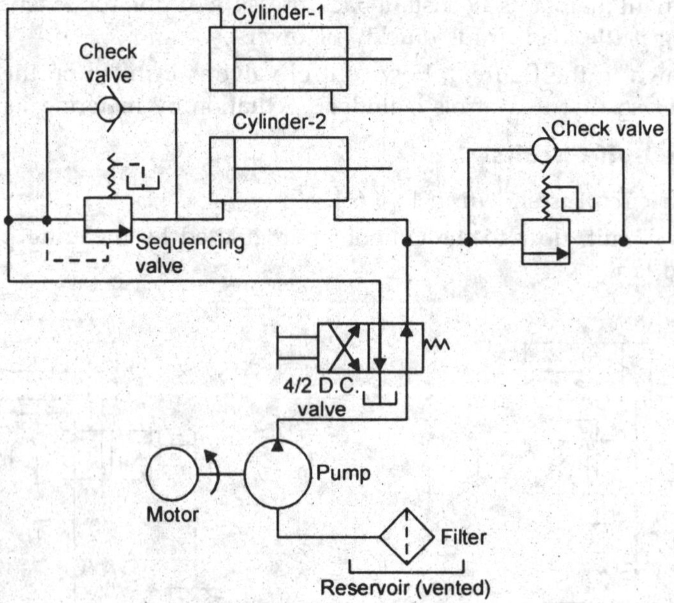

Fig. 6.6. Sequencing circuit.

6.4.9. Regenerative Circuit

This circuit is employed to speed up the extending speed of a double-acting hydraulic cylinder.

Fig. 6.7. shows schematically a regenerative circuit controlled with two solenoid valves, a pressures switch and two check valves.

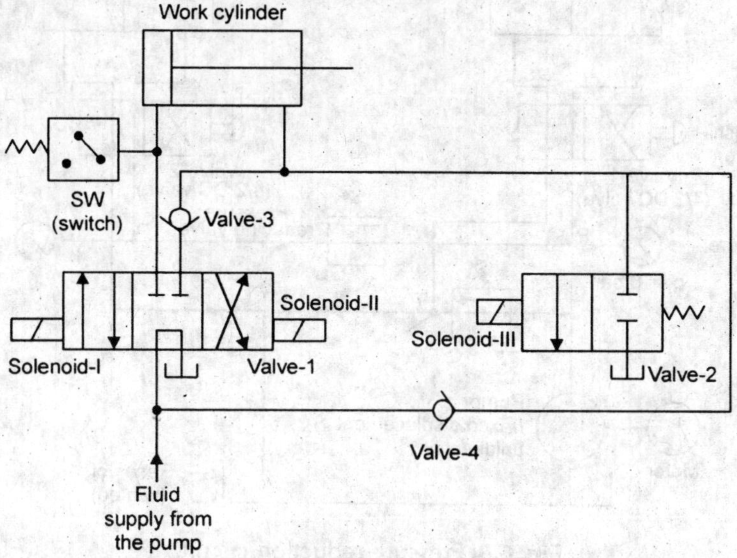

Fig. 6.7. Regenerative circuit.

— When the switch (SW) is manually turned to the extend position, solenoid-I gets energized causing the cylinder to extend. Oil from the rod end flows through check valve-4 to *join the incoming oil from the pump to provide a rapid cylinder extension in the regenerative condition.*

— Owing to loadings on the cylinder the oil pressure builds up, and as a consequence, normally open pressure switch actuates and solenoid-III is energised. Rod end oil is vented directly back to the oil tank through the valve-2. As a result, the *cylinder extends slowly as it drives a high resistive load.* During the slow extension of the cylinder, solenoid-III *is energised continuously.*

— On turning the switch (SW) to the retract position, *solenoid-II becomes energised while the solenoid-III gets de-energised.* Consequently, the *cylinder retracts in a normal manner* to its full retractable position.

— On turning the switch to the unload position, all the solenoids (and the relay coils) are de-energised, subsequently the valve-1 is placed in its spring centered position *to unload the pump.*

6.4.10. Pressure Reduction Circuit

Fig. 6.8. shows a pressure reduction circuit :

A "pressure reducing valve" is used in supply line joining two subsystems.

— Let supply from pump is taken at 8 bar pressure and is sent directly to *actuator-1* (double acting cylinder) via valve-1 (4/2 manually operated spring return DC valve).

— In order to drive *actuator*-2 in low-pressure, fluid has to pass through pressure reducing valve where its pressure is reduced to 4 bar and then sent to actuator-2 via valve-2.

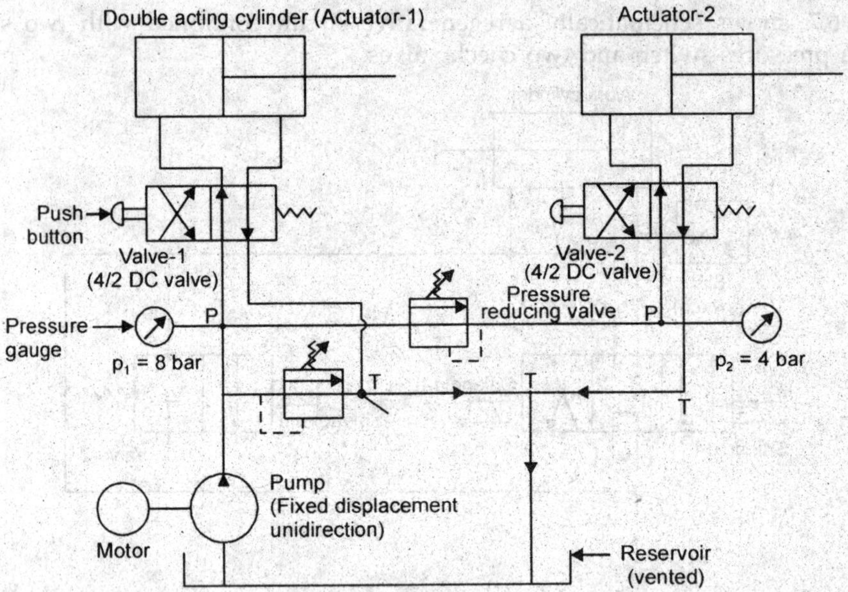

Fig. 6.8. Pressure reduction circuit.

6.4.11. Hydraulic Circuit for a Robot Arm

The robot arm can be put to desired motion by providing a power source. A hydraulic power source is used where the *pay load is more*; it may also be used in volatile, hazardous and explosive environments like a *spray painting booth*.

Fig. 6.9. Illustrates an hydraulic circuit for arm extension of non-servo robot :

— When solenoid-I is energised, the oil goes to the blank end of the cylinder and the arm extends to the desired position. It *touches the limit switch and signal is sent back to solenoid-II to energise*, causing the arm to retract.

— The *'accumulator' helps the system to send additional fluid* to the cylinder if there is a demand for the fluid and it also *acts as a shock absorber.*

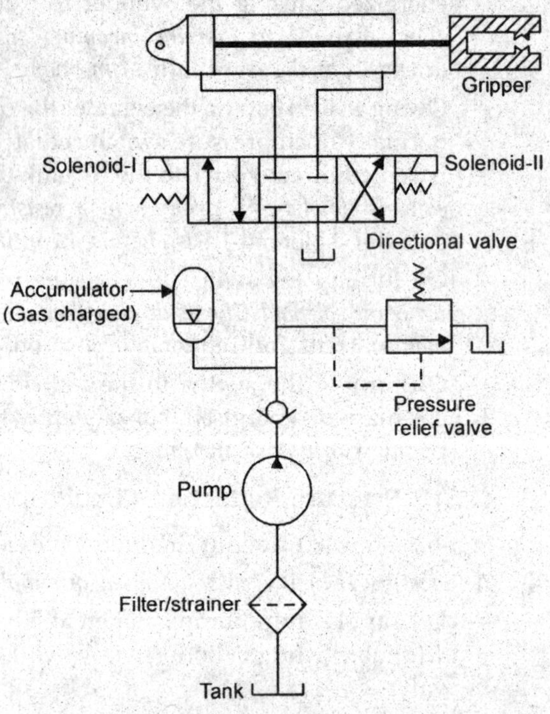

Fig. 6.9. Hydraulic circuit for a robot (non-servo) arm.

6.5 PNEUMATIC CIRCUITS

- These circuits are similar to hydraulic circuits; the difference being that *no return lines are used in pneumatic circuits because the exhaust air is released directly into the atmosphere.* This is shown by a short *dashed line leading* from the exhaust part of each valve or by a *smaller triangle.*
- *No input device is shown because most pneumatic circuits employ a centralised compressor as their source of energy.*
- The input to the circuit is fixed at some conveniently located *manifold which leads directly into the FRL* (Filter regulator lubricator) unit.

6.5.1. Basic Pneumatic Circuit Control of Double Acting Cylinder

Fig. 6.10. shows the schematic arrangement of basic pneumatic circuit control of double acting cylinder. A 4/2 (4-way 2 position) valve is used because the forward and return motions of the cylinder are controlled by the air pressure.

— When the push button of the 4/2 valve is operated, air moves to the piston side.

— When the valve is not actuated, the valve position is determined by the spring which makes the air to move to the rod side of the cylinder, thus the cylinder is retracted.

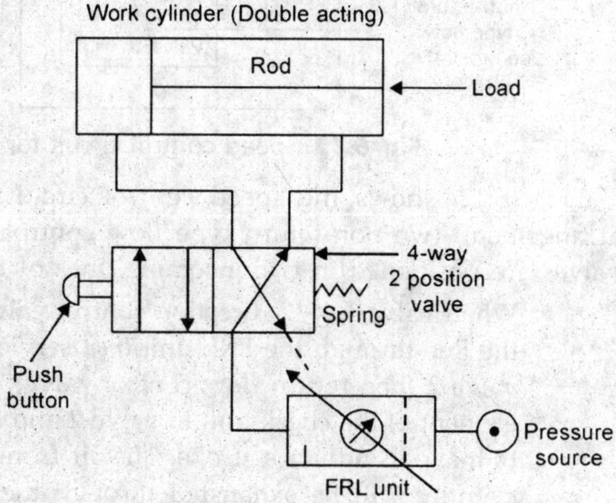

Fig. 6.10. Basic pneumatic circuit control of a double acting cylinder.

- This type of circuit can be used in the shifting of simple jigs, clamps etc., in a manually controlled system. However, if more versatility, controllability and flexibility in operation are required, a pilot control system (Fig. 6.11.) is a better choice.

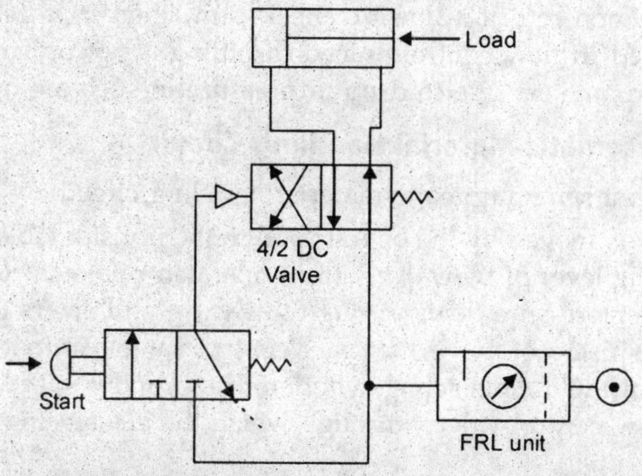

Fig. 6.11. Pilot controlled double acting cylinder.

6.5.2. Speed Control Circuit for a Double Acting Cylinder

In a pneumatic system, the *actuator speed is controlled by using a flow control valve.* This is also known as a *"throttle valve"* or a *"flow restrictor"*. In pneumatic systems, the non-return type flow control valves are more popular than flow restrictors.

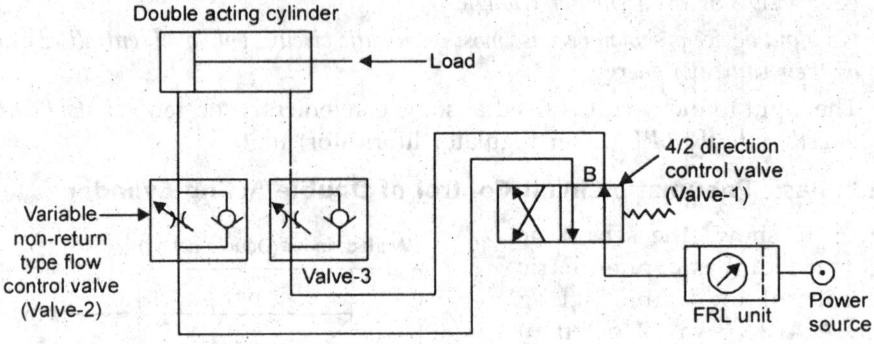

Fig. 6.12. Speed control circuit for a double acting cylinder.

Fig. 6.12. shows the speed control circuit for a double acting cylinder. In this arrangement, two non-return type flow control valves are used; the two flow control valves are positioned in two incoming lines of the double acting cylinder.

- When valve-1 (4/2 direction control valve) is actuated, the pressurised air from the line through the FRL unit (*i.e.*, service unit) will move to part B and then to valve-2 (non-return flow control valve). As a result, the air will lift the valving element of the check unit in valve-2 and fall amount of air will be fed to cylinder to make its advance stroke. The air from the earlier stroke in the rod side of the cylinder will be exhausted through the feed check valve-3. Here, the valving element in the check unit will remain closed and hence the air from the cylinder has to pass only through the *'throttle'* chamber, thereby importing the *desired flow control to cylinder.*
  - — *The above arrangement of circuit design with non-return flow control valve is called* **Exhaust throttling** or **Throttle-out**.
  - — If the air to cylinder is fed in a the throttled chamber and exhaust is effected via the non-return valve or check unit, then it is termed as **Infeed** or **Throttled-in** flow control. Infeed throttling is not preferred for cylinders of bigger diameters due to drop in pressure in such an arrangement.

6.5.3. Semi-automatic Material Handling Circuit

Fig. 6.13. shows a semi-automatic material handling circuit :

— A large box is moved by the operator on to the transfer table. He then depresses the operating lever of the valve-4 (cam operated pilot control valve). As a result the air under pressure flows through this valve and enters the pilot chamber (I) shifting the piston of valve-1 (master control valve), so that the air flows to the blank end of the double acting cylinder. The cylinder speed is controlled by the valve-3 (flow control valve), through which the air is being exhausted from the rod end.

— When the box is pushed on to the gravity conveyor, valve-4 (cam operated pilot control valve) is *tripped* and air flows to the pilot chamber (II) of the shifting piston so that air flows to rod end of the double acting cylinder and piston retracts to the original position. Now, the operator is ready to place the next box on the transfer table.

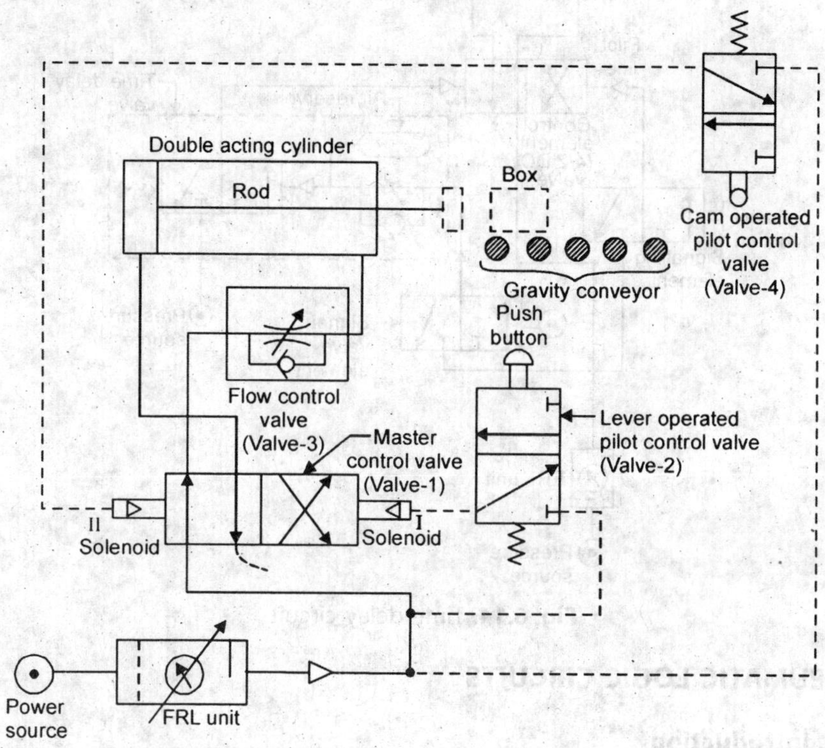

Fig. 6.13. Semi-automatic material handling circuit.

6.5.4. Time Delay Circuit

Fig. 6.14. illustrates a time delay circuit, used *to delay an already started signal.*

— When the manual push button of *signalling element* is pressed the airflow starts in the circuit.

— The control element is actuated by the pilot line (X). As a result, the air enters cylinder and pushes the piston in the outward direction; air pressure is directed to *time delay valve,* which can delay the signal for 5 to 15 seconds depending upon the requirement.

— The signal emanating from time delay valve actuates 4/2 DC (direction control) valve (*control element*), consequently the position of DC valve is changed and fluid is permitted to enter rod side of the cylinder and to discharge from the other side.

- Time delay valves find use in certain special applications (*e.g.*, bonding of two workpieces by filling adhesives in between and applying heat and pressure on it for a short time.)

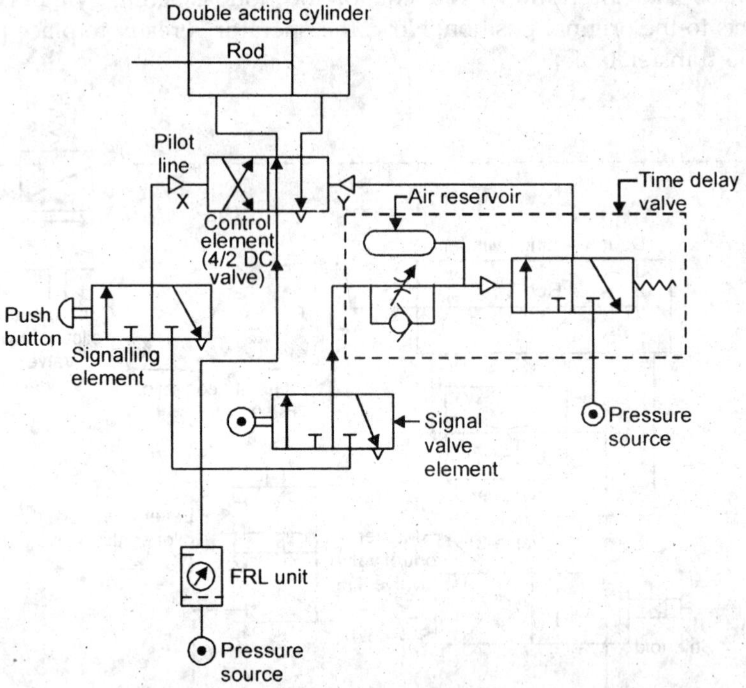

Fig. 6.14. Time delay circuit.

6.6 PNEUMATIC LOGIC CIRCUITS

6.6.1. Introduction

Arkototle (Greek philosopher) was the first to examine in details the form of reasoning called *deductive logic.*

George Boole (an Irish logician and mathematician) worked out a mathematical system (in 1847) to solve problems in logic, probability and engineering.

Logic was first applied to electrical relay logic and switching. Now, this logic is applied to pneumatics.

The basic component of pneumatic logic controllers are "sensors" and "switching elements" called binary elements. This means that, at any given moment, they can only be in *one of the two positions, "ON-OFF".*

Pneumatic logic control systems are available in varying degrees of sphistication.

- *One specific advantage of pneumatic control system is that air may perform the work function as well as control function and does not require additional equipment to convert its energy into motive power.*
  - — A *"control system"* may be defined as an integrated *complex device that governs or regulates a process or an operation.* Pneumatic control is one of the control devices used in modern industries for automation and rationalisation.
- To evolve a methodical pneumatic circuit diagram basic knowledge of functional diagram, logic function, binary arithmetic, Boolean algebra, etc. may be of great help.

6.6.2. Advantages of Pneumatic Logic Controls

Pneumatic logic controls offer the following *advantages* :

1. Systems are simple in design.
2. The life expectancy of such systems is longer than their electrical counterparts.
3. The response of pneumatic switching elements is faster than electrical systems.
4. Employed in specialised applications, *e.g.*, explosive environments.

6.6.3. Pneumatic Sensors

The different sensing elements include the following.

1. **Pneumatic sensors :**
   - (*i*) Pneumatic limit valves.
   - (*ii*) Back pressure sensors.
   - (*iii*) Proximity sensors.
   - (*iv*) Gap sensor or interruptible jet sensor.
   - (*v*) Limit switch.
2. **Pressure sensors :**

- The different types of pneumatic switching elements (in the form of logic gates like OR, NOR) are :
  - (*i*) Pneumatic valves.
  - (*ii*) Pneumatic moving-part logic (MPL) :
    - (*a*) AND MPL element;
    - (*b*) OR MPL element;
    - (*c*) NOT MPL element.
  - (*iii*) Fluidic element :
    - (*a*) Bistable flip-flop;
    - (*b*) OR/NOR gate device;
    - (*c*) AND/NAND gate
    - (*d*) Diaphragm amplifier.

— These elements are being used with power pneumatic circuits to offer better control to the pneumatic systems.

6.6.4. Common Methods of Circuit Design for Automatic System

The following are the common method used for circuit design of automatic system :

1. Intuitive based on knowledge of elements.
2. Intuitive based on logics.
3. Use of Karnaugh – Veitch diagram, programme generators etc.
4. Cascade system.
5. Other methods.

6.6.5. Common Logic Functions

Some common logic functions are :

1. *AND – function* : Here all the incoming impulses or pneumatic signals must be activated and added up before there could be an output signal.

2. *OR – Function* : Here, any one of the input signals will produce an output signal for schematic arrangement.
3. *Not – function* : In this case, the input signal goes through to the output signal until an impulse stops the flow. In other words, impulse tells the input signal NOT to proceed to the output signal.
4. *NAND – function* : (Negative AND) – Here, all input signals are removed to have an output.
5. *Memory – function* : A momentary impulse causes an output to commence and continue unless a second impulse is fed to release the state of continuity.

   An impulse operated (pilot operated) valve is a good example.
6. *LIMITED Memory* : Here, the impulse is momentary and action between the input signal to output signal is stopped until a specified time elapses. The device 'remembers' what to do and does it.

6.6.6. Description of Some Pneumatic Logic Circuits

Some pneumatic logic circuits (*design problems*) are enumerated and described below :

1. Pneumatic logic cylinder sequencing circuit.
2. Pneumatic circuit for a clamping device having variable clamping forces.
3. Pneumatic braking circuit for serial braking action in a number of wagons placed one after the other.
4. Pneumatic circuit for control of cylinder using flip-flop.

1. Pneumatic logic cylinder sequencing circuit :

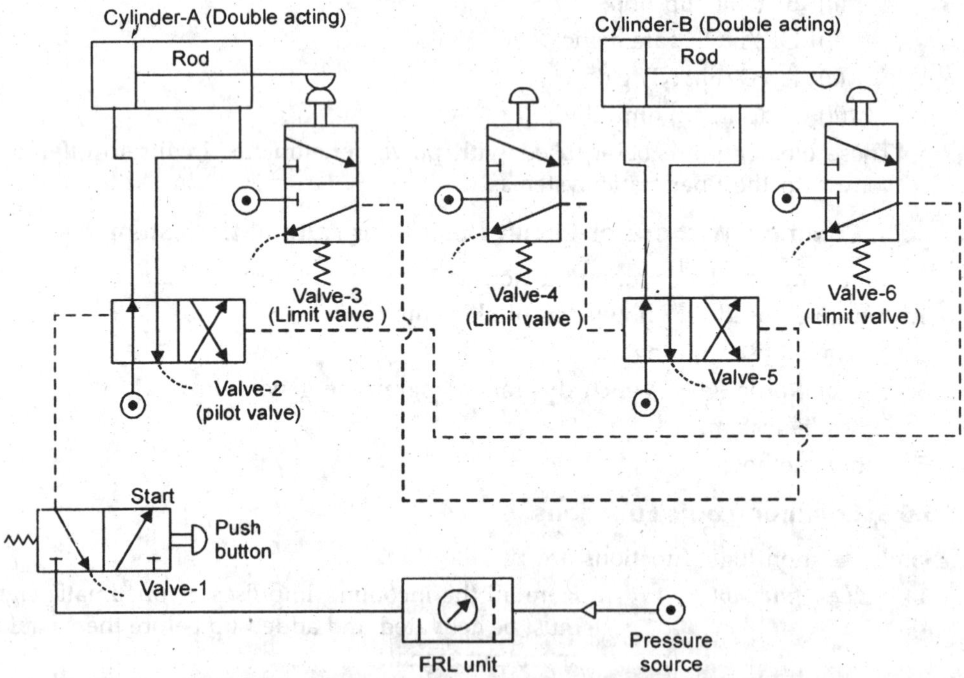

Fig. 6.15. Pneumatic logic cylinder sequencing circuit.

Fig. 6.15. shows a pneumatic logic circuit for controlling the extension and retraction of two double acting cylinders A and B; the sequence is as follows :

— Cylinder – A extends (A^+);
— Cylinder – B extends (B^+);
— Cylinder – A retracts (A^{-1});
— Cylinder – B retracts (B^{-1}).

The operation is as following, assuming that both cylinders are *fully retracted* :

— When the valve-1 is momentarily depressed, the pilot signal shifts valve-2 (pilot valve), causing the cylinder-A to *extend*.
— At full extension (of cylinder-A), valve-4 (*limit valve*) is actuated to shift valve-5 and *extend* cylinder-B.
— Upon full extension (of cylinder-B), valve-6 (*limit valve*) is actuated; this shifts valve-2 to *retract* cylinder-A.
— Upon full retraction (of cylinder-A) valve-3 (limit valve) is actuated, shifting valve-5 to fully *retract* cylinder-B.

2. Pneumatic circuit for a damping device having variable damping forces :

Fig. 6.15. shows a pneumatic circuit for a clamping device having variable clamping forces :

— The air energy to the single acting cylinder is led through the valve-6 (pressure regulating valve) which can be adjusted to the required pressure to achieve the desired clamping force.
— The D.C. valves-1, 3 and 4 provide a self-holding arrangement to the circuit which can be made *'on'* and *'off'* by valve-1 and valve-3 respectively.

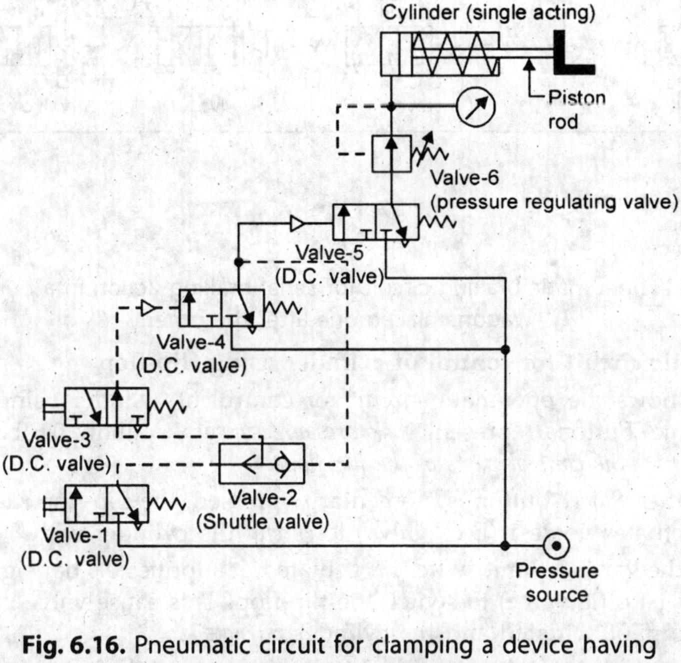

Fig. 6.16. Pneumatic circuit for clamping a device having variable clamping forces.

3. Pneumatic braking circuit for serial braking action in a number of wagons placed one after the other :

Fig. 6.17. shows a pneumatic braking circuit for serial braking action in a number of wagons placed one after the other :

— A number of double acting braking cylinders are actuated in series.

— Restrictors 1.3, 2.3 and 3.3 are used with varying adjustment so that the cylinders are actuated one after another in a sequence after predetermined timings.

— The quick exhaust valves 1.2, 2.2, 3.2 and 4.2 are used to release the return air from the cylinders quickly.

The following points may be noted :

(*i*) Instead of restrictors, time delays valves may also be used.

(*ii*) The cylinders need not always be double acting, single acting cylinders may also serve the purpose.

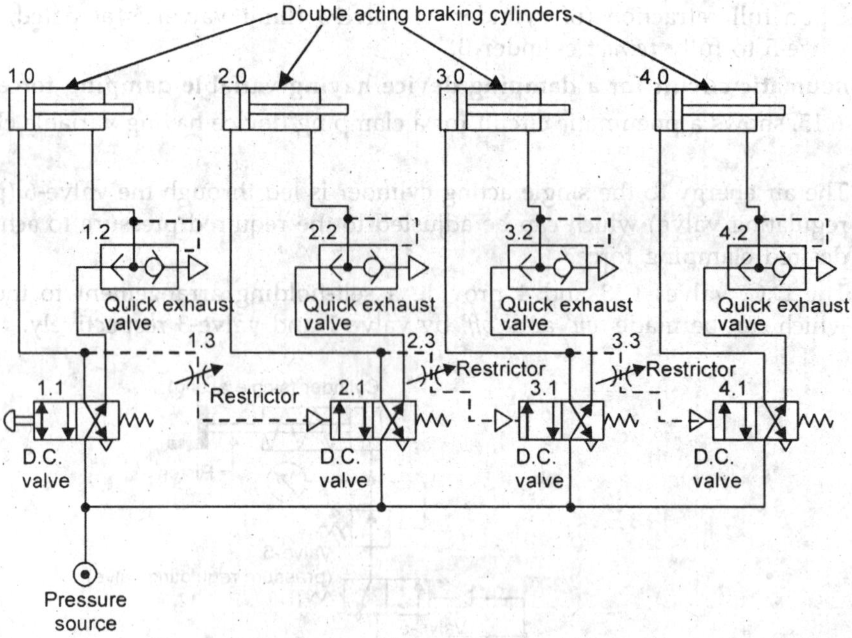

Fig. 6.17. Pneumatic braking circuit for serial braking action in a number of wagons placed one after the other.

4. Pneumatic circuit for control of cylinder using flip-flop

Fig. 6.18. shows the pneumatic circuit for control of cylinder using flip-flop, here 'start' and 'panic' buttons represent *back pressure sensors* to allow 0.6 bar restricted air supplies *to activate the control parts of the flip-flop* :

— When the "*Start*" button is momentarily pressed, the flip-flop (using 0.6 bar air) pilot actuates valve-1 (D.C. valve) to *extend* the cylinder with 8 bar air.

— When the valve-3 (limit switch) is actuated, it supplies 0.6 bar signal through the valve-4 (shuttle valve) to switch the flip-flop. This cause valve-2 (D.C. valve) to become pilot actuated and the cylinder *retracts.*

— When the "panic" button is momentarily pressure while the cylinder is extending, it supplies signal through the valve-4 to switch the flip-flop. As a result, the valve-2 is actuated, causing the cylinder to *retract instantly.*

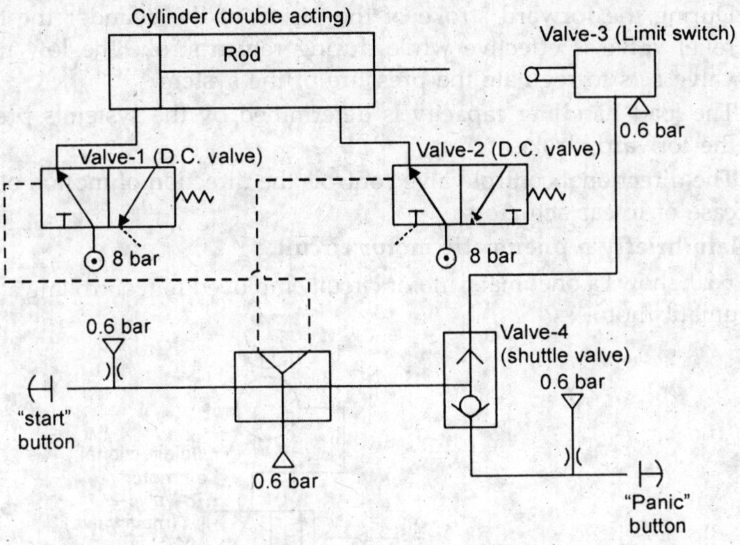

Fig. 6.18. Pneumatic circuit for control of cylinder using flip-flop.

QUESTIONS WITH ANSWERS

Q. 6.1. Explain briefly with a neat sketch a simple hydraulic circuit.

Ans. Fig. 6.19. shows symbolically the *hydraulic system of linear actuator in a simplest arrangement* :

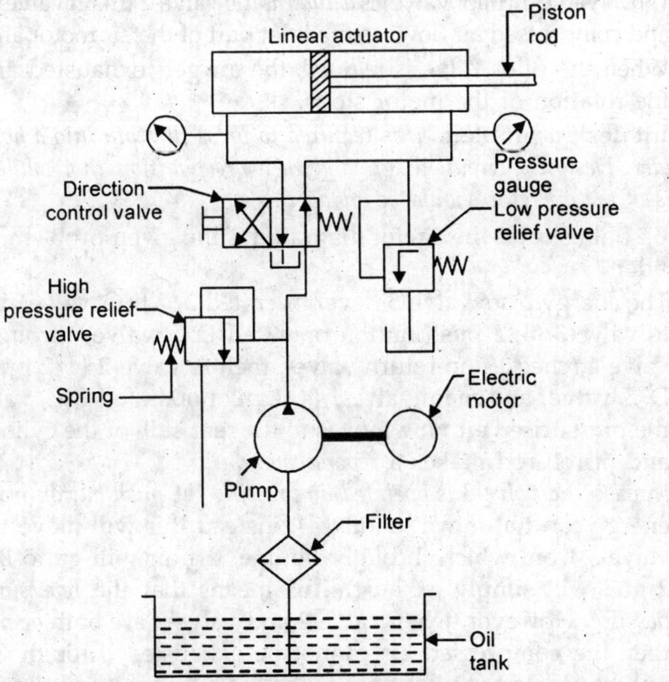

Fig. 6.19. Typical hydraulic circuit.

— During the forward stroke of the piston of the cylinder the high pressure relief valve is effective while during return stroke the low pressure relief valve acts to regulate the pressure of the system.

— The load handling capacity is determined by the system's pressure during the forward stroke.

— The directional control valve controls the direction of motion of the piston in case of linear actuators.

Q. 6.2. Explain briefly a pneumatic motor circuit.

Ans. Fig. 6.20. shows a pneumatic motor circuit employed for controlling a unidirectional pneumatic motor :

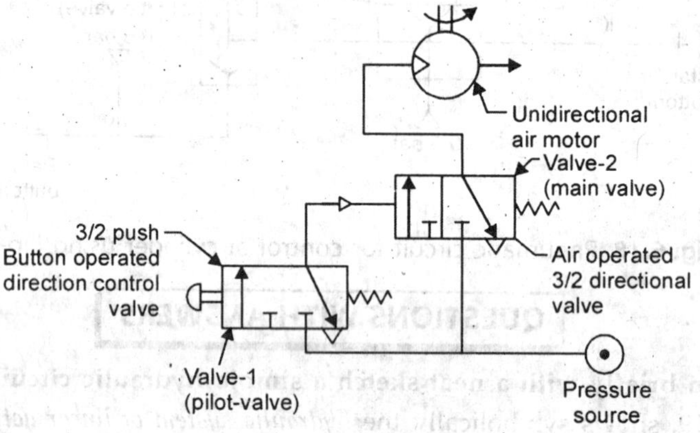

Fig. 6.20. Pneumatic motor circuit.

— When valve-1 (pilot valve) is *actuated,* the valve-2 (main valve) changes its position and compressed air flows to the inlet part of the air motor and the motor rotates.

— When the pilot valve is *released*, the air gets exhausted and as a conquence, the rotation of the motor stops.

Q. 6.3. Circuit design problem. *It is required to press-fit a pin into a hole using a pneumatic cylinder. Design a circuit diagram with a precondition that while actuating, both the hands of the operator should be engaged.*

Ans. Fig. 6.21. illustrates the circuit diagram for the given problem. This system works as under :

— The energy from valve-5 (lever operated 2/2 D.C. valve, manual start) moves to valve-3 (4/2 push button operated D.C. valve) through which it goes to valve-4 (check/non-return valve), then to valve-2 (4/2 push button operated D.C. valve) and main valve-1 (4/2 pilot operated D.C. valve). Consequently, the pressurised air now moves to the rear side of the cylinder (double acting) and piston-rod makes a *reverse motion.*

— In case the *valve-3 is operated by pressing the push button manually,* then the air energy can not move to valve-4, instead this will move to the B port of the valve-3 from which through the line, the air will go to B port of the valve-2 and will simply exhaust. This means that the *reversing of cylinder is not possible.* However, when valve-2 and valve-3 are both *operated simultaneously,* then the compressed air cannot be exhausted through valve-2 and energy will be fed to the pilot port of the main valve-1 and the *desired motion of the cylinder will be possible.*

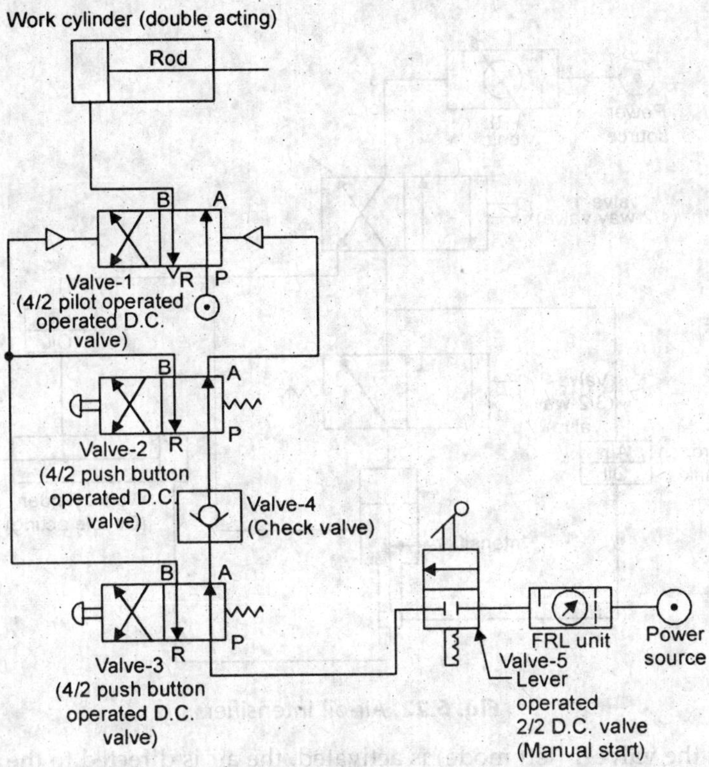

Fig. 6.21. Logic circuit- To press-fit a pin to a hole; both hands of the operator being engaged.

Q. 6.4. (*a*) What do you understand by 'hydropneumatics' ?

(*b*) How are hydro-pneumatic appliances categorised ?

(*c*) Explain briefly air-oil intensifier circuit.

Ans. (*a*) ***Hydropneumatics :*** Owing to the compressibility characteristics of air, the air cylinder tends to slow down on meeting an increase in load and to accelerate or jump forward when working against a load which suddenly decreases. Thus, pneumatic actuators are not suitable for the finer movement of machine tool, elements. The solution of these problems lie in the combination of two fluids–*air and the oil*. By using these two media, the quick action of air and the smooth high pressure action of oil blend ideally to provide a solution to several fluid power problems. It should be understood clearly that although the two fluids are used in the circuit they are *not to be mixed.*

(*b*) ***Classification :*** All hydro-pneumatic appliances may broadly be classified as follows :

(*i*) Air-oil reservoir ; (*ii*) Air-oil cylinder; (*iii*) Air-oil intensifier.

(*c*) **Air-oil intensifier circuit :**

Fig. 6.22. shows an air-oil intensifier circuit. *Such a system drives a cylinder over a large distance at a low pressure and then over a small distance at a high pressure.* Shop air can be used to extend or retract the cylinder during the low pressure portion of the cycle.

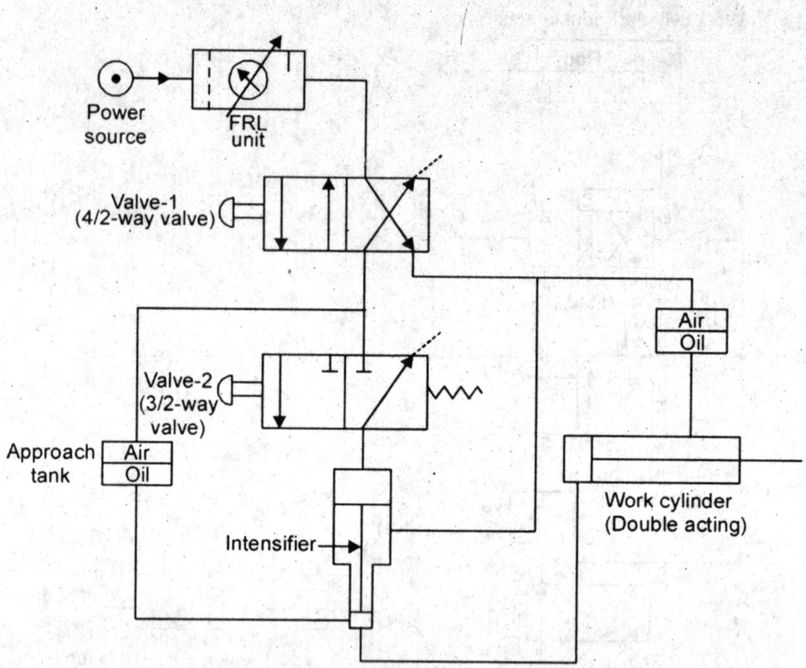

Fig. 6.22. Air-oil intensifier.

— When the valve-1 (left mode) is activated, the air is directed to the approach tank and consequently, the oil is forced through the bottom of the intensifier to the blank end of the cylinder.

— When the cylinder is loaded, valve-2 is actuated (left mode) which sends air to the top end of the intensifier. The high pressure oil cannot return to the approach tank because this part is blocked off by the downward motion of the intensifier piston. As a result, the cylinder receives the high-pressure oil at the blank end to carry out the high pressure punching operation.

— On releasing the valve-2 (right mode), the shop air is blocked and the top end of the intensifier is vented to the atmosphere. When valve-1 is released (right mode), the air is directed to the return tank. This delivers oil to the rod end of the cylinder causing it to retract. The return oil from the blank end of the cylinder enters the bottom end of the intensifier and flows back to the approach tank.

HIGHLIGHTS

1. The major difference between hydraulic and pneumatic circuits is that hydraulic elements are designed for high pressures (20 bar to 400 bar) and pneumatic circuits are designed for low pressures (5 bar to 20 bar).
2. A *hydraulic circuit* is an arrangement of interconnected components, selected to achieve the desired work output.
3. The basic components of pneumatic logic controllers are *sensors* and *switching elements* called binary elements.

OBJECTIVE TYPE QUESTIONS

Fill in the Blanks or Say "Yes" or "No".

1. .... diagrams are prepared by arranging various components into a sequence according to their functions.
2. Hydraulic elements are designed for low pressures.
3. The weight-to-pressure ratio in case of hydraulic system is very small.
4. In hydraulic systems, there is no cavitation problem.
5. While drawing a circuit, it must begin with .... and end with ....
6. In general, lower the number of components .... is the cost and .... is the cost effectiveness.
7. Meter-in circuit is ideal in applications where a load always offers a positive resistance to flow during a controlled stroke.
8. Bleed-off circuit is less efficient than meter-in and meter-out circuits.
9. A pressure reducing valve is used in supply line joining two subsystems.
10. Return lines are not used in pneumatic circuits because this air is exhausted directly into the atmosphere.

ANSWERS

1. Circuit	2. No	3. Yes	4. No	5. Pump, actuator
6. lower, higher	7. Yes	8. No	9. Yes	10. Yes.

THEORETICAL QUESTIONS

1. What is a fluid circuit ?
2. What is the major difference between hydraulic and pneumatic circuits ?
3. Draw symbols for the following hydraulic components :
   (*i*) Fixed displacement (unidirectional) pump.
   (*ii*) Intensifier.
   (*iii*) Hydraulic oscillator.
   (*iv*) Pilot operated check valve.
   (*v*) Complete valve assembly.
   (*vi*) 6/3 direction valve.
   (*vii*) Pressurised reservoir.
4. Draw symbols for the following pneumatic components :
   (*i*) Variable displacement compressor.
   (*ii*) Oscillating motor.
   (*iii*) 5/3-way valve.
   (*iv*) Differential pressure regulator.
   (*v*) Pressure switch adjustable.
   (*vi*) Pneumatic-electrical converter.
5. Give a comparison of hydraulic, pneumatic and hydro-pneumatic systems.
6. How are the elements of hydraulic and pneumatic circuits categorised ?
7. With the help of a neat diagram, explain the sections of a hydraulic circuit.
8. What are the main criteria in the design of any system ?

9. Enlist the main parameters which affect the selection of a particular type of pump.
10. Name at least six basic hydraulic circuits available for simple applications.
11. Explain briefly any two of the following basic hydraulic circuits :
    (*i*) Meter-out circuit;
    (*ii*) Bleed-off circuit;
    (*iii*) Sequencing circuit;
    (*iv*) Regenerative circuit.
12. Explain with a neat sketch a hydraulic circuit for a robot (non-servo) arm.
13. Describe briefly basic pneumatic circuit control of double acting cylinder.
14. Explain the following :
    (*i*) Speed control circuit for a double acting cylinder.
    (*ii*) Semiautomatic material handling circuit.
15. What are the advantages of pneumatic logic controls ?
16. Name the common methods of circuit design for automatic system.
17. Enlist some common logic functions.
18. Explain with a neat sketch the pneumatic logic cylinder sequencing circuit.

CHAPTER

7 Electrical and Electronic Controls

7.1 Sensors and transducers – Introduction – Definition of transducer – classification of transducers – Transducer sensitivity – Specifications for transducers; **7.2 Control systems** – Introduction – System – Control system – Classification of control systems – Open-loop control systems – Closed loop control systems – Automatic control systems – Servo-mechanism – Regulator – Representation through model – Analogous system – Block diagram – Mathematical block diagram – Signal flow graph – Time response of control system – Stability– Frequency response – Error defector – LVDT – Servo-amplifier – Sampled data system – Industrial controllers – Pneumatic control systems – Hydraulic control system; **7.3 Electronic devices** – General aspects – Semiconductor – Intrinsic semiconductor – Extrinsic semiconductor – P-N junction diodes – Zener diode – Biopolor junction transistor (BJT) – Field effect transistor (FET) – Unijunction transistor (UJT) – Thyristor; **7.4 Digital electronics** – Introduction – Advantages and disadvantages of digital electronics – Digital circuit – Logic gates – Universal gates – Half adder (HA) – Full adder (FA) – Boolean algebra – De Morgan's Theorem – Flip-Flop Circuits – Counters – Integrated circuits (ICs) – Operational amplifiers (Op-amp); **7.5 Microprocessors** – Micorprocessor-General aspects – Microprocessor systems – Intel soft microprocessor – Programmable logic controllers (PLCs). ***Questions with Answers*** – Highlights – Objective Type Questions – Theoretical Questions.

7.1 SENSORS AND TRANSDUCERS

7.1.1. Introduction

The *primary sensing element* (*sensor*) is the first and foremost requirement for measurement and automatic controls. The *sensors sense the condition, state or value of the process variable and produce an output which reflects this condition, state or value. The* **transducers** *transform the energy of the process variable to an output of some other type of energy which is able to operate some control device.* Sometimes a *secondary transducer* may be employed to transform the output of the primary sensor to still another type of energy.

Examples :

- In the ordinary *dial indicator* the indicating spindle acts as a sensor/detector for displacement. It simply performs the function of sensor/detector and nothing else.
- The function of a *Bourdon tube* of a pressure gauge is *twofold*: Firstly to sense the pressure and secondly to give the resulting effect or output in the form of displacement. Here the tube acts a sensor/detector transducer.
- In a *compressive load cell,* the platform detects the force and gives an output in the form of deflection. This deflection may be further converted into an electrical output by strain gauges (called *secondary transducer*).

For the measurement of particular quantity, different types of sensors and transducers are available and the choice of a suitable unit depends upon the static and dynamic performance characteristics.

7.1.2. Definition of Transducer

A broad definition of a transducer is as follows:

"A **transducer** *is a device which converts the energy from one form to another".*

Most of the transducers either convert electrical energy into mechanical displacement and/or convert some non-electrical physical quantity (*e.g.*, force, sound, temperature etc.) to an electrical signal.

A transducer performs the following *functions* in an electronic instrumentation system :

1. *Detects or senses the presence, magnitude and changes in physical quantity being measured.*
2. *Provides a proportional electrical output signal (see Fig. 7.1.)*

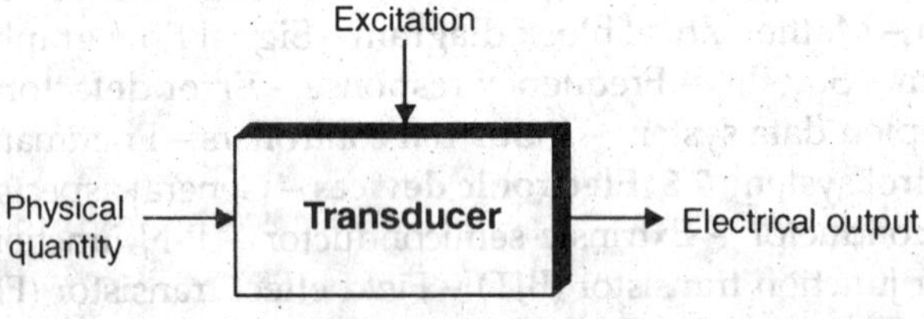

Fig. 7.1. Transducer.

- A **transducer** can be broadly defined as a *device which converts a non-electrical quantity into an electrical quantity.*

An **inverse transducer** *is defined as a device which converts an electrical quantity into a non-electrical quantity.* It is a precision actuator which has an electrical input and a low power non-electrical output. A *piezoelectric* crystal acts as an inverse transducer because when a voltage is applied across its surfaces, it changes its dimensions causing a mechanical displacement.

7.1.3. Classification of Transducers

A. Transducers are broadly *classified* into two groups as follows:

1. ***Active transducers.*** They are also known as *self-generating type transducers.* These transducers *develop their own voltage or current. The energy required for production of an output signal is obtained from the physical phenomenon being measured.*

   Examples: *Thermocouples and thermopiles, piezoelectric pick-up, photovoltaic cell.*

2. ***Passive transduces.*** *They are known as externally-powered transducer.* These transducers *derive the power required for the energy conversion from an external power source.* However, they *may absorb some energy from the physical phenomenon under study.*

   Examples: *Resistance thermometers and thermistors, potentiometric devices, differential transformer, photoemission cell etc.*

B. Classification based on the type of output :

1. ***Analogue transducers.*** These transducers convert the input physical phenomenon into an analogous output which is a continuous function of time.

   Examples: *Strain gauge, a thermocouple, a thermistor or an LVDT* (linear voltage differential transformer).

2. ***Digital transducers.*** These transducers convert the input physical phenomenon into an electrical output which may be in form of pulse.

C. Classification based on electrical principle involved :

1. ***Variable-resistance type :***
   (*i*) Strain and pressure gauges.
   (*ii*) Thermistors, resistance thermometers.
   (*iii*) Photoconductive cell etc.
2. ***Variable-inductance type :***
   (*i*) Linear voltage differential transformer (LVDT).
   (*ii*) Reluctance pick-up.
   (*iii*) Eddy current gauge.
3. ***Variable-capacitance type :***
   (*i*) Capacitor microphone.
   (*ii*) Pressure gauge.
   (*iii*) Dielectric gauge.
4. ***Voltage-generating type :***
   (*i*) Thermocouple.
   (*ii*) Photovoltaic cell.
   (*iii*) Rotational motion tachometer.
   (*iv*) Piezoelectric pick-up.
5. ***Voltage-divider type :***
   (*i*) Potentiometer position censor.
   (*ii*) Pressure-actuated voltage divider.

Table 7.1. shows the measurements versus transduction methods.

- While *describing a particular transducer* the information must be available about the following aspects :
  (*i*) The measurand.
  (*ii*) The sensing element which responds directly to the measurand.
  (*iii*) The principle of operation of the transducer and where the output of the transducer originates.
  (*iv*) The useful range.

Table 7.1. Measurements versus Transduction Methods

S. No.	Quantity to be measured	Type of transducer	S. No.	Quantity to be measured	Type of transducer
1.	***Displacement***	— Resistive — Inductive — Capacitive — Piezoelectric — Magnetoelectric — Radioactive — Electron tube.	7.	***Pressure***	— Resistive — Inductive — Capacitive — Piezoelectric — Thermoelectric — Magnetoelectric — Magnetostrictive — Radioactive — Electron tube.
2.	***Thickness***	— Inductive — Capacitive — Piezoelectric — Photoelectric — Radioactive.	8.	***Flow***	— Resistive — Inductive — Capacitive

3.	*Velocity*	— Resistive — Inductive — Capacitive — Piezoelectric — Photoelectric — Magnetoelectric — Radioactive — Electron tube.			— Piezoelectric — Magnetoelectric — Radioactive.
			9.	*Level*	— Resistive — Capacitive — Piezoelectric — Photoelectric — Radioactive.
4.	*Acceleration*	— Resistive — Inductive — Capacitive — Piezoelectric — Magnetoelectric — Electron tube.	10.	*Temperature*	— Resistive — Photoelectric — Thermoelectric — Radioactive.
			11.	*Humidity*	— Resistive — Capacitive.
5.	*Mass*	— Inductive — Piezoelectric — Magnetoelectric — Radioactive.	12.	*Viscosity*	— Resistive — Capacitive — Piezoelectric — Magnetostrictive.
6.	*Force*	— Resistive — Inductive — Piezoelectric — Radioactive.			

- While *selecting a dector-transducer element*, the following major considerations need to be looked into:
  - (*i*) Mechanical suitability in terms of
    - — Physical size, weight and shape;
    - — Mounting arrangement;
    - — Ruggedness.
  - (*ii*) Electrical suitability in terms of:
    - — Sensitivity ;
    - — Frequency response ;
    - — Ease of signal transmission.
  - (*iii*) Environmental suitability in terms of
    - — Sensitivity to temperature and self-heating effects;
    - — Magnetic fields;
    - — Vibration; dust and humidity;
    - — Supply frequency etc.
  - (*iv*) Transducer performance in terms of calibration accuracy.
  - (*v*) Desired measurement accuracy and range, power requirements, overload protection and vulnerability to sudden failure.
  - (*vi*) Purchase aspects.

7.1.4. Transducer Sensitivity

The relationship between the measurand and the transducer output signal is referred to as "transducer sensitivity".

$$i.e., \text{ Transducer sensitivity} = \frac{\text{Output signal increment}}{\text{Measurand increment}}$$

Sensitivity of a transducer should be *usually as high as possible* since then it becomes easier to take the measurements.

7.1.5. Specifications for Transducers

While selecting the proper transducer for any application, or ordering the transducers, the following specifications should be thoroughly considered:

(*i*) Ranges available.
(*ii*) Squaring system.
(*iii*) Sensitivity.
(*iv*) Maximum working temperature.
(*v*) Method of cooling employed.
(*vi*) Mounting details.
(*vii*) Maximum depth.
(*viii*) Linearity and hysteresis.
(*ix*) Output for zero input.
(*x*) Temperature coefficient of zero drift.
(*xi*) Natural frequency.

7.2. CONTROL SYSTEMS

7.2.1. Introduction

Automatic control has played a significant role in the advance of engineering science. Besides its extreme importance in space-vehicle systems, missile-guidance systems, etc., automatic control has become an important and integral part of modern manufacturing and industrial processes. Automatic control, for example, is essential in :

— Design of auto pilot systems in *aero space industries* ;
— Design of cars and trucks in the *automobile industries* ;
— Industrial operations as controlling pressure, temperature, humidity, viscosity, and flow in the *process industries*.

7.2.2. System

A system may be defined as follows :

- *"A **system** is an arrangement, set or collection of things connected or related in such a manner as to form an entirely or whole".*

  Or

  *"A **system** is an arrangement of physical components connected or related in such a manner as to form and / or act as entire unit."*

- A system consists of a sequence of components in which *each component has some cause as input and its effect will be its output. Broadly it is a sequential set of cause and effects.*

Each system may have a *large number of subsytems; "Examples"* :

(*i*) This universe is iteself a system consisting of large number of subsystems.
(*ii*) Human body as a system has digestive system, respiratory system etc.

7.2.3. Control System

A *control system* is an *arrangement of physical components connected or related in such a manner as to command, direct or regulate itself or another system.*

Elements of a control system:

The elements of a control system are enumerated and defined as follows :

Element	Definition
1. *Controlled variable*	The quantity or condition of the controlled system which can be directly measured and controlled is called *controlled variable.*
2. *Indirectly controlled variable*	The quantity or condition related to controlled variable, but cannot be directly measured is called *indirectly controlled variable.*
3. *Command*	The input which can be independently varied is called *command.*
4. *Reference input*	A standard signal used for comparison in the close-loop system.
5. *Actuating signal*	The difference between the feedback signal and reference signal is called *actuating signal.*
6. *Disturbance*	Any signal other than the reference which affects the system performance is called *disturbance.*
7. *System error*	The difference between the actual value and ideal value is called *system error.*

Examples of control system applications:

Following are some examples of control system applications:

1. Steering control of automobile.
2. Print wheel control system.
3. Industrial sewing machine.
4. Sun-tracking control of solar collectors.
5. Speed control system.
6. Temperature control of an electric furnace.

7.2.4. Classification of Control systems

Control systems are classified into the following *two basic types* :

1. Open-loop control systems (Unmonitored or non–feedback control systems)
2. Closed–loop control systems (Monitored or feedback control systems).

Comparison between Open-loop and Closed-loop Systems

Open-loop	*Closed-loop*
1. Less accurate.	1. More accurate.
2. Generally build easily.	2. Generally complicated and costly.
3. Stability can be ensured.	3. May become unstable at times.
4. Presence of non-linearities cause malfunctioning.	4. It usually performs accurately even in the presence of non-linearities.
5. Any change is system component cannot be taken care of automatically.	5. Change in system component is automatically taken care of.
6. Input cammand is the sole factor responsible for providing the control action.	6. The control action is provided by the difference between the input command and the corresponding output.

7. The control adjustment depends upon human judgement and estimate. *Examples :* (*i*) Automatic washing machine. (*ii*) The electric switch. (*iii*) An automatic toaster. **Note:** All control systems *operated by present timing mechanisms are open-loop.*	7. The control adjustment depends on output and feedback element. *Examples :* (*i*) Liquid level control system. (*ii*) Traffic signal system. (*iii*) Human being reaching for an object.

7.2.5. Open-loop Control Systems (Non-feedback Systems)

- *An* ***Open-loop control system*** *is one in which the control action is independent of the desired output.* The actuating signal depends only on the input command and output has *no control* over it.
- The ***elements*** of an open-loop control system can usually be divided into the following two parts (Refer to Fig. 7.2):
  (*i*) Controller;
  (*ii*) Controlled process.

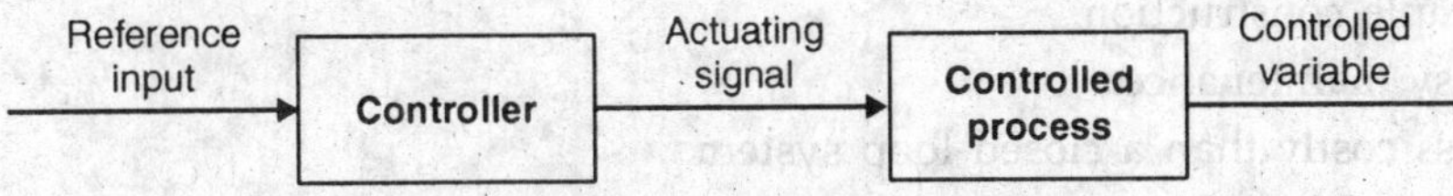

Fig. 7.2. Elements of an open-loop control system.

— An input signal or command is applied to the controller, whose output acts as the actuating signal; the actuating signal then controls the controlled process so that the controlled variable will perform according to prescribed standards.

— *In simple cases,* the controller can be an *amplifier, mechanical linkage, filter, or other control element,* depending on the nautre of the system. In *more sophisticated cases,* the controller can be a computer such as a *microprocessor.*

— Because of the *simplicity and economy* of open-loop control systems we find this type of system in many non-critical applications.

Examples :

1. ***Idle-speed control system :***

- The following are the main objectives of the idle-speed control system of automobile :
  (*i*) To eleminate or minimise the speed drop when engine loading is applied.
  (*ii*) To maintain the engine speed at a desired value.
- Fig. 7.3 shows an idle-speed control system from the stand point of inputs-system outputs. In this case the throttle angle and the load torque (due to the application of air conditioning, power steering, transmission, power brake, etc.) are the inputs, and the engine speed is the output. The engine is the controlled process of the system.

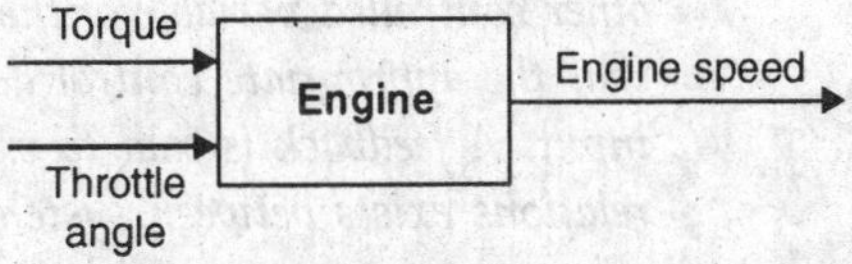

Fig. 7.3. Idle-speed control system.

2. ***Print wheel control system:***

Fig. 7.4 shows an example of the printwheel control system of a word processor or electronic typewriter (and also shows a typical input-ouput set for the system).

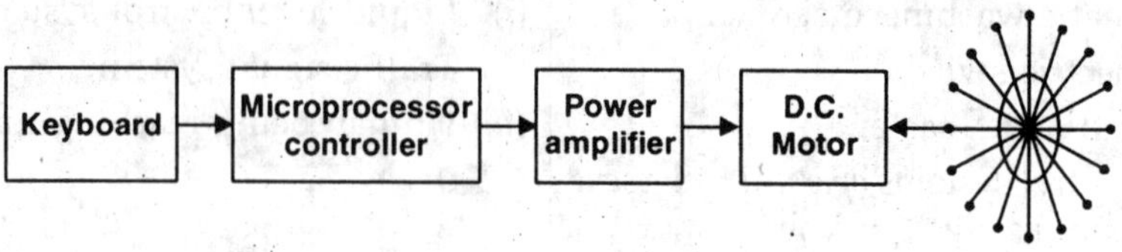

Fig. 7.4. Open-loop word processor control system.

— When a reference command input is given, the signal is represented as a step function. Since the electric windings of the motor have inductance and the mechanical load has inertia, the printwheel cannot respond to the input instantaneously. Typically it will follow the response and settle at the new position after sometime. Printing should not begin until the printwheel has come to complete stop; otherwise, the character will be smeared.

Advantages and limitations of open-loop control system :

Advantages :

1. Simple construction.
2. Easy maintenance.
3. Less costly than a closed-loop system.
4. No stability problem.
5. Convenient when output is difficult to measure or measuring the output precisely is economically not feasible.

Limitations/Disadvantages :

1. Since the system is affected by internal and external disturbances, the *output may differ from the desired value.*
2. For getting accurate results, this system needs frequent and careful calibrations.
3. Any change in system component cannot be taken care of automatically.
4. Presence of non-linearities causes malfuctioning.

7.2.6. Closed-loop Control System (Feedback Control System)

- **A closed-loop system** is one in *which control action is somehow dependent on the output.* In this case the controlled output is fed back through a *feedback element* and compared with the reference input. Thus *the actuating signal is the difference of desired output and reference input.*
- **Feedback** *is that property of a closed-loop system which permits the output or some other controlled variable of the system, to be compared with the input to the system, so that the appropriate control action may be formed as some function of the output and input. A feedback is said to exist in system when a closed sequence of cause and effect relations exists between system variables.*

The ***Characteristics of feedback*** *are as follows :*

(*i*) Increased bandwidth.

(*ii*) Increased accuracy.

(*iii*) Tendency towards oscillation or instability.

(*iv*) Reduced effects or non-linearities and distortion.

(*v*) Reduced sensitivity of the ratio of output to input to variations in system characteristics.

- *A closed-loop idle-speed control system* is shown in Fig. 7.5.

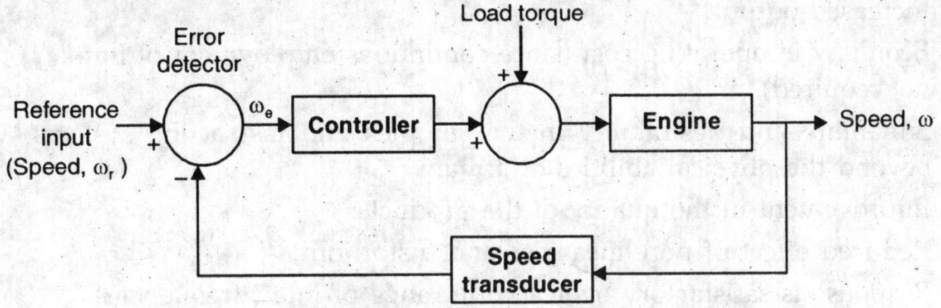

Fig. 7.5. Closed-loop idle-speed control system.

— The reference input (ω_r) sets the desired idling speed. The engine speed at idle should agree with reference value (ω_r), and any difference such as load torque is sensed by the speed transducer and the error detector. The controller will operate on the difference and provide a signal to adjust the throttle angle to correct the error.

Advantages and limitations :

Advantages :

1. More accurate comparatively.
2. Usually performs accurately even in the presence of non-linearities.
3. Change in system component is automatically taken care of.
4. The use of feedback system response is relatively insensitive to external distrubanes and internal variations in system parameters. It is thus *possible to use relatively inaccurate and inexpensive components to obtain the accurate control of a given plant* (whereas doing so is impossible in the open-loop case).

Limitations/Disadvantages :

1. Generally complicated in construction.
2. Generally higher in cost and power.
3. May become unstable at times.

7.2.7. Automatic Control Systems

- *A closed-loop control system operating without human operator is called an* **automatic control system.**

Examples :

(*i*) Centrifugal watt governor, where the lift of the rotating balls is used as speed monitor. The supply of steam is automatically controlled as speed tends to increase or decrease beyond a set point.

(*ii*) *A pressure control system* where the pressure inside the furnace is automatically controlled by affecting changes in the position of the damper.

(*iii*) *The level control system* where the inflow of water to the tank is dependent on the water level in the tank. The automatic controller maintains the liquid level by comparing the actual level with a desired level and correcting any error by adjusting the opening of the control valve.

Advantages and limitations :

Advantages :

1. Increased output.
2. Economy in operating cost (since continuous employment of human operator is not required).
3. Suitability and desirability in the complex and fast acting systems which are beyond the physical abilities of a man.
4. Improvement in the quality of the products.
5. Reduced effect of non-linearities and distortions.
6. Response is satisfactory over a wide range of input frequencies.

Limitation : Automatic control system has a tendency to *overcorrect errors* which may *result in oscillations of constant or changing amplitude.*

7.2.8. Servo-Mechanism

A servo-mechanism is a feedback control system used to *control position or its derivative.* It has the following essential *features :*

1. It is a closed-loop system.
2. It is used to control position, velocity or acceleration.
3. Its characteristics include :
   — automatic control;
   — remote operation;
   — high accuracy.
4. It has high power amplifying stages to operate the system from very small error to signal.

7.2.9. Regulator

A regulator is a system employed to *control quality which is to be kept constant for a fairly long interval.*

Example : Voltage regulator or speed regulator.

7.2.10. Representation Through Model

In order to solve a system problem, the specifications or description of the system configuration and its components must be put into a form amenable to analysis, design and evolution. Following three basic models may be used for various system :

1. Differential equations and other mathematical solutions.
2. Block diagrams.
3. Sign flow graphs (SFG).

7.2.11. Analogous Systems

For mathematical relations analogies are drawn between features of a system and features of some known elements or properties; some analogous systems are given below:

Table 7.2. Force-Current Analogy

S.No.	Mechanical System		Electrical System
	Translational	*Rotational*	
1.	Force, F	Torque, T	Current, I
2.	Mass, M	Moment of inertia, M.I.	Capacitance, C
3.	Displacement, x	Angular displacement, θ	Magnetic flux linkage, f
4.	Velocity, V	Angular velocity, ω	Voltage, E
5.	Viscous friction co-efficient, f	Viscous friction co-efficient, f	Reciprocal of resistance, $\frac{1}{R}$
6.	Spring stiffness, K	Torsional spring stiffness, K	Reciprocal of inductance, $\frac{1}{L}$

Table 7.3. Force-Voltage Analogy

S.No.	Mechanical System		Electrical System
	Translational	*Rotational*	
1.	Force, F	Torque, T	Voltage, E
2.	Mass, M	Moment of inertia, M.I.	Inductance, L
3.	Displacement, x	Angular displacement, θ	Charge, q
4.	Velocity, U	Angular velocity, ω	Current, I
5.	Spring stiffness, K	Torsional spring stiffness, K	Reciprocal of capacitance, $\frac{1}{C}$
6.	Viscous friction co-efficient, f	Viscous friction co-efficient, F	Resistance R

Table 7.4. Electrical, Thermal, Liquid level and Pneumatic Systems

S.No.	Electrical system	Thermal system	Liquid-level system	Pneumatic system
1.	Charge, coloumbs (C)	Heat flow, joules (J)	Liquid flow cum. (m^3)	Air flow, cum. (m^3)
2.	Current, amperes (A)	Heat flow rate, joules/sec. (J/s)	Liquid flow rate, cum/sec (m^3/s)	Air flow rate, cum/sec. (m^3/s)
3.	Voltage, volts (V)	Temperature, °C	Head, meters (m)	Pressure, N/m^2
4.	Resistance, ohms (Ω)	Resistance, $°CsJ^{-1}$	Resistance, $m^{-2}s$	Resistance N-ms^{-1}
5.	Capacitance, farad (F)	Capacitance, J/°C	Capacitance, m^3/m	Capacitance, m^3/Nm^2

7.2.12. Block Diagram

A ***block diagram*** is the *diagrammatic representation of a physical system. The follwing steps* are worth noting :

— Firstly a functional block diagram is drawn to represent the functions of the system.

— Then it is *converted into a mathematical block diagram* by expressing the transfer function for each block.

— Finally is *reduced to an equivalent simpler block diagram* for system analysis.

Fig. 7.6 shows a block diagram of the feedback control system.

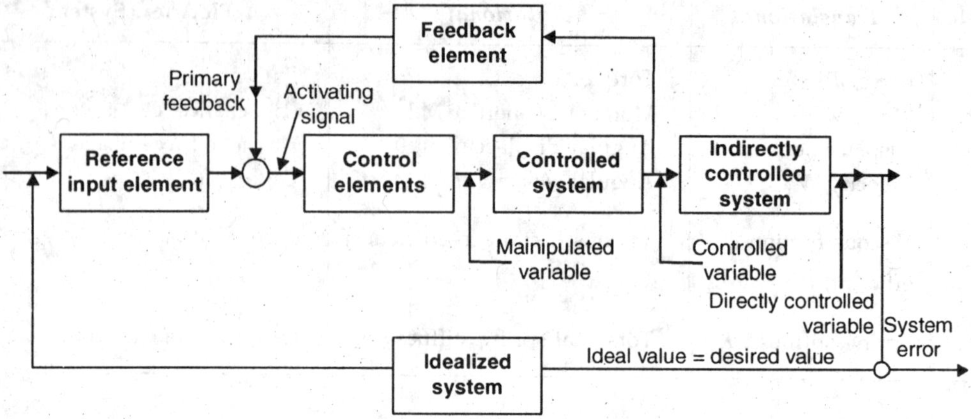

Fig. 7.6. Block diagram of the feedback control system.

7.2.13. Mathematical Block Diagram

Fig. 7.7 shows the block diagram of a closed-loop system. The various quantities shown are as follows :

$R(s)$ = Laplace transform of the reference input

$C(s)$ = Laplace transform of the output

$H(s)$ = Transfer function of the feedback path

$B(s)$ = Laplace transform of the feedback signal

= $C(s)\,H(s)$

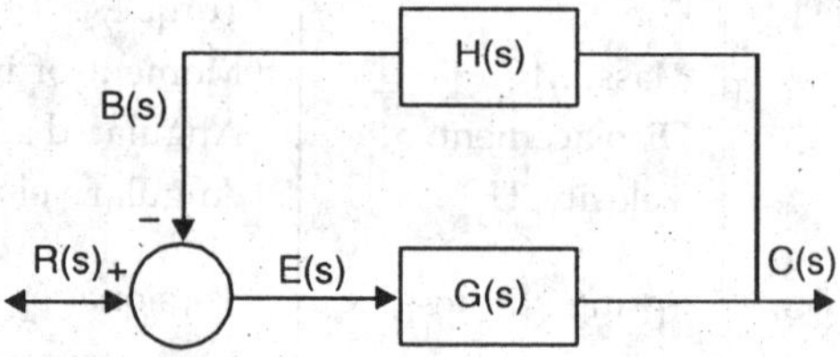

Fig. 7.7. Closed-loop system.

$E(s)$ = Laplace transform of the actuating signal

= $R(s) - B(s) = R(s) - C(s)\,H(s)$

$G(s)$ = Laplace transform of the formed path

$$\therefore \quad C(s) = G(s)\,E(s) = G(s)\,R(s) - G(s)\,H(s)\,C(s)$$

or, $$C(s) + G(s)\,H(s)\,C(s) = G(s)\,R(s)$$

or, $$C(s)\,[1 + G(s)\,H(s)] = G(s)\,R(s)$$

or, $$\frac{C(s)}{R(s)} = \frac{G(s)}{1+G(s)\,H(s)}$$

Hence the transfer function of the system,

$$M_s' = \frac{C(s)}{R(s)} = \frac{G(s)}{1+G(s)\,H(s)}$$

In the above equation the following points are worth noting :

(*i*) Product of transfer function of forward path and feedback path $G(s) \times H(s)$, sometimes expressed as $GH(s)$.

(*ii*) The system performance depends on its *characteristic equation* (it is a key equation in the control system analysis) which is given as under :

$$1 + G(s)H(s) = 0.$$

Block reductions :

By using the rules (derived by simple algebraic manipulation of the equations representing the blocks) of block diagram algebra, a complex block diagram configuration can be simplified by certain rearrangements of block diagrams; such rules are given in the Table 7.5.

Table 7.5. Rules for Block Diagram Algebra

S. No.	*Rule*	*Original diagram*	*Equivalent diagram*
1.	Combination of block in cascade	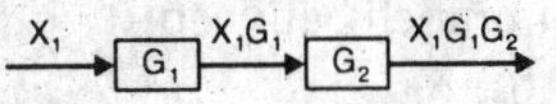	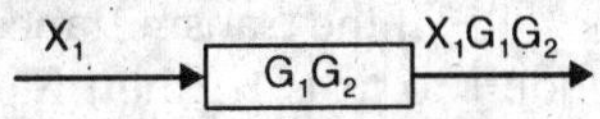
2.	Moving a summing point after a block	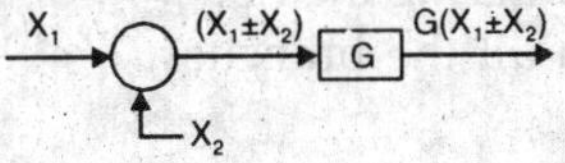	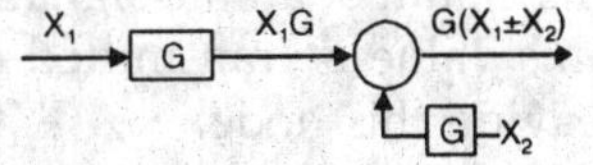
3.	Moving a summing point *ahead* of a block	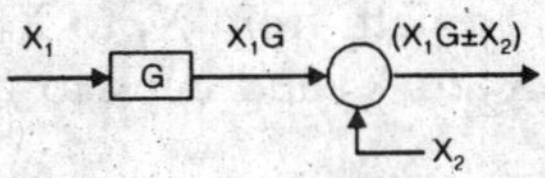	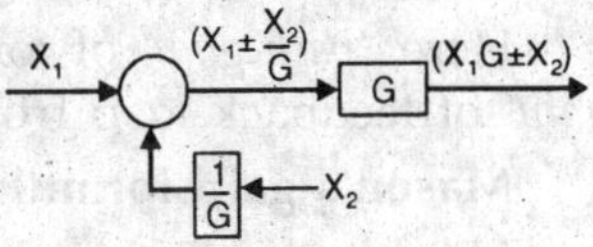
4.	Moving a take off point *after* of a block	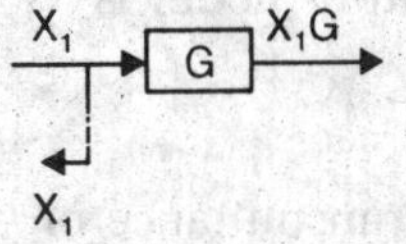	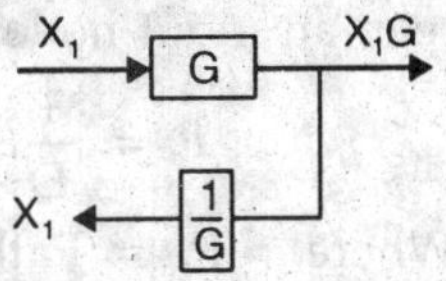
5.	Moving a take off point *ahead* of a block	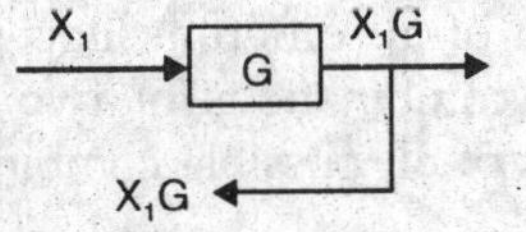	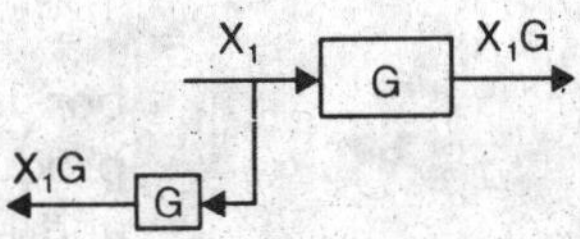

Fig. 7.8. Rules for block diagram algebra.

7.2.14. Signal Flow Graph

The block diagram reduction process, for complicated systems, becomes tedious and time consuming. For this purpose signal flow graphs (developed by S.J. Mason) are used.

A signal flow graph is a pictorial representation of the simultaneous equations describing a system.

Some important definitions relating to *signal flow graph* are given below :

1. Input and output nodes. A node having only outgoing branches is called *input node* while a node having only incoming branches is called *output node or sink.*

2. Path. Any continuous unidirectional succession of branches traversed in the indicated direction of branch is called *path.*

3. Forward path. The path from input node to the output node along which no anode is traversed twice is called *forward path.*

4. Loop or feedback path. It is defined as the path originating from a node and after traversing terminates at the same node without traversing any other node twice.

5. Path gain and loop gain. The product of transmittances along a path is called *path gain* while along the loop is called *loop gain.*

Rules for drawing a 'Signal Flow Graph' :

A signal flow graph can be drawn following the rules given as under :

1. The system variables are represented by nodes and arranged from left to right to represent cause and effect through the system.

2. The direction of the branch from node to X_t to X_m represents that the variable X_m depends upon the variable X_t and not the reverse.

3. Signals travel along the direction of arrows only.

4. If the transmittance of branch between bode l to m is t_{lm} then $X_m = t_{lm} X_l$.

5. The value of the variable at the node is the sum of all the signals entering the node. This value is transmitted on all the branches leaving this node.

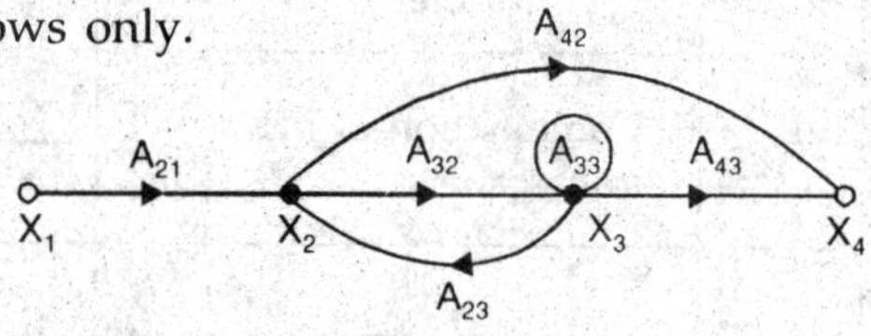

Fig. 7.9

A typical example of signal flow graph is shown in Fig. 7.9

Here, *'path gain'* of forward path from X_1 to X_2 to X_3 to X_4 is A_{21}. $A_{32}.A_{43}$. The *'loop gain'* of feedback loop from X_2 to X_3 and back to X_2 is $A_{32}.A_{23}$.

Mason's gain formula :

Mason's gain formula, which is applicable to the overall gain (the transmittance between an input node and output node) is

$$P = \frac{1}{D} \text{å}_{k} P_k D_k$$

Where, P_k = Path gain transmittance of k_{th} forward path,

D = Determinant of the graph

= 1 – (sum of all different loop gains) + (sum of gain products of all possible combinations of two non-touching loops) – (sum of gain products of all possible combination of three non-touching loops), and

D_k = Co-factor of k_{th} path determinant of this graph with loops touching the k_{th} forward path removed.

7.2.15. Time Response of Control System

Some important definitions :

1. **Delay time.** It is the time required for the response to reach line half the final value.
2. **Rise time.** It is the time required for the response to rise from 10 to 90% of its final value.
3. **Peak time.** It is the time required for the response to reach the first peak of the overshoot.
4. **Maximum overshoot.** It is the maximum value of the response curve measured from unity.
5. **Overshoot.** It is the maximum difference between the transient and steady state solutions for unit-step function input. It is a measure of relative stability and is represented as percentage of final value of the output.

6. **Setting time.** It is the time required for the response to a unit-step function input to reach and remain within a specified percentage of its final value.

Test functions :

Following are the commonly used typical test signals :

(*i*) Step function.

(*ii*) Ramp function.

(*iii*) Impulse function.

(*iv*) Sinusoidal function.

7.2.16. Stability

The response of a system to input or disturbances determines its *stability*. A stable system is one that will remain at rest unless excited by an external source and will return to rest if all the excitations are removed. A system is stable if its impulse response approaches zero as time approaches infinity. The *necessary condition for the system to be stable is that the real part of the roots of the characteristic equation have negative real parts. This ensures that the impulse response will decay exponentially with time.*

When the system has some roots with real parts equal to zero, but none with positive real parts, the system is said to be *"marginally stable"* which in unstable.

Routh stability criterion :

Routh stability criterion is a method for determining system stability that can be applied to an *n*th order characteristic equation of the form

$$a_n S^n + a_{n-1} S^{n-1} + \ldots\ldots + a_1 S + a_0 = 0$$

The Routh table is prepared as defined below :

s_n	a_n	a_{n-2}	a_{n-4}
S_{n-1}	a_{n-1}	a_{n-3}	a_{n-5}
:	b_1	b_2	b_3
:	c_1	c_2	c_3

After the array is completed the following criterion is applied :

"The number of changes in sign for the terms in the first column equals the number of roots of the characteristic equation with positive real parts.

Hence by the Routh criterion, for a system to be stable the array resulting from its characteristic equation must have a first column with terms of the same sign.

Deficiencies of Routh's criterion :

1. It does not provide the facility for selecting in a simple and direct fashion the parameters of a system component to stabilize the system when it is found to be absolutely unstable.
2. It assumes that characteristic equation is available in polynomial form; which is not necessarily always true.
3. The Routh array may show no change in sign in the first column but the ensuing dynamic response may be characterised by overshoots so excessive as to render the system useless for control purposes. Thus the system may be relatively unstable inspite of the fact that it is absolutely stable.
4. Although this criterion gives information about absolute stability, it conveys little or no information about how close the system may be to become unstable.

7.2.17. Frequency Response

The analysis of the *system* whose input is frequency and amplitude is dealt under frequency response. *The system is actuated by a sinusoidal input and allowed to settle. The output amplitude and its phase with respect to input are measured. The phase difference and amplitude change indicate the nature of the system.*

Graphical methods :

The following four graphical methods are available to control systems analyses which are simpler and more direct than the time domain method for practical linear models of feedback control systems :

1. Bode's-Plot-Representation
2. Nyquist Diagrams
3. Nichols Charts
4. The Root Locus method.

The first three are *frequency-domain techniques.*

Bode's Plot. This method has the following *advantages :*

(*i*) It is the simplest method.

(*ii*) The multiplication of magnitudes can be converted into addition.

(*iii*) Transfer function can be determined easily.

Nyquist method :

- This method *handles* systems with time delays without the necessity of approximations and hence yields exact results about both absolute and relative stability of the system.
- It is also useful for obtaining information about transfer functions of components or systems from experimental frequency response data.

Root locus method :

This method permits accurate computations of the time-domain response in addition to yielding readily available frequency response information.

7.2.18. Error Detector

An **error detector** is *a sensor to sense the error between the reference input and the desired output.*

- It gives an input to the amplifier and actuator in proportion to the error.
- Its output should be directly electrical or a transducer should be cascaded to give electrical output.
- An error-cum-transducer is obtained by connecting two potentiometers in parallel to a voltage source. Their movable points are brought out to give output voltage in proportion to the difference between the positions of the movable contacts.

7.2.19. LVDT

LVDT (Linear-Variable-Differential Transformer) is a transformer having one primary, and two secondary windings and movable core. The secondary windings are connected in series opposition, so as to have output which is difference of the two induced secondary voltages. The movable core is connected to the shaft and a normal position output voltage is zero. When the core moves the *output voltage is a function of the shaft position.*

7.2.20. Servo-Amplifier

A servo-amplifier is the amplifier used to *amplify the small output of the error detector to directly operate the actuator.*

- It can be electronic, magnetic or rotating.
- It should have high input impedance, low output impedance, frequency response curve should be flat in the range of operating frequencies, phase sensitive, small residual voltage and minimum noise.

7.2.21. Sampled Data Systems

These systems (also called *discrete time systems*) are *dynamic systems, in which one or more variables change at the discrete instant of time.* The time interval between two discrete instants is very small so that the data during this interval can be approximated by interpolation.

These systems find application in :

(*i*) Numerically controlled machine tool operations.

(*ii*) Pulse control or digital control of electric drives.

(*iii*) High speed tin plate rolling mill using quantized data for control.

(*iv*) Large complex systems employing telemetry links based on pulse modulation transmission of data.

7.2.22. Industrial Controllers

Industrial controllers may be classified *according to their control action* as follows :

1. Two-position or on-off controllers.
2. Proportional controllers.
3. Integral controllers.
4. Proportional-plus-integral controllers.
5. Proportional-plus-derivative controllers.
6. Proportional-plus-integral-plus-derivative controllers.

— Most industrial controllers use pressurised fuel such as oil or air or electricity as power sources. Consequently, controllers may also be classified according to the kind of power employed in the operation, such as *"pneumatic controllers"*, *"hydraulic controllers"* or *"electronic controllers"*. However, the kind of controllers to be used must be decided based on the nature of the plant and operating conditions, including such considerations as safety, cost, availability, reliability, accuracy, weight, and size.

7.2.23. Pneumatic Control Systems

- Pneumatic controllers use air control medium to provide an output signal which is a function of an input error signal.
- Fig 7.10 shows the schematics of a pneumatic control system, the major components are :

  Error detector; Flopper nozzle (controller mechanism); Amplifier or Pilot relay.

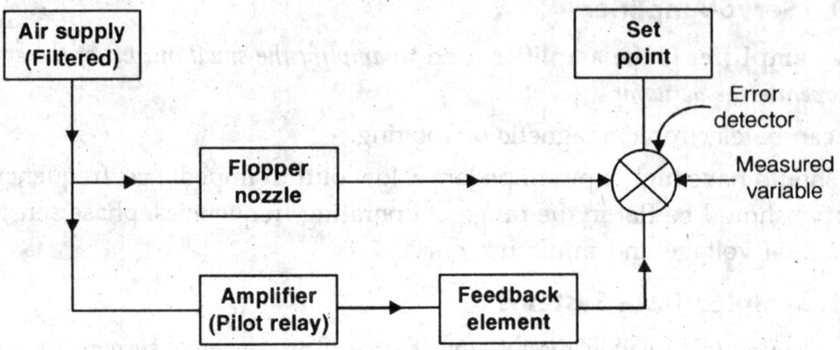

Fig. 7.10. Schematics of a pneumatic control system.

— The controller mechanisms are of two types : *Free balance* and *motion balance.*

Advantages :

1. Simple construction and easy maintenance.
2. Relatively high power amplification for operating the final control elements.
3. Relatively inexpensive power system.
4. No return pipes are required when air is used.
5. Insensitive to temperature changes.
6. Fire-and explosion-proof.
7. The normal operating pressure of pneumatic system is very much lower than that of hydraulic systems.

Limitations/Disadvantages :

1. Output powers are considerably less (than those of hydraulic systems).
2. Accuracy of pneumatic actuators is poor at low velocities.
3. Slow response of final control elements, and transmission lag.
4. Operation difficult under freezing conditions.
5. Lubrication of the mating parts is difficult.

Uses :

- The pneumatic systems are employed for majority of the plant and process control actions in petroleum, petrochemical, chemical, paper, textile and food industries.
- They are also sometimes used in the aircraft systems and guided missiles.

7.2.24. Hydraulic Control System

- Compressed air has seldom been used (except for low-pressure controllers) for the continuous control of the motion of devices having significant mass under external load forces. For such a case, hydraulic controllers are generally preferred.
- The widespread use of hydraulic circuitry in "*machine tool applications*", "*aircraft control systems*" and "*similar operations*" occurs because of such factors as *accuracy, positiveness, flexibility, high power-to-weight ratio, fast starting, stopping, and reversal with smoothness and precision and simplicity of operations.*
  — The operating pressure in hydraulic systems lies between 1 and 35 MPa; in some special applications the operating pressure may go upto 70 MPa.

— For the same power requirement, the weight and size of the hydraulic unit can be made smaller by increasing the supply pressure. *Very large force can be obtained with hydraulic systems.*

— With *hydraulic systems, rapid-acting, accurate positioning of heavy loads is possible.*

— A combination of electronic and hydraulic systems is widely used because it combines the advantages of both electronic control and hydraulic power.

- **Hydraulic controllers** *employ a liquid control medium to provide an output signal which is a function of an input error signal.*

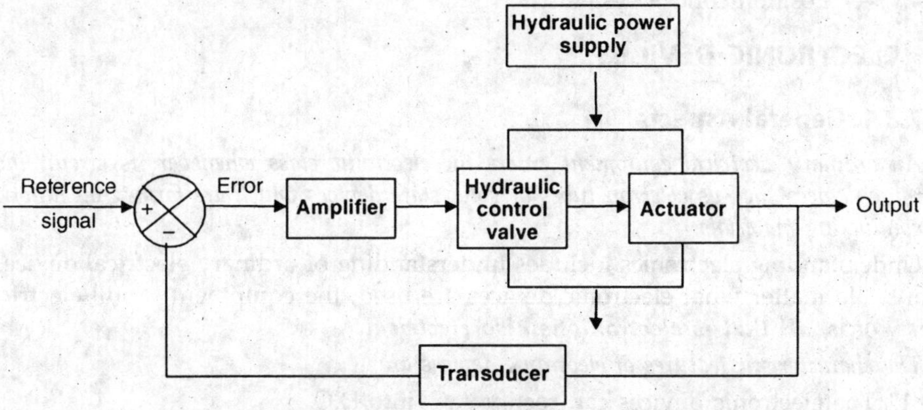

Fig. 7.11. Schematics of an hydraulic control system.

Fig. 7.11 shows the schematics of a hydraulic control system; the major components are :

Error detector; an amplifier; a hydraulic control valve; an actuator.

— Hydraulic power supply system is of the following two types : *"Constant flow arrangement"* and *"Constant pressure arrangement"*

Advantages :

1. Because of low leakages in hydraulic actuators, *speed drop* when loads are applied is *small*.
2. Hydraulic actuators have a *higher speed of response with fast starts, stops, and speed reversals.*
3. Availability of both linear and rotary actuators gives *flexibility in design*.
4. Simplicity of actuator system.
5. Operation of hydraulic actuators under continuous, intermittent, reversing and stalled conditions without damage is possible.
6. Large forces or torques can be developed by the comparatively small sized hydraulic actuators.
7. Long life due to self lubricating properties of the hydraulic liquids.

Disadvantages/Limitations :

1. In order to prevent the leakage of hydraulic fluid, the proper seals and connections are needed.
2. Unless fire-resistant fluids are used, fire and explosion hazards exist.
3. For keeping the fluid clean and pure careful maintenance of the system is required.

4. As a result of the non-linear and other complex characteristics involved, the design of sophisticated hydraulic systems is quite complicated.
5. Contaminated oil may cause failure in the proper functioning of a hydraulic system.

Uses : The hydraulic systems, because of their high power-to-weight ratio find a wide range of use in :

— Machine tools;
— Speed governing systems;
— Position control systems.

7.3 ELECTRONIC DEVICES

7.3.1. General Aspects

An ordinary electrical equipment enters the electronic class whenever its circuit includes electronic devices such as electron tubes or solid state devices which are formed by junctions of semiconducting materials.

Understanding electronics includes understanding of ordinary electrical devices and circuits. No matter what electronic devices are used, the equipment is still electrical. In other words, ***all that is electronic is also electrical.***

The characteristic features of electronic devices are

1. The electronic devices can rectify A.C. into D.C.
2. They can amplify input signals.
3. They can respond at speeds far beyond the speeds that one comes across in electrical and mechanical devices.
4. Some electronic devices are photosensitive. Some of these devices can produce radiations such as X-xays

7.3.2. Semiconductors

Semiconductors *are solid materials, either non-metallic elements or compounds, which allow electrons to pass through them so that they conduct electricity in much the same way as a metal.*

7.3.2.1 Characteristics of semiconductors

Semiconductors possess the following *characteristics* :

1. The resistivity is usually high.
2. The temperature coefficient of resistance is *always negative.*
3. The contact between semiconductor and a metal forms a layer which has a higher resistance in one direction than the other.
4. When some suitable metallic impurity (e.g., Arsenic, Gallium, etc.) is added to a semiconductor, its *conducting properties change appreciably.*
5. They exhibit a rise in conductivity in the increasing temperature, with the decreasing temperatures their conductivity falls off, and at low temperatures semiconductors become dielectrics.
6. They are usually metallic in appearance but (unlike metals) are generally hard and brittle.

Both the resistivity and the contact effect are as a rule very sensitive to small changes in physical conditions, and the *great importance* of semiconductors for a wide range of uses apart from rectification depend on the *sensitiveness.*

Examples of semiconducting materials :

Of all the elements in the periodic table, *eleven* are *semiconductors* which are listed below :

S. No.	*Element*	*Symbol*	*Group in the periodic table*	*Atomic No.*
1.	Boron	B	III	15
2.	Carbon	C	IV	6
3.	Silicon	Si	IV	14
4.	Germanium	Ge	IV	32
5.	Phosphorus	P	V	15
6.	Arsenic	As	V	33
7.	Antimony	Sb	V	51
8.	Sulphur	S	VI	
9.	Sellinum	Se	VI	
10.	Tellurium	Te	VI	
11.	Iodine	I	VIII	

Examples of semiconducting *compounds* are given below :

(*i*) Alloys : Mg_3Sb_2, ZnSb, Mg_2Sn, CdSb, AlSb, InSb, GeSb.

(*ii*) Oxide : ZnO, Fe_3O_4, Fe_2O_3, Cu_2O, CuO, BaO, CaO, NiO, Al_2O_3, TiO_2, UO_2, Cr_2O_3, WO_2, MoO_3.

(*iii*) Sulphides : Cu_2S, Ag_2S, PbS, ZnS, CdS, HgS, MoS_2.

(*iv*) Halides : AgI, CuI.

(*v*) Selenides and Tellurides.

PbS is *used in photo-conductive devices, BaO in oxide coated cathodes, caesium antimonide in photomultipliers, etc.*

7.3.3. Intrinsic Semiconductor

A *pure semiconductor is called "intrinsic semiconductor"*. Here no free electrons are available since all the covalent bonds are complete. A *pure semiconductor, therefore behaves as an insulator.* It exhibits a peculiar behaviour even at room temperature or with rise in temperature. The *resistance of a semiconductor decreases with increase in temperature.*

When an electric field is applied to an intrinsic semiconductor at a temperature greater than 0°K, conduction electrons move to the anode and, the holes (when an electron is liberated into the conduction band a positively charged hole is created in valence band) move to cathode. *Hence semiconductor current consists of movement of electrons in opposite direction.*

Fig. 7.12 shows the energy diagram for intrinsic (pure) semiconductor at absolute zero.

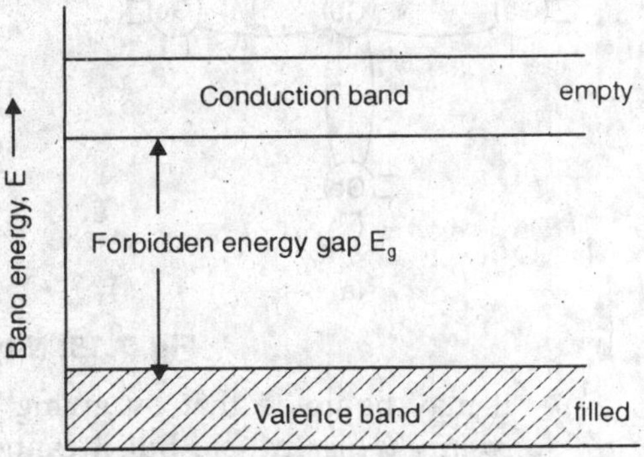

Fig. 7.12. Energy diagram for intrinsic (pure) semiconductor at absolute zero.

7.3.4. Extrinsic Semiconductor

In a pure semiconductor, which behaves like an insulator *under ordinary conditions,* if small amount of certain *metallic impurity* is *added* it attains *current conducting properties.* The impure semiconductor is then called *"impurity semiconductor"* or *"extrinsic semiconductor". The process of adding impurity (extremely in small amounts, about 1 part in 10^8) to a semiconductor to make it extrinsic (impurity) semiconductor is called* **Doping.**

Generally following doping agents are used :

(*i*) *Pentavalent atom* having five valence electrons (arsenic, antimony, phosphorus) ... called *donor atoms.*

(*ii*) *Trivalent atoms* having three valence electrons (gallium aluminium, boron) ... called *acceptor atoms.*

With the addition of suitable impurities to semiconductor, two type of semiconductors are :

(*i*) N-type semiconductor.

(*ii*) P-type semiconductor.

N-type semiconductor :

The presence of *even a minute quantity of impurity,* can produce N-type semiconductor. If the impurity atoms has *one valence electron more* than the semiconductor atom which it has substituted, this *extra electron* will be loosely bound to the atom. For example, an atom of *Germanium* possesses *four valence electrons* ; when it is replaced in the crystal lattice of the substance by an impurity atom of antimony (Sb) which has *five valence electrons,* the fifth valence electron (free electron) produces extrinsic N-type conductivity *even at room temperature.* Such an impurity into a semiconductor is called *donor impurity* (or donor). The conducting properties of germanium will depend upon the *amount of antimony* (*i.e.,* impurity) *added.* This means that controlled conductivity can be obtained by proper addition of impurity. Fig. 7.13 (*a*) shows the loosely bound excess electron controlled by the donor atom.

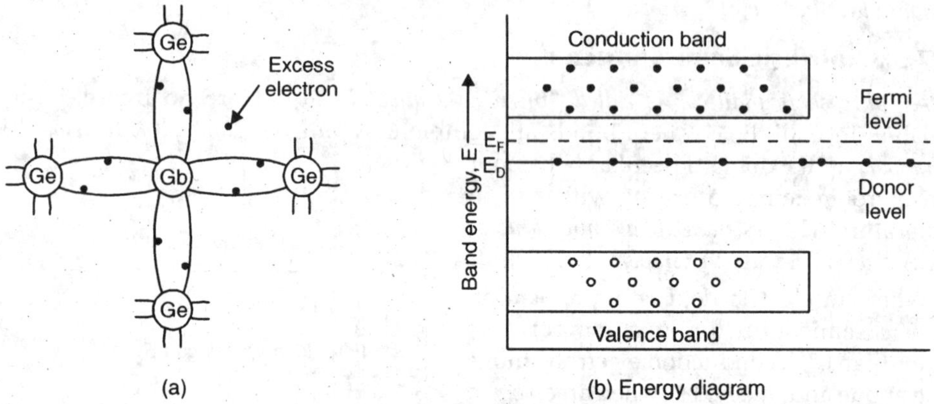

Fig. 7.13. N-type semiconductor.

- It may be noted that by giving away its one electron, the *donor atom becomes positively charged ion.* But it cannot take part in conduction because it is firmly fixed or tied into the crystal lattice. In addition to the electrons and holes *intrinsically available in germanium,* the addition of antimony greatly increases the number of conduction electrons. Hence, *concentration of electrons in the conduction band is increased and exceeds the concentration of the holes in the valence band.* Consequently, *Fermi level shifts upwards towards the bottom of the conduction band* as shown in Fig. 7.13 (*b*). [Since the number of electrons as compared to the number of holes increases with temperature, the *position of Fermi level also changes considerably with temperature*].

- It is worth noting that even though N-type semiconductor has excess of electrons, still it is *electrically neutral.* It is so because by addition of donor impurity, number of electrons available for conduction purposes becomes more than the number of holes available intrinsically. But the total charge of the semiconductor does not change because the donor impurity brings in as much negative charge (by way of electrons) as positive charge (by way of protons).

Note. In terms of energy levels, the fifth antimony electron has as energy level (called donor level) just below the conduction band. Usually, the donor level is 0.01 eV below conduction band for germanium and 0.054 eV for silicon.

P-type semiconductor :

- P-type extrinsic semiconductor can be produced if the impurity atom has *one valence electron less* than the semiconductor atom that it has replaced in the crystal lattice. This impurity atom cannot fill all the *interatomic* bonds, and the free bond can accept an electron from the neighbouring bond ; leaving behind a vacancy of *hole.* Such an impurity is called an *acceptor impurity* (*or acceptor*). Fig. 7.14 (*a*) shows structure of P-type semiconductor (Germanium and Boron).

- In this type of semiconductor, conduction is by means of holes in the valence band. Accordingly, *holes form the majority carriers whereas electrons constitute minority carriers.* The process of conduction is called *deficit conduction.*

- Since the concentration of holes in the valence band is more than the concentration of electrons in the conduction band, Fermi level shifts nearer to the valence band [Fig. 7.14 (*b*)]. The acceptor level lies immediately above the Fermi level. *Conduction is by means of hole movement at the top of valence band, the acceptor level readily accepting electrons from the valence band.*

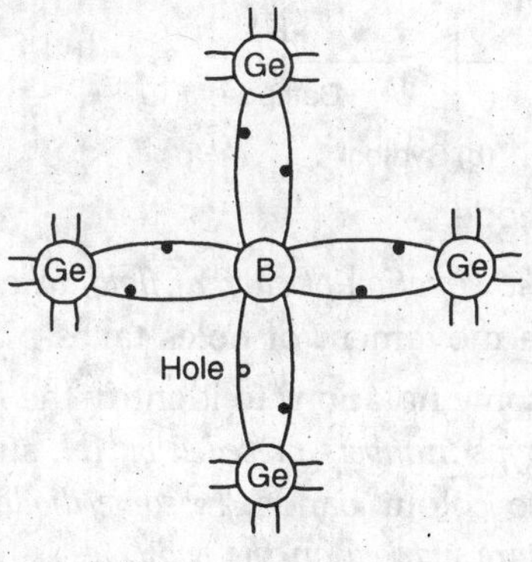

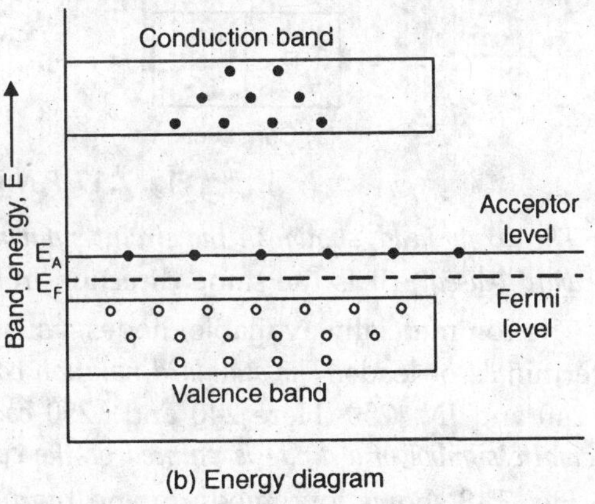

Fig. 7.14 P-type semiconductor.

It may be noted again that even though P-type semiconductor has excess of holes for conduction purposes, as a whole it is *electrically neutral* for the same reasons as discussed earlier.

7.3.5. P-N Junction Diode

In an **N-type material** (Fig. 7.15) *the electron is called the majority carrier and the hole as the minority carrier.*

In a **P-type material** (Fig. 7.16) *the hole is the majority carrier and the electron is the minority carrier.* The N- and P-type materials represent the *basic building blocks of semiconductor devices.*

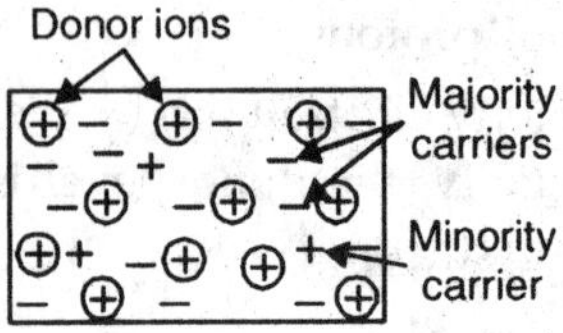

Fig. 7.15 N-type material.

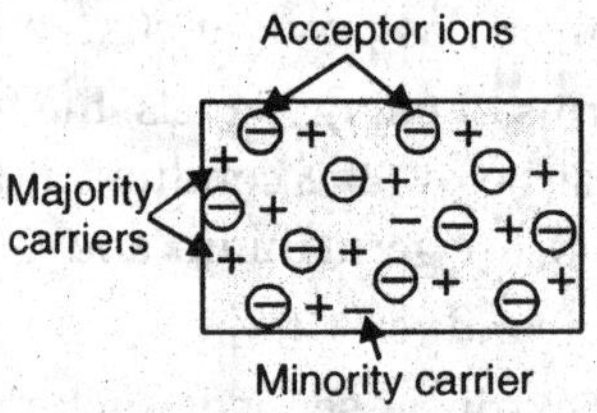

Fig. 7.16 P-type material.

The semiconductor diode is simply bringing these materials together (constructed from the same base-Ge or Si). At instant the two materials are *"joined"* the electrons and holes in the region of the junction will combine resulting in a lack of carriers in the region near the junction. *This region of unconvered positive and negative ions is called the* ***depletion region*** *due to the depletion of carriers in this region.*

Construction and types of P-N junction diodes :

The most extensively used elements in the manufacture of junction diodes are *germanium and silicon* (although some other materials are also assuming importance in recent years).

A *P-N* junction diode (known as a semiconductor or crystal diode) consists of a *P-N* junction, formed either in germanium or silicon crystal. The diode has two terminals namely *anode and cathode.* The anode refers to the *P*-type region and cathode refers the *n*-type region as shown in Fig. 7.17 (*a*)

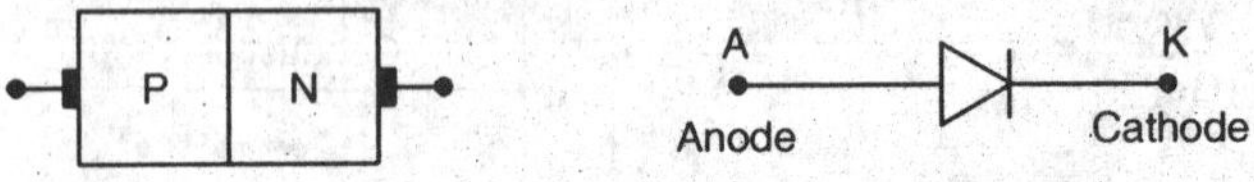

(a) Construction (b) Symbol

Fig. 7.17 *P-N* junction diode.

The arrow head, shown in the circuit symbol, points the direction of current flow, when it is "forward biased" (It is the same direction in which the movement of holes takes place).

The commercially available diodes, usually have some notations to identify the *P* and *N* terminals or leads. *The standard notation consists of type numbers preceded by IN,* such as *IN* 240 and *IN* 1250. Here 240 and 1250 *correspond* to colour bands. *In some diodes, the schematic symbol of a diode is painted or the colour dots are marked on the body.*

Fig. 7.18 shows low, medium and high current diodes.

— *Refer to Fig. 7.18* (*a*). The diode shown has a colour band located near one of the ends. The end, which is near the colour band, is identified as cathode, and other end, obviously, is the anode (*A*).

— *Refer to Fig 7.18 (b).* The diode has a schematic symbol actually painted at its cathode (*K*) and the other end as anode.

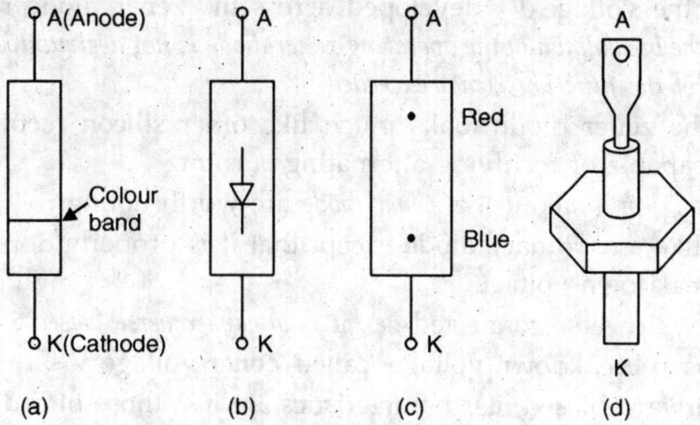

(*a*), (*b*) = Low current diodes ; (*c*) = Medium current diode ;
(*d*) = High current or power diode.

Fig. 7.18 Low, medium and high current diodes.

The diodes of Fig. 7.18 (*a*) and (*b*) *can pass a forward current of 100 mA and are known as* ***low current diodes.***

— *Refer to Fig. 7.18* (*c*). The diode has colour dots marked on its body. The end lying near the blue dot is a cathode, while the other end is anode. Sometimes this diode is *shown bigger in size* than that of diodes shown in Fig. 7.18 (*a*) and (*b*). The diodes of this size can *pass a forward current of 500 mA and are known as* ***medium current diodes.***

— *Refer to Fig. 7.18* (*d*). It shows a diode, which *can pass a forward current of several amperes.* Therefore it known as a ***power diode*** or a ***high current diode.***

- *The outstanding property of P-N junction / crystal diode to conduct current in one direction only permits it to be used as a* **rectifier.**

7.3.6 Zener Diode

A properly doped P-N junction crystal diode which has a sharp breakdown voltage is known as **Zener diode.**

The voltage-regulator diode is commonly called a ***'Zener' diode.*** It is a *voltage limiting diode* that has some applications in common with the older voltage-regulator gas tubes but serves a much wider field of application, because the devices cover a wide spectrum of voltages and power levels.

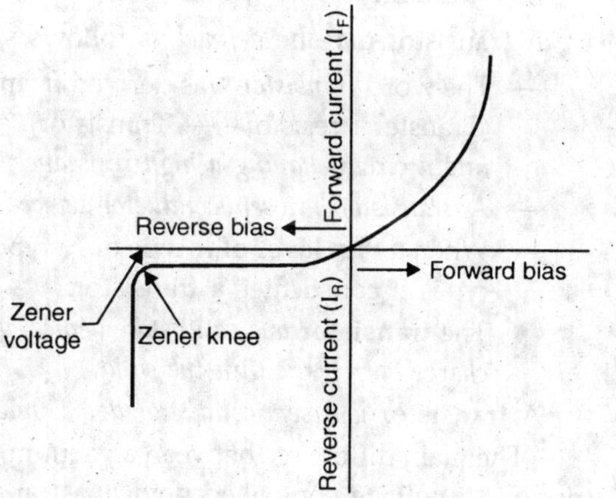

Fig. 7.19 Zener diode characteristics.

Performance/Operation :

The electrical performance of a zener diode is based on the *avalanche characteristics* of the *P-N* junction. When a source of voltage is applied to a diode in the *reverse direction* (negative to anode), a reverse current I_R is observed (see Fig 7.19). As the reverse potential is increased beyond the *"Zener knee"* avalanche breakdown becomes well developed at zener voltage V_Z. At voltage V_Z, the high counter resistance drops to a low value and the junction current increases rapidly. The current must of necessity be limited by an external resistance, since the voltage V_Z developed across the zener diode remains essentially constant. *Avalanche breakdown of the operating zener diode is not destructive as long as the rated power dissipation of the junction is not exceeded.*

Externally, the zener diode looks much like other silicon rectifying devices, and electrically it is capable of rectifying alternating current.

The following points about the *Zener diode* are worth noting :

(*i*) It looks like an ordinary diode except that it is properly doped so as to have a sharp breakdown voltage.

(*ii*) It is always reverse connected *i.e.*, it is *always reverse biased.*

(*iii*) It has sharp breakdown voltage, called Zener voltage V_Z.

(*iv*) When forward biased, its characteristics are just those of ordinary diode.

(*v*) It is not immediately burnt just because it has entered the breakdown region (The current is limited only by both external resistance and power dissipation of Zener diode).

- The location of Zener region can be controlled by varying the doping levels. An *increase in doping, producing an increase in the number of added impurities, will decrease the Zener potential.*
- Zener diodes are available having Zener potentials of 1.8 to 200 V with power ratings from $\frac{1}{4}$ to 50 W. *Because of its higher temperature and current capability, silicon is usually preferred in the manufacture of Zener diodes.*

7.3.7. Biopolor Junction Transistor (BJT)

Introduction :

A transistor may be *defined* as follows :

— The word *transistor* was derived from the two word combination, *transfer-resistance* (Transfer + resistor ⟶ Transistor). A **transistor** *is a device to transfer a low resistance into a circuit having a high resistance.*

— A *'transistor' is a semiconductor device in which current flows in semiconductor materials.*

— When a thin layer of *P*-type or *N*-type semiconductor is between a pair of opposite types it constitutes a transistor.

- The **transistor** *is a solid state device, whose operation depends upon the flow of electric charge carriers within the solid.*

A **transistor** *is a semiconductor device having both rectifying and amplifying properties.*

"The main difference between a vacuum triode and a transistor is that while a vacuum triode is a voltage controlled device, a transistor is a current controlled device".

The transistor was invented by a team of three scientists at Bell Laboratories, USA in 1947. Although the first transistor was not a bipolar junction device, yet it was the beginning of a technological revolution that is still continuing in the twenty first century. All of the complex electronic devices and systems developed or in use today, are an outgrowth of early developments in semiconductor transistors.

The two basic types of transistors are :

1. Bipolar junction transistor (BJT)
2. Field-effect transistor (FET)

The *bipolar junction transistor is used in the following two broad area of electronics :*

(*i*) *As a **linear amplifier** to boost an electric signal.*

(*ii*) *As an **electronic switch.***

P-N-P and N-P-N transistors. To understand the basic mechanism of transistor operation the following facts need to be kept in mind.

1. Since *emitter is to provide charge carriers, it is always "forward biased".*
2. *First letter of transistor type indicates the polarity of the emitter voltage with respect to base.*
3. Collector's job is *to collect or attract those carriers through the base, hence it is always "reverse-biased".*
4. *Second letter* of transistor type *indicates the polarity of collector voltage with respect to the base.*

The above points apply both to *P-N-P* and *N-P-N* transistors.

Working of P-N-P tansistor. Fig. 7.20 shows a ***P-N-P transistor*** connected in the common-base (or grounded-base) configuration (it is so called because both the emitter and collector are returned to the base terminals). The *emitter junction is forward-biased whereas the collector junction is reverse-biased.* The holes in the emitter are repelled by the positive battery terminal towards the *P-N* or emitter junction. The potential barrier at the junction is reduced due to the forward-biased, hence holes cross the junction and enter the *N*-type base. Because the base is thin and lightly-doped, majority of the holes (about 95%) are able to drift across the base without meeting electrons to combine with. The balance of 5% of holes are lost in the base region due to recombination with electrons. The holes which after crossing the *N-P* collector junction enter the collector region are swept up by the negative collector voltage V_C.

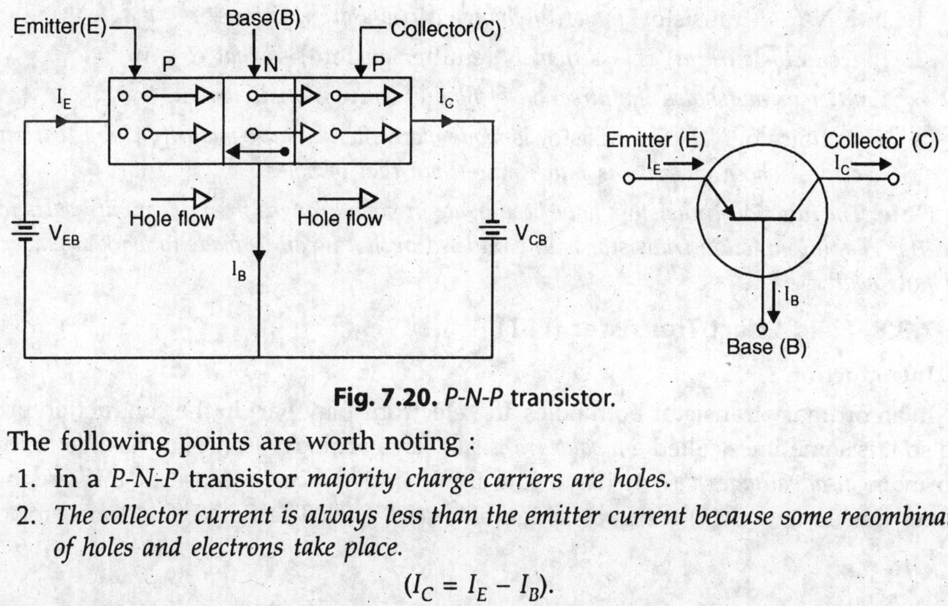

Fig. 7.20. *P-N-P* transistor.

The following points are worth noting :

1. In a *P-N-P* transistor *majority charge carriers are holes.*
2. *The collector current is always less than the emitter current because some recombination of holes and electrons take place.*

$$(I_C = I_E - I_B).$$

3. The *current amplification* (α) (or gain of P-N-P transistor) for steady conditions when connected in common base configuration is expressed as :

$$\alpha = \frac{I_C \text{ (collector current)}}{I_E \text{ (emitter current)}} < 1.$$

4. *Emitter arrow shows the direction of flow of conventional current.* Evidently, electron flow will be in the opposite direction.

Working of N-P-N transistor. Fig 7.21 shows a ***N-P-N junction transistor.*** The emitter is forward-biased and the collector reverse-biased. The electrons in the emitter region are repelled by the negative battery terminal towards the emitter or *N-P* junction. The *electrons cross over into the P-type base region because potential barrier is reduced due to forward bias.* Since the base is thin and lightly doped, most of the electrons (about 95%) cross over to the collector junction and enter the collector region where they are readily swept up by the positive collector voltage V_C. Only about 5% of the emitter electrons combine with the holes in the base and are lost as charge carriers.

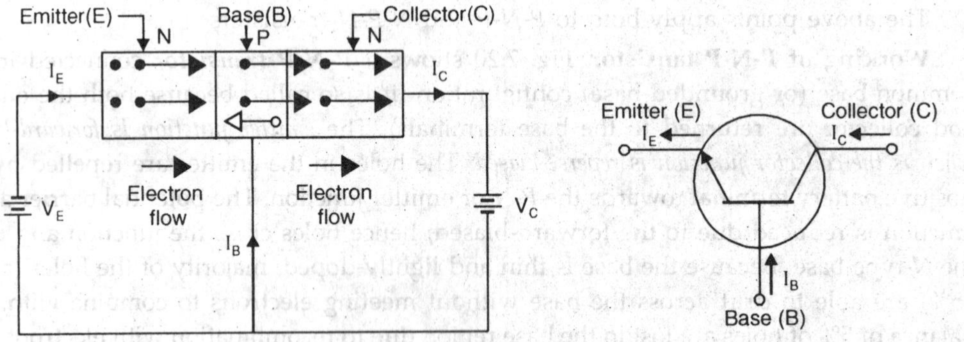

Fig. 7.21. *N-P-N* transistor.

The following points are worth noting :

1. In a *N-P-N* transistor, *majority change carriers are electrons.*
2. I_C *(collector current) is less than* I_E (emitter current) *so that* $\alpha < 1$.
3. *Emitter arrow shows the direction of flow of conventional current.*

- The choice of *N-P-N* transistor is *made more often because majority charge carriers are electrons whose mobility is much more than that of holes.*

Note. *The junction transistors have been made in power ranges from a few milliwatts to tens of watts. The tiny junction transistor is unparalleled in that it can be made to work at power level as I microwatt.*

7.3.8 Field-Effect Transistor (FET)

Introduction :

In an ordinary transistor both holes and electrons play part in the conduction process and so it is sometimes called a *bipolar transistor.* This ordinary transistor has the following two main *disadvantages* : (*i*) It has a low input impedance (because of forward biased emitter junction), and (*ii*) It has considerable noise level. The field-effect transistor (FET) has, by virtue of its construction and biasing, large input impedance (which may be more than 100 MΩ. The FET is generally *much less noisy* than the BJT).

Types of Field-effect Transistors :

A **field-effect transistor** *(FET) is a three terminal (namely* **drain, source** *and* **gate**) *semiconductor device in which current conduction is by only one type of majority carriers (electrons in case of an N-channel FET or holes in a P-channel FET).* It is also sometimes called the *unipolar transistor.*

In a broad sense, following are two main types of field-effect transistors :

1. *Junction field-effect transistor (JFET)*
   (*i*) *N*-channel
   (*ii*) *P*-channel.
2. *Metal oxide semiconductor field-effect transistor (MOSFET) or insulated gate field-effect transistor (IGFET).*
   (*i*) Depletion type :
      (*a*) *N*-channel
      (*b*) *P*-channel
   (*ii*) Enhancement type :
      (*a*) *N*-channel
      (*b*) *P*-channel

1. Junction field-effect transistors (JFET) :

The junction field-effect transistors (JFETs) can be divided depending upon their structure into the two following categories :

1. *N*-channel JFET
2. *P*-channel JFET

Construction :

- The basic construction of a *N*-channel JFET is as shown in Fig. 7.22 (*a*). It consists of an *N-type semiconductor bar with two P-type heavily doped regions diffused on opposite sides of its middle part.* The *P*-type regions form two *P-N* junctions. The *space between the junctions (i.e., N-type regions) is called a* **channel.** Both the *P-type* regions are connected internally and a single wire is taken out in the form of a terminal callled the **gate (G).** The electrical connections (called *ohmic contacts*) are made to both ends of the **N**-type semiconductor and are taken out in the form of two "terminals called **drain** (D) and **source (S).** *The "drain (D)" is a terminal through which the electrons leave the semiconductor and "source (S)" is a terminal through which the electrons enter the semiconductor.*

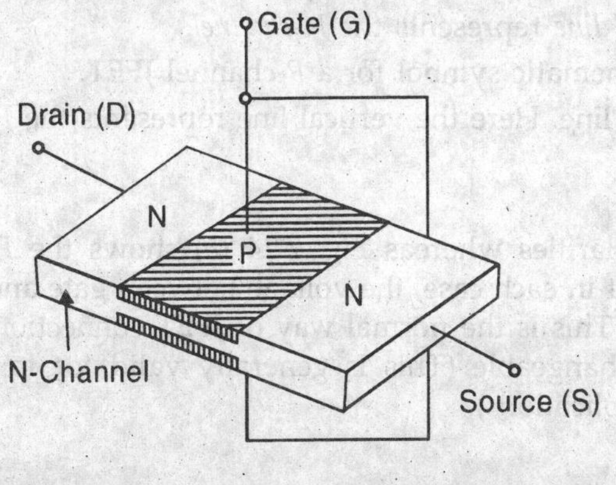

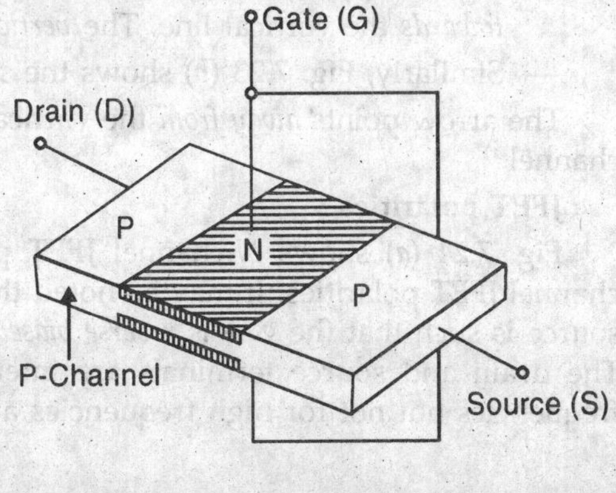

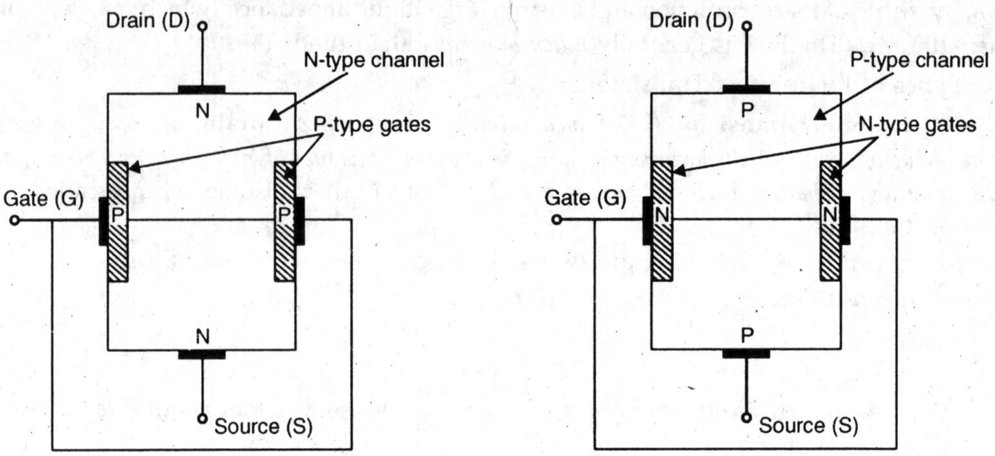

(a) N-channel JFET (b) P-channel JFET

Fig. 7.22. JFETs.

Whenever a voltage is applied across the drain and source terminals, a current flows through the *N*-channel. The current consists of only one type of carriers (*i.e.*, electrons), therefore, the FET is called a *unipolar device.* (This distinguishes FET from BJT where the current consists of the flow of both the electrons and holes).

- A *P*-channel JFET is shown in Fig. 7.22 (*b*). Its construction is similar to that of *N*-channel JFET, except that it consists of a *P-channel and N-type junctions.* The current carriers in JFET are the *holes,* which flow through the *P*-type channel.

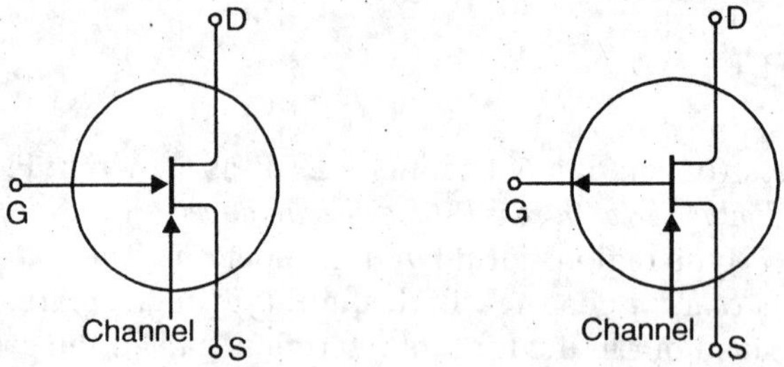

N-channel JFET (a) P-channel JFET (b)

Fig. 7.23. Symbols for JFETs.

— Fig. 7.23 (*a*) shows the schematic symbol for a *N*-channel *JFET.* The arrow points *towards* the vertical line. The *vertical line* represents the *N-channel.*

— Similarly, Fig. 7.23 (*b*) shows the schematic symbol for a *P*-channel JFET.

The arrow points *away from* the vartical line. Here the vertical line represents the *P*-channel.

JFET polarities :

Fig. 7.24 (*a*) shows *N*-channel JFET polarities whereas Fig. 7.24 (*b*) shows the *P*-channel JFET polarities. It may be noted that in each case, the voltage between gate and source is such that the *gate is reverse biased.* This is the normal way of JFET connection. The drain and source terminals are interchangeable (This is generally valid for low frequencies but not for high frequencies applications).

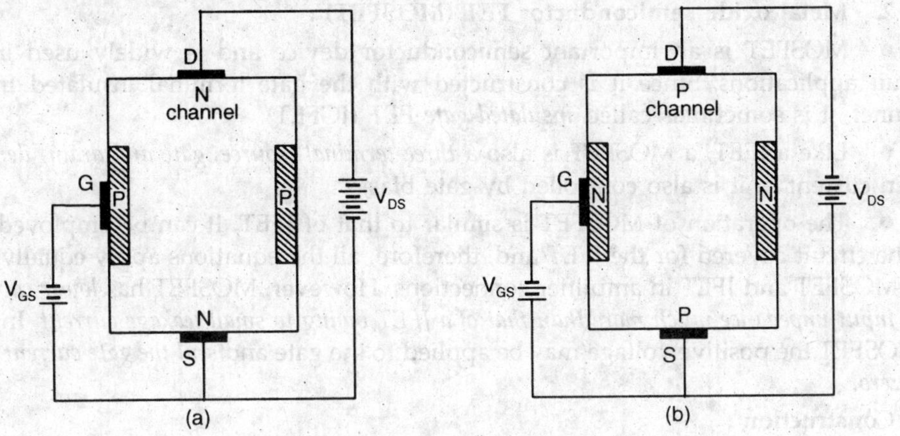

Fig. 7.24. JFET polarities.

Working :

Fig. 7.25 shows the circuit of *N*-channel JFET with normal polarities. The circuit action is as follows :

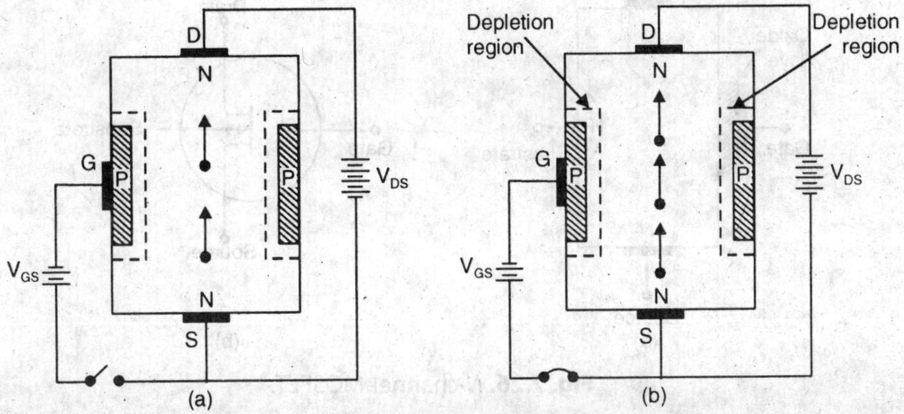

Fig. 7.25

(*a*) When a voltage V_{DS} is applied between drain and source terminals and voltage on the gate is zero [Fig. 2.83 (*a*)], the two *P-N* junctions at the sides of the bar establish depletion layers. The electrons will flow from source to drain through a channel between the depletion layers. *The size of these layers determines the width of the channel* and hence the current conduction through the bars.

(*b*) When a reverse voltge V_{GS} is applied between the gate and source [Fig. 7.25 (*b*)], the *width of the depletion layers is increased*. This *reduces* the width of conducting channel, thereby increasing the resistance of *N*-type bar. Consequently, the current from the source to drain is *decreased*. On the other hand if the reverse voltage on the gate is decreased, the width of depletion layers also decreases. This increases the width of the channel and source to drain current increases.

From the above discussion it is evident that current from source to drain can be controlled by the application of potential or electric field on the gate. It is due to this reason that this device is called field-effect transistor. Note that a *P-channel* JFET operates in the same manner as an *N*-channel JFET *except that channel current carriers will be the holes instead of electrons and the polarities of* V_{GS} *and* V_{DS} are reversed.

2. **Metal oxide semiconductor FET (MOSFET) :**

- MOSFET is an important semiconductor device and is widely used in many circuit applications. Since it is constructed with the gate terminal insulated from the channel, it is sometimes called *insulated gate FET* (IGFET).
- Like a JFET, a MOSFET is also a *three-terminal (source, gate and drain) device* and drain current in it is also controlled by gate bias.
- The operation of MOSFET is similar to that of JFET. It can be employed in any of the circuit covered for the JFET and, therefore, all the equations apply equally well to the MOSFET and JFET in amplifier connections. However, MOSFET has *lower capacitance and input impedance much more than that of a JFET owing to small leakage current.* In case of a MOSFET the positive voltage may be applied to the gate and *still the gate current remains the zero.*

Construction :

Fig. 7.26 (*a*) shows constructional details of *N*-channel MOSFET. It is similar to FET except with the following modifications :

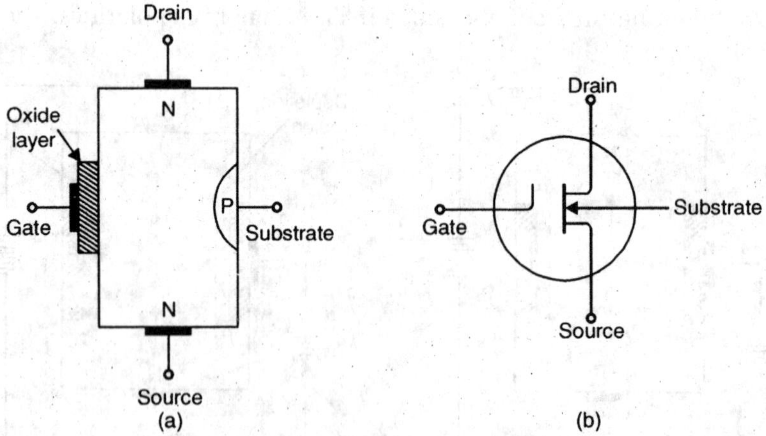

Fig. 7.26. *N*-channel MOSFET.

(*i*) There is only a single *P*-region. This region is called ***Substrate.***

(*ii*) A thin layer of metal oxide (usually silicon oxide) is deposited over the left side of the channel. A *metalic* ***gate*** *is deposited over the oxide layers.* As silicon dioxide is an insulator, therefore, *gate is insulated from the channel.*

(*iii*) Like JFET, a MOSFET has three terminals *viz, source, gate* and *drain.*

Fig. 7.26 (*b*) shows the symbolic symbol of *N*-channel MOSFET.

Working :

Fig. 7.27 shows the MOSFET circuit. Instead of gate diode as in JFET, here gate is formed as a small capacitor. *One plate of this capacitor is the gate and the other plate is the channel with metal oxide as the dielectric.*

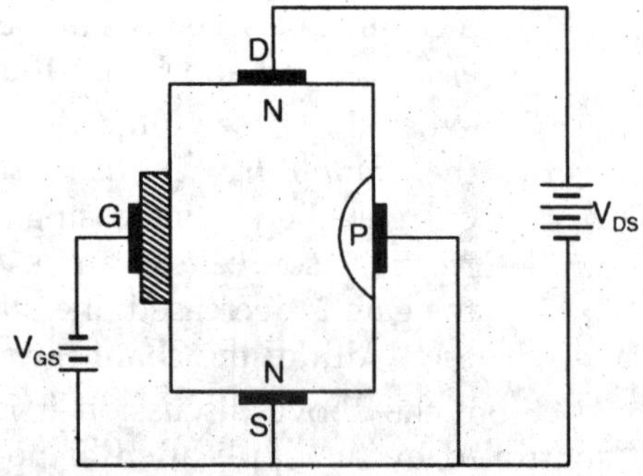

Fig. 7.27. MOSFET circuits.

— When *negative voltage* is applied to the gate, electrons accumulate on it. These electrons repel the conduction band electrons in the *N*-channel. Therefore, lesser number of conduction electrons are made available for current conduction through the channel. The greater the negative voltage on the gate, the lesser is the current conduction from source to drain.

— When the gate is given *positive voltage*, more electrons are made available in the *N*-channel. Consequently, current from source to drain *increases*.

Regarding MOSFET, the following *points* are worth noting :

- **A MOSFET,** *unlike the JFET, has* ***no gate diode.*** Therefore, it is possible *to operate the device with positive or negative gate voltage.*
- In a MOSFET, *the source to drain current is controlled by the electric field of capacitor formed at the gate.*
- As the gate forms a capacitor, therefore, negligible currents flows whether + ve or – ve voltage is applied to the gate. Consequently, the input impedance of MOSFET is *very high* (varying from 10^4 MΩ to 10^6 MΩ).

7.3.9 Unijunction Transistor (UJT)

A unijunction transistor, unlike a bipolar transistor has *only one junction*. Like other conventional transistors, it also processes the transistor action and *works like a switch.*

Its characteristics are similar to those of a silicon unilateral switch (SUS) and a complementary silicon controlled rectifier (CSCR). Its construction is, however, different and it does *not* belong to the thyristor family.

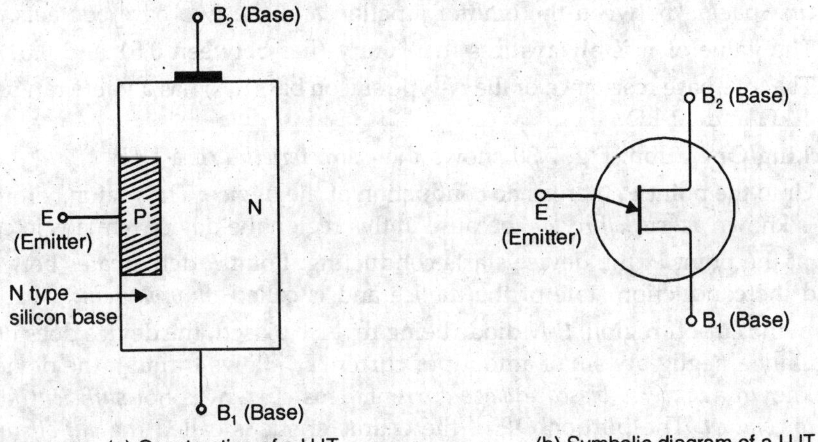

(a) Construction of a UJT (b) Symbolic diagram of a UJT

Fig. 7.28. Unijunction Transistor (UJT).

- Basically, a UJT is a *three-terminal silicon diode.* As its name indicates, it has only one *P-N* junction. It differs from an ordinary diode in that it has *three* leads and it differs from a FET in that it has *no ability to amplify*. However, it has the ability to *control a large A. C. power with a small signal.* It also exhibits a *negative resistance characteristic which makes it useful as an oscillator.*

Construction : Refer to Fig. 7.28

A unijunction transistor (UJT) consists of a *lightly doped silicon* bar with a *heavily doped P-type* material alloyed to its one side (closer to B_2) for producing single *P-N* junction. There are three terminals : one emitter, E and two bases B_1 and B_2 at the bottom and top of the silicon bar.

Interbase resistance (R_{BB}): Refer to Fig. 7.29.

The interbase resistance (R_{BB}) is the total resistance of the silicon bar from one end to the other with emitter terminal open; from equivalent circuit (see Fig. 2.87), we have

$$R_{BB} = R_{B_2} + R_{B_1}$$

The point C is such that $R_{B_1} > R_{B_2}$ (usually R_{B_1} is 60 percent of R_{BB}). R_{B_1} has been shown as a variable resistor because its value varies inversely as I_E.

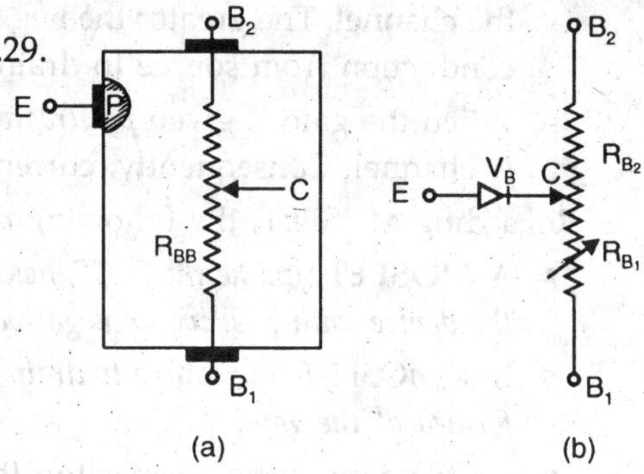

Fig. 7.29

Let the voltage drop across R_{B_1} is V_C. Then,

$$V_C = V_{BB} \times \frac{R_{B_1}}{R_{B_1} + R_{B_2}} \quad \text{...using voltage binder relationship}$$

$$= \eta . V_{BB}$$

where, $$\eta = \frac{R_{B_1}}{R_{B_1} + R_{B_2}}$$

η is called the ***instrinsic stand ratio.***

— The value of η depends on two factors namely : (*i*) Construction of the UJT, and (*ii*) Spacing between the emitter junction and the two base contacts.

— The value of η is always less than unity (lies between 0.51 and 0.81)

— The interbase resistance of the *N*-type silicon bar (R_{BB}) has a value ranging between 4 kΩ and 12 kΩ.

Working/Operation. Fig. 7.30 shows the *characteristics* of a UJT.

— Upto the point *P*, there is no conduction of the device. The region before this point is known as *'cut-off region'* because in this region the device remains in cut-off state.

Just at the point *P*, the device starts conducting. Point *P* demarcates between cut-off state and the conduction state of the device and is called its *peak point.*

— In the cut-off region, *P-N* diode being reverse biased, the device does not conduct. Only a negligibly small amount of current I_{EO} flows through the device which is known as *reverse biased leakage* current. This current is not *sufficient for the device to conduct.* The portion *OP* of the characteristic is called the *cut-off region* of the device.

— When the peak point *P* is reached, the increase in charge carriers causes decrease in resistance R_{B_1} and the device starts *conducting.* In the conduction state, the device depicts a negative resistance characteristics. This means, as the emitter voltage V_E is further increased, the voltage across the device decreases, but the current increases. This region of conduction is called the *negative resistance region.* This region continues till the valley point *V* is reached in the characteristic. The portion *PV* of the characteristic is called the *negative resistance region.*

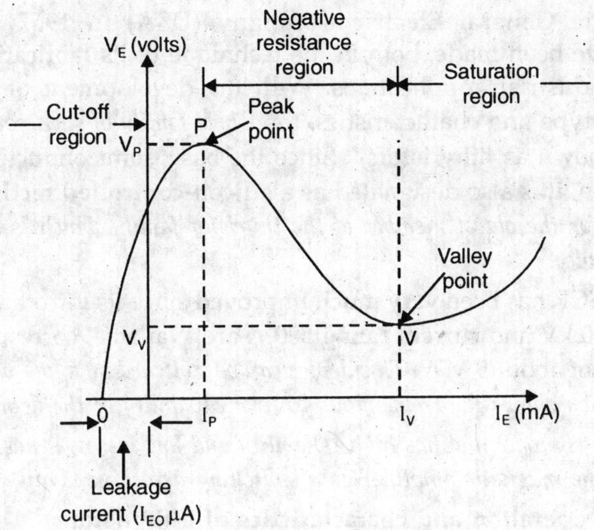

I_{EO} = Leakage current ; V_P = Peak point voltage; I_P = Peak point current; V_V = Valley point voltage; I_V = Valley point current; V_E = Emitter voltage ; I_E = Emitter current.

Fig. 7.30. Characteristics of UJT.

— After reaching the valley point, the device goes to its saturation state where further fall in the voltage across the device does not take place. The device voltage and current both reach standard values and do not change any more even if the emitter voltage is changed. This portion of the characteristic beyond the valley point is called *'saturation region'*.

A set of *V-I* characteristic for UJI can be obtained for different values of interbase voltage V_{BB}.

- It is seen that only terminals *E and* B_1 *are active terminals whereas* B_2 *is the bias terminal i.e.,* it is meant only for applying external voltage across UJT.
- Generally, *UJT is triggered into conduction by applying a suitable positive pulse at its emitter.* It can be brought *back to OFF state by applying a negative trigger pulse.*

Applications. One significant property of UJT is that it *can be triggered by* (or an output can be taken from) *any one of its three terminals.* Once triggered, the emitter current I_E of the UJT increases regeneratively till it reaches a limiting value determined by the external power supply. Owing to their particular behaviour, UIT is used in variety of circuit applications; some of these are :

1. Pulse generation;
2. Sine wave generator;
3. Saw tooth wave generator;
4. Switching;
5. Phase control;
6. Voltage or current regulated supplies;
7. Timing and trigger circuits.

7.3.10. Thyristor

Introduction :

Ample pioneering work on theory and fabrication of the *power-switching device,* which later came to be known as a *thyristor* (because its characteristics are *similar to those of the gas-tube thyratron*), was done at the Bell Laboratories in the U.S.A. The first prototype was introduced by the General Electric Company (USA) in 1957. Since then, many improvements have been made, both in the technique of its fabrication and in adapting it to numerous industrial appplications. With the development of a number of other devices of similar type and characteristics, the *whole family of such power-switching devices* has come to be known as **"thyristors"**. Since the basic semiconductor material used for the device is silicon, it is also designated as a **silicon-controlled rectifier (SCR).** *The term SCR is often used for the oldest member of the thyristor family which is the most widely used power-switching device.*

The rating of SCR has been very much improved since its introduction and now SCRs of voltage rating 10 kV and current rating 500 A are available, corresponding to a power-handling capacity of about 5 MW. *This device can be switched by a low-voltage supply of about 1 A and 10 W, which shows the tremendous control capability of the device.*

Because SCR is compact and has high reliability and low losses, it has more or less replaced the thyratron and the magnetic amplifier as a switching device in many applications.

Construction, operation and characteristics of a thyristor :

Construction :

- The cross-sectional view of a typical SCR is shown in Fig. 7.31. Basically, the SCR consists of a four-layer pallet of *P*-type and *N*-type semiconductor materials. Silicon is used as the intrinsic semiconductor to which the proper impurities are added. The junctions are either *diffused* or *alloyed.*
- The *planer construction* shown in Fig 7.31 (*a*) is used for *low-power* SCRs. This technique is useful for making a number of units for a single silicon wafer. Here, all the junctions are *diffused.*

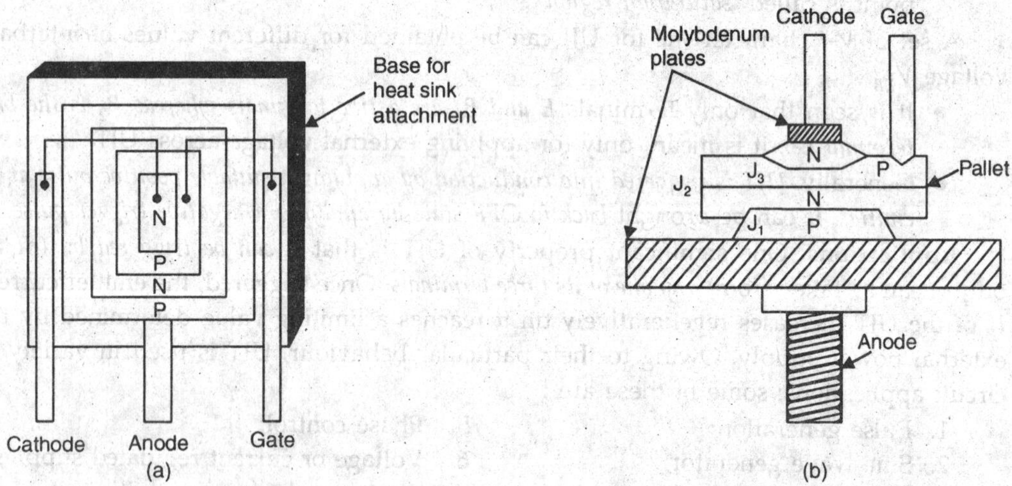

Fig. 7.31. (*a*) Planer type (all diffused), (*b*) Mesa type (alloy diffused).

- The other technique the *mesa construction* is shown in Fig. 7.31 (*b*). This is used for *high-power* SCRs. Here, the inner junction J_2 is obtained by diffusion, and then the outer two layers are alloyed to it. Because the *PNPN* pallet is required to handle large currents, it is properly braced with tungsten or molybdenum plates to provide greater mechanical strength. One of these plates is handsoldered to a copper or an aluminium stud, which is threaded for attachment to a heat sink. This provides an efficient thermal path for conducting the internal losses to the surrounding medium. The use of hand solder between the pallet and back up plates *minimises thermal fatigue* when the SCRs are subjected to temperature-induced stresses. For medium and low-power SCRs, the pallet is mounted directly on the copper stud or casing, using soft-solder which absorbs the thermal stresses set-up by differential expansion and provides a good thermal path for heat transfer. When a larger cooling arrangement is required for highpower SCRs, the press-pack or hockey pack construction is used, which provides for double-sided air or water cooling.

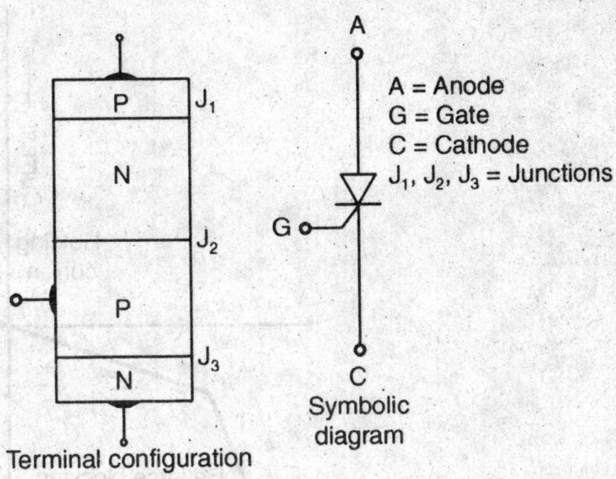

Fig. 7.32. Schematic diagram of an SCR.

- Fig. 7.32 shows the terminal configuration and symbolic diagram of an SCR.
- Fig. 7.33 shows a thyristor. It has four layers alternately of *P* and *N* with *outer layers heavely doped.* The control current is generally applied *via* middle *P* layer and *N* emitter. *Junctions J_1 and J_3 are forward biased while junction J_2 is reverse biased. If the voltage gradient between anode and cathode is too high then due to increased inherent current the thyristor may be switched on.*
- Fig. 7.34 shows the *forward* and *reverse characteristics* of a thyristor.

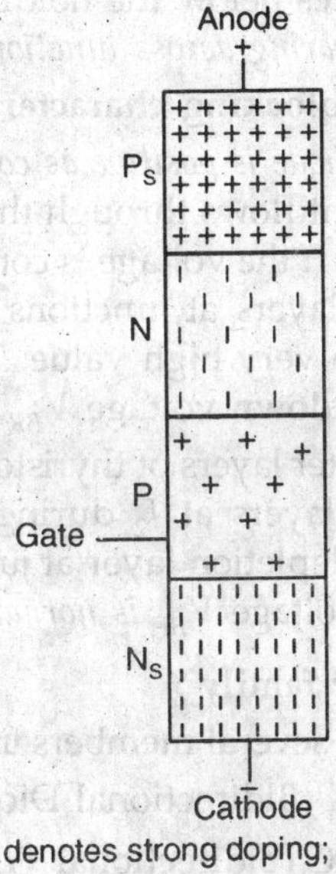

Suffix 's' denotes strong doping; J_1, J_2, J_3-junctions.

Fig. 7.33. Thyristor.

Forward characteristics :

When the *anode is positive compared to the cathode,* junctions J_1 and J_3 are forward biased and J_2 is reverse biased. Under such circumstances *a very small current* due to inherent conductivity will flow through the device which is in *forward blocking state.*If the voltage at anode is continuously increased the inherent current is increased and at one stage *it switches on the device.* The 'switching on' of the device results *because of the breakdown of reverse biased junction J_2 due to high voltage gradient.* The 'switching on' condition of thyristor is referred to as *conducting state* and during this state *current through the device is only limited by external resistance.*

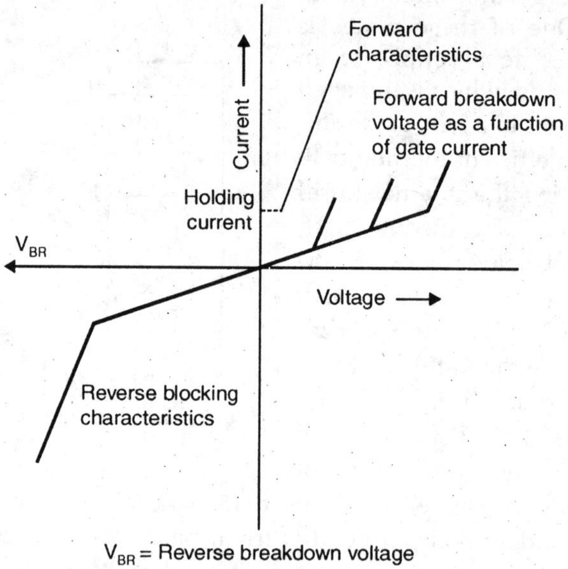

Fig. 7.34. Thyristor characteristics.

If the anode voltage is reduced, the device will continue to conduct till the forward current reduces below the holding current I_h. *Below this value of current the depletion layer will start appearing across junction* J_2 *and the device will be 'blocked'.*

Reverse blocking characteristics :

If the *cathode is positive as compared to anode,* the junction J_2 *is reverse biased* and only a small current flows through the device and the characteristics are called *reverse blocking characteristics.* If the voltage is continuously increased at one stage it may result in breaking of depletion layers at junctions J_1 and J_3 and the current through the device suddenly increases to a very high value. This is called *reverse breakdown* and the voltage is called reverse breakdown voltage V_{BR}.

As the outer layers of thyristor are highly doped compared to inner layers, the thickness of depletion layers at J_2 during forward bias is much more as compared to the total thickness of depletion layer at junctions J_1 and J_3 during reverse bias. Hence, the forward breakdown voltage V_{BF} *is normally greater than reverse breakdown voltage* V_{BR}.

Thyristor family :

There are several members in the thyristor family, some of them are mentioned below:

1. DIAC (Bidirectional Diode Thyristor)
2. TRIAC (Bidirectional Triode Thyristor)
3. SCR (Silicon Controlled Rectifier)
4. SUS (Silicon Unilateral Switch), also known as complementry SCR (CSCR)
5. SBS (Silicon Bilateral Switch)
6. SCS (Silicon Controlled Switch)
7. LASCR (Light Activated SCR)
8. LASCS (Light Activated SCS).

7.4. DIGITAL ELECTRONICS

7.4.1 Introduction

The *branch of electronics which deals with digital circuits is called* **digital electronics.**

- A *continuously varying signal (voltage or current) is called an* **"analog signal".**

Example. A *sinusoidal voltage.*

In an analog electronic circuit, the output voltage changes continuously according to the input voltage variations *i.e.,* the output voltage can have an *infinite number of values.*

- A *signal (voltage or current) which can have only two discrete values is called a* **"digital signal".**

Example. A *square wave.*

- *An electronic circuit that is designed for two-state operation is called a* **digital circuit.**

These days digital circuits are being used in many electronic products such as *video games, microwave ovens, oscilloscopes etc.*

7.4.2. Advantages and Disadvantages of Digital Electronics

The advantages and disadvantages of digital electronics are listed below :

Advantages :

1. Digital system can be normally *easily designed.*
2. Digital circuits are less *affected by noise.*
3. Storage of information is easy with digital circuits.
4. Digital circuits provide greater accuracy and precision.
5. More digital circuitry can be fabricated on integrated chips.

Disadvantages :

1. The digital circuits can handle only digital signals ; it requires encoders and decoders, due to which cost *of the equipment is increased.*
2. Under certain situations the use of only the analog techniques is simpler and economical (*e.g.* the process of signal amplification).

However, since the advantages outweigh the disadvantages, therefore, we are switching to digital techniques at a faster pace.

7.4.3. Digital Circuit

An electronic circuit that handles only a digital signal is called a **digital circuit.**

Or

An electronic circuit in which a state switches between the two states with time or with the change of the input states, and it is its state at the inputs and the outputs which has a significance is called a **digital circuit.**

"Digital" is derived from *"digitus"*. In Latin, the latter means "finger". A finger is either up or down. Similarly an electronic circuit may have one of the states as :

(*i*) 'ON' (conduction) or 'OFF' (poor conduction), or

(*ii*) *'High'* voltage or *'Low'* voltage between two terminals, or

(*iii*) *'High'* current through a circuit or *'Low'* current through a circuit, or

(*iv*) *'High'* frequency signal or *'Low'* frequency signal, or

(*v*) 'Negative' potential difference or *'Positive'* potential difference, or (*vi*) "1" or "0" etc.

Therefore, *digital circuit is one that expresses the values in digits 1's or 0's, hence the name "digital'*. The number concept that uses only the two digits 1 and 0 is the *binary numbering system.*

- As a digital circuit is based upon the two states, it is used in dealing with binary numbers ; digital circuit is therefore used in *computers.*

Advantages of digital circuit :

1. *Noise free* as output is measured in terms of its state, not in terms of a voltage, or a current, or a frequency. A *state has a difiniteness.*
2. *Capabilities of logical decision, arithmetic and Boolean operation on the binary numbers.*

Disadvantages :

1. *Slower speed* due to greater number of components to represent a state.
2. The circuits have *complexities* also. To represent a big decimal number, a *large number of components needed.*

Advantages of "Analog circuit" :

1. More close to physical system values.
2. A voltage level may represent temperature, wind, speed etc.

Disadvantages :

Lack of definiteness, preciseness and reliability.

7.4.4. Logic Gates

General aspects :

A *digital circuit with one or more input signals but only output signal is called a* **logic gate.** A ***logic gate*** *is an electronic circuit which makes logic decision.*

- Logic gates are the *basic building blocks* from which most of the digital systems are built up. They implement the hardware logic function based on the logical algebra developed by George Boolean which is called *Boolean algebra* in his honour.
  - A unique characteristic of Boolean algebra is that variables used in it can assume only one of the two values *i.e., either 0 or 1.* Hence, every variable is either a 0 or a 1 (Fig. 7.35–limits on TTLIC's).
- Each gate has *distinct graphic symbol and its operation can be described by means of Boolean algebraic function.*

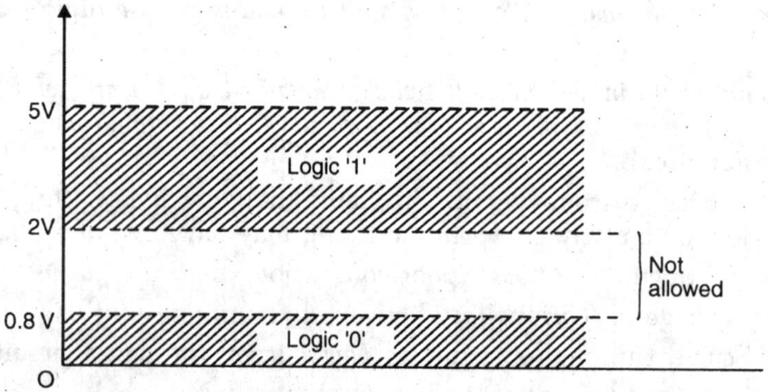

Fig. 7.35. Voltage assignment in a digital system.

- The *table which indicates output of gate for all possible combinations of input is known as a* ***truth table.***
- These gates are available today in the form of various IC families. The *most popular families are :*

(*i*) Transistor-transistor logic (TTL)

(*ii*) Emitter-coupled logic (ECL)

(*iii*) Metal-oxide-semiconductor (MOS)

(*iv*) Complementary metal-oxide-semiconductor (CMOS).

Applications of logic gates :

The following are the *fields of application of logic gates :*

1. Calculators and computers.
2. Digital measuring techniques.
3. Digital processing of communications.
4. Musical instruments.
5. Games and domestic appliances, etc.
6. The logic gates are also employed for decision making in *automatic control of machines and various industrial processes and for building more complex devices such as binary counters* etc.

Positive and negative logic :

The number symbols 0 and 1 represent, in computing systems, two possible states of a circuit or device. It does not make any difference if these two states are referred to as *'ON'* and *'OFF'*, *'Closed'* and *'Open'*, *'High'* and *'Low'*, *'Plus'* and *'Minus'* or *'True'* and *'False'* depending upon the situations. The main point is they must be symbolized by *two opposite conditions.* In *positive logic* a **'1'** *represents* : an *'ON circuit'* ; a *'Closed switch'*; a *'High voltage'*, a *Plus sign'*, *'True statement'*. *Consequently,* a **0** *represent* : an 'OFF circuit' ; an *'Open switch'*, a *'Low voltage'* ; a *'Minus sign'*, a 'False statement'.

In negative logic, the just opposite conditions prevail.

Example. A digital system has two voltage levels of 0 V and 5 V. If we say that symbol 1 stands for 5 V and symbol 0 for 0 V, then we have *positive logic system.* If on the other hand, we decide that a 1 should represent 0 V and 0 should represent 5 V, then we will get negative logic system.

Main point is that in *'positve logic'* the ***more positive*** of the two voltage levels represents the 1 while in *'negative logic'* the ***more negative*** voltage represents the 1.

Types of Logic Gates : Refer to table 7.6 (page 330)

In the *complex circuits,* the following *six* different digital electronics gates are used as *basic elements :*

1. NOT Gate
2. NAND Gate
3. AND Gate
4. OR Gate
5. NOR Gate
6. XOR Gate.

— A truth table has 2^n rows. It gives in each of its row *m* outputs for a given combination of *n* inputs.

1. NOT Gate :

- *Not operation means that the output is the complement of input.* If input is logic '1', the output is logic '0' and if input is logic '0', the output is logic '1'.

Table 7.6. Symbols, Boolean Expressions and Truth Tables of various Logic Gates

Logic	NOT	NAND	AND	OR	NOR	EX.OR	COINCIDENCE
Symbol	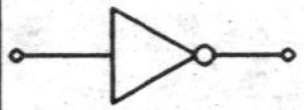 Fig. 7.36	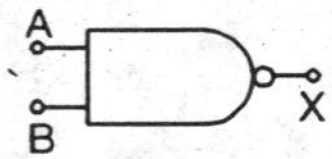 Fig. 7.37	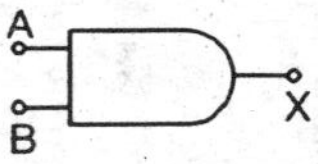 Fig. 7.38	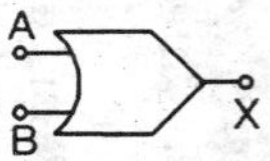 Fig. 7.39	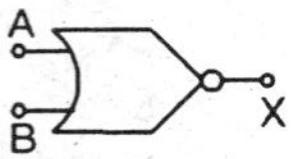 Fig. 7.40	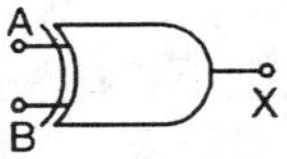 Fig. 7.41	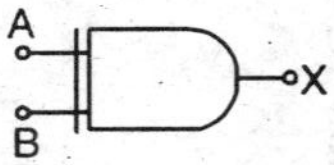 Fig. 7.42
Boolean expression	$X = \overline{A}$	$X = \overline{A \, . \, B}$	$X = A.B$	$X = A + B$	$X = \overline{A + B}$	$X = A \, . \, \overline{B} + \overline{A} \, . \, B = A \oplus B$	$X = \overline{A} \, . \, \overline{B} + A.B$
Truth table	A X 0 1 1 0	A B X 0 0 1 0 1 1 1 0 1 1 1 0	A B X 0 0 0 0 1 0 1 0 0 1 1 1	A B X 0 0 0 0 1 1 1 0 1 1 1 1	A B X 0 0 1 0 1 0 1 0 0 1 1 0	A B X 0 0 1 0 1 1 1 0 1 1 1 0	A B X 0 0 1 0 1 0 1 0 0 1 1 1
Definition	Output available – when there is no input.	Output available – to all states except when all the inputs are available.	Output available– when all inputs available.	Output available– when only one or more inputs available.	Output available– when no input is available.	Output available– when the inputs are *not* identical.	Output available– to those states when the inputs are identical.

- Fig. 7.36 shows the *symbol of NOT Gate.* It is generally represented by a *triangle followed by a bubble* (or a bubble followed by a triangle).
- NOT gate is *used when an output is desired to be complement of the input.*
- If all inputs of NAND gates are joined it shall act as NOT gate.
- NOT gate is also *called 'inverting logic circuit'*. It is also *called a 'complementing circuit'*.

2. NAND Gate :

- A NAND gate can said to be *basic building block* of the *all digital TTL logic gates and other digital circuits.*
- It is represented by the symbol shown in Fig. 7.37.
- Its *unique property is that output is high '1' if any of the input is at low '0' logic level.*

Let us consider two inputs with the states *A* and *B* at the NAND gate. The answer (output) $X = \overline{A.B}$. Bar denotes a NOT log operation on A.B. The meaning of A.B, called AND operation, is given in 3 below.

3. AND Gate :

- *A NAND gate followed by a NOT gate gives us AND gate.*
- It is represented by a symbol in Fig. 7.38. Its symbol differs from NAND only by *omission of a bubble (circle).*
- Its *unique property is that its output is '0' unless all the inputs to it are at the logic 1's.*
- A two inputs, AND gate has X = *A.B. Dot* between the two states indicates 'AND' logic operation using these.

4. OR Gate :

- An 'OR' operation means that the *output is '0' only if all the inputs are '0s'.*
- It is represented by a symbol shown in Fig. 7.39.
- If any of the inputs is '1' the output is '1'. A two inputs 'OR' gate has $X = A + B$. Sign + between the two states indicates an 'OR' logic operation.

5. NOR Gate :

- An 'OR' circuit followed by a NOT circuit gives a 'NOR' gate (Fig. 7.40).
- Its *unique property is that its output is '0' if any of input is '1'.*
- A NOR gate is a *basic building block for other types of the logic gates than TTLs.* In the TTL circuits, a NOR is fabricated in an IC by the several NANDs.

A two input NOR has $X = \overline{A + B}$.

6. XOR Gate :

- A XOR gate (Fig. 7.41) is called 'Exclusive OR' gate.
- Its *unique property is that the output is '1' only if odd number of the inputs at it are '1's.*
- The 'Exclusive OR' can be written as : $X = A . \overline{B} + \overline{A} . B$ or $A \oplus B$.
- Exclusive OR gate is important in the circuits *for addition of two binary numbers.*

7. Coincidence Gate :

- This gate (Fig. 7.42) can be written as : $X = \overline{A} . \overline{B} + A . B$.
- Output available to those states when the inputs are identical.

Basic building blocks. AND, OR and NOT gates are called *basic* building blocks or basic gates because they are *essential to realize any boolean expression.*

Universal gates. NAND and NOR gates are known as *universal gates* because any logic gate can be constructed either by using NAND gates only or by using NOR gates only.

7.4.5. Universal Gates

NAND and NOR gates are known as **universal gates.**

The AND, OR, NOT gates can be realized using only NAND or NOR gates.

— Demorgan's theorem afford a convenient method to use these two gates in *logic design*. The entire logic system can be implemented by using any of these two gates.

— These two gates are *easier to realize and consume less power than other gates.*

(*i*) **Realization of logic gates using NAND gates :**

Fig. 7.43 (*a*), (*b*), (*c*) shows realization of NOT, AND, OR gates using NAND gates respectively, which is self explanatory.

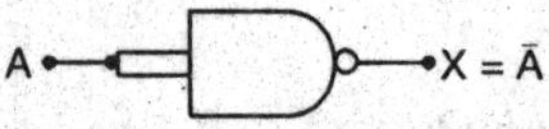

(a) Realization of NOT gate using NAND gate

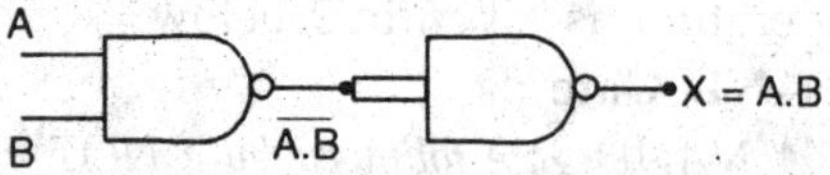

(b) Realization of AND gate using NAND gate

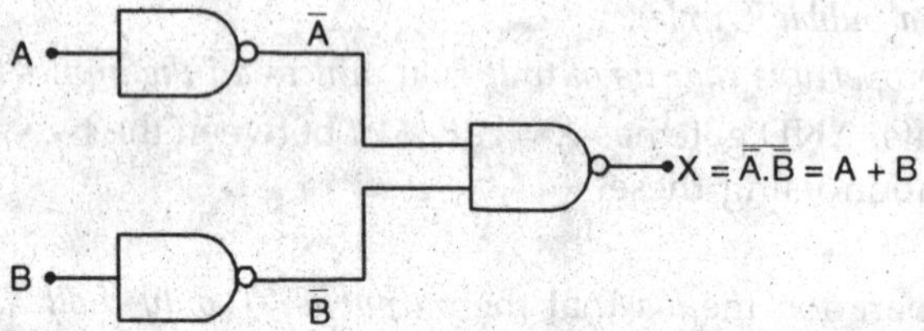

(c) Realization of OR gate using NAND gate

Fig. 7.43. Realization of NOT, AND and OR gates using NAND gates.

(*ii*) **Realization of logic gates using NOR gates :**

The realization of NOT, OR and AND gates using NOR gates is shown in Fig. 7.44 (*a*), (*b*), (*c*) respectively.

(a) Realization of NOT gate using NOR gate

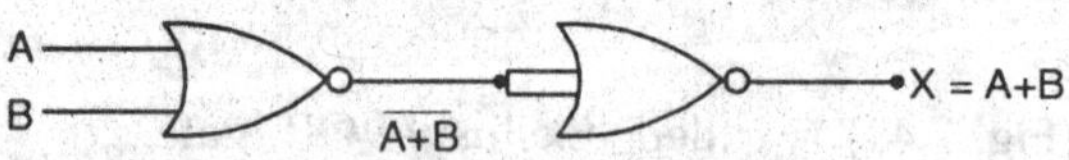

(b) Realization of OR gate using NOR gate

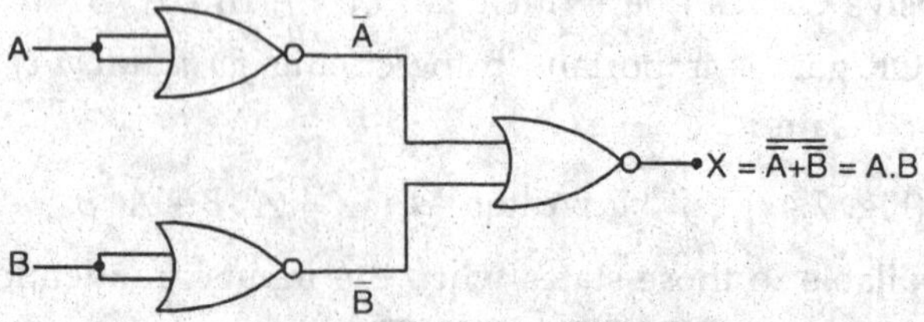

(c) Realization of AND gate using NOR gate

Fig. 7.44. Realization of NOT, OR and AND gates using NOR gates.

7.4.6. Half Adder (HA)

It is a *1-bit adder* and carries out binary addition with the help of XOR and AND gates. It has *two inputs and two outputs.*

It can add 2 binary digits at a time and produce a 2-bit data *i.e.,* 2-bit data *i.e.,* SUM and CARRY according to binary addition rules.

The circuit of a half adder is shown in Fig. 7.45. (*a*). It consists of an Ex-OR gate and AND gate. The outputs of the Ex-OR gate is called the SUM (S), while the output of the AND gate is known as CARRY-(C). As the AND gate produces a *high* output only when both inputs are *high* and Ex-OR gate produces a *high* output if either input (not both) is high, the truth table of a half adder is developed by writing the truth table output of AND gate in the CARRY column and the output truth table of Ex-OR gate in SUM column. Truth table for half adder is given in table 7.7.

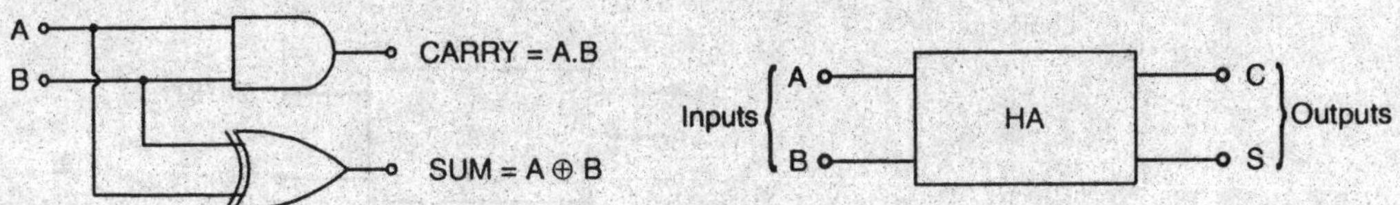

(a) Logic circuit (b) Logic symbol

Fig. 7.45. Half adder.

Table 7.7. Truth table for Half Adder

Inputs		*Outputs*	
A	*B*	*C*	*S*
0	0	0	0
0	1	0	1
1	0	0	1
1	1	1	0

The logical expressions for CARRY and SUM can be written from the truth table for a half adder as follows :

$$\text{CARRY, C} = \text{A.B}$$
$$\text{SUM, S} = \text{A} \oplus \text{B}$$

- This circuit is called *half-adder*, because it *cannot* accept a CARRY-IN from previous additions. Owing to this reason the half-adder circuit can be *used for binary addition of lower most bit only.*

For higher-order columns, 3-input adder called *full adder* is used.

7.4.7. Full Adder (FA)

A full adder has *three inputs* and *two outputs.* It *can add 3 digits (or bits) at a time.* The bits A and B which are to be added come from the two registers and the third input C comes from the '*carry*' generated by the previous addition. It produces two outputs, SUM and CARRY-OUT (going to next higher column).

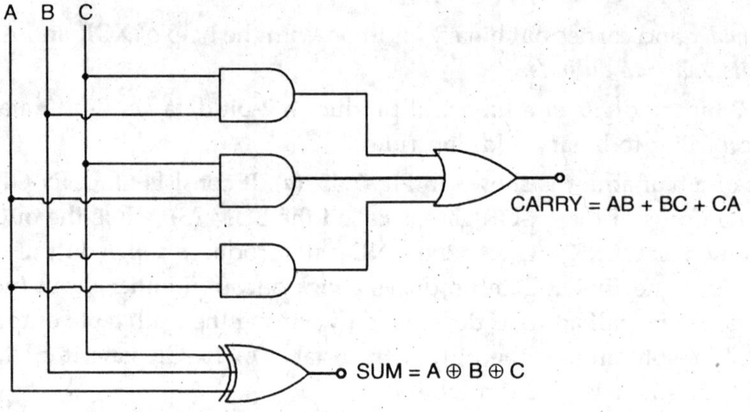

(a) Logic circuit

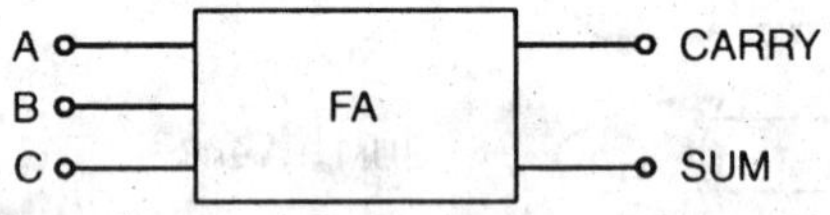

(b) Logic symbol

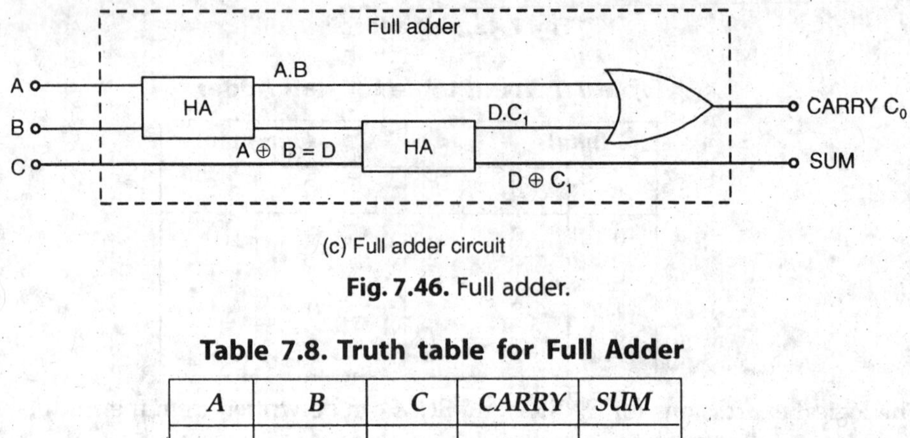

(c) Full adder circuit

Fig. 7.46. Full adder.

Table 7.8. Truth table for Full Adder

A	*B*	*C*	*CARRY*	*SUM*
0	0	0	0	0
0	0	1	0	1
0	1	0	0	1
0	1	1	1	0
1	0	0	0	1
1	0	1	1	0
1	1	0	1	0
1	1	1	1	1

A simple circuit of a full adder is shown in Fig. 7.46 (*a*), though other designs are also possible. It uses 3 AND gates, one Ex-OR gate and one OR gate. The final CARRY is given by the OR gate while the final SUM is given out by the Ex-OR gate.

Fig. 7.46(*b*) shows the logic symbol for a full adder.

Truth table for full adder for all possible inputs/outputs is given in Table 7.8. Truth table can be checked easily for its validity.

A full adder can be made by using two half adders and an OR gate. The circuit is shown in Fig. 7.46(*c*).

- The full adder can do more than a million additions per second. Besides that, it never get tired or bored or asks for a rest.

Note : *Binary additions :* Following are the four rules/cases for addition of binary numbers :

(1) $0 + 0 = 0$

(2) $0 + 1 = 1$

(3) $1 + 0 = 1$

(4) $1 + 1 = 10$ (This sum is *not ten* but *one-zero*).

7.4.8. Boolean Algebra

George Boolean in 1854 developed a mathematics now referred as *Boolean algebra.* It is the *algebra of logic presently applied to the operation of computer devices.* The rules of this algebra are based on human reasoning.

Digital circuits perform the binary arithmetic operations with binary digits 1 and 0. These operations are called logic functions or logic operations. *The algebra used to symbolically describe logic functions is called* **Boolean algebra.** Boolean algebra is *a set of rules and theorems by which logical operations can be expressed symbolically in equation form and be manipulated mathematically.*

Boolean algebra differs from ordinary algebra in that Boolean constant and variables *can have only two values :* '0' and '1' :

In Boolean algebra the following *four connecting symbols* are used :

1. **Equal sign (=).** In Boolean algebra the 'equal sign' refers to the standard mathematical equality. In other words, the logical value on one side of the sign is identical to the logical value on the other side of the sign.

Example. We are given two logical variables such that $A = B$. Then if $A = 1$, then $B = 1$ and if $A = 0$ then $B = 0$.

2. **Plus sign (+).** In Boolean algebra the 'plus sign' refers to logical *OR operation.*

The statement $A + B = 1$ means *A* ORed with B equals 1. Consequently, either $A = 1$ or $B = 1$ or both equal to 1.

3. **Multiply sign (.).** In Boolean algebra the 'multiply sign' refers to *AND operation.*

The statement $A.B = 1$ means *A* ANDed with *B* equals 1. Consequently, $A = 1$ and $B = 1$.

The function $A . B$ often written as *AB*, *omitting the dot* for convenience.

4. **Bar sign (–).** In Boolean algebra the 'bar sign' refers to *NOT operations.* The NOT has the effect of inverting (complementing) the logic value.

Thus if A = 1, then $\overline{A} = 0$.

Boolean Laws (For Outputs from Logic Inputs) :

The following Laws can said to be associated with Boolean algebra :

1. **'OR' Laws**

The 'OR' Laws are described by the following equations :

$$A + 1 = 1 \qquad ...[7.1(a)]$$

$$A + 0 = A \quad ...[7.1(b)]$$
$$A + A = A \quad ...[7.1(c)]$$
$$A + \bar{A} = 1 \quad ...[7.1(d)]$$

- An 'OR' operation is denoted by *plus sign*.
- 'OR' Law means :

(*i*) Any number (0 or 1) is a first input to an OR gate and another member at the second input is 1 then answer is 1,

(*ii*) If another is 0 then answer is as first input, and

(*iii*) If two inputs to an OR gate *complement* then output is '1'.

2. **'AND' Laws**

'AND' operation is denoted by the *dot* sign.

- True and true make true
- True and false make false
- False and false make false.

$$A \,.\, 1 = A \quad ...[7.2(a)]$$
$$A \,.\, 0 = 0 \quad ...[7.2(b)]$$
$$A \,.\, A = A \quad ...[7.2(c)]$$
$$A \,.\, \overline{A} = 0 \quad ...[7.2(d)]$$

3. **'NOT' Laws (Laws of Complementation)**

A NOT operation is denoted by putting a *bar* over a number.

- The NOT true means false.
- The NOT false means true.

$$\bar{1} = 0 \quad ...[7.3(a)]$$
$$\overline{\overline{A}} = A \quad ...[7.3(b)]$$

Eqn. [7.3(*b*)] means that if *A is inverted (complemented) and then again inverted, we get the original number.*

4. **Commutative Laws**

These Laws mean that *order of a logical operation is immaterial.*

$$A + B = B + A \quad ...[7.4(a)]$$
$$A \,.\, B = B \,.\, A \quad ...[7.4(b)]$$

5. **Associative Laws**

These laws *allow a grouping of the Boolean variables.*

$$A + (B + C) = (A + B) + C \quad ...[7.5(a)]$$
$$A \,.\, (B \,.\, C) = (A \,.\, B) \,.\, C \quad ...[7.5(b)]$$

6. **Distributive Laws**

These laws *simplify the problems in the logic disigns.*

$$A \,.\, (B + C) = (A \,.\, B) + (B \,.\, C) \quad ...[7.6(a)]$$
$$A + (B \,.\, C) = (A + B) \,.\, (A + C) \quad ...[7.6(b)]$$
$$A + (\overline{A} \,.\, B) = A + B \quad ...[7.6\ (c)]$$

The last two equations are typical to the Boolean algebra, and are not followed in the usual algebra.

7.4.9. De Morgan's Theorems

First theorem shows an equivalence of a NOR gate with an AND gate having bubbled inputs (Fig. 7.47), and is given by the equation :

$$\overline{A+B} = \overline{A}\,.\,\overline{B} \qquad ...(7.7)$$

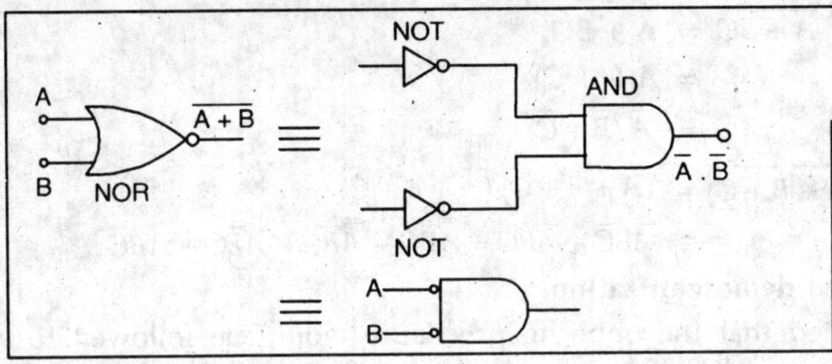

A	B	X
0	0	1
1	0	0
0	1	0
1	1	0

Fig. 7.47. De Morgan's First theorem showing an equivalence of a NOR gate (same holds for multiple inputs).

Second theorem shows an equivalence of a NAND gate with an OR having bubbled inputs as shown in Fig. 7.48 and is given by the equation :

$$\overline{A\,.\,B} = \overline{A} + \overline{B} \qquad ...(7.8)$$

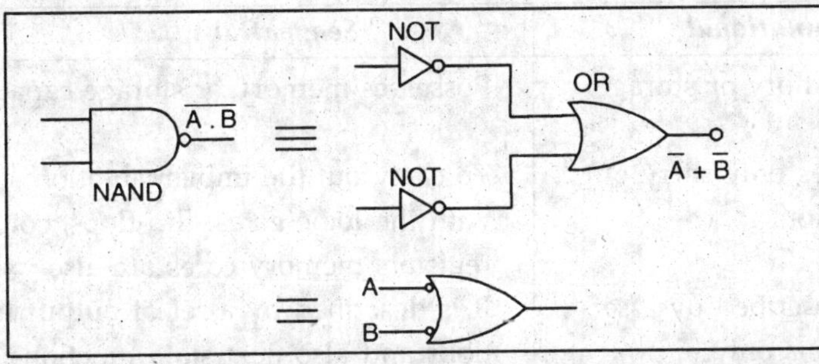

A	B	X
0	0	1
1	0	1
0	1	1
1	1	0

Fig. 7.48. De Morgan's Second theorem showing the equivalence of a NAND gate (same holds for the multiple inputs).

In fact the eqns. (7.7) and (7.8) also hold for the cases of the multiple (more than two) inputs.

$$\overline{A+B+C+.....} = \overline{A}\,.\,\overline{B}\,.\,\overline{C} \qquad ...[7.9(a)]$$

$$A\,.\,B.\,C\,..... = \overline{A}+\overline{B}+\overline{C} \qquad ...[7.9(b)]$$

The *purpose of these theorems is to enable digital circuit designers to implement all the other logic gates with the help of either NOR gates only or NAND gates only*. For example, a NOT gate is implementable by a NAND or a NOR as shown in the left part or lower right part of Fig. 7.47 respectively. This theorem finds wide use in the digital logic circuits as these are implementable on one single basic logic gate considered as a basic building unit.

- The *'first statement'* (De Morgan's) says that *the complement of a sum equals the product of the complements.* The *'second statement'* says that the *complement of a product equals the sum of the complements.* In fact, it allows transformation from a sum-of-products form to a product-of-sum form.

— The procedure required for *taking out an expression from under a NOT sign is as follows :*

1. *Complement the given expression i.e., remove the overall NOT sign.*
2. *Change all and ANDs to ORs and all the ORs to ANDs.*
3. Complement or negate all individual variables.

Examples : (*i*) $\overline{A + BC} = A + BC$ ...*Step 1*

$= A(B + C)$ ...*Step 2*

$= \bar{A}(\bar{B} + \bar{C})$ ...*Step 3.*

(*ii*) $\overline{(\bar{A} + B + \bar{C})(\bar{A} + B + C)} = (\bar{A} + B + \bar{C})(\bar{A} + B + C)$

$= \bar{A}B\bar{C} + \bar{A}BC = \bar{\bar{A}}\bar{B}\bar{\bar{C}} + \bar{\bar{A}}\overline{BC} = A\bar{B}C + A\overline{BC}$

This *process is called* **demorganization.**

— It may be noted that the opposite procedure would be followed to *bring an expression under the NOT sign :*

Example : $\bar{A} + \bar{B} + \bar{C} = \bar{\bar{A}} + \bar{\bar{B}} + \bar{\bar{C}}$ ...*Step 3*

$= A + B + C$

$= ABC$ ...*Step 2*

$= \overline{ABC}$ ...*Step 1*

Combinational	***Sequential***
1. Possesses no memory or storage capacity.	Possesses memory or storage capacity.
2. Simple logic gates only carry out the implementation.	To carry out the implementation along-with the logic gates, flip-flops, counters, registers memory cores are also used.
3. The system is described by a set of output functions only.	It is described by a set of output funct-tions and also next state functions.
4. Output of the system depends only on the present input.	Output of the system depends on the present input as well as on the present state of the system.

Combinational circuits :

A ***combinational circuit*** *consists of logic gates whose outputs at any time are determined directly from the combination of inputs without regard for previous input.*

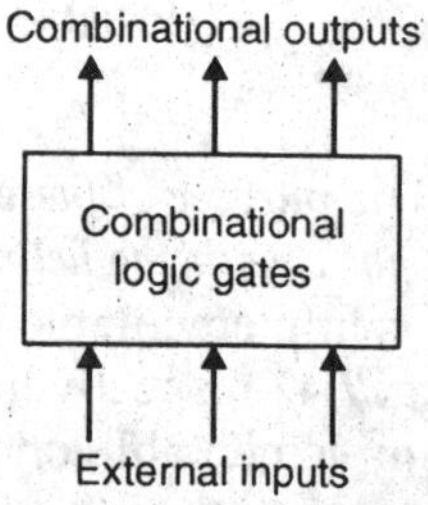

Fig.7.49. Combinational logic circuit.

The circuit possesses a set of *inputs*, a *memoryless logic network* to operate on the inputs and a *set of outputs* as shown in Fig. 7.49. Moreover, output combinational networks are used to make logical decisions and control the operation of different circuits in digital electronic systems. For a given set of input conditions, the output of such a circuit is the same. Consequently, truth table can fully describe the operation of such a circuit.

Examples. Examples of a combinational circuit are :

(*i*) Decoders (*ii*) Adders

(*iii*) Multiplexers (*iv*) Demultiplexers etc.

- **Multiplexers and demultiplexers :**
  — *Transmission of a large number of information units over a small number of lines is known as* **"Multiplexing".**
  — **"Demultiplexing"** is a *reverse operation and denotes receiving information from a small number of channels and distributing it over a large number of destinations.*

Design procedure of combinational circuit :

Following *operations* are involved in the design procedure :

1. To state the problem.
2. To determine the number of available input variables and required output variables.
3. To assign letter symbol to each input and output variable.
4. To derive the truth table that defines the required relationship between inputs and outputs.
5. To obtain the simplified Boolean function for each output.
6. To draw the logic diagram.

- A circuit that adds two bits is called a ***half adder.***
- A ***full adder*** consists of three inputs and two outputs. The outputs are designated by the symbol S for sum and C for carry.
- A ***two bit subtractor*** has two inputs X (minuend) and Y (subtrahend).
- A ***full subtractor* (FS)** is a *combinational circuit that performs a subtraction between two bits.* This circuit has *three inputs and two outputs.*

Code conversion :

- A variety of codes are used by different digital systems. It is sometimes necessary to use the output of one system as input to the other.
- A conversion circuit must be inserted between the two systems if each uses different codes for the same information.
- To convert from binary code A to binary code B, the input lines must supply the bit combinations of elements as specified by code A and the output lines must generate the corresponding bit combination of code B. A combination circuit performs this transformation by means of *logic gates.*

Comparator. A comparator is a combinational circuit that compares two number A and B and determine their relative magnitude. The outcome of the comparison is displayed in three outputs that indicate $A > B \equiv X$, $A = B = Y$, $A < B \equiv Z$.

Decoders and encoders :

— A ***decoder*** is a combination circuit that converts a binary code of n variables into m output lines, one for each discrete element of information.

— An ***encoder*** is a combination circuit that accepts m input lines, one for each element of information, and generates a binary code of n output lines.

Sequential Circuits :

Such circuits have *inputs, logic network, outputs* and a *memory,* as shown in Fig. 7.50. Their present output depends not only on their present inputs but also on the pevious logic states of the outputs.

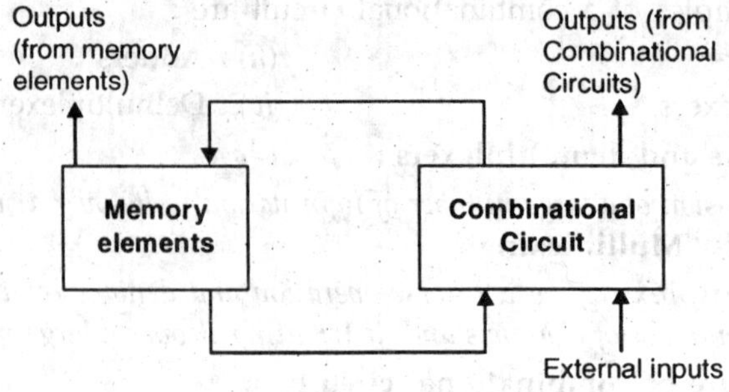

Fig. 7.50. Block diagram of a sequential circuit.

Examples. Examples of sequential circuits are :

(*i*) *Latches* (*ii*) *Flip-flops.*

The two main *types* of sequential circuits are :

1. Synchronous sequential circuits ..... referred to as *clocked-sequential circuits*
2. Asynchronous sequential circuits.

- The ***synchronous sequential circuits*** are *built to operate at a clocked rate whereas* **asynchronous ones** *are without clocking.*

7.4.10. Flip-Flop Circuits

The memory elements used in clocked sequential circuits are called **flip-flops.** These circuits are binary cells capable of storing one bit of information. It has two *outputs, one for the normal value and one for the complement value of the bit stored in it.* Binary information can enter a flip-flop in a variety of ways. Hence there are different types of flip-flops.

A number of flip-flops are available in *IC* form. Some of these are *SR* (Set-Reset), *J-K* and *D* flip-flops. They are widely used as *switches, latches, counters, registers* and *memory cells* in computers.

- A salient feature of the flip-flop is that *output can exist in one of the two stable states, logic 1 and logic 0, simultaneously.* This is ensured by the appropriate crossed feedback connections associated with the most elementary form of the flip-flop known as a *latch.*

The following flip-flops will be discussed in the following articles :

1. *R-S* flip-flop.
2. Clocked *R-S* flip-flop.
3. *D* flip-flop.
4. *J-K* flip flop.
5. *T* flip-flop.

R-S flip-flop :

Fig. 7.51 shows a *R-S flip-flop using* **NOR** *gates.* There are two inputs to the flip-flops called *S* (set) and *R* (*reset*). *The* cross-coupled connection from the output of one gate and input of the other constitutes a feedback path. For that reason, the *circuit is classified as synchronous circuit.*

- A low *R* and a high *S* results in the *set* state.
- A high *R* and a low *S* give the *reset* state.
- If both *R* and *S* are high, the output becomes *indeterminate* and this is known as *'race condition'.* This condition is *avoided by proper design.*

The truth table is shown in Table 7.9.

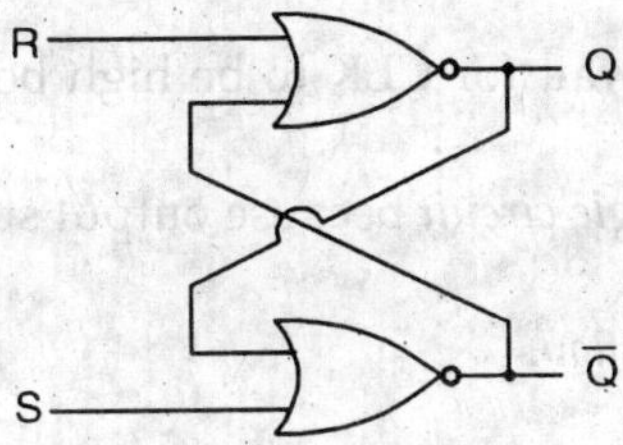

Table 7.9. Truth table for NOR latch

R	*S*	*Q*	*Comment*
0	0	NC	No change
0	1	1	Set
1	0	0	Reset
1	1		Race

(*a*) Circuit diagram (*b*) Truth table

Fig. 7.51. *R-S* flip-flop using NOR gates.

Fig. 7.52 shows a *R-S* flip-flop using **NAND** *gates.* Table 7.10 shows the truth table. It is seen that the inactive and race conditions are **reversed.**

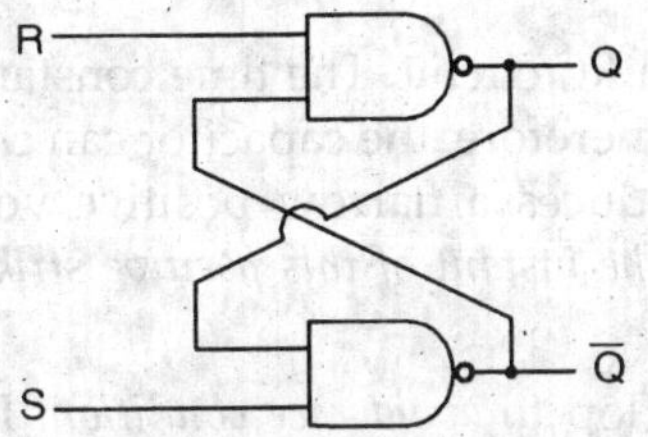

Table 7.10.Truth table for NAND latch

R	*S*	*Q*	*Comment*
0	0		Race
0	1	1	Set
1	0	0	Reset
1	1	NC	No change

(*a*) Circuit diagram (*b*) Truth table

Fig. 7.52. *R-S* flip-flop using NAND gates.

- When *R* is *low,* output *Q* is *high.*
- When *R* is *high,* output *Q* is *low.*
- When both *R* and *S* are low, we get race condition which must be *avoided.*
- When both *R* and *S* are high - *no change condition.*

Clocked R-S flip-flop.

A large number of flip-flops are used in a computer. In order *to coordinate their working a square wave signal* known as *clock* is applied to the flip-flop. This *clock signal (indicated as CLK) pevents the flip-flop changing state till the right instant occurs.*

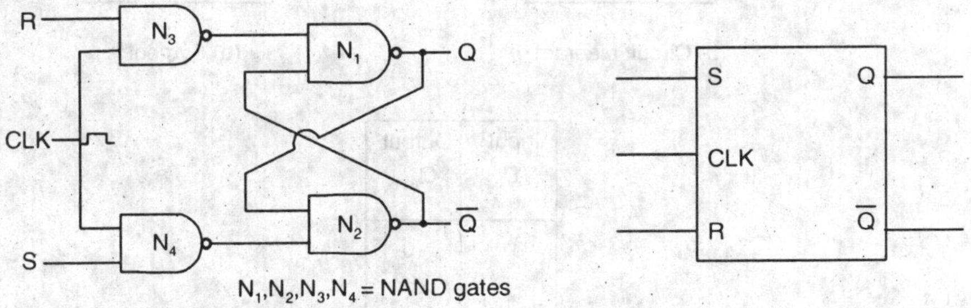

(a) Circuit diagram (b) Symbol

Fig. 7.53. Clocked *R-S* flip-flop

Fig. 7.53(*a*) shows a clocked R-S flip-flop using NAND gats (N_1 and N_2). This circuit uses two NAND gates N_3 and N_4 to apply CLK signal.

- When CLK is *low* the flip-flop *output Q indicates no change.*

- If S is high and R is *low*, the *flip-flop must wait till CLK becomes high before Q can be set on 1.*
- If S is low and R is low, the flip-flop must wait for CLK to be high before Q is reset to low (0).

Clocked *R-S* flip-flop is a *synchronous sequential logic circuit* because output state of the circuit changes at *discrete clocked instant of time.*

Fig. 7.53(*b*) shows a symbol for clocked *R-S flip-flop.*

Level clocking and edge triggering :

In a clocked flip-flop, the output can change state when CLK is high. When CLK is low, the output remains in the same state. Thus, the output can change state during the entire half cycle when CLK is high. This may be a *disadvantage in several situations. It is necessary that the output should change state only at one instant in the positive half cycle of the clock.* This *is known as* ***edge triggering*** *and the resulting flip-flop is known as* ***edge triggered flip-flop.***

Edge triggering can be made feasible by the use of an *RC* circuit. The time constant *RC* is made much smaller than the width of the clock pulse. Therefore, the capacitor can charge fully when CLK is high. The exponential charging produces a narrow positive voltage strike across the resistor. *The input gates are activated at the instant of this positive strike.*

D flip-flop :

A *D* flip-flop is an improvement over the *R-S* flip-flop to *avoid race condition.* It can be *level clocked or edge triggered.* The edge triggered one causes the change in output state at a unique instant.

In a *clocked R-S* flip-flop two input signals are required to drive the flip-flop which is a disadvantage with many digital circuits. In some events, both input signals become *high* which is again an undesirable condition. So these shortcomings/drawbacks of clocked *R-S* flip-flop are overcome in *D* flip-flop.

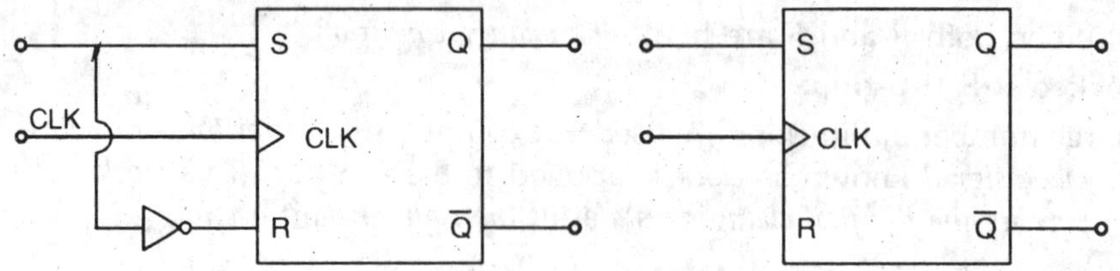

(b) Circuit diagram (b) Symbol

Input D_n	Output Q_{n+1}
0	0
1	1

(c) Truth table

Fig. 7.54. D flip-flop

Fig. 7.54(*a*), (*b*), (*c*) show the circuit diagram, symbol and truth table of *D* flip-flop respectively. It may be observed that only single data bit, *D* is required to drive the flip-flop.

— When the clock signal is at low level, data bit *D* is prevented to reach at output *Q* until clock signal becomes *high* at next pulse.

— It may be noted from the truth table that when data bit D_n is *high*, output Q_{n+1} gets at *high* level and when data bit D_n is low, Q_{n+1} gets at low level. Thus *D* flip-flop transfers the data bit *D* to *Q* as it is, and *Q* remains in the same state until the next pulse of the clock arrives.

- The flip-flop is named (*D*) flip-flop since *the transfer of data from the input to output is delayed.*
- The *D*-type flip-flop is *either used as a delay device* or *as a latch to store 1-bit of binary information.*

Edge triggered D flip-flop :

Fig. 7.55.(*a*) shows the circuit diagram and symbols of an edge triggered *D* flip-flop. The *clock provides the square wave signal. RC* circuit *converts* this signal *into strikes* so that triggering occurs at the instant of positive strike. The data bit *D drives one of the inputs.* Because of inverter, the complement $\overline{D}$ *drives the other output.* At the instant of positive strike, input *D* and its complement $\overline{D}$ cause the output *Q* to set or reset. Fig. 7.55(*b*) shows the truth table.

- When CLK is 0 or 1, the *D* input is not *there* and there is no change in state of *Q*.
- On the *negative* edge of the clock (marked ↓) the output remains in the same state.
- On the *positive* edge of the clock (marked ↑) *Q* changes to 0 if *D* is 0 and to 1 if *D* is 1.

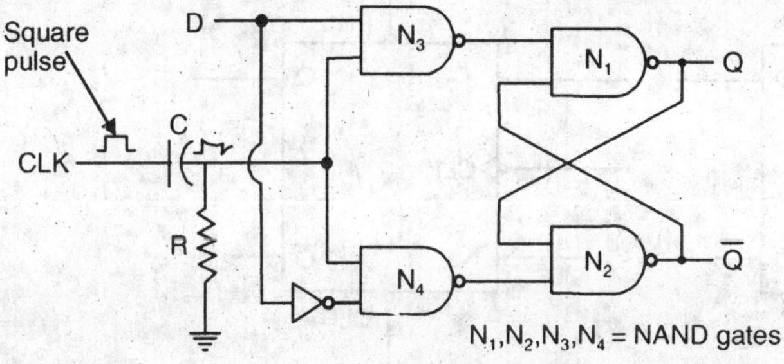

(a) Circuit diagram

CLK	D	Q
0	×	No change
1	×	No change
↓	×	No change
↑	0	0
↑	1	1

(b) Truth table

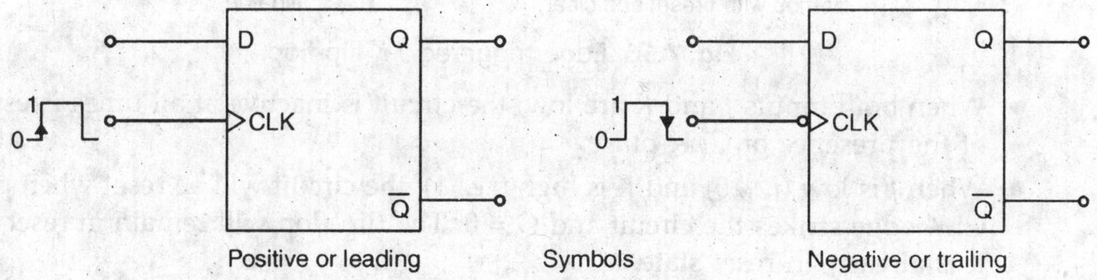

Fig. 7.55. Edge triggered *D* flip-flop.

Edge triggered J-K flip-flop :

— *J-K* fip-flop is *very versatile* and is perhaps the most *widely used type of flip-flop.*

— The *J* and *K* designations for the inputs have no known significance except that they are adjacent letters in the alphabet.

— *J-K* flip-flop *functions identically to R-S flip-flop.*

The difference is that the *J-K* flip-flop has no *invalid state* as does the *R-S* flip-flop.

— It is widely used in digital devices such as *counters, registers, arithmetic logic units, and other digital systems.*

Fig. 7.56(*a*) shows the circuit diagram of a edge triggered *J-K* flip-flop used in *digital counters.* The CLK input is through an *RC* circuit with a short time constant. The *RC* circuit converts the rectangular clock pulse to narrow spikes as shown. Due to double inversion through NAND gates, the circuit is positive edge triggered.

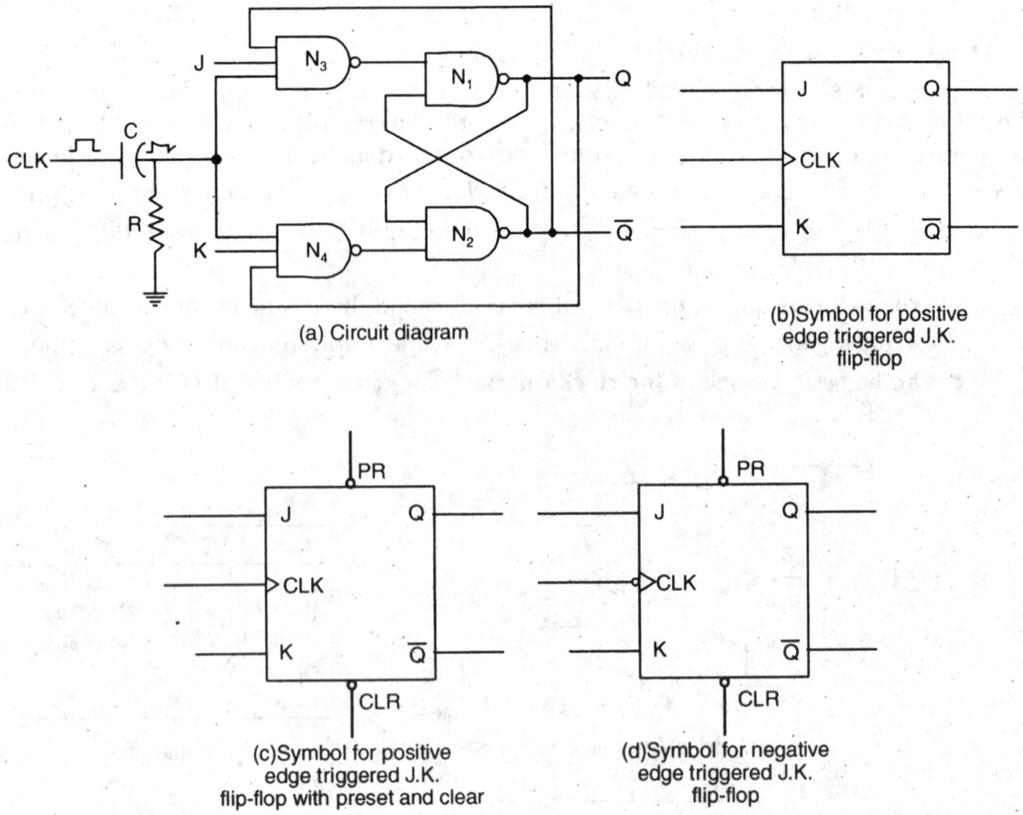

Fig. 7.56. Edge triggered *J-K* flip-flop.

- When both inputs *J* and *K* are low, the circuit is inactive at all times irrespective of the presence of CLK pulse.
- When *J* is low (*i.e.*, 0) and *K* is high (*i.e.*, 1), the circuit will be reset when positive CLK edge strikes the circuit and $Q = 0$. The flip-flop will remain in reset state if it is already in reset state.
- When $J = 1$ and $K = 0$, the circuit sets at the arrival of next positive clock edge.
- When $J = 1$ and $K = 1$, the flip-flop will toggle (means to switch to opposite state) on the next positive CLK edge. The action is illustrated in the table 7.11 :

Table 7.11. Positive edge triggered J-K flip-flop

CLK	*J*	*K*	*Q*
0	×	×	No change
1	×	×	No change
↓	×	×	No change
X	0	0	No change
↑	0	1	0 (reset)
↑	1	0	1 (set)
↑	1	1	toggle

Using of *RC* circuit for edge triggering is not very convenient for fabrication. Actual circuits use additional NAND gates for edge triggering, such circuits are known as *direct coupled circuit.*

— Fig. 7.56(*b*) shows the symbol for positive edge triggered *J-K* flip-flop.

— Fig. 7.56(*c*) shows a positive edge *J-K* flip-flop with present (*PR*) and clear (CLR).

— Fig. 7.56(*d*) shows the symbol for negative edge triggered *J-K* flip-flop with *PR* and CLR.

The *small bubble at CLR indicates negative triggering.*

T flip-flop :

T flip-flop is basically a *J-K* flip-flop, in this circuit input terminals *J* and *K* are connected with each other and this input is named as *T*.

Fig. 7.57(*a*) and (*b*) show the circuit diagram and symbol respectively of a trailing edge triggered *T* flip-flop.

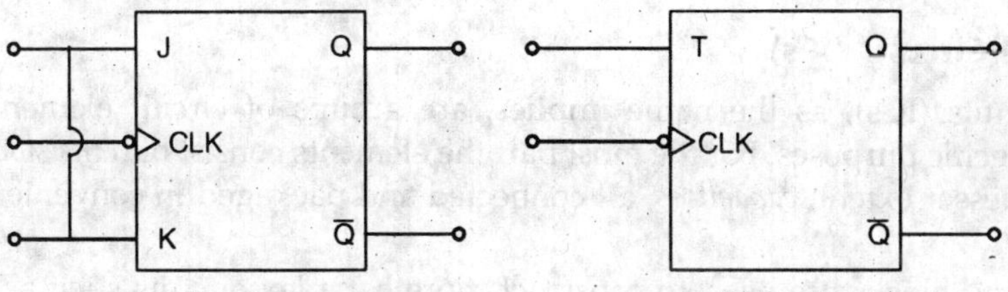

Input T_n	Output T_{n+1}
0	Q_n
1	$\bar{Q}_n$

(a) Circuit diagram (b) Symbol (c) Truth table

Fig. 7.57. Trailing edge triggered T flip-flop.

— When ***low** level signal* is applied to the input terminal *T*, then *initial state of output of flip-flop remains the same.*

— When ***high*** level signal is applied to the input terminal *T*, output of the flip-flop toggles after arrival of every new clock pulse. So the frequency of output signal is *half* of the clock signal frequency.

This flip-flop can be treated as frequency divider or a device which takes the input frequency at the clock terminal and divide it by *two.*

7.4.11. Counters

— A ***counter** is a sequential circuit that goes through a prescribed sequence of states upon the application of input pulses.*

— The input pulses, called *count pulses,* may be clock pulse or may originate from an external source and may occur at prescribed intervals of true or random.

— The sequence of states in a counter may follow a binary count or any other sequence of states.

— They are used for counting the number of occurrences of an event and are useful for generating time sequences to control operations in a digital system.

Straight binary sequence counter. It is the simple and most straight forward. An n-bit binary counter has n flip-fops and can count in binary from 0 to 2^{n-1}.

Binary ripple counter. It consists of a series connections of T flip-flops without any logic gates. Each flip-flop is triggered by the output of its preceding flip-flop goes from 1 to 0. The signal propagates through the counter in a *Ripple* manner, *i.e.*, the flip-flop essentially changes once at a time in rapid succession. It is the most simplest and most straight forward. It, however, has speed limitations ; an increase in speed can be obtained by the use of a parallel or a synchronous counter.

Synchronous 3-bit binary counter. In this all flip-flops are triggered simultaneously by count pulse. The flip-flop is complemented only if its T input is equal to 1.

Counter-decoder circuits :

— Counters together with decoders are used to generate timing and sequencing signals that control the operation of digital systems.

— The counter-decoder can be designated to give any desired number of repeated timing sequence.

Applications of counters :

The fundamental applications of counters are given below :

1. Measurement of time interval.
2. Direct counting.
3. Measurement of speed.
4. Measurement of frequency.

7.4.12. Integrated Circuits (ICs)

The integrated circuits (ICs), as the name implies, are groups of circuit elements combined to perform specific purposes. For the most part the elements consist of *transistors, diodes, resistors,* and, to lesser extent, *capacitors,* all connected and packaged in convenient plug-in units.

ICs from the building blocks are used to construct more complex circuits such as :

— Differential amplifiers;

— Mixers (for combining signals);

— Timers;

— Filters;

— Audio preamps;

— Auto-power amplifiers;

— Voltage references;

— Regulators and comparators;

— Several digital devices.

7.4.13. Operational Amplifiers (Op-amp)

An ***operational amplifier (Op-amp)*** *is a linear integrated circuit (IC) that has a very high voltage gain, a high input impedance and a low output impedance.*

It is so called because it can be employed to carry out many different mathematical operations like *"addition"*. *"subtraction"*, *"multiplication"*, *"division"*, *"integration"*, *"differentiation"* etc.

- Operational amplifiers are linear integrated circuits that work on *relatively low supply voltage.*
- They are *reliable and inexpensive.*
- An *ideal operational amplifier* is device of infinite voltage gain, infinite bandwidth, infinite input impedance (open) and zero output impedance.
- An Op-amp may contain two dozen transistors, a dozen resistors and one or two capacitors.

  Examples : μA 709, LM 108-LM 208, CA 741 CT and CA741T.

7.4.13.1. Specifications/Characteristics of an Op-amp

While selecting an Op-amp, the following characteristics need to be considered :

1. *Input offset voltage.* It is the voltage that must be applied at the input terminals to make the output voltage zero (This is about 2 mV for a 741 amplifier). The offset voltage changes with temperature.
2. *Input offset current.* It is defined as the net difference in current that must be applied at the input terminals to make the output voltage zero (This is 20 nA for a 741 amplifier).
3. *Input check current.* It is the mean of the two input currents to make the voltage zero.
4. *Slew rate.* It is the maximum rate at which the output can change. It is expressed as volts/microseconds.
5. *Unity gain frequency.* This is the frequency at which the open loop gain of the amplifier becomes unity.
6. *Common mode rejection ratio (CMRR).* It is the ratio of desirable signals to undesirable signals.

- An Op-amp is the *basic building block* for:

— Amplifiers
— Integrators
— Summers
— Differentiators
— Comparators
— A/D and D/A converters
— Active filters
— Sample and hold amplifiers.

7.4.13.2. Op-amp description

Fig. 7.58(*a*) shows a standard symbol (a triangle having two input labelled differently and a single output) for an Op-amp, the one shown in Fig. 7.58(*b*) is also oftenly used.

One input terminal is designated by *–ve* sign, it is called *inverting end* while other input terminal is designated by a *+ve* sign, it is called *non-inverting end.*

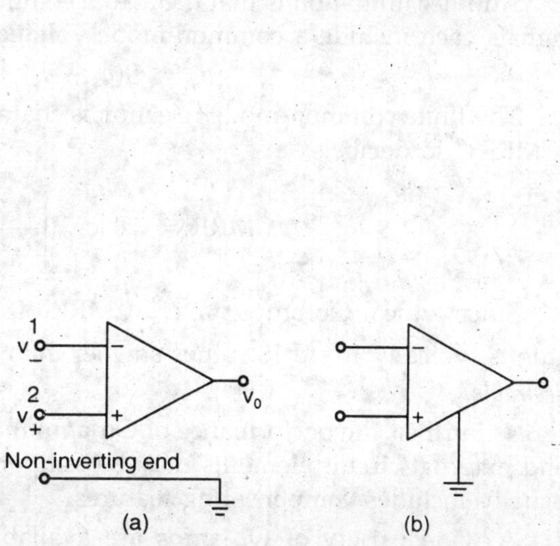

Fig. 7.58. Op-amp symbol.

These plus (+) and minus (–) polarities indicate *phase reversal* only. It does not mean that v_1 and v_2 [Fig. 7.58(*a*)] are negative and positive respectively. Additionally, it also *does not imply* that a positive input voltage has to be connected to the plus-marked non-inverting terminal 2 and negative input voltage to negative-marked inverting terminal 1. *In fact amplifier can be used either way up so to speak.* It may also be noted that all input and output voltages are referred to a common reference usually the ground shown in Fig. 7.58(*a*).

Operational amplifier operating with –ve feedback possesses stable closed loop gain and also insensitivity to variation of supply voltage and ambient temperature.

- The voltage at the output terminal v_o, is the product of the amplifier gain G, and the voltage difference :

$$v_o = G(v_+ - v_-) \quad \text{...(7.10)}$$

The output voltage is roughly limited to the power supply voltages V_{cc} and V_{ee}, as the voltage difference increases; if the voltage difference becomes too large, the output *saturates* near one of these values and remains constant.

- *The differential characteristic of op-amp has great importance in instrumentation because it eliminates offset voltages and noise signals common to both input terminals.* For example, nearby power lines may induce 50-cycle noise in the exterior circuitry leading to the amplifier. Such line noise is often present in identical form at both input terminals. This behaviour is known as *common-mode rejection.* If, instead, an op-amp receives the output of a voltage-sensitive Wheatstone bridge, the common offset voltages of the two voltage dividers are cancelled, and only the desired difference voltage is applied.

Limitations of Op-amp : Most of the Op-amps have a nonideal characteristic according to which they *do not completely satisfy the differential amplifying property.* With *both inputs grounded residual output voltage remains.*

— The multitude of transistors, resistors, and other elements within the Op-amp are *never perfectly matched,* so the amp output actually reaches zero at some small *non-zero* input voltage. To accommodate this input offset voltage, the common op-amp is provided with pins marked "offset null" or "balance" which provides a means for adjusting the unwanted offset voltage towards zero.

Another limitation is that the actual common-mode rejection is finite. If the two input signals each include a common-mode voltage v_{cm}, the Op-amp's actual voltage will be,

$$v_o = G(v_+ - v_-) + G_{cm}\, v_{cm} \quad \text{...(7.11)}$$

The finite common-mode rejection is characterised by the "*common-mode rejection ratio* (CMRR)" in decibels:

$$\text{CMRR} = 20 \log_{10}\left(\frac{G}{G_{cm}}\right) \text{dB} \quad \text{...(7.12)}$$

Since typical Op-amps have a CMRR of 60 to 120 dB, therefore, the common-mode gain is typically 10^3 to 10^6 times smaller than the differential gain, hence a *high CMRR is desirable.*

- Further, the performance of Op-amp may be limited by thermal drift. Both internal and external circuit elements may be temperature sensitive, and design of each circuit usually includes compensating features.

A wide variety of Op-amps are available, and their differences largely represent attempts to *improve:*

— Thermal stability;
— CMRR;
— Offset voltages;
— Frequency response .... These refinements, however, increase the cost.

7.4.13.3. Applications of Op-amp

Operational amplifiers may be used as the *basic components of* :

- Linear voltage amplifiers;
- Differential amplifiers;
- Integrators and differentiators;
- Voltage comparators;
- Function generators;
- Filters;
- Impedance transformers;
- Many other devices.

7.4.13.4. Op-amp circuits used in Instrumentation

Some of the commonly used Op-amp circuits are described below:

1. Inverter;
2. Adder;
3. Subtracter;
4. Multiplier and divider;
5. Integrator;
6. Differentiator;
7. Buffer amplifier;
8. Differential amplifier.

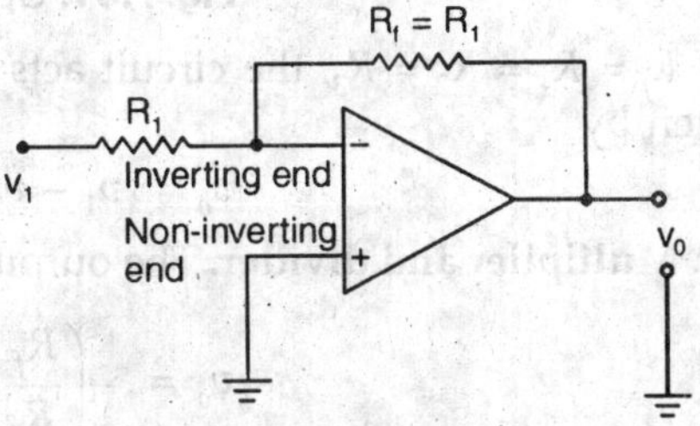

Fig. 7.59. Op-amp as an inverter.

1. Inverter. Fig. 7.59, shows the circuit of an Op-amp used as *inverter*. The feedback resistance R_f is made equal the resistance R_1, connected to the inverting end of the amplifier,

$$\text{Output voltage, } v_o = -\frac{R_f}{R_1}v_1 = -v_1 \qquad ...(7.13)$$

$$(\because R_f = R_1)$$

Obviously, the output voltage is 180° out of phase with the input voltage.

2. Adder. Fig. 7.60 shows an Op-amp circuit that performs the signals with amplification (if desired); using superposition theorem, we get

Output voltage,

$$v_o = -\left(\frac{R_f}{R_1}v_1 + \frac{R_f}{R_2}v_2 + \frac{R_f}{R_3}v_3\right) \qquad ...(7.14)$$

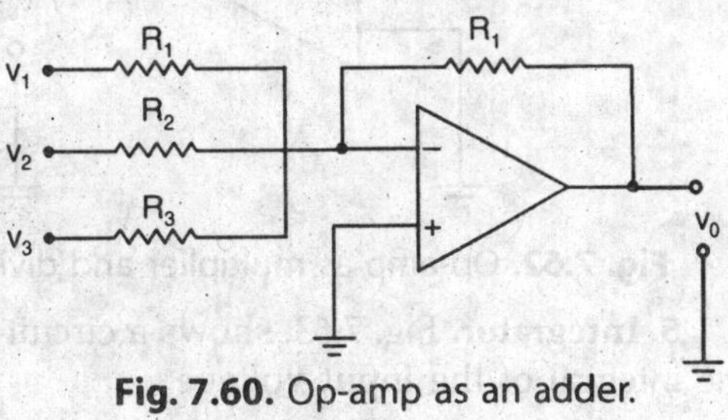

Fig. 7.60. Op-amp as an adder.

If $R_1 = R_2 = R_3 = R_f$, then

$$v_o = -(v_1 + v_2 + v_3) \qquad ...(7.15)$$

i.e., sum of the individual input voltages. The inversion that occurs cannot be avoided.

3. Subtracter. The Op-amp circuit used for subtraction of two input signals is shown in Fig. 7.61. The output of the 2nd Op-amp is given by :

$$v_o = -\left(-v_1 \frac{R_f}{R_1} \cdot \frac{R_f}{R_3} + v_2 \frac{R_f}{R_2}\right) \qquad ...(7.16)$$

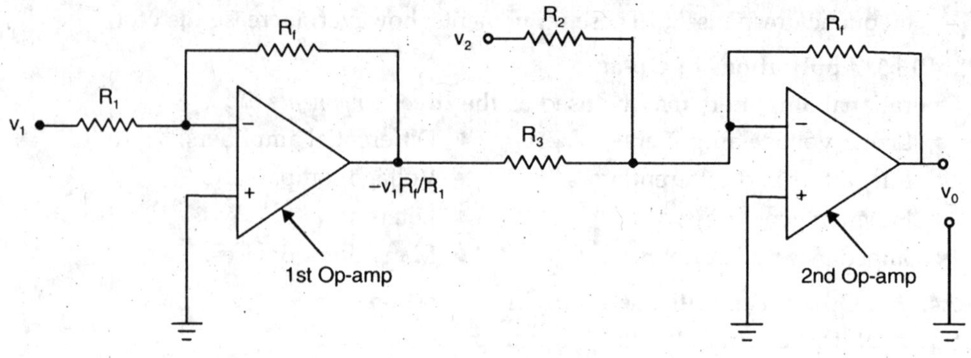

Fig. 7.61. Op-amp as a subtractor.

If $R_f = R_1 = R_2 = R_3$, the circuit acts as a *pure subtractor* and the output, in this case, is given by

$$v_o = v_1 - v_2 \qquad ...(7.17)$$

4. Multiplier and divider. The output of an Op-amp in the inverting mode is given by

$$v_o = -\left(\frac{R_f}{R_1}\right) \times v_1$$

In case $R_f > R_1$, the circuit shown in Fig. 7.62 acts as a multiplier and in case $R_f < R_1$ it acts as a divider.

Thus, by choosing the values of R_f and R_1, the multiplier and the divider circuits can be designed for multiplication and division by any number.

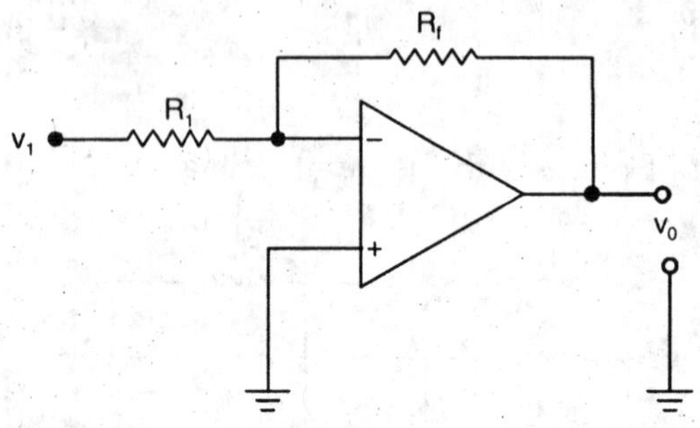

Fig. 7.62. Op-amp as multiplier and divider.

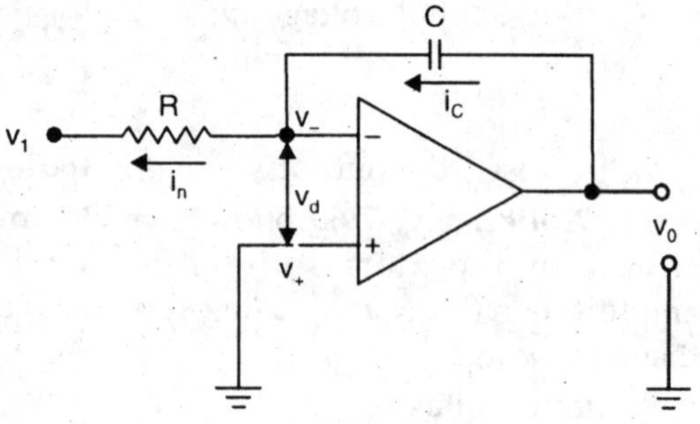

Fig. 7.63. Op-amp as an integrator.

5. Integrator. Fig. 7.63. shows a circuit in which the output voltage is proportional to the integral of the input voltage.

In order to show that the circuit shown in Fig. 7.63. acts as an integrator using KCL (Kirchoff's Current Law) at node v_-, we have,

$$i_n = i_C$$

or, $$\frac{v_- - v_1}{R} = C\frac{d}{dt}(v_o - v_-)$$

For *infinite differential gains*, $v_- = 0$

$$\therefore \quad -\frac{v_1}{R} = C\frac{d}{dt}v_o$$

By integration, we have

$$v_o = -\frac{1}{RC}\int v_1\, dt \quad ...(7.18)$$

The convenient values of R and C are MΩ and μF range respectively.

6. Differentiator. The differential amplifier circuit is obtained by *interchanging* the positions of resistance R and capacitor C as shown in Fig. 7.64.

At node v_-, we have :

$$i_C = i_R$$

$$C\frac{d}{dt}(v_- - v_1) = \frac{v_o - v_-}{R}$$

Now, $\quad v_- = 0$

$$\therefore \quad -C\frac{d}{dt}(v_1) = \frac{v_o}{R}$$

or, Output voltage, $v_o = -RC\frac{d}{dt}(v_1)$ ...(7.19)

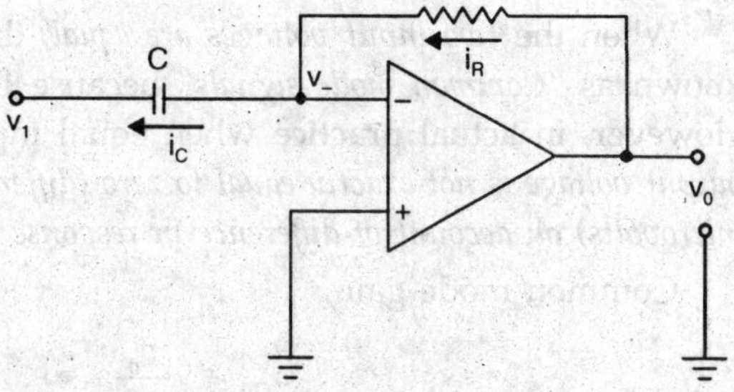

Fig. 7.64. Op-amp as differentiator.

Thus, the output voltage is equal to the differentiated input voltage.

- The *Op-amps are normally used as differentiators as they tend to decrease the signal to noise (S/N) ratio.*

7. Buffer amplifier. The buffer amplifier is essentially an impedance transformer which converts a voltage at high impedance to the same voltage at low impedance. The circuit of a unity gain buffer amplifier also called a *"voltage follower"* is shown in Fig. 7.65.

- *The use of unity gain buffer amplifiers greatly reduces the loading effects in measurements systems.*

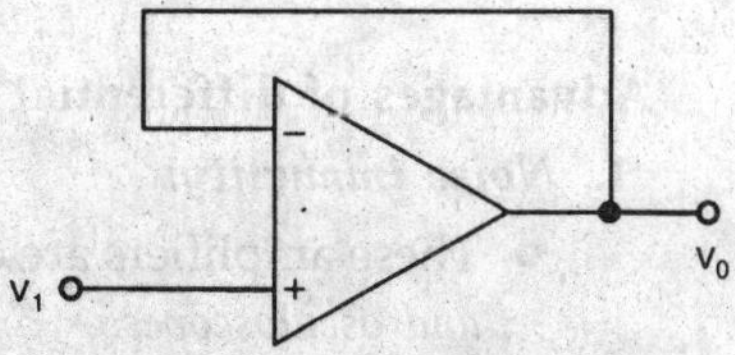

Fig. 7.65. Unity gain buffer amplifier voltage follower.

8. Differential amplifier. A differential amplifier (an Op-amp) is of significant importance in an instrumentation system.

In its basic form it has two *inputs* and *outputs.* The signals available to the two outputs are identical except that the two are 180° out-of-phase with each other.

The output voltage of the amplifier is proportional to the difference between the two input voltages.

Fig. 7.66, shows an Op-amp used as a differential amplifier.

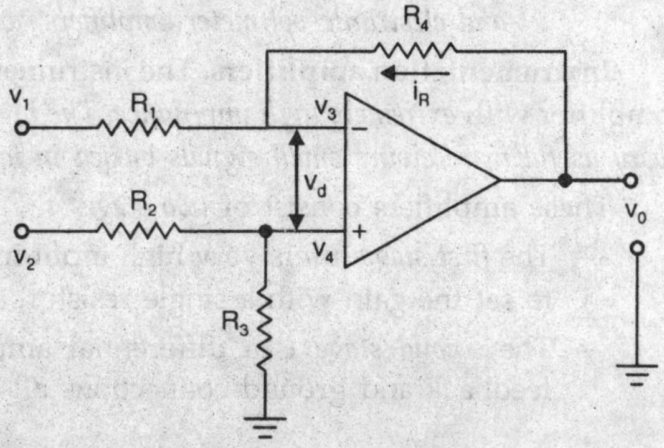

Fig. 7.66. Op-amp used as a differential amplifier.

Here, $v_o = G_d(v_+ - v_-)$

where, G_d = Differential gain.

The signal $v_d = (v_+ - v_-)$ is called *"Difference Mode Signal"* or simply *"Difference Signal"*

$$G_d = \frac{v_o}{v_+ - v_-} = \frac{v_o}{v_d} \quad ...(7.20)$$

It can be proved that,

$$v_o = G_d(v_2 - v_1) \quad ...(7.21)$$

When the *two input voltages are equal,* the output voltage is zero. Equal inputs are known as *"Common mode signals"* because the input signal is common to both inputs. However, in actual practice when equal input voltages are applied to the inputs, the *output voltage is not exactly equal to zero (difference is typically of the order of several hundred microvolts) on account of difference in response of the two inputs to common mode signals.*

Common mode gain,

$$G_{cm} = \frac{v_o}{v_{cm}} \quad ...(7.22)$$

where, G_{cm} = Common mode gain, and

v_{cm} = Common mode input signal.

The *"Common mode rejection ratio (CMRR)"* is defined as

$$CMRR = \frac{G_d}{G_{cm}} \quad ...(7.23)$$

Also,

$$CMRR = 20 \log_{10}\left(\frac{G_d}{G_{cm}}\right) \text{dB} \quad ...(7.23a)$$

... when expressed in *dB.*

Advantages of differential amplifiers :

1. ***Noise immunity:***
   - These amplifiers are extensively used in equipment such as *electronic voltmeters and oscilloscopes.*
2. ***Drift immunity :***
   - The differential amplifier *has inherent capabilities of eliminating problem of drift.*
   - The differential amplifier construction is used for the *early stages of oscilloscope and electronic voltmeter amplifiers, where low drift is extremely important.*

Instrumentation amplifiers. The instrumentation amplifier is a dedicated differential amplifier with *extremely high impedance. The high common mode rejection makes this amplifier very useful in receiving small signals buried in large common-mode offsets and noise.*

These amplifiers consist of *two stages:*

— The *first stage* offers very high input impedance to both input signals and allows to set the gain with a single resistor.

— The *second stage* is a differential amplifier (unity gain) with output, negative feedback and ground connections all throughout.

7.5 MICROPROCESSORS

7.5.1. Microprocessor—General Aspects

7.5.1.1. Definition and brief description

A ***microprocessor*** *is a large scale integration (LSI) chip that is capable of performing arithmetic and logic functions as defined by a given programme.* This system by itself does not form an operational computer, and additional circuit for memory and input/output must be supplied and interfaced with the system. The software (it is the programme for controlling the operation of the microprocessor itself) is also to be provided.

Or

A state machine on a single IC chip with very large scale integration, capable at a desired instant of working as per programme or an instruction of a programme, and which is driven by a clock of frequency of 1 MHz or more, is called a ***microprocessor***. Such machine is also called a central processing unit (CPU). A CPU forms main part of a computer.

The ***microprocessor*** consists of the following *three segments* (See Fig. 7.67).

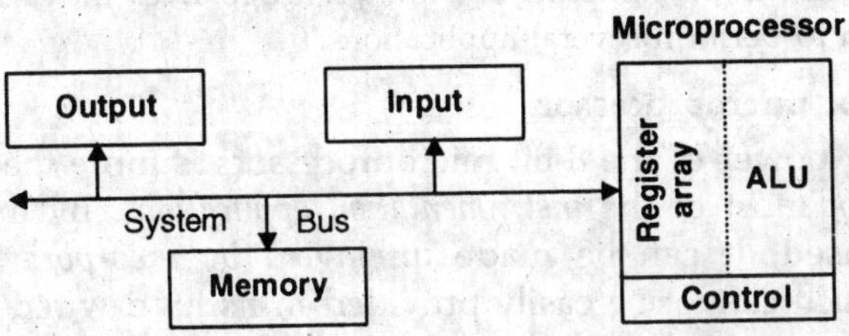

Fig. 7.67. Block diagram of a microcomputer.

1. **Arithmetic/Logic Unit (ALU).** In this area of the microprocessor, *computing functions are performed on data*. The ALU performs arithmetic operations such as addition and subtraction, and logic operations such as AND, OR and exclusive OR. Results are stored *either* in registers or in memory or sent to output devices.
2. **Register unit.** This area of the microprocessor consists of various registers. The registers are used primarily *to store data temporarily during the execution of a program.* Some of the registers are accessible to the user through instructions.
3. **Control unit.** The control unit provides the necessary timing and control signals to all the operations in the microcomputer. It *controls the flow of data between the microprocessor and peripherals including memory.*

In short a *microprocessor performs the following functions* :

- Communicates with all peripherals (memory and I/O) using system.
- Controls timing of information flow.
- Performs the computing tasks specified in a programme.

7.5.1.2. Characteristics of microprocessor

In nearly every type of design, with any complexity at all, microprocessors have potential for *drastically reducing component count and shortening design time.* In fact a microprocessor is considered to represent long-awaited next generation of digital building blocks, and that microprocessor will provide the best single approach to the system-level digital integrated circuit.

Some of the characteristics of a microprocessor are listed below :

1. It *handles shorter words than other computers,* usually 4 to as many as 16 bits.

2. It consists of *integrated circuits* from 1 to 30 in number.
3. It contains arithmetic logic unit (ALU), registers, control, random access memory (RAM), data buses and read only memory (ROM) with programmes.

7.5.1.3. Important features

The important features of the microprocessors are :

1. Low cost
2. Small size
3. Low power consumption
4. Versatile (The versatility of a microprocessor results from its 'stored programme' mode of operation).
5. Extremely reliable.

Note: Probably the term 'micro' in the name of the device can be contributed to its *low cost, small size and low power consumption.* The processing capability of a microprocessor should not, however, be underestimated. Currently available 32-bit microprocessors have a processing power similar to that of the mainframe computer of a few years ago. Even the early 8-bit microprocessors are powerful enough to perform several applications.

7.5.1.4. Uses of microprocessor

The processing power of the 8-bit microprocessors is more than adequate to satisfy the requirements of most of the *instrumentation applications.* By making an instrument microprocessor-based, it can be made *intelligent by incorporating new features like programmability* which cannot be easily provided in its hard-wired counterpart.

Some important *uses* of microprocessors in instrumentation area are listed below :

1. Frequency meters.
2. Function generators.
3. Frequency synthesizers.
4. Spectrum synthesizers.
5. *Intelligent instruments CRT terminals*
6. Digital millimeters.
7. Oscilloscopes.
8. Counters.
9. **Process control**
   — Instrumentation
   — Monitoring and control
   — Data acquisition
   — Logging and processing.
10. **Medical Electronics**
    — Patient-monitoring in intensive care unit
    — Pathological analysis
    — Measurement of parameters like blood pressure and temperature.

Under this heading the following instruments/machines are included:

(*i*) Microprocessor based medical instrument.
(*ii*) Microprocessor based ECG machines.
(*iii*) Microprocessor based EEG machines etc.

Other applications of microprocessors :

(*i*) High level language computers.

(*ii*) Replacing hard-wired logic by a microprocessor.

(*iii*) Control of automation and continuous processes.

(*iv*) Computer peripheral controllers.

(*v*) Home entertainment and games.

(*vi*) Inventory control system, pay roll banking etc.

7.5.2. Microprocessor Systems

Microprocessor systems consist of the following *three* parts :

1. *Central processing unit (CPU) :*
   — This part uses the microprocessor.
   — It recognises and carries out program instructions.
2. *Input and output interfaces :*
   — These interfaces handle communications between the computer and the outside world.
   — For the interface, the term *port* is used.
3. *Memory :*
   — Its function is to hold the program instructions and data.

- *"Microprocessors" which have memory and various input/output arrangements on the same chip are called* ***microcontrollers.***

7.5.2.1. The microprocessor

The microprocessor (generally referred to as CPU) is that part of the processor system which carries out the following functions :

(*i*) Processes the data ;

(*ii*) Fetches instructions from memory;

(*iii*) Decodes and executes the instructions.

Following are the various parts of a microprocessor :

1. *Arithmetic and logic unit (ALU):* This part of the microprocessor *manipulates the data.*

2. *Registers:* Registers are memory locations within the microprocessor and are employed to store information involved in program execution. The various types of registers used are :

(*i*) *Accumulator register :*

— It is a temporary holding register for data to be operated on by the arithmetic and logic unit and also, after the operation, the register for holding the results.

— It deals with all data transfers associated with the execution of arithmetic and logic operations.

(*ii*) *Status register :*

— The status register (also called flag register or condition code register) contains information concerning the result of the latest process carried out in the ALU.

— It contains individual bits with each bit having special significance; the bits are called flags.

— The status of the latest operation is indicated by each flag with each flag being set or reset to indicate a specific status.

(*iii*) *Program counter register (PC) :*

— This register, also called instruction pointer (IP), contains the address of the memory location that contains the next program instruction.

— This register is *updated* after the execution of each instruction, so that it contains the memory location where the next instruction to be executed is stored.

(*iv*) *Memory address register (MAR) :*

— This register contains the address of data.

(*v*) *Instruction register (IR) :*

— This register stores an instruction.

— The control processing unit (CPU), after fetching an instruction from the memory via the data bus, stores it in the instruction register. The microprocessor, after each such fetch, increments the program counter by one with the result that the program counter points to the next instruction waiting to be fetched. The instruction can then be decoded and used to execute an operation. This sequence is known as *fetch-execute cycle.*

(*vi*) *General purpose registers :*

— These registers serve as temporary storage for data or addresses and *used in operations involving transfers between other registers.*

(*vii*) *Stack pointer register (SP) :*

— The *stack* is a special area of the memory in which program counter values can be stored when a subroutine part of the program is being used.

— The contents of this register form an address which defines the top of the stack in RAM.

The number and form of the registers depends on the microprocessor concerned.

7.5.2.2. Buses

***Buses** are the paths along which digital signals move from one section to another.*

— A bus is just a number of conductors along which electrical signals can be carried. It might be tracks on a printed circuit board or wires in a ribbon cable.

In a microprocessor system there are the following three forms of bus :

1. Data bus;
2. Address bus;
3. Control bus.

1. *Data bus :*

— The data bus carries the data associated with the processing function of the CPU.

— Word lengths used may be 4, 8, 16, 32 or 64.

— Each wire in the bus carries a binary signal, *i.e.,* a0 or a1.

— The more wires the data bus has the longer the word length that can be used.

- The earliest microprocessors were 4-bit (word length : $2^4 = 16$) devices and such 4-bit microprocessors are still used in such devices as toys, washing machines etc. They were followed by 8-bit microprocessors (*e.g.,* Motorola 6800, the Intel 8085 A and the Zilog Z80). Now 16-bit, 32-bit and 64-bit microprocessors are available.

2. *Address bus :*

— It carries signals which indicate where data is to be found and so the selection of certain memory locations or input or output ports.

— Each storage location within a memory device has a unique identification, termed its address, so that system is able to select a particular instruction or data item in the memory.

— Each input/output interface also has an address.

— When a particular address is selected by its address being placed on the address bus, *only that location is open to the communications from the CPU.* The CPU is thus *able to communicate with just one location, at a time.*

— A computer with an 8-bit data has typically a 16-bit wide address bus, *i.e.*, 16 wires. This size of address enables 2^{16} locations to be addressed. 2^{16} is 65 536 locations and is usually written as 64 K, where K is equal to 1024.

3. *Control bus :*

— This bus *carries the signals relating to control actions.*

— It is also used *to carry the system clock signals;* these are to synchronise all the actions of the microprocessor system.

7.5.2.3. Memory

In a microprocessor, the memory unit stores binary data and takes the form of one or more integrated circuits (ICs).

— The data may be program instruction codes or numbers being operated on.

— *The size of the memory is determined by the number of wires in the address bus.*

Following are the various forms of memory unit :

1. ROM
2. PROM
3. EPROM
4. EEPROM
5. RAM.

1. *ROM :*

— ROM (*Read Only Memory*) is a memory device in which *data is stored permanently.*

— While manufacturing the integrated circuit ROMs are programmed with the required contents. *No data then can be written into this memory chip in the computer.* The data *can only be read and is used for fixed programs;* memory is not lost when power is removed.

● *Programs stored in ROM are termed as **firmware**.*

2. *PROM :*

— PROM (Programmable ROM) is used for ROM chips that *can be programmed by the user.*

3. *EPROM :*

— EPROM (Erasable and programmable ROM) is employed for *ROMs that can be programmed and their contents altered.*

— A typical EPROM chip contains a series of small electronic circuits, cells, which can store charge.

— The program is stored by applying voltages to the integrated circuit connecting pins and producing a pattern of charged and uncharged cells. The pattern remains permanently in the chip until *erased by shining ultraviolet light through a quartz window on the top of the device.*

4. ***EEPROM :***

— EEPROM (Electrically erasable PROM) is similar to EPROM; erasure, however, is done by the application of a relatively high voltage rather than using ultraviolet light.

5. ***RAM :***

— RAM (Random-access memory) is a read/write memory in which data currently being operated on (temporary data) is stored.

— Such a memory can be read or written to.

— When RAM is used for program storage then such programs are referred to as ***software.*** When the system is switched on, software may be loaded into RAM from some other peripheral equipment such as a keyboard or hard disc or floppy disc.

Difference between a software of a computer and a microprocessor :

In computer software is loaded into the computer at the beginning of each computation, software in microprocessor is stored within the computer itself in a ROM chip. The modification of the program is achieved by merely replacing ROM IC with another ROM IC containing a different control program. This is a very notable advantage of software implementation in microprocessor.

7.5.2.4. Input/Output

*The transfer of data between the microprocessor and the external world is termed as the **input/ output operation.***

— The pieces of equipment that exchange data with a microprocessor systems are called ***peripheral devices.***

— In *input operations* the input device places the data in the data register of the interface chip; this holds the data until it is read by the microprocessor. In *output operations* the microprocessor places the data in the register until it is read by the peripheral.

7.5.3. Intel 8085 Microprocessor

7.5.3.1. Brief history

— Intel Corporation in early seventies introduced the first microprocessor, *Intel 4004.* This microprocessor was a single chip device which was capable of performing simple arithmetic and logic operations such as addition, subtraction, comparison, AND, and OR. Its control unit could perform various functions such as fetching of an instruction from the memory, decoding it and generating control pulses for executing it. It was a 4-bit microprocessor operating upon 4-bits of data at a time.

— Intel introduced *4040* as modified version of microprocessor 4004.

— Intel, later on, introduced 8-bit microprocessors called 8008 and 8080 which could perform arithmetic and logic operations on 8-bit words.

— These days, modified and better version of 8-bit microprocessor is **Intel 8085** which is most widely used and most popular microprocessor.

Now-a-days 12-bit, 16-bit and 32-bit microprocessors are also available.

Fig. 7.68, *shows the block diagram of Intel 8085 microprocessor.*

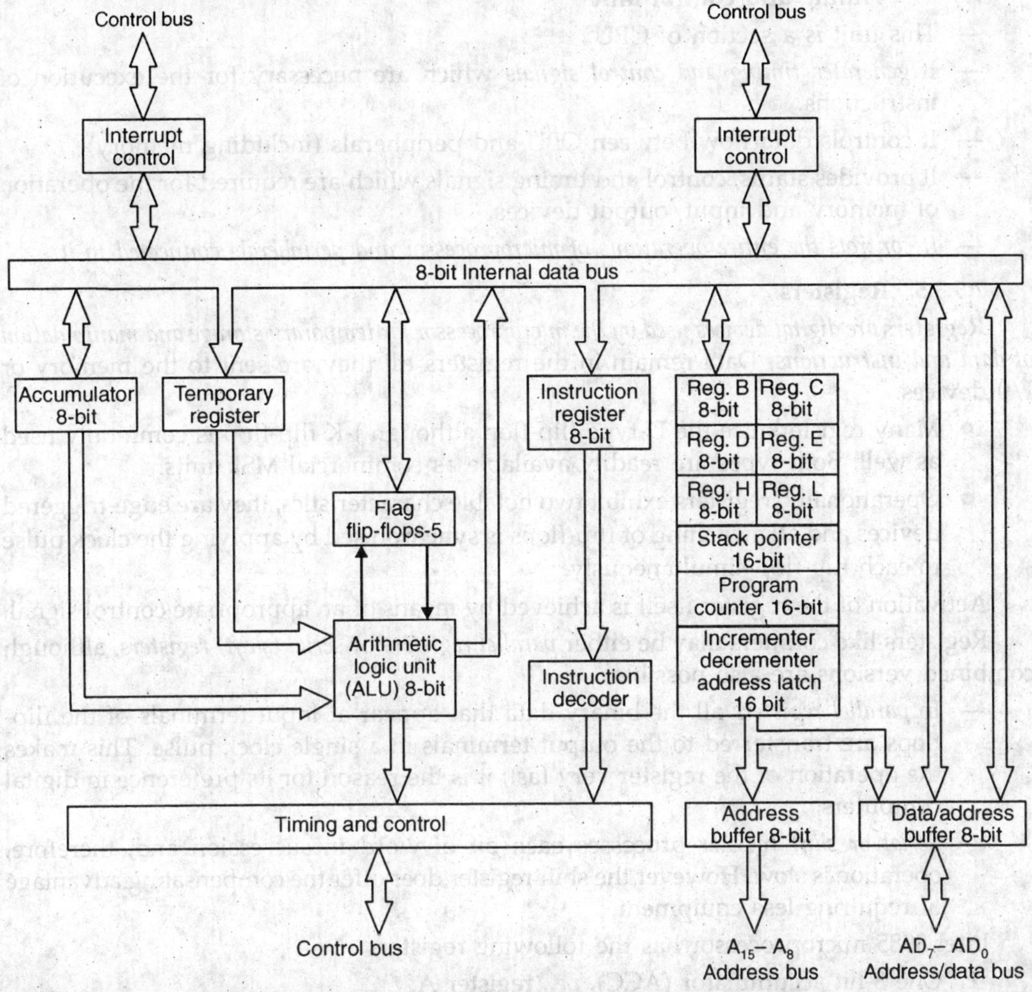

Fig. 7.68. Block diagram of Intel 8085 microprocessor.

7.5.3.2. Introduction

- Intel 8085 is an 8-bit, NMOS microprocessor.
- It is a 40 pin I.C. package fabricated on a single LSI chip.
- It uses a single +5 $V_{D.C.}$ supply for its operation.
- Its clock speed is about 3 MHz. The clock cycle is of 320 ns. The time for the clock cycle of the Intel 8085 AH-Z version is 200 ns. It has *80 basic instructions and 246 op-codes.*

7.5.3.3. Arithmetic and logic unit (ALU)

ALU performs the following arithmetic and logical operations :

1. Addition
2. Subtraction
3. Logical AND
4. Logical OR
5. Logical EXCLUSIVE OR
6. Complement (logical NOT)
7. Increment (add 1)
8. Decrement (subtract 1)
9. Left shift, Rotate left, Rotate right.
10. Clear etc.

7.5.3.4. Timing and control unit

— This unit is a section of CPU.

— It *generates timing and control signals* which are necessary for the execution of instructions.

— It controls data flow between CPU and peripherals (including memory).

— It provides status, control and timing signals which are required for the operation of memory and input/output devices.

— *It controls the entire operations of microprocessor and peripherals connected to it.*

7.5.3.5. Registers

Registers are digital devices used by the microprocessor for temporary storage and manipulation of data and instructions. Data remain in the registers till they are sent to the memory or I/0 devices.

- Many registers use the D-type flip-flop although J-K flip-flop is commonly used as well. Both types are readily available as commercial MSI units.
- Operationally, registers exhibit two notable characteristics: they are edge-triggered devices and all switching of flip-flops is synchronised by applying the clock pulse to each flip-flop simultaneously.

Activation of the register itself is achieved by means of an appropriate control signal.

Registers like counters may be either *parallel registers* or *serial (shift) registers*, although combined versions are also possible.

— In *parallel registers* all the binary data that appear at input terminals of the flip-flops are transferred to the output terminals in a single clock pulse. This makes the operation of the register very fast; it is the reason for its preference in digital computers.

— *Serial or shift register* processes each bit of word in succession and, therefore, operation is slow. However the shift register does offer the compensating advantage of requiring less equipment.

Intel 8085 microprocessor has the following registers :

1. One 8-bit accumulator (ACC), *i.e.*, register A.
2. Six 8-bit general purpose registers (B, C, D, E, H and L).
3. One 16-bit stack pointer, SP.
4. One 16-bit program counter, PC.
5. Instruction register.
6. Temporary register.

These registers are described below :

1. ***Accumulator (ACC) :***
   - It is an 8-bit special purpose register that is a part of the ALU. It is also identified as register A.
   - In arithmetic and logical operations the accumulator may store the operand, execute an instruction with the help of other registers, and memory and finally store the result of the operation. In the former case it acts as a source, and in the latter a destination.
2. ***General purpose registers :***
   - The 8085 microprocessor contains six 8-bit general purpose registers. These are identified as B, C, D, E, H and L as shown in Fig. 7.68.

- These registers are used in microprocessor for temporary storage of operands or intermediate data in calculations.
- These registers can be used either simply for storage of 8-bit data or in pairs for storage of 16-bit data. When used in pairs, only selected combination can be used for pairing, *i.e.*, B-C, D-E and H-L. When registers are used in pairs the high order byte resides in the first register and low order byte in the second register.

3. ***Stack pointer (SP) :***
   - It is a 16-bit special function register.
   - The *stack* is a sequence of memory locations set aside by a programmer to store/retrieve the contents of accumulator, flags, program counter and general purpose registers during the execution of a program. Any portion of the memory can be used as a stack.
   - In this register, data is stored temporarily on first come and last go basis.
4. ***Program counter (PC) :***
   - It is a 16-bit special-purpose register and is used to hold the memory address of the next instruction to be executed.
   - The contents of the PC are automatically updated by the microprocessor during the execution of an instruction so that at the end of execution it points to the address of the next instruction in the memory.
   - The microprocessor uses the PC for sequencing the execution of instructions.
5. ***Instruction register :***
   - During the execution of a program, microprocessor addresses some memory, which supplies an 8-bit data of instruction code to the data bus which gets stored in the register called the *instruction register.*
   - The instruction register holds the op-code (operation code or instruction code) of the instruction which is being decoded and executed.
6. ***Temporary register :***
   - It is an 8-bit register associated with ALU.
   - It holds data during an arithmetic/logical operation.
   - It is used by the microprocessor and is not accessible to programmer.

Flags. There are five-flops in Intel 8085 microprocessor to serve as status flags; these are : (*i*) Carry Flag (CS), (*ii*) Parity Flag (P), (*iii*) Auxiliary Carry Flag (AC); (*iv*) Zero Flag (Z), and (*v*) Sign Flag (S).

— The flip-flops are set or reset according to the conditions which arise during an arithmetic or logical operation.

— If a flip-flop for a particular flag is set, it indicates 1. When it is reset, it indicates 0.

Instruction decoder. Data from the instruction register is sent to the instruction decoder, where microprocessor decodes it and then translates into specific actions.

7.5.3.6. Data and address bus

The data bus of Intel 8085 microprocessor is 8-bit wide and hence, 8 bits of data can be transmitted in parallel from or to the microprocessor. This microprocessor requires a 16-bit wide address bus as the memory addresses are of 16-bits. The 8 *most significant* bits of the address are transmitted by the address bus, A-bus (pins A_8 to A_{15}). The 8 *least significant* bits of the address are transmitted by address/data bus, AD-bus (pins AD_0–AD_7).

The address/data bus transmits data and address at different movements. At particular moment it transmits either data or address. Thus *AD-bus operates in time shared mode. This technique is called **multiplexing**.*

7.5.3.7. Pin configuration

Fig. 7.69. shows the schematic diagram of Intel 8085 microprocessor.

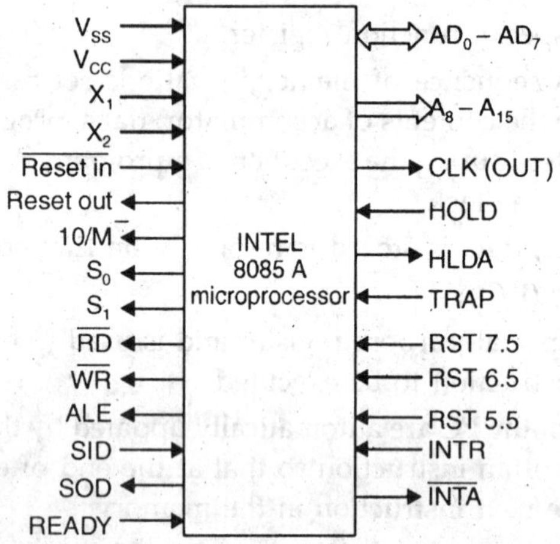

Fig. 7.69. Schematic diagram of Intel 8085 microprocessor.

ADo–AD$_7$* (*Input/Output): They are used for the least significant 8-bits of the memory address of I/O address during the first clock cycle of a machine cycle. Again they are used for data during second and third clock cycles.

A_8–A_{15} (output): These are address bus and are used for the most significant bits of the memory address of 8-bits of I/O address.

IO/$\bar{M}$ (Output): It is a status signal which distinguishes whether the address is for memory or I/O. When it goes high the address on the address bus is for an I/O device, whereas, when it goes low the address on the address bus is for a memory location.

ALE (Output): It is an *address latch enable signal.* It goes high during first clock cycle of a machine cycle and enables the lower bits of the address to be latched either into the memory or external latch.

S_0, S_1 (Output): These are status signals sent by the microprocessor to distinguish the various types of operations given in the table below,

S_1	S_0	*Operations/Machine cycle*
0	0	HALT
0	1	WRITE
1	0	READ
1	1	FETCH

$\overline{RD}$ (Output) :

— It is a signal to control READ operation.

— When it goes low the selected memory or I/O device is read.

$\overline{WR}$ (Output) :

— It is a signal to control WRITE operation.

— When it goes low the data on the data bus is written into the selected memory or I/O location.

READY INPUT :

— It is used by the microprocessor to sense whether a peripheral is ready to transfer data or not.

— If READY is high the peripheral is ready, if it is low the microprocessor waits till it goes high.

— A slow peripheral may be connected to the microprocessor through READY line.

HOLD (Input) :

— It indicates that another device is requesting for the use of the address and data bus.

— The microprocessor after having received a HOLD request relinquishes the use of the buses as soon as the current machine cycle is completed. Internal processing may continue.

— The processor regains the bus after the removal of the HOLD signal.

HLDA (Output) :

— This signal indicates that the HOLD request has been received.

— After the removal of a HOLD request the HLDA goes low. The CPU takes over the buses half clock cycle after the HLDA goes low.

INTR (Input); $\overline{INTA}$:

— INTR signal indicates an interrupt request.

— The INTR line is sampled in the last state of the last machine cycle of an instruction. The microprocessor acknowledges the interrupt signals and issues an $\overline{INTA}$ signal.

$\overline{RESET\ IN}$ (Output) :

— When this signal is applied the program counter is set to zero and resets the Interrupt Enable and HLDA flip flops. Except the instruction register no other register or flag is affected.

— The CPU remains in the reset condition as long as reset is applied.

RESET OUT :

— This is an output signal which shows that CPU is being reset. This can be used as a system RESET.

— The signal is synchronised to the microprocessor clock.

X_1, X_2 (Input) :

— X_1 and X_2 are the terminals to be connected to an external crystal oscillator which drives an internal circuitry of the microprocessor to produce a suitable clock for the operation of microprocessor.

CLK (Output) :

— It is a clock output for user, which can be used for other digital ICs.

— Its frequency is same at which processor operates.

SID and SOD line :

— SID is an input line and it is for serial input data. The serial input data at SID can be identified by an instruction called RIM and serial data can be an instruction called SIM.

— In Intel 8085 microprocessor only serial transmission facility is available. The SID and SOD lines pursuit the input and output serial data. The actual transfer of the data is accomplished by software using the RIM and SIM instructions. Both these instructions are single byte and are also used to read or set/reset interrupt masks.

7.5.3.8. Opcode and operands

Each instruction contains the following *two parts* :

(*i*) Operation code (opcode);

(*ii*) Operand.

- ***Opcode*** specifies the task to be performed by the computer.
- ***Operand*** is the data to be operated on.
  - — The operand (or data) given in the instruction may be in various forms such as 8-bit or 16-bit data, 8-bit or 16-bit address, internal registers or a register or memory location.
  - — In some instructions the operand is implicit. When the operand is a register it is understood that data is the content of the register.

7.5.3.9. Instruction cycle

An ***instruction*** *is a command given to the computer to perform a specified operation on given data.*

In order to perform a particular task a programmer writes a sequence of instructions, called a *program*. Program and data are stored in the memory. The CPU fetches one instruction from the memory at a time and executes it. It executes all instructions of a program one by one to produce the final result.

An ***instruction cycle*** consists of a *fetch cycle* and *execute cycle*. The total time required to execute an instruction is the some of time required to fetch an opcode and the time required to execute it.

— The opcode (the Ist byte of an instruction is its opcode; the instruction may be more than one byte long) fetched from the memory goes to the data register, DR (data/address buffer in Intel 8085 microprocessor) and then to instruction register, IR. From the instruction register it goes to the decode circuitry which *decodes* the instruction. The decoded circuitry is within the microprocessor.

— After the instruction is decoded, *execution begins.* If the operand is in the general purpose registers, execution is immediately performed. The time taken in decoding and execution is one clock cycle. If an instruction contains data or operand address which are still in the memory, the CPU has to perform some read operations to get the desired data. After receiving the data it performs execute operation.

A *read cycle* is similar to a fetch cycle. In case of a read cycle the quantity received from the memory are data or operand address instead of an opcode. In some instructions write operation is performed.

In write cycle data are sent from CPU to the memory or an output device.

In some cases an execute cycle may involve one or more read or write cycles or both.

— The necessary steps carried out to perform a fetch, a read or write operation constitute a *"Machine cycle"*. An instruction cycle consists of several machine cycles.

7.5.3.10. Microprocessor programming

Prior to performing to task, a microprocessor has to be programmed (*Program* is a sequence of instruction that operates the microprocessor on a certain data to achieve desired results).

- The program written for performing a particular task, is stored in the semiconductors memory that is accessible to the microprocessors.
- During the execution of the program, microprocessor fetches one instruction at a time from the memory and executes it.
- Microprocessor understands only instructions written in sequence by using 0s and 1s, and this type of program is known as *machine language program.* These types of programs are very difficult to write. So *first of all programs are written is assembly language using mnemonic operation codes and symbolic addresses.* After that this program is *translated into machine language programme* manually or by using some special translator known as an *assembler.*

7.5.4. Programmable Logic Controllers (PLCs)

7.5.4.1. Introduction

***PLCs** are specialised industrial devices for interfacing to and controlling analog and digital devices.*

— They are designed with a small instruction set suitable for industrial control applications.

— They are usually programmed with *"ladder logic"*, which is graphical method of laying out the connectivity and logic between system inputs and outputs.

— They are designed with industrial control and industrial environments specifically in mind. Therefore, in addition to being *flexible and easy to program, they are robust and relatively immune to external interference.*

- A programmable logic controller (first conceived in 1968), is shown in Fig. 7.70. *It is a "digital electronic device" that uses a programmable memory to store instructions and to implement functions such as logic sequencing, timing, counting and arithmetic in order to control machines and processes.*

It has been specifically designed to *make programming easy.*

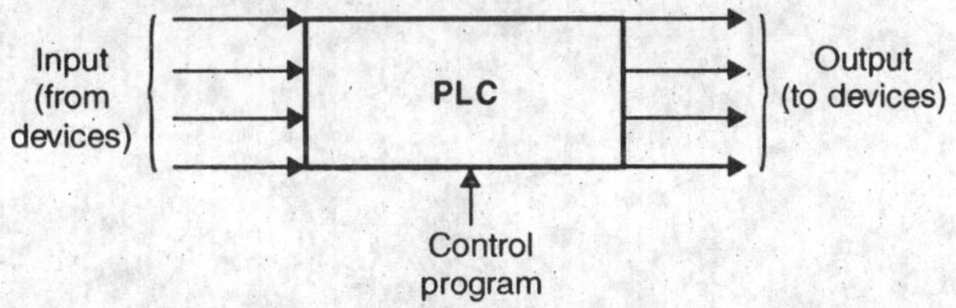

Fig. 7.70. Programmable logic controller

Advantages :

(*i*) The primary ***advantage*** of the PLCs is that it is *possible to modify a control system without having to rewire the connections to the input and output devices,* the only requirement being that an operator has to *key in different set of instructions.*

(*ii*) PLCs are also *much faster than relay-operated systems.*

Uses. PLCs are widely used and extend from small-contained units for use with perhaps 20 digital inputs/outputs to modular systems which can be used for large numbers of inputs/outputs, handle digital or analog inputs/output, and also carry out PID control modes.

7.5.4.2. Special features

Although PLCs are similar to computers, yet they have the following *specific features* to their use as controllers:

1. The interfacing for inputs and outputs is *inside* the controller.
2. *Easily programmable.* They have an easily understood programming language.
   — Programming is mainly concerned with *logic* and *switching operation.*
3. *Rugged* and designed to withstand vibrations, temperatures, humidity and noise.

7.5.4.3. Architecture basic structure

Fig. 7.71 shows the architecture/internal structure of a programmable logic controller (PLC):

A PLC consists of the following *main components* :

1. Central processing unit (CPU);
2. Memory;
3. Input/Output circuitry.

1. *Central processing unit (CPU) :*

- It controls and processes all the operations within the PLC.

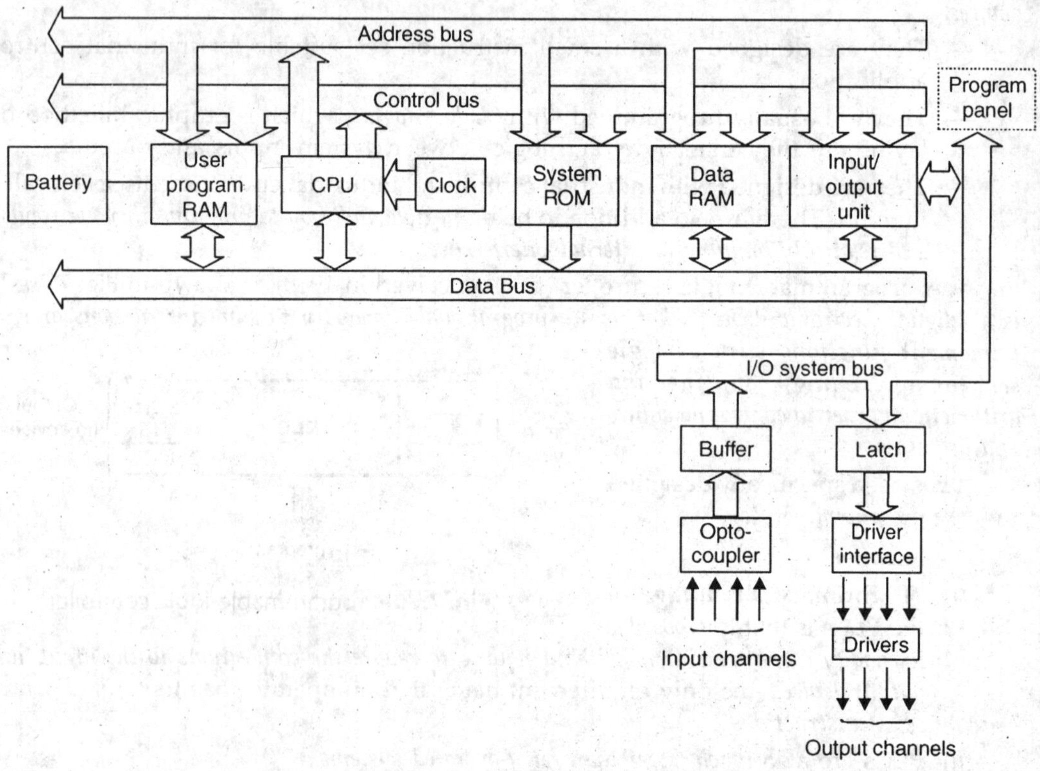

Fig. 7.71. Architecture of a programmable logic controller (PLC).

- It is provided with a *"clock"* with a frequency of typically between 1 and 8 MHz. This frequency determines the operating speed of the PLC and provides the timing and synchronisation for all elements in the system.

- A *"bus system"* carries information and data to and from the CPU, memory and input/output units.

2. ***Memory:*** The various memory elements available in a PLC are:
   (*i*) A system ROM to give permanent storage for the operating system and fixed data.
   (*ii*) RAM for user's program.
   (*iii*) Temporary buffer stores for input/output channels.
   - The programs in RAM can be changed by the user. However, to prevent the loss of these programs when the supply is switched off a battery is likely to be used in the PLC to maintain the RAM contents for a period of time.
   - Specifications for small PLCs often specify the program memory size in terms of the number of program step (A program step is an instruction for some event to occur) that can be stored. Typically the number of steps that can be handled by a small PLC is of the order of 300 to 1000, which is generally more than adequate for most control situations.

3. ***Input/Output (I/O) circuitry:*** The I/O unit provides the interface between the system and outside world.
   - Programs are entered into the I/O unit from a panel which can vary from small keyboards with liquid crystal displays to those using a visual display unit with keyboard and screen display. The programs, alternatively, can be entered into the system by means of a link to a personal computer which is loaded with an appropriate software package.
   - The I/O channel provides signal conditioning and isolation functions so that sensors and actuators can be generally directly connected to them without the need for other circuitry.
   - The basic form of programming commonly used with PLCs is *ladder programming.* This involves each program task being specified as though a rung of a ladder.

   Following methods can be used for *I/O processing :*

   1. Continuous updating.
   2. Mass I/O copying.

Timers: The timers are commonly regarded as relays with coils which, when energised, result in the closing or opening of input contacts after some preset time.

A *timing circuit* is specified by stating the interval to be timed and the conditions or events that are to start and/or stop the timer.

- PLCs are generally provided with only *delay-on timers, i.e.,* a timer which comes on after a time delay.

Internal relays:

- These relays are often used when there are programs with multiple input conditions.
- The internal relays are also used for the starting of multiple outputs.

Counters: The use of counters is restored to when there is *a need to count a specified number of contact operations.*

- Counter circuits are supplied as an internal feature of PLCs.

Shift registers: Several internal relays can be grouped together to form a *register* which can provide a storage area for a series sequence of individual bits. Thus a 4-bit and a 8-bit registers would be formed by using four and eight internal registers respectively.

The term *shift register* is used because the bits can be shifted along by one bit when there is a suitable input to the register.

Shift registers have *three inputs* :

- One to load data into the first element of the register (OUT);
- One as the shift command (SFT);
- One for resetting (RST).

7.5.4.4. Selection of a PLC

For selection of a PLC, the following criteria need to be considered:

1. *Types of inputs/outputs* required, such as:
   — Isolation;
   — Out-board power supply for inputs/outputs;
   — Signal conditioning.
2. *Input/Output capacity* required.
3. *Size of memory* required. This is linked to the number of inputs/outputs and the complexity of program used.
4. *Speed and power* required for CPU—This is linked to the number of types of instructions that can be handled by a PLC.

Difference between PLCs and Microcomponents :

Differences between PLCs and microcomputers :

S.NO	*Aspects*	*PLCs*	*Microcomputers*
1.	*Interfacing*	• They come with I/O modules for different voltage levels, especially designed to connect in the industrial environment. • These are easily interfaced with hundreds of input and output lines.	• They can be interfaced with external equipment using special circuit cards and interfaces. • It is difficult to interface hundreds of input and output lines in microcomputers.
2.	*Programming*	• These are ususally programmed using relay ladder logic diagrams and high–level language.	• These are programmed using machine language, assembly language and high–level language.
3.	*Data processing and calculation*	• These are able to accept analog data and perform simple arithmetical operations.	• These are not able to accept analog data; they required A/D converters. • They can perform complicated mathematical operations.
4.	*Application*	• These are especially designed for *industrial environments* where they can be exposed to heat, corrosive atmosphere, mechanical shock, vibration etc.	• They are not built to operate reliably under industrial conditions. • They are preferable where a great deal of data processing is required, such as *robot control and where the number of I/O lines is limited.*

QUESTION WITH ANSWERS

Q. 7.1. Enlist the characteristics which should be kept in mind while selecting a transdurcer.

Ans. A transdurcer should possess the following *characteristics* :

1. Ability to recoginse and sense the desired input signal and be sensitive to other signals.
2. Good accuracy
3. Good precision
4. Amplitude linearity.
5. Environment capability.

Q. 7.2. State the advantages of PLCs.

Ans. The *advantage* of PLCs are :

1. Fast operational speed.
2. Low cost.
3. Easy to install on the system.
4. Small space requirements.
5. High reliability.
6. Ability to withstand harsh environments.
7. Expandability.
8. Consistency in manufacturing.

Q. 7.3 What are the applications of microprocessors?

Ans. Microprocessors are widely used in contolling the devices; some of the typical examples are:

1. Temperature control.
2. Strain measurement.
3. Speed control of D.C. motors.
4. Protection relays.
5. Frequency measurement.
6. Traffic light control.
7. Tuning control.
8. Firing angle measurement.
9. On-off of power main supply.

Q. 7.4 Name common input-output devices and standard interfaces.

Ans. ***Common input / output devices :***

- Digital–to–analog converters (DAC).
- Analog–to–digital converters (ADC).
- Magnetic tape drives.
- Magnetic disk drives.
- VDU terminals.
- Line printers.

Standard interfaces :

- RS–232 interface.
- IEEE–488 interface.

Q. 7.5 (*a*) What is mechatronics? Explain briefly.

(*b*) What are the components of a mechatronic system?

(*c*) What are the advantages and disadvantages of mechatronics?

Ans. (*a*) "***Mechatronics***" may be defined as follows :

"The synergistic combination of precision mechanical engineering, electronic control and systems thinking in the design of products and manufacturing processes."

Or

"Integration of electronics, control engineering and mechanical engineering".

Or

"The interdisciplinary field of engineering dealing with the design of products whose function relies on the synergistic integration of mechanical and electronic components co-ordinated by a control architecture."

- *"Mechatronics"* involves a number of technologies such as :
  - — Mechanical engineering ;
  - — Electronic engineering ;
  - — Electrical engineering ;
  - — Computer technology ;
  - — Control engineering.

  This can be *considered to be the application of computer-based digital control techniques, through electronic and electric interfaces to mechanical engineering problems.*
- It represents the *next generation machines,* **robots** and *smart mechanisms for carrying out work in a variety of environments–predominantly factory automation, office automation and home automation.*

(*b*) **Components of a mechatronic system :**

The term *mechatronic system* (sometimes referred to as 'smart device') encompasses a myriad of devices and systems. Increasingly, *microcontrollers are embedded in the electromechanical devices, creating much more flexibility and control possibilities in system design.*

Fig. 7.72 shows all components in a typical *"mechatronic system"*.

- The **actuators** *produce motion or cause some action ;*
- The **sensors** *detect the state of the system parameters, inputs and outputs ;*
- **Digital devices** *control the system ;*
- **Conditioning and interfacing circuits** *provide connection between the control circuits and the input/output devices ;*
- **Graphical displays** *provide visual feedback to users.*

(*c*) **Advantages and disadvantages of mechatronics**

Following are the *advantages* and *disadvantages* of mechatronics :

Advantages :

1. The products produced are cost effective and of very good quality.
2. The performance characteristics of mechatronics products are such which are otherwise very difficult to achieve without the synergistic combination
3. High degree of flexibility.
4. A mechatronics product can be better than just sum of its parts.

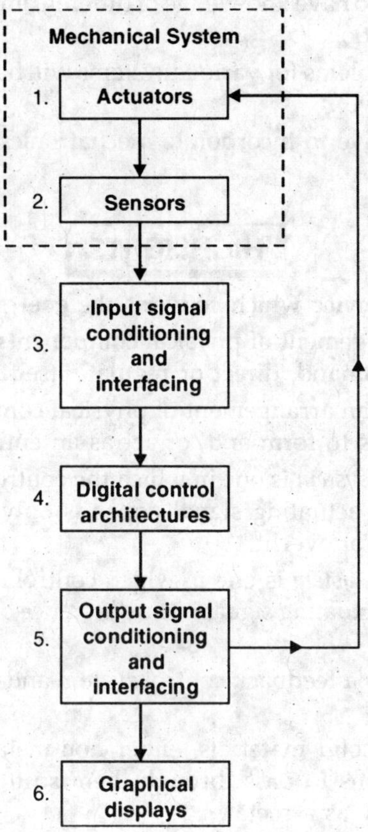

1. **Actuators :** Solenoids, voice coils ; D.C. motors ; Stepper motors ; Servomotor ; hydraulics; pneumatics.
2. **Sensors :** Switches ; Potentiometer ; Photoelectrics ; Digital encoder ; Strain guage ; Thermocouple ; accelerometer etc.
3. **Input signal conditioning and interfacing :** Discrete circuits ; Amplifiers, Filters ; A/D, D/D.
4. **Digital control architectures :** Logic circuits ; Microcontroller ; SBC ; PLC ; Sequencing and timing ; Logic and arithmetic ; Control algorithms ; Communication.
5. **Output signal conditioning and interfacing :** D/A, D/D ; Amplifiers ; PWM ; Power transistors ; Power Op-amps.
6. **Graphical displays :** LEDs ; Digital displays ; LCD ; CRT.

Fig. 7.72. Components of a typical "mechatronic system".

5. Greater extent of machine utilisation.
6. Due to the integration of sensors and control systems in a complex system, capital expenses are reduced.
7. Owing to the incorporation of intelligent, self correcting sensory and feed-back systems, the *mechatronic approach results in* :
   — greater productivity ;
   — higher quantity and producing reliability.

Disadvantages :

1. High initial cost of the system.

2. Imperative to have knowledge of different engineering fields for design and implementation.
3. Specific problems for various systems will have to be addressed separately and properly.
4. It is expensive to incorporate mechatronics approach to an existing/old system.

HIGHLIGHTS

1. A transducer is a device which converts the energy from one form to another.
2. A *system* is an arrangement of physical components connected or related in such a manner as to command, direct or regulate itself or another system.
3. An *control system* is an arrangement of physical components connected or related in such a manner as to form and/or act as an entire circuit.
4. An *open-loop control system* is one in which the control action is independent of the desired ouput. The actuating signal depends only on the input command and output has no control over it.
5. A *closed-loop control system* is one in which control action is somehow dependent on the output. The actuating signal is the difference of desired output and reference input.
6. A *servo-mechanism* is a feedback control system and used to control position or its derivative.
7. *Semiconductors* are solid materials, either non-metallic elements or compounds, which allow electrons to pass through them so that they conduct electricity in much the same way as a metal.
8. A pure semiconductor is called *instrinsic semiconductor.*
9. The process of adding impurity (extremely in small amounts, about 1 part in 10^8) to a semiconductor to make it extrinsic (impure) semiconductor is called *doping*.
10. The *N*- and *P*-type materials represent the basic building blocks of semiconductor devices.
11. The outstanding property of *P-N* junction diode to conduct current in one direction only permits it to be used as a *rectifier.*
12. *P-N* junction diodes usually made of germanium or silicon, are commonly used as *power rectifiers.*
13. A properly doped *P-N* junction diode which has a sharp breakdown voltage is known as *Zener diode.*
14. *Tunnel diode* is a heavily doped *P-N* junction type germanium having an extremely narrow junction.
15. A *"transistor"* is a semiconductor device having both rectifying and amplifying properties.
16. The two basic types of transistors are :
    (*i*) Bipolar junction transistor (BJT)
    (*ii*) Field-effect transistor (FET).
17. *Transistor circuit configurations :*
    (*i*) Common-base (CB) configuration.

(*ii*) Common-emitter (CE) configuration.

(*iii*) Common-collector (CC) configuration.

18. A *FET* is a three terminal (namely drain, source, gate) semiconductor device in which current conduction is by only one type of majority carriers (electrons in case of an *N*-channel FET or holes in a *P*-channel *FET*).
19. In a broad sense, following are two main types of FETs :
    (*i*) JFET (*ii*) MOSFET
20. Metal oxide semiconductor FET (MOSFET) is an important semiconductor device and is widely used in many circuit applications. It is also called insulated gate FET (IGFET).
21. An electronic circuit that handles only a digital signal is called a *digital circuit.* In digital circuits the following four systems of arithmetic are often used : *Decimal, Binary, Octal, Hexadecimal.*
22. A digital circuit with one or more input signals but only one output signal is called a *logic gate.*

    In the complex circuits, the following *six* different digital electronics gates are used as basic elements :

    (*i*) NOT gate (*ii*) NAND gate
    (*ii*) AND gate (*iv*) OR gate
    (*v*) NOR gate (*vi*) XOR gate.

OBJECTIVE TYPE QUESTIONS

Choose the Correct Answer :

1. LVDT is a
   (*a*) capacitive transducer (*b*) resistive transducer
   (*c*) inductive transducer (*d*) none of them
2. The size of air-cored transducers in comparison to their iron-cored counter parts is
   (*a*) smaller (*b*) bigger
   (*c*) same (*d*) unpredictable.
3. LVDT windings are wound on
   (*a*) steel sheets (laminated) (*b*) aluminium
   (*c*) ferrite (*d*) copper.
4. Piezoelectric crystals are used for measurement of .... changes.
   (*a*) static (*b*) dynamic
   (*c*) static and dynamic (*d*) any of these.
5. Piezoelectric crystals produce an e.m.f.
   (*a*) When external mechanical force is applied
   (*b*) when external magnetic field is applied
   (*c*) when radiant energy stimulates the crystal
   (*d*) when the junction of two such crystals is heated.
6. Piezoelectric crystals are used for the measurement of
   (*a*) temperature (*b*) velocity
   (*c*) sound (*d*) none of the above.
7. The transducers used for measurement of linear displacement are
   (*a*) strain gauges and resistive potentiometers

(*b*) LVDTs, capacitive transducers and Hall effect transducers
(*c*) thermocouples, thermistors and RTDs
(*d*) both (*a*) and (*b*).

8. Rotational displacement can be measured by
(*a*) strain gauges (*b*) resistive potentiometers
(*c*) reluctance transducers (*d*) both (*b*) and (*c*).

9. Capacitive transducers operate upon the principle (s) of
(*a*) variation of over-lapping area of plates
(*b*) variation of separation of plates
(*c*) variation of relative permittivity of dielectric material between two plates
(*d*) all of the above.

10. Capacitive transducers are normally employed for .... measurements.
(*a*) static (*b*) dynamic
(*c*) both static and dynamic (*d*) transient.

11. The thermo-electric effect was first observed by
(*a*) Seebeck (*b*) Thomas Young
(*c*) Pirani (*d*) Thermus.

12. Which of the following is valid for both P-N-P as well as N-P-N transistors?
(*a*) The emitter injects holes into the base region
(*b*) The electrons are the minority carriers in the base region
(*c*) The EB junction is forward biased for active operation
(*d*) When biased in the active region, current flows into emitter terminal.

13. In a transistor with normal bias
(*a*) the emitter junction offers high resistance (*b*) the emitter junction is reverse biased
(*c*) the emitter junction has a low resistance (*d*) none of the above.

14. Which region of a transistor is lightly doped?
(*a*) Collector (*b*) Base
(*c*) Emitter (*d*) All regions are equally doped.

15. A *P-N* junction diode has ....
(*a*) one *P-N* junction (*b*) two *P-N* junctions
(*c*) three *P-N* junctions (*d*) none of these.

16. FETs have similar properties to
(*a*) thermionic valves (*b*) unijunction transistor
(*c*) P-N-P transistors (*d*) N-P-N transistors.

17. A JFET can operate in
(*a*) depletion mode only (*b*) enhancement mode only
(*c*) depletion and enhancement modes
(*d*) neither enhancement nor depletion mode.

18. In a JFET ..... is usually the point of reference.
(*a*) gate (*b*) drain
(*c*) source (*d*) either (*b*) or (*c*).

19. The primary control on drain current in a JFET, is exerted by which of the following?
(*a*) Gate reverse bias
(*b*) Channel resistance
(*c*) Voltage drop across channel
(*d*) Size of depletion regions.

20. For the operation of enhancement-only N-channel MOSFET, value of gate voltage has to be
(*a*) zero
(*b*) low positive
(*c*) high positive
(*d*) high negative.

21. A FET can be used as a variable
(*a*) inductor
(*b*) capacitor
(*c*) resistor
(*d*) voltage source
(*e*) current source.

22. In FET the drain voltage above which there is no increases in the drain current is called ..... voltage
(*a*) pick off
(*b*) pinch off
(*c*) breakdown
(*d*) critical.

23. A resistance thermometer is basically a/an
(*a*) active transducer
(*b*) passive transducer
(*c*) potentiometer
(*d*) none of these.

24. Platinum resistance thermometer can be used upto
(*a*) 200°C
(*b*) 850°C
(*c*) 1200°C
(*d*) 1500°C.

25. Which of the following should be incorporated in the RTD to make a temperature sensitive bridge most sensitive to temperature ?
(*a*) Platinum
(*b*) Nickel
(*c*) Thermistor
(*d*) Copper.

26. Piezo-electric transducers are .... transducers.
(*a*) active
(*b*) passive
(*c*) inverse
(*d*) both (*a*) and (*c*).

27. Piezo-electric transducers work when we apply to it
(*a*) heat
(*b*) mechanical force
(*c*) vibrations
(*d*) illumination.

28. A Zener diode has ....
(*a*) one *P-N* junction
(*b*) two*P-N* junctions
(*c*) three *P-N* junctions
(*d*) none of these.

29. A Zener diode is always .... connected.
(*a*) reverse
(*b*) forward
(*c*) either reverse or forward
(*d*) none of these.

30. A Zener diode is used as ....
(*a*) an amplifier
(*b*) a voltage regulator
(*c*) a rectifier
(*d*) a multivibrator.

31. A transducer is basically a device which converts
(*a*) mechanical energy into electrical
(*b*) energy or information from one form to another
(*c*) mechanical displacement into electrical
(*d*) none of these.

32. A strain gauge is a passive transducer and is employed for converting
    (*a*) mechanical displacement into a change of resistance
    (*b*) pressure into a change of resistance
    (*c*) force into displacement
    (*d*) none of the above.
33. In a transistor highly doped part is
    (*a*) emitter (*b*) base
    (*c*) collector (*d*) none of these.
34. In a transistor lightly doped part is
    (*a*) emitter (*b*) base
    (*c*) collector (*d*) none of these.
35. When a diode is reverse biased, it is equivalent to
    (*a*) an OFF switch (*b*) an ON switch
    (*c*) a high resistance (*d*) none of these.
36. The reverse current in a diode is of the order of ....
    (*a*) kA (*b*) mA
    (*c*) μA (*d*) A.
37. Regarding MOSFET which of the following statements is *incorrect*?
    (*a*) It can operate in depletion mode.
    (*b*) It can operate in enhancement mode.
    (*c*) It can operate in depletion and enhancement modes.
    (*d*) It can operate in depletion-only-mode.
    (*e*) It can operate in enhancement-only mode.
38. An SCR may be considered to be diodes back-to-back consisting of an anode, cathode and .....
    (*a*) two, plate (*b*) three, plate
    (*c*) three, gate (*d*) four, base.
39. On stopping the gate pulse to a SCR
    (*a*) it will stop conduction
    (*b*) it will continue conduction in the same direction
    (*c*) it will continue conduction in the opposite direction
    (*d*) none of the bove will happen.
40. UJT is also called a
    (*a*) transistorized junction (*b*) voltage controllable device
    (*c*) current controllable device (*d*) relaxation oscillator.
41. In the symbols of P-N-P transistors and N-P-N transistor the arrow on the emitter shows the direction of flow of
    (*a*) electrons, electrons (*b*) holes, holes
    (*c*) holes, electrons (*d*) electrons, holes.
42. In P-N-P transistor, base will be of
    (*a*) P material (*b*) N material
    (*c*) either of the above (*d*) none of these.
43. A crystal diode is used as ....
    (*a*) an amplifier (*b*) a rectifier
    (*c*) an oscillator (*d*) A voltage regulator.

44. SCR is
(*a*) three layer two terminal device
(*b*) three layer three terminal device
(*c*) four layer two terminal device
(*d*) four layer three terminal device.

45. Hall effect can be used to measure
(*a*) carrier concentration
(*b*) electric field intensity
(*c*) magnetic field intensity
(*d*) none of the above.

46. The strain gauges should have low
(*a*) gauge factor
(*b*) resistance
(*c*) resistance temperature coefficient
(*d*) all of the above.

47. ..... is the best electronic device for fast switching.
(*a*) MOSFET
(*b*) JFET
(*c*) BJT
(*d*) Triode.

48. LVDT can be used for
(*a*) vibration measurement
(*b*) angular velocity measurement
(*c*) force measurement in beam
(*d*) load measurement on a column.

49. A Zener diode is .... device.
(*a*) a non-linear
(*b*) a linear
(*c*) an amplifying
(*d*) none of these.

50. Bipolar transistor is a
(*a*) three terminal semiconductor device
(*b*) three layer semiconductor device
(*c*) three junction semiconductor device
(*d*) none of these.

ANSWERS

1. (*c*)	2. (*b*)	3. (*c*)	4. (*b*)	5. (*a*)	6. (*a*)	7. (*d*)
8. (*d*)	9. (*d*)	10. (*c*)	11. (*a*)	12. (*c*)	13. (*c*)	14. (*b*)
15. (*a*)	16. (*a*)	17. (*a*)	18. (*c*)	19. (*a*)	20. (*c*)	21. (*c*)
22. (*b*)	23. (*b*)	24. (*b*)	25. (*c*)	26. (*d*)	27. (*b*)	28. (*a*)
29. (*a*)	30. (*b*)	31. (*b*)	32. (*a*)	33. (*a*)	34. (*b*)	35. (*a*)
36. (*c*)	37. (*d*)	38. (*c*)	39. (*b*)	40. (*b*)	41. (*b*)	42. (*b*)
43. (*b*)	44. (*d*)	45. (*c*)	46. (*c*)	47. (*c*)	48. (*c*)	49. (*a*)
50. (*b*).						

THEORETICAL QUESTIONS

1. What is a transducer? Explain briefly.
2. How are transducers classified?
3. Define a 'system'.
4. What is a 'Control system'?
5. Enumerate and define the elements of a control system.
6. List four examples of control system applications?
7. How are control systems classified?
8. What is an'open-loop' control system?
9. What are the elements of an 'open-loop' control system?
10. Explain briefly two examples of 'open-loop' control system.
11. State the advantages and disadvantages of open-loop control system.

12. What is 'closed-loop' control system?
13. Define the term 'feedback'.
14. State the characteristics of 'feedback'.
15. Explan briefly a 'closed-loop' control system with an example.
16. State the advantages and limitations/disadvantages of a 'closed-loop' control system.
17. What is an 'automatic control system'? What are its advantages and limitations?
18. What is a block diagram?
19. What is a signal flow graph?
20. What do you understand by the term 'stability'?
21. How are industrial controllers classified?
22. Explain briefly a 'Pneumatic control system'. State its advantages and disadvantages.
23. Describe briefly 'Hydraulic control system'. State its advantages and disadvantages.
24. Define a 'semiconductor'.
25. List the important characteristics of semiconductors.
26. Give examples of semiconducting materials.
27. What is the difference between a semiconductor and an insulator?
28. What is an intrinsic semiconductor?
29. What do you mean by the term *doping*?
30. How does an extrinsic semiconductor differ from an intrinsic semiconductor?
31. Explain the structure of a *P*-type semiconductor with help of neat sketches.
32. What is a P-N junction diode? How its terminals are identified?
33. What is a Zener diode ? Draw its equivalent circuit.
34. Explain briefly the applications of a Zener diode.
35. What do you understand by Zener voltage?
36. Explain why Zener diode is always operated in reverse biasing.
37. Explain how a Zener diode can stabilize the voltage across the load.
38. Explain the process of Zener breakdown.
39. Define the term 'Transistor'.
40. What are the various types of transistors?
41. Explain the function of emitter in the operation of a junction transistor.
42. What is the significance of arrow in the transistor symbol?
43. Why is emitter wider than collector and base?
44. Why is base made thin?
45. Draw *N-P-N* and *P-N-P* transistors.
46. Explain the working of a *P-N-P* transistor.
47. Differentiate between *P-N-P* and *N-P-N* transistors. Why are collector and emitter currents nearly equal in these transistors?
48. Explain the construction and working of a JFET.
49. What is the difference between a JFET and a BJT?
50. How will you determine the drain characteristics of JFET? What do they indicate?
51. What are the advantages and disadvantages of JFET?
52. What are the applications of FETs?
53. What is the difference between MOSFET and JFET?
54. Draw the *V-I* characteristics of an *N*-channel FET.
55. Discuss briefly, the construction, working, characteristics and applications of SCR.
56. Explain the forward and reverse characteristics of a thyristor.
57. List the applications of thyristor?

58. What is 'digital electronics'?
59. State the advantages and disadvantages of digital electronics?
60. What is a 'digital circuit'?
61. Draw the diagram of a clocked R-S flip-flop and give the truth table.
62. Show that a R-S flip-flop results when two NOR gates are cross-coupled.
63. What is a flip-flop? Explain the principle of operation of S-R flip-flop with truth table.
64. With the aid of a neat sketch, explain the operation of J-K flip-flop.
65. Briefly describe J-K, D- and T-type flip-flops.
66. Write short notes on "logic families".
67. What is a "Microprocessor"?
68. Draw the block diagram of a microcomputer and explain briefly the three segments (ALU, Register and Control unit) of a microprocessor.
69. What are the characteristics of microprocessor?
70. Mention the important features of the microprocessors.
71. What are the uses of microprocessors?
72. Explain briefly the various parts of a microprocessor system.
73. Explain briefly the following registers:
Accumulator register; Status register; Program counter register (PC); memory address register; Instruction register; General purpose registers; Stack pointer register.
74. What are 'buses'? Explain briefly the following buses:
Data bus, Address bus, Control bus.
75. Explain briefly the following forms of memory unit:
ROM; PROM; EPROM; EEPROM; RAM.
76. Explain briefly Intel 8085 microprocessor with the help of a block diagram.
77. What are programmable logic controllers? Explain.
78. State the advantages and uses of PLCs.
79. State the special features of a PLC.
80. Discuss briefly with a neat sketch the architecture of a PLC.
81. What criteria should be considered while selecting a PLC?

CHAPTER

8 Fluidics

8.1 Introduction; 8.2 Advantages, disadvantages and applications of fluidic divices/ fluidics; 8.3 Fluidic (or fluid logic) elements – General aspects – Coanda effect – Classification of fluidic devices – Fluid logic devices – Fluidic sensors – Fluidic amplifiers; **8.4 Comparison among different switching elements.** Highlights – Objective Type Questions – Theoretical Questions.

8.1 INTRODUCTION

The term **fluidics** relates to the combination of the two functions namely *fluid amplification and fluid logic.*

Fluidics is *defined as a control technology which makes use of fluids interaction to produce useful signals.*

- The *"field of fluidics" is the study of the performance and response characteristics of control systems, computing devices and logical switch gears based on the fluidic elements.*
- In finer control engineering, non-moving logic elements find a prominent place. *Irrespective of the development of electronics, low pressure pneumatics and fluidic elements have certain specific characteristics which put them at par with electronic controls even for modern sophisticated machines.* Various fluidic elements have been developed conforming to the need of logic functions in the industrial automation.
  - — The *basic principle is derived from the "Tesla's fluid-diode" and theory of "wall attachment" discovered by Coanda.*
  - — More and more fluid-logic elements in the form of logic gates like OR, NOR, etc. are being used along with power pneumatic circuits to offer better control and feedback to the pneumatic system. One of the major areas of their application is in the *field of sensors.* The present day state of art of pneumatic sensors is quite competitive with other form of sensors *e.g.*, fine mechanical, opto-electrical, inductive, hydraulic, ultra sound and magnetic devices etc. and are therefore widely used in various engineering tools and instruments.

History of fluidics :

- 1904 : *L. Prandtl* (German aerodynamist) suggested that in a wide angled diffuser the flow separation could be controlled by applying suction to the boundary layer.
- 1916 : — Nikola *Tesla* filed a patent for a "Valvular Conduit" for fluids in which there was an easy direction of flow and a difficult direction, owing to the interference caused by the divided branch flow opposing the intended flow-direction as can be seen in Fig. 8.1. which was granted in 1920.

— Whill Tesla claimed to have invented the first fluid device having no moving parts.

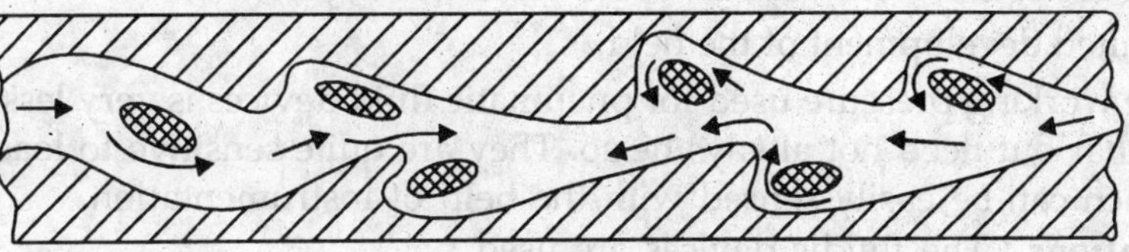

Fig. 8.1. Tesla's tube.

- 1930 : Henri *Coanda* (a Rumanian engineer) discovered that a free jet would follow an adjacent curved or inclined surface. Coanda's theory of *"wall attachment"* was a major stepping stone in the development of this field. Later this effect was used for the development of fluidic logic components.
- 1958 : Moore and Klive, working with wide angled diffusers, discovered that a jet could have two or three stable stages, depending on the angle of diffuser through which the jet was flowing.
- 1962 : Ray Auger discovered a fluidic logic element called *"Turbulence amplifier"*.

Boolean algebra and truth tables : Refer to Art. 7.4

8.2 ADVANTAGES, DISADVANTAGES AND APPLICATIONS OF FLUIDIC DEVICES/FLUIDICS

During the last few decades, the technology of electronic control system has leapt through milestones of innovations with various forms of transistors, ICs and the like, production costs have drastically reduced, and reliability has increased many a fold. Inspite of these developments, fluidic devices are finding wide applications and claim the following *advantages* :

Advantages :

1. More reliable.
2. Simpler in construction.
3. Smaller in size, mass or weight.
4. Offer exceptional physical and thermal stability and ruggedness.
5. Noise free.
6. Hazard free.
7. Mode of energy feeding to a fluidic system is very simple.
8. Fuidic elements are easily adaptable to logic functions in engineering applications.
9. Have good response and performance characterisitics (0.001 second).
10. Since they contain no moving parts, not much maintenance problems are encountered.

Disadvantages :

1. Unsuitable for incompressible fluids.
2. Slow speeds and low power outputs.
3. Complex systems are impracticable.

4. Inefficient in operation; they cannot be used in high-speed switching operations.
5. Not suitable for intermittent operation control systems.
6. Limited development of the field.

- The working pressure used for pneumatic fluid devices is very less within 0.05 to 0.1 bar but need not always be so. They are quite sensitive to load and position which can be easily sensed with the help of instrumentation.

Applications : The fluidic devices are used :

1. To measure flow rates;
2. To provide on-off controls;
3. To check weights;
4. To operate various types of machinery, etc.

- If used with a chemically inert gas such as nitrogen or helium, these devices are especially advantageous in cases where fire or electrical hazards are present, as in the manufacture of explosives.

8.3 FLUIDIC (OR FLUID LOGIC) ELEMENTS

8.3.1. General Aspects

The fluidic elements (also called *fuid logic elements*) were developed in the early sixties. The primary merit of these elements over all other forms of control elements is that they have a *minimum number of mechanical moving parts;* because of this, these elements are also known as *"non-moving logic"* controllers.

These element claim the following *advantages :*

1. Quite insensitive to temperature, vibration, shock, electric noise and radiation.
2. Insensitive to electromagnetic interferences etc.
3. Need no actuating force.
4. No wear and tear of elements.
5. Need very little space for mounting.

8.3.2. Coanda Effect

Physically a fluidic device is a block of material having an internal network of passages. One of the basic underlying principle of the functioning of these flow passages was given by Henri Marie Coanda (a Ramanian engneer) and is known as *"Coanda effect"* or *"Wall attachment effect".*

This effect is explained below : Refer to Fig. 8.2.

- **Fig. 8.2. (*i*)** : A free jet of air is emitted into a confined region or orifice at a velocity high enough to *produce turbulent flow.*

  Fig. 8.2. (*ii*) : The free jet of air will continue in a given direction, pulling in with it the available air from its surroundings as it leaves the orifice. If there is greater availability of this entraining air from one side, a small vortex area (low pressure area/region) is created near the nozzle exit. This low pressure area then tends to attract the free jet, distorting it and pulling it towards the wall, because the atmospheric pressure on the other side forces the jet to cling to the surface.

  Fig. 8.3. (*iii*) : The free jet attachment continues until a small air supply is fed to the low pressure area, thus relieving the attraction of the jet to the wall. When this signal is injected, the free jet then detaches itself from the wall and resumes its normal uninterrupted flow path.

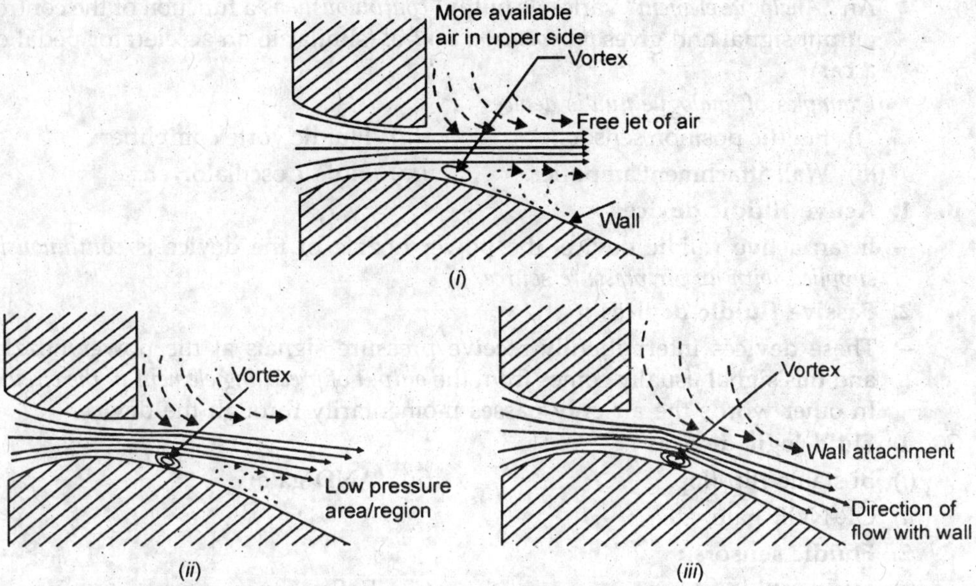

Fig. 8.2. Coanda/Wall attachment effect.

8.3.3 Classification of Fluidic Devices

Fluidic devices may be *classified* as follows :

A. 1. Digital fluidic devices :

(*i*) Wall attachment. (*ii*) Turbulence.

(*iii*) Vortex feedback.

- A *digital element""* has two outputs, and flow takes place from one or other output depending upon the presence or absence of control signal. The flow never splits between two outputs; it is either fully out of one or other. It can be compared to a simple "ON-OFF" switch.
- The digital devices are of the following two categories :

(*a*) ***Monostable :***

— In the absence of a signal the device always selects one of its two output states. The other state can only be selected by applying and maintaining the signal *i.e.,* when the applied signal is removed the device returns to its former state.

These devices do *not* have any *memory*.

Example : 3/2 normally operated, spring return D.C. valve.

(*b*) ***Bistable :***

— The output of this device in any direction is stable, irrespective of the fact whether applied signal is present or absent.

— These devices are *equipped with memory*.

Example : 4/2 pilot operated D.C. valve.

Examples of digital fluidic devices :

(*i*) Bistable flip-flops; (*ii*) Logic gates, etc.

2. Analogue fluidic devices :

(*i*) Stream or beam deflection. (*ii*) Impact modulator.

(*iii*) Vortex.

- An "*Analogue element*" varies its output *continuously* as a function of the control output signal and gives *proportional control* (similar to an accelerator pedal of a car)

  Examples of analogue fluidic devices :

  (*i*) Fluidic position sensors; (*ii*) Fluidic vortex amplifier;

  (*iii*) Wall attachment amplifier; (*iv*) Fluidic oscillator, etc.

B. 1. **Active fluidic devices :**

— In an active fluidic device, the power nozzle of the device is *continuously supplied with an air pressure source.*

2. **Passive fluidic devices :**

— These devices intermittently receive pressure signals at the power nozzle, and this signal usually comes from the *output of preceding element in the circuit.* In other words the air only passes momentarily through the device.

C. 1. **Fluid logic devices :**

(*i*) Bi-stable flip-flop; (*ii*) AND gate;

(*iii*) OR-NOR gate, etc.

2. **Fluidic sensors :**

(*i*) Interruptible jet sensor; (*ii*) Reflex sensor;

(*iii*) Back pressure sensor.

3. **Fluidic amplifiers :**

(*i*) Turbulence amplifier; (*ii*) Vortex amplifiers.

8.3.4. Fluid Logic Devices

The following terms are frequently used in connection with fluid logic devices :

1. ***Logic function or gate.*** *It is defined as an assembly of one or more logic elements which produce a desired effect on satisfying certain conditions, i.e., if we apply correct input signals, we will get the required output.*
2. ***Input signal.*** *It is the pressure or flow which is directed into the input part to control the logic function or an element.*
3. ***Output signal.*** *It is the pressure or flow which is leaving the output part of a logic function or an element.*

8.3.4.1. Bi-stable flip-flop

A bi-stable flip-flop is the most common fluid logic device. It works on the principle of *Coanda effect.*

Fig. 8.3. shows this device :

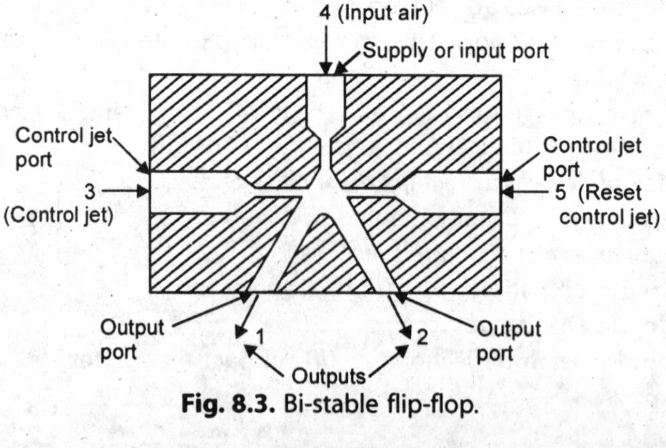

Fig. 8.3. Bi-stable flip-flop.

— In general, it consist of five ports, out of which one is supply port; two control jet ports and two output ports.

— Supply is always present on the input port (4) whereas output is dependent on the control jets (3, 5), *i.e.*, when fluid is passed from control jet 3 there will be an output – 1 and similarly when there is a supply at control jet–5 the output will be an output – 2.

- The device is suited for binary logic functions, as the output either exists or does not. The output from this device can be used as "pilot signal" for actuating various valves with low pressure actuating element.
- This device can be manufactured easily out of glass or plastic, not much larger than the size of a coin, the orifice size in such devices is of the order of 0.25 mm in diameter and the working pressure is about 0.05 to 0.1 bar. The operational speed is about 1000 cycles per second.

8.3.4.2. ADN gate

Fig. 8.4. shows a fluidic AND gate/element; its operation is based on Coanda's wall attachment theory.

(*i*) Under no-control condition, output is at 0 (output–0).

(*ii*) When control jet C_1 is on, output is at 1 (output–1).

(*iii*) *When both the control jets C_1 and C_2 are on, only then there is an output at the AND gate,* *i.e.*, output is at 2 (output–2).

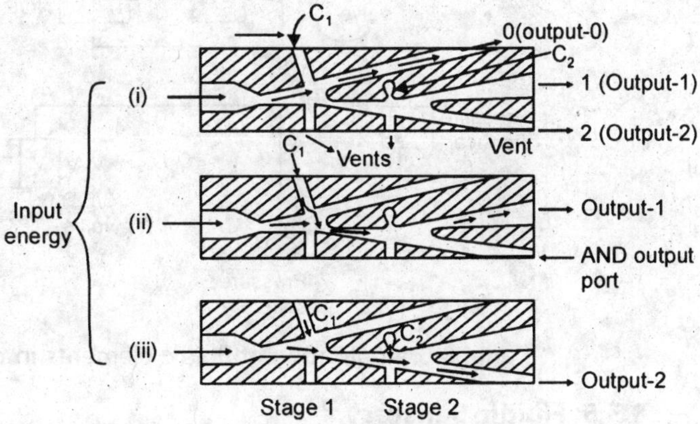

Fig. 8.4. AND gate.

8.3.4.3. OR–NOR gate

Fig. 8.5. shows fluidic OR–NOR gate.

— It consists of supply port S from which the fluid is supplied all the time. There are two control ports C_1 and C_2 and two output ports X and X_1. X represents the OR gate output while X_1 represent NOR gate output.

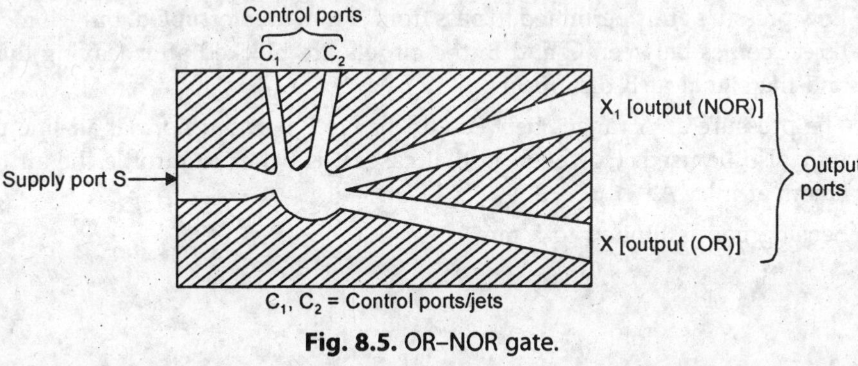

Fig. 8.5. OR–NOR gate.

— When the control jets C_1 and C_2 are absent, supply S is passed through the gate via output X_1. It implies that when there is no control input, there is an output. This is the characteristics of NOR gate and X_1 is considered as NOR output.

— When fluid stream is present at either of control jet C_1 or C_2 or both at C_1 and C_2, there will be output at X. It implies that even with one control input there is an output. This is the logic characteristic of OR gate and X output is considered as OR output.

The complete device is called as OR–NOR gate.

- Fig. 8.6. shows a circuit diagram illustrating the application of fluidic pressure for actuating a low pressure diaphragm valve.

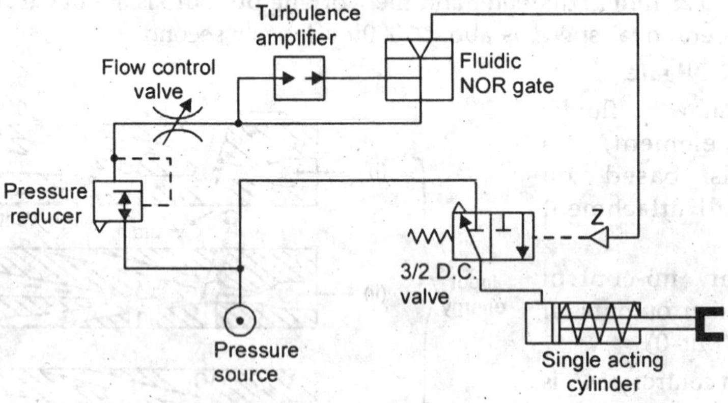

Fig. 8.6. Application of fluidic elements in pneumatic circuit.

8.3.5. Fluidic Sensors

These sensors (generally of very small size) are mainly used *to detect the presence of objects.* They are *designed to provide a signal in the form of fluid jet, to indicate the presence of an object.*

- The output of these sensors can be directly used as a *control signal for pneumatic logic circuits.* These control signals, however, need to be amplified when employed in pneumatic and hydraulic devices.

A few important fluidic sensors are described in the following subarticles.

8.3.5.1. Interruptible jet sensor

Fig. 8.7. shows an interruptible jet sensor :

— Low pressure air is permitted to pass from A to B uninterrupted. But if a mechanical object comes between C and B, the air jets are blocked after leaving the port C and the signal at B disappears.

— The pressure at A ranges between 0.1 bar to 2 bar, but normal air-line pressure may also be used. However, in that case, it is better to throttle the air before it enters at inlet A.

— Sensing gap is limited to 5 mm.

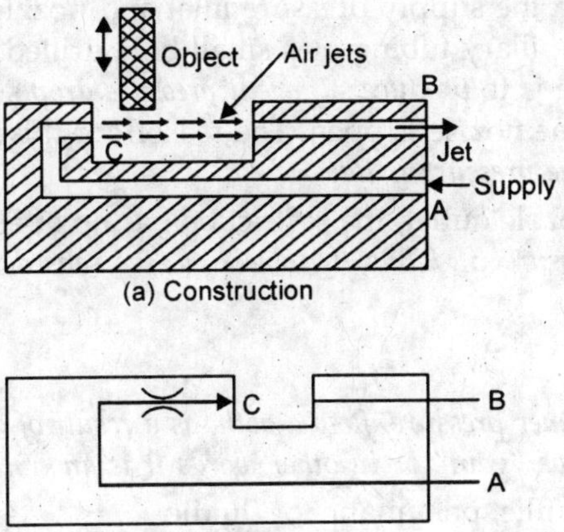

(a) Construction

(b) Symbol

Fig. 8.7. Interruptible jet sensor.

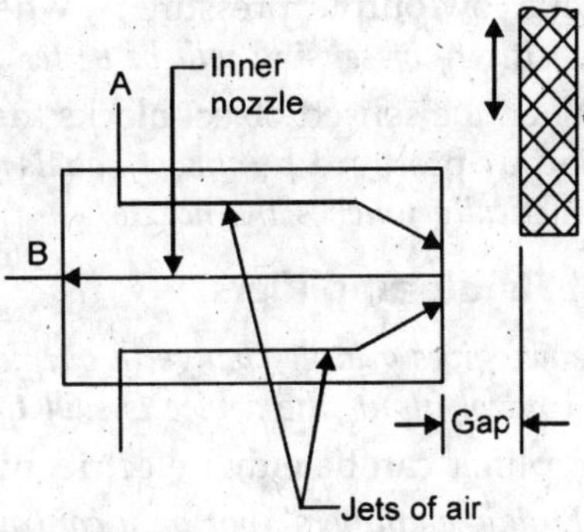

Fig. 8.8. Reflex sensor-symbol.

8.3.5.2. Reflex sensor

Fig. 8.8. shows the symbol of a reflex sensor :

— It *works on the principle of creating a back pressure when the pressure flowing out to the atmosphere is blocked.*

— Two jets of air flow out through an annular opening. If an object disrupts the jets, a back pressure is created which flows back into the inner nozzle and controls other valves.

— Here, the inlet pressure is limited to 0.1 to 0.2 bar (gauge).

8.3.5.3. Back pressure sensor

Fig. 8.9. shows a back pressure sensor :

— It consists of a small nozzle exhausting air to the atmosphere.

— When nozzle is blocked, the pressure backs up and this *increased back pressure signals the presence of an object.*

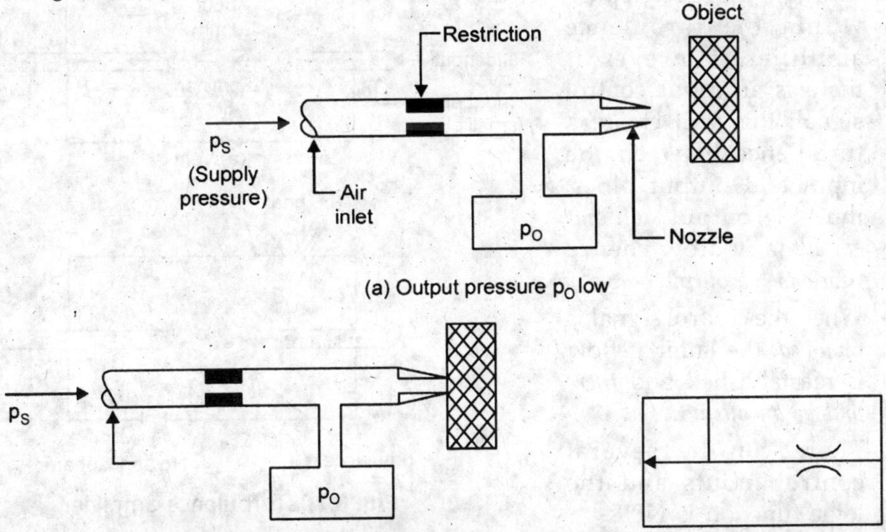

(a) Output pressure p_O low

(b) Output pressure p_O High

(c) Symbol

Fig. 8.9. Back pressure sensor.

— A restriction must be placed between the supply pressure inlet and nozzle. This restriction could be a short length capillary tubing or a small hole drilled into a plug inserted in the tube. Its purpose is to *produce sufficient pressure drop* so as to get a low output pressure p_o, when the nozzle is open. The reason for this is that *sensitivity of sensing will be better if the pressure is low.*

— When the sensed object blocks the nozzle outlet, the restriction has no effect, and output pressure p_o *almost equals pressure* $\mathbf{p_s}$. *Thus the object can be sensed only if practically touches the nozzle.*

8.3.6. Fluidic amplifiers

If a device gives a large change in output either pressure, flow a both, as a result of a small change in control input, the device is said to have "gain" or in other words it is an ***amlifier.***

An amplifier can be either electric, hydraulic, pneumatic, or fluidic.

- *"Fluidic amplifiers" not only control but also provide certain amplification to the fluid signals.*

8.3.6.1. Turbulence amplifier

Fig. 8.10. Illustrates a turbulence amplifier (T.A.)

— Low pressure fluid is conducted through a long, small bore pipe to achieve *laminar flow.*

— This laminar flow issued from the inlet pipe transits the space of about 20 mm and is then collected by the outlet pipe. The space between input pipe and output pipe is protected by a cylindrical shield of diameter 20 to 30 times that of input/output pipe. This shield houses the *'control inputs'* and it is vented to atmosphere as shown in the figure.

— When there is no control input or signal, the laminar flow proceeds from input to output through the open space within the turbulence amplifier. However if there is an input control signal, this will create a turbulence between the input and output pipes; thus the output *will not collect* any flow, hence there is *No output.*

— When the control signal is *removed,* the laminar flow is reestablished and *there will be an output.*

— There can be several control inputs and the logic function is NOR.

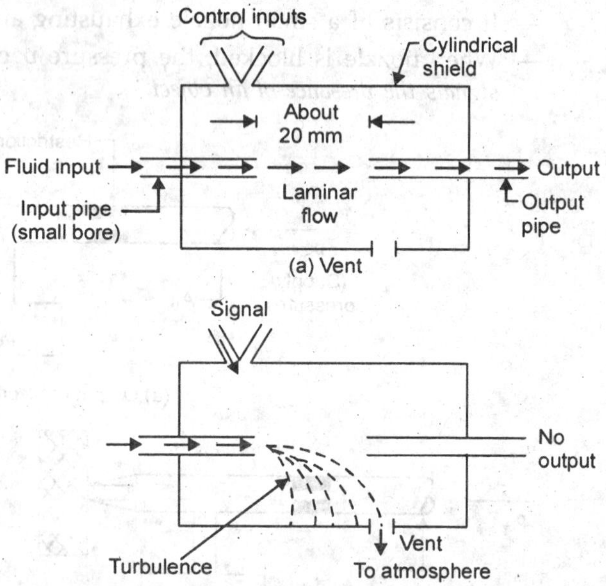

Fig. 8.10. Turbulence amplifier.

In these devices, the switch off time is about 4 ms (milli seconds) and the total cycle time may not exceed 6 to 7 ms.

- Following are the examples of the application of the "turbulence amplifier" to some industrial problems :

  (*i*) Sensing; (*ii*) Counting;
  (*iii*) Discrimination; (*iv*) Timing;
  (*v*) Dispensing; (*vi*) Liquid level sensing.

8.3.6.2. Vortex amplifier

This fluidic device is used to regulate the flow of fluid by *utilising the properties of a vortex.*

Fig. 8.11. shows a vortex amplifier :

— It consists of a cylindrical disc like container, which is divided by a *cylindrical porous element* into two chambers : *Outer* chamber and *vortex* chamber.

— Supply port S is provided for fluid inlet to the cylindrical disc.

— Control jet C is provided to generate vortex in the vortex chamber. Several control jets along the circumference can be provided depending upon capacity, which throw streams of fluid in *tangential direction* and generate vortex motion in fluid.

— At the centre there is an output port X from where signal is transmitted.

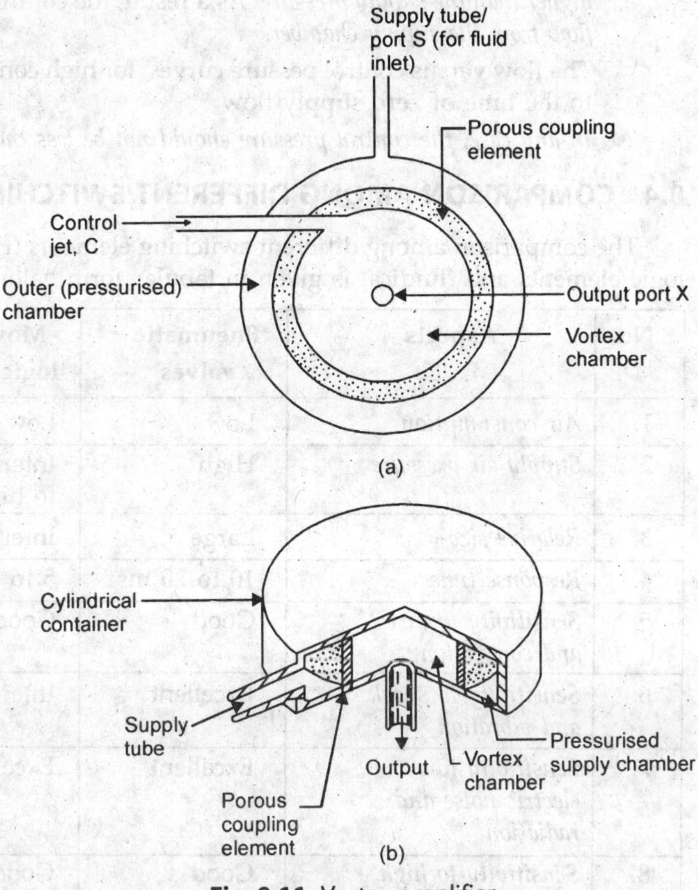

Fig. 8.11. Vortex amplifier.

Working :

— When a fluid element enters the vortex chamber through the porous coupling element and it flows towards the output port its *radial velocity must increase.* Again, when the same fluid element enters through the porous coupling with tangential velocity imparted to it as it leaves coupling to conserve angular momentum, the *angular velocity of fluid must increase* as it approaches the output port X. Thus there are two amplifying properties of vortex amplifier : (*i*) Increase in radial velocity; (*ii*) Increase in angular velocity.

— When the supply pressure is regulated, if the vortex motion of the fluid produces a centrifugal pressure drop across the vortex chamber, there is less pressure at the exit to expel the liquid, thus *less fluid flows from the exit*. The introduction of vortex motion into the chamber effectively throttles the fluid through the chamber.

It has been observed from the curves of output flow as function of the control pressure applied to vortex chamber that, the *flow is reduced with the increase of control pressure.*

— *When the control pressures are sufficiently high, the flow from supply is zero or even slightly reversed.* Centrifugal pressure increases the pressure at the inside surface of the porous coupling element so that the total pressure is *equal to or slightly higher than the supply pressure.* As a result, the control flow *can cut-off completely the flow from the supply chamber.*

The flow versus control pessure curves, for high control pressures, become tangent to the time of zero supply flow.

- *In any case, the control pressure should not be less than the supply pressure.*

8.4 COMPARISON AMONG DIFFERENT SWITCHING ELEMENTS

The comparison among different switching elements (Pneumatic valves, moving part logic elements and fluidics) is given in tabular form below :

S.No.	Aspects	Pneumatic valves	Moving part logic elements	Fluidics
1.	*Air consumption*	Low	Low	High
2.	*Supply air pressure*	High	Intermediate to high	Low
3.	*Relative size*	Large	Intermediate	Intermediate
4.	*Response time*	10 to 20 ms	5 to 10 ms	1-2 ms
5.	*Sensitivity to dirt and conversion*	Good	Good	Poor
6.	*Sensitivity to shock and vibration*	Excellent	Intermediate	Excellent
7.	*Sensitivity to electric noise and radiation*	Excellent	Excellent	Excellent
8.	*Sensitivity to high temperatures*	Good	Good	Excellent
9.	*Life expectancy*	10^7 to 10^8 cycles	10^7 to 10^8 cycles	Unlimited (with clean air)

HIGHLIGHTS

1. *Fluidics* is defined as a control technology which makes use of fluids interaction to produce useful signals.
2. Fluid devices may be of following types :

   (*i*) Digital; Analogue; (*ii*) Active; passive; (*iii*) Logic devices; sensor; amplifiers.
3. Fluidics amplifiers not only control but also provide certain amplification to the fluid signals.

OBJECTIVE TYPE QUESTIONS

Fill in the Blank or Say "Yes" or "No".

1. 'Valvular conduit' was patented by ....
2. The control circuit element has a limited life.
3. A digital element has one output.
4. .... element varies its output continuously as a function of the control output signal and gives proportional control.
5. The digital devices are least important for industrial applications.
6. A device having bistable properties is also said to have ....
7. In .... device the air only passes momentarily.
8. A logic gate is a device which has inputs and outputs.
9. .... devised the first NOT gate.
10. Coanda effect was discovered in 1930 by ....
11. Bi-stable flip-flop works on the principle of .... effect.
12. The output from Bi-stable flip-flop can be used as a pilot signal for actuating various valves with low pressure actuating element.
13. Fluidic sensors are primarily used for detecting the presence of objects.
14. Vortex amplifier is used to regulate the flow of fluid by utilizing properties of a vortex.
15. The power required to turn off the turbulence amplifier is much more than the power in the output line.

ANSWERS

1. Nikola Tesla	2. No	3. No	4. Analogue	5. No
6. Memory	7. passive	8. Yes	9. Prandtl	10. Henri Coanda
11. Coanda	12. Yes	13. Yes	14. Yes.	15. No.

THEORETICAL QUESTIONS

1. Define the term 'Fluidics'.
2. Enlist the advantages, disadvantages and applications of fluidic devices/fluidics.
3. Explain briefly "Coanda effect".
4. Give the classification of fluidic devices.
5. Differentiate between the following fluidic devices :
   (*i*) Digital and analogue. (*ii*) Active and passive.
   (*iii*) Mono-stable and bi-stable.
6. Define the following terms :
   (*i*) Logic function or gate. (*ii*) Input signal.
   (*iii*) Output signal.
7. Explain briefly any two of the following fluid logic devices :
   (*i*) Bi-stable flip-flop. (*ii*) AND gate.
   (*iii*) OR-NOR gate.
8. Describe briefly the following fluidic sensors :
   (*i*) Interruptible jet sensor. (*ii*) Back pressure sensor.
9. Explain briefly air following fluidic amplifiers :
   (*i*) Turbulence amplifier. (*ii*) Vortex amplifier.
10. Give the comparison among pneumatic valves, moving part logic elements and fluidics.

CHAPTER

9 Industrial Automation

9.1 General aspects; 9.2 Advantages and limitations of automation; 9.3 Applications of automation; 9.4 Elements of automation; 9.5 Aims of automation; 9.6 Mechanisation and automation; 9.7 Types of automation – Fixed automation – Programmable automation – Flexible automation; **9.8 Low cost automation; 9.9 Assembly automation equipment – Transfer devices and feeders** – Introduction – Classification of material handling systems – Transportation system – Feed system – AGVs – Automated storage systems; **9.10 Flexible manufacturing systems (FMS)** – Introduction – FMC and FMS – Components of a FMS – Requirement for FMS – Advantages and limitations of FMS; **9.11 Computed integrated manufacturing (CIM); 9.12 Mechatronics and concurrent engineering; 9.13 Computer aided process planning (CAPP) systems; 9.14 Group technology (GT)** – Introduction – Advantages and limitations of group technology – Part families – Formation and establishment of component family – Collection of production data. ***Questions with Answers.*** Highlights – Objective Type Questions – Theoretical Questons.

9.1 GENERAL ASPECTS

Automation may be *defined* as follows :

"It is the process of following a predetermined sequence of operations with little or no human labour, using specialised equipment and devices that perform and control manufacturing processes."

Or

"It is a system in which many or all of the processes in the production, movement, and inspection of parts and materials are automatically performed or controlled by self-operating devices."

Or

"It is technology concerned with performing a process by means of programmed commands combined with automatic feedback control to ensure proper execution of the instructions."

— ***Partial automation*** means replacement of human activities or involvement by automatic means only *partially*.

— In ***full automation***, the human involvement is totally eliminated and the process is entirely carried out and controlled through automatic means alongwith a proper feedback system.

Mechanisation. It means that something is done or operated by *machinery* and not by hand. Feedback is *not* provided and thus one deals with *open-loop systems*. The *automatic screw machine* with its cam actuated tools and sequencing represents a high degree of mechanisation. *"Automation" is a step beyond "mechanisation"*, where human operators are provided with machinery to help them in their jobs. *Industrial robots are said to be the most visible part of automation.*

- The history of industrial automation is characterised by periods of rapid change in popular methods. Either as a cause, or, perhaps, an effect, such periods of change in automation techniques seem closely tied to world economics. Use of *industrial robot*, which became identifiable as a unique device in the 1960s alongwith computer-aided design (CAD) systems and computer-aided manufacturing (CAM) systems, characterises the latest trends in the automation of the manufacturing process. These technologies are leading industrial automation through another transition, the scope of which is still unknown.
  - — *A primary reason for the growth in the use of "industrial robots" is their declining cost.* Also, robots are not just getting cheaper, they are becoming more effective – *faster, more accurate, more flexible.* If we factor these quality adjustments into the numbers, the cost of using robots is dropping even faster than their price tag is.
  - — As robots become more cost effective at their jobs, and as human labour continues to become more expensive, more and more industrial jobs become candidates for robotic automation. This is the single most important trend *propelling growth of the industrial robot market.* Another trend is that, economics aside, as *robots become more capable they become able to do more and more tasks that might be dangerous or impossible for human worked to perform.*
- *"Manufacturing automation"* deals with the production of goods and includes the following :
  - (*i*) Industrial robots;
  - (*ii*) Automatic machine tools to process parts;
  - (*iii*) Automatic assembly machines;
  - (*iv*) Automatic material handling;
  - (*v*) Automated storage and retrieval systems;
  - (*vi*) Automatic inspection systems;
  - (*vii*) Feckback control system;
  - (*viii*) Computer systems for automatically transforming design into parts;
  - (*ix*) Computer systems for planning and decision making to support manufacturing.

9.2 ADVANTAGES AND LIMITATIONS OF AUTOMATION

Following are the *advantages* and *limitations* of automation :

Advantages :

1. Increased productivity.
2. Reduced overall production cost.
3. Less floor area required.
4. Increased overall profits of the manufacturing concern.
5. Human fatigue is greatly minimised.
6. Reduced maintenance requirements.
7. The workers avail better working conditions.
8. Owing to the use of standardized parts and assemblies the inventory requirement is minimised.
9. Uniform components are produced.
10. There is an effective control over the production process.
11. The workpiece is tidy and safe.

12. With the use of group technology a considerable saving is effected in terms of design cost, material cost, tooling cost etc.
13. There is great improvement in the quality and reliability of the products.
14. Human safety is fully ensured.

Limitations :

1. *Initial investment.* The considerations of cash flow may make an investment in automation difficult even if the estimated rate of return is high.
2. *Labour resistance.* Although the use of robots increases productivity, makes the firm more competitive and preserves jobs, yet some jobs are lost. Therefore, people look at robots and manufacturing automation as a *cause of unemployment.*
3. *Cost of upgraded lobour.* It is very easy to automate routine monotous tasks. But skill is required to automate the difficult tasks, this requires the upgradation of the manufacturing labour, which involves cost.

9.3 APPLICATIONS OF AUTOMATION

Applications of automation include the following :

1. Numerical control.
2. Automated production lines.
3. Automated assembly.
4. Robots in manufacturing.
5. Flexible manufacturing systems.
6. Building Automation System. (BAS)
7. CAD/CAM and Computer Integrated Manufacturing (CIM).
8. Automation in daily life.

9.4 ELEMENTS OF AUTOMATION

The three basic building blocks (elements) of automation are :

1. **Source of power** .... to perform some function.
2. **Feedback controls** .... widely used in modern automated systems.

(*i*) *Sensors* :

— Manual switches; — Limit switches;
— Proximity switches; — Photoelectric sensors;
— Infrared sensors; — Fiber optics;
— Lasers.

(*ii*) *Analysers* :

— Computers; — Counters;
— Timers; — Bar code readers;
— Optical encoders.

(*iii*) *Actuators* :

— Cylinders; — Solenoids;
— Relays.

(*iv*) *Drives* :

— Motors; — Stepper motors;
— D.C. servomotors; — Kinematic linkages;
— Walking beams; — Genevas.

3. **Machine programming :**
   — The set of actions that is to be accomplished automatically by the system, is determined by the *programmed instructions.*
   — The program specifies what the automated system should do and how its various components must function in order to accomplish the desired result.
   — The content of the program varies considerably from one system to the next.

9.5 AIMS OF AUTOMATION

The *aims* of automation are as follows :

1. To improve product quality.
2. To enhance labour productivity.
3. To reduce labour cost.
4. To check the effects of labour shortages.
5. To reduce or eliminate routine manual clerical tasks.
6. To improve work safety.
7. To reduce manufacturing lead time.
8. To accomplish process that cannot be done manually.
9. To avoid the high cost of not automating.
10. To economise on floor space in manufacturing plant by arranging the machines, material movement, and related equipment more efficiently.

9.6 MECHANISATION AND AUTOMATION

The **mechanisation** (*the precursor to automation*) *can be defined in its simplest sense as the transfer of skills and manual activities to machine operations.*

The primary difference between mechanisation and automation includes *feedback for controlling an automated system.*

"Microautomation" differs from "macroautomation" in that the latter pertains to the *low-level control system commonly envisioned for industrial automation.*

1. **"Microautomation"** is primarily concerned with logical control focused on *individual machines* and the logical linkage between machines and devices. In particular microautomation pertains to *servomechanism, hydraulic and pneumatic devices* and associated low-level software used for physical movement of parts through a production system which typically results in "islands" of automation.
2. **"Macroautomation"** as the term indicates, has a larger scope and *deals with the coordination and supervision of various smaller scooped islands of automation.* This has led to various models and landings for communications and control of large-scale manufacturing (automated) system. Pertinent issues include refractive models and corresponding architectures for implementation of large scale systems.

9.7 TYPES OF AUTOMATION

The various types of "automation" are :

1. Fixed automation (Hard automation).
2. Programmable automation.
3. Flexible automation (Soft automation).

9.7.1. Fixed or Hard Automation

- It is characterized by having the sequence of operations necessary to manufacture or assemble a product *fixed by the equipment configuration.* As such, there is typical equipment which is *inflexible to product changes.*
- The production lines are designed to produce a standardized product, such as engine blocks, valves, gears spindles.
- It is typically associated with *high production rates.*
- This type of automation makes sense only when *product designs are stable and product life cycles are long.*
- The machines employed in fixed automation applications are usually built on the 'building block' or 'modular' principle; they are generally called *"transfer machines"* and consists of the two major components : (*i*) Power led production units; (*ii*) Transfer mechanisms.

  Advantages : (*i*) The movement of parts is very fast and efficient.
  (*ii*) Very high efficiency.
  (*iii*) Unit cost is low.

  Disadvantages : (*i*) High initial investment.
  (*ii*) inflexibility.

9.7.2. Programmable Automation

- The word *"programmable"* means that one set of task can be easily switched over to another set by changing the computerised instructions.
- ***Programmable automation*** *uses information technology and numerical engineering to provide coordination, machine control and communication through computers in the most effective way. It attempts to bridge the gap between consistency and flexibility.*
- Programmable automation equipment is highly reprogrammable to accommodate high product variety but has *low production rates relative to fixed automation.*
- Parts are typically loaded into programmable automated production system in *batches.* Each batch consists of a different part; and the machines comprising the systems to manufacture the part are reprogrammed for each batch. Change-over from one batch to the next also requires a change in physical set-up of the machine tools, that is their fixing and tooling. Such change-over *results in a loss of production time.*

  Example : Numerically Controlled (NC) machine tools and industrial *robots.*
- The *principal programmable automation technologies are :*

  A. ***Computer Aided Design (CAD) :***
  (*i*) Computer Aided Design and Drafting (CADD).
  (*ii*) Computer Aided Engineering (CAE)

  B. ***Computer Aided Manufacturing (CAM) :***
  (*i*) Computer Numerically Controlled machine tools (CNC).
  (*ii*) Direct Numerically Controlled machine tools (DNC).
  (*iii*) Robots.
  (*iv*) Flexible Manufacturing System (FMS).
  (*v*) Automated Material Handling System (AMH).
  (*vi*) Automated Storage and Retrieval System (AS/RS).

C. ***Computer Integrated Manufacturing (CIM) :***

(*i*) Management Information Systems (MIS).

(*ii*) Computer Aided Planning (CAP).

And

Computer Aided Process Planning (CAPP).

Advantages : (*i*) For large batches the unit cost is low.

(*ii*) The changes in product can be dealt with flexibility.

Disadvantages : (*i*) Unit cost, as compared to fixed automation, is high.

(*ii*) Long set up time is required for a new product.

9.7.3. Flexible (or soft) Automation

- The birth of flexible automation has been possible with the *advent of microelectronics, microcomputers and programmable controllers.*
- Conceptually, a flexible automated system has the *capability of producing a variety of parts with minimal change over time from one part to the next.* The ability to change part programs and to change the physical setup of the production system *with little or no loss of production time is the primary difference between flexible automation and programmable automation.*
- To accomplish this, one strategy has been to formulate flexible manufacturing class based on *group technology principles.* The overall objective is to increase productivity by properly designing a flexible system that offers advantages similar to fixed automation schemes of mass production but has capability for handling part variety.
- Flexible automation gives the manufacturer the ability *to produce multiple products cheaply in coordination than separately.*

Advantages : (*i*) Customised prducts.

(*ii*) Design variations can be dealt with flexibility.

(*iii*) Parts with complex shapes can be produced.

Disadvantages : (*i*) High initial investment.

(*ii*) Unit cost is relatively high.

9.8 LOW COST AUTOMATION

Low cost automation *is a technology that creates some degree of automation around the existing equipment, tools, methods, people, etc., using mostly standard components available in the market with low investment, so that payback period is 'short'.*

- Low cost automation is one of the very practical, safe, economical and rewarding strategies in the present regime of globalisation and liberalisation to enhance productivity through *quality enhancement* and *cost reduction.*
- Low cost automation system may be adopted for the following activities :

— Loading;	— Feeding;
— Clamping;	— Machining;
— Forming;	— Welding;
— Gauging;	— Assembly and packing etc.

- In low cost automation systems, the various systems which can be employed are: mechanical, hydraulic, pneumatic, electrical and electronics. Any of these systems can be used in simple situations, however, in more difficult practical applications hybrid systems are used.

Advantages :

1. Low investment required.
2. Better utilisation of the material.
3. Small batch sizes.
4. Increased labour productivity.
5. Changes are gradual, smooth and very cost effective.
6. Less rejections due to consistency in process.
7. The hardware components are flexible and reusable and very adaptable to changes in product, market conditions etc.

9.9 ASSEMBLY AUTOMATION EQUIPMENT – TRANSFER DEVICES AND FEEDERS

9.9.1. Introduction

- The automatic assembly equipment include the *"technical devices"* by which assembly processes can either be undertaken fully automatically or semi-automatically with the help of a person. These are available in the market in the form of *mechanical assembly equipment, assembly machines, assembly lines and automatic modules* in several different designs, performance ranges and fields of application.
- In the field of material conveying technology, a Fundamental knowledge of *"handling"* is a prerequisite for correct selection. ***Handling*** is *the realisation, defined change or temporary maintenance of a specified 3-dimensional arrangement of geometrical bodies in a reference coordinate system.*
- In order to increase the productivity of a manufacturing organisation proper *transportation and feeder solutions* are required. The main requirement of such a system is to utilise the resources effectively.
- The analysis and design of a material handling system is complex job. The transportation has to be solved not only between manufacturing cells but between the manufacturing unit to the warehouse. *A computer offers a wide range of solutions for material handling. Some simulation tools are also available to plan the transportation of the material as transfer or feeder device.*

9.9.2. Classification of Material Handling Systems

Material handling systems may be classified as follows :

1. Transportation system.
2. Feeding system.
3. Random access transport systems (New trends) - Automated Guided Vehicles (AGVs).

9.9.3. Transportation System

9.9.3.1. Transfer Systems

The various types of transfer systems are :

1. ***Continuous transfer system* :**
   — In such systems the work carriers move at constant speed while the work head index back and forth.

— The assembly operations are carried out during the period the work heads are moving *forward,* keeping pace with the work carriers.

2. ***Intermittent or synchronous transfer system* :**
   — In this type of system, the work carriers are transferred *intermittently,* while the workheads remain *stationary.*
   — It is named as synchronous system since all the parts are moved at the same time.
3. ***Asynchronous transfer system* :**
   — *This system permits each part to advance to the next station when processing at the current station has been completed;* as a result the *flexibility is increased.*
   — It is suitable for both manual and automated operations.
4. ***Stationary system* :**
   — In this system, the part remains in a fixed location as located during the whole assembly process.
   — It is employed when the assembly products are bulky or difficult to handle.

9.9.3.2. Transfer machines

Transfer machines are invariably found to be most suitable method for continuous flow of identical or very similar components in mass production of consumer goods. A wide variety of such machines with different models, designs and sizes are available in the market.

Some of these machines available in the market are :

(*i*) Computer numerically controlled (CNC) rotary transfer machine;
(*ii*) Economy precision transfer machine;
(*iii*) Rotary transfer machines for brass goods;
(*iv*) Automatic transfer line machine;
(*v*) Transfer machines for production of valve bodies;
(*vi*) Six-station dial rotary transfer machines, etc.

Advantages :
1. Higher output.
2. Higher accuracy.
3. Lesser number of operators required.
4. Reduced cost of component.
5. Reduced production time.
6. Better utilization of floor space.
7. Possible to machine heavy and irregular parts.

Disadvantages :
1. High initial cost.
2. Suitable for high production industries only.
3. Control system is complex.
4. Breakdown of one machine leads to stoppage of all machines.

9.9.3.3. Transfer devices

The following transfer devices are used in transfer machines :

1. Roller conveyors.
2. Band and belt conveyors.
3. Chain conveyors.
6. Screw conveyor.
7. Slat conveyor.
8. Bucket conveyor.

4. Pneumatic conveyors.
5. Chute Conveyors.
9. Vibrating conveyor.
10. Magnetic belt conveyor.

1. ***Roller conveyors*** :

These can take any article that has a flat firm surface. Roller conveyors may be of two types :

1. Gravity-roller conveyors.
2. Powered type roller conveyors.

- The *gravity type roller conveyors* must have adequate fall to move the work. These conveyors can be quickly set up and are suitable for temporary handling arrangements such as loading of motor vehicles. They are used for inter-operational movement and provide line storage for limited quantities of work. Fig. 9.1. shows a gravity roller conveyor.

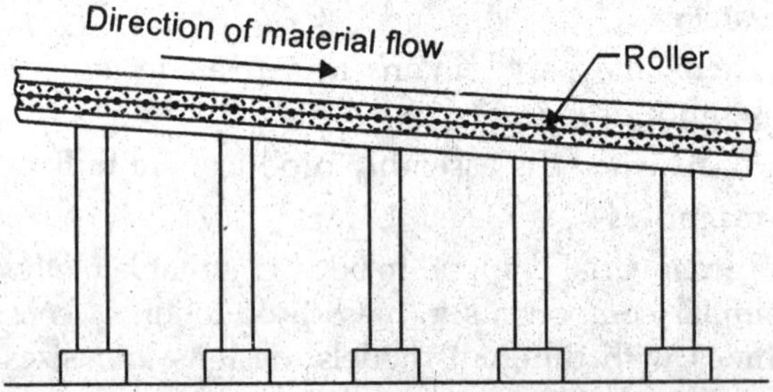

Fig. 9.1. A Gravity roller conveyor.

- When it is necessary to carry parts over horizontal runs or an incline, a section of *power driven roller conveyor* may be interposed. The whole length of a roller conveyor may of course be power driven, or power driven sections may alternate with gravity sections. The rollers may be driven by a belt which is held in contact with their undersides by idling rollers, or each roller may be fitted with a sprocket and driven by an endless chain. An alternative to the long chain driven is to drive each roller from its predecessor by a short endless chain, the first roller in the conveyor being similarly driven from the motor sprocket.

2. ***Band and belt conveyors*** :

- One of the most commonly used forms of power driven conveyors is the belt or band type in which the load is carried on an endless belt driven by the rotation of one or both of its terminal rollers (or pulleys). The terms 'belt' and 'band' include types made from rubber, fabric, impregnated with rubber or bitumen and endless bands of steel and wire-mesh.
- A conveyor belt must be flexible so that it may pass easily round its supporting terminal roller or pulley. As it may be called upon to withstand heat, a steel band or a belt woven of wire may be the solution to the problem. Resistance to corrosion and ease of cleaning when conveying plastic or sticky masses are other features that are desirable in some conveyor installation.

  The belt conveyor may serve both as a means of transport and as a support for a load while it is being worked upon or is under treatment. Although comparatively simple, belt conveyors vary somewhat according to their specific uses. They may be installed at almost any level and can almost be completely enclosed under a flow or arranged with the upper strand of the belt at floor level and the other strand beneath the floors when heavy loads are carried and it is necessary to eliminate lifting. A conveyor with its upper strand at bench or table height is perhaps the more useful type of installation, the lower strand sometimes conveying parts such as rejects. Suspended installations are used having the advantage of conserving floor space.

Components of a belt conveyor :

A belt conveyor system (Fig. 9.2.) has the following important parts :

1. Belt
2. Prime mover to drive the belt
3. Support for the belt rollers or idlers
4. Arrangement to maintain the belt tension
5. Arrangement for loading and unloading the belt
6. Arrangement to clean and protect the belt.

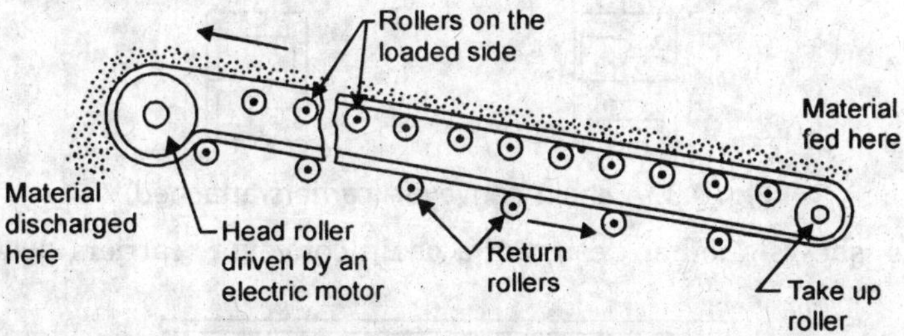

Fig. 9.2. A belt conveyor system.

Advantages : A belt conveyor possesses the following advantages :

1. High efficiency.
2. Low frictional loss.
3. Simple in construction.
4. Maintenance is easy and economical.
5. Low power requirement.
6. Can carry material either up or down an incline.
7. Capable of handling almost all types of bulk materials.
8. Large load carrying capacity.
9. Continuous in operation.
10. Silent in operation.

Uses : In chemical plants and in mass production assembly lines the materials are moved on conveyors. Ovehead conveyors between shops and production shops and between stages of manufacture and assembly are often planned to act as buffer storage units. The modern trend has been to integrate handling with processing operations.

3. ***Chain conveyors*** :

For the continuous transport of goods and component parts in warehouses and factories, the overhead system of chain conveyors is more advantageous than any other system of mechanical handling.

The use of this type of conveyor makes it possible to carry goods or components through a complete circuit, bringing the parts to the points where the operations are to be carried out and moving them out of the way where they are not required–turning around corners; communicating from one building to another without impeding ground traffic, and generally performing any movement which may be necessary.

The conveyor consists of a steel rack, curved in the vertical and horizontal planes as required and suspended from the roof of the factory or from brackets attached to the walls or the floor. On the lower flange of the girder, forks are fitted with rollers. Each fork carries a hook or an arm from which the parts or a platform, bin, wire basket, etc., are hung. Motive power is provided in by linking together with forks by an endless chain or wire driven by geared electric motor.

Fig. 9.3. shows a typical chain conveyor (carriers attached)

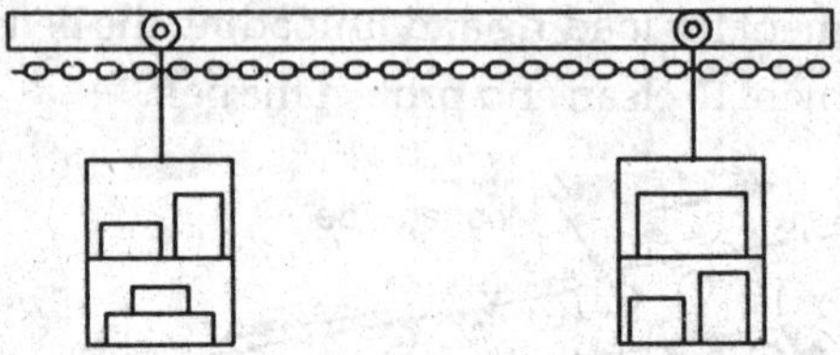

Fig. 9.3. A chain conveyor (carriers attached).

Fig. 9.4. shows another design of a chain conveyor (carriers detached).

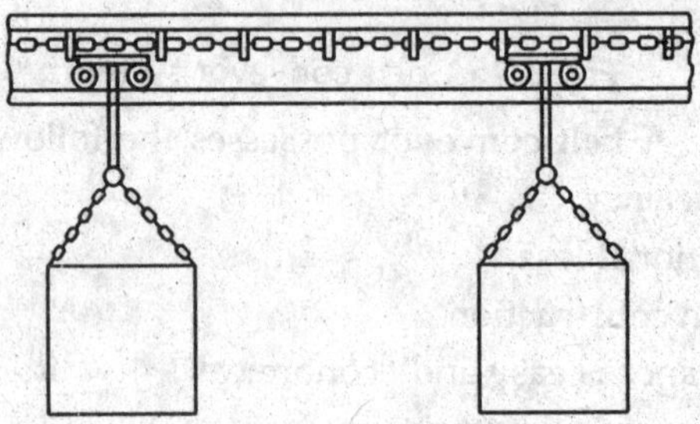

Fig. 9.4. A chain conveyor (carriers detached).

4. ***Pneumatic conveyors*** :

The pneumatic tube system, which is familiar feature in many department stores, is used in a number of factories for the transport, at a speed of about 9 meters per second, of small tool samples, matrices and similar parts between different buildings. Samples of iron or steel from a batch of castings, may be passed from the foundry to the works laboratory with minimum of delay.

A complete system consists of a series of pneumatic tubes connecting the various shops to a control desk. An impeller or turbine maintains a constant vacuum or pressure in the tubes, so that when a tubular carrier is inserted in the tube, it is carried by air pressure. By means of colour codes or numerals on the body of the carrier, the operator at the control desk is able to re-route the carrier to the correct department.

In more elaborate systems a number of metallic bonds on the body of the carrier operate switches. Which in turn energise relays to divert the carrier into the correct tube. This system, analogous to a dial telephone system, renders an operator unnecessary at the central exchange desk.

5. ***Chute conveyor*** : Refer to Fig. 9.5.

— It is used to *link* two handling devices.

— It is *inexpensive.*
— These conveyors are employed to provide accumulation in shipping areas and to convey items betwen floors.
— Their *limitation* lies in controlling the position of the items.

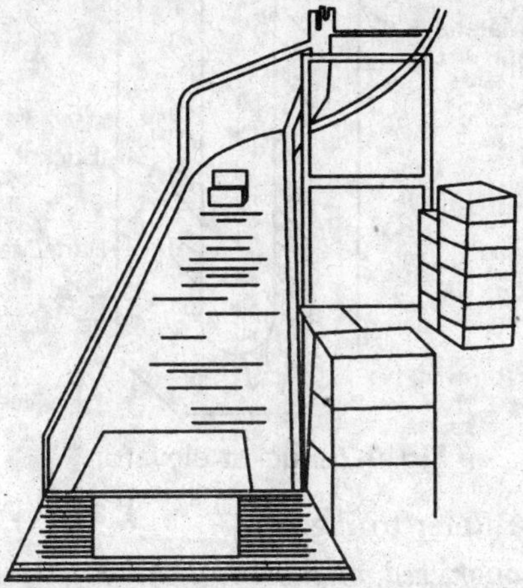

Fig. 9.5. Chute conveyor.

6. ***Screw conveyor*** :

A screw conveyor or spiral conveyor (Fig. 9.6.) consists of a revolving shaft provided with continuous or broken spiral flighting. The shaft operates within an enclosed or open casing or trough, and material is thus transferred from one place to another. Usually the screw is designed in such a way that it operates unidirectionally. An electric motor drives the screw at a suitable speed (through a gear train). Screw conveyors are commonly employed for conveying non-abrasive and fine materials. They can, however, be used to transfer abrasive materials if the casting is fitted with hardened liners and screw tipped with hard steel.

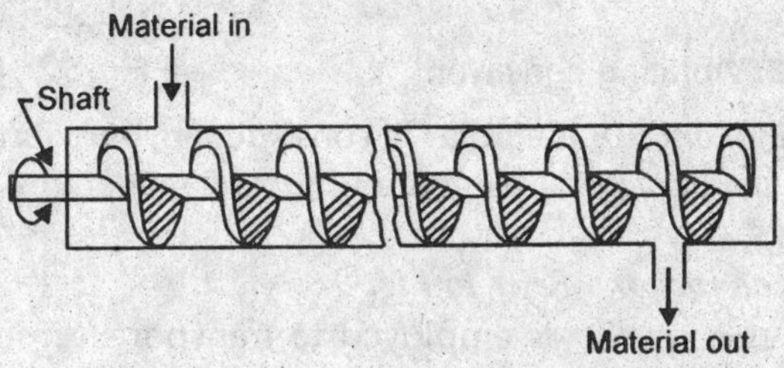

Fig. 9.6. A screw conveyor.

7. ***Slat conveyor*** :
— This type of conveyor uses discretely slats connected to a chain. The unit being transported retains its position, like a belt conveyor.
— The orientation and placement of the load is controlled.
— It is used for heavy loads or loads that may damage a belt.

8. ***Bucket conveyor*** : Refer to Fig. 9.7.
— It is used to move bulk materials in a vertical or inclined path. Buckets are attached to a cable, chain, or belt.

— The buckets are automatically unloaded at the end of the conveyor run.

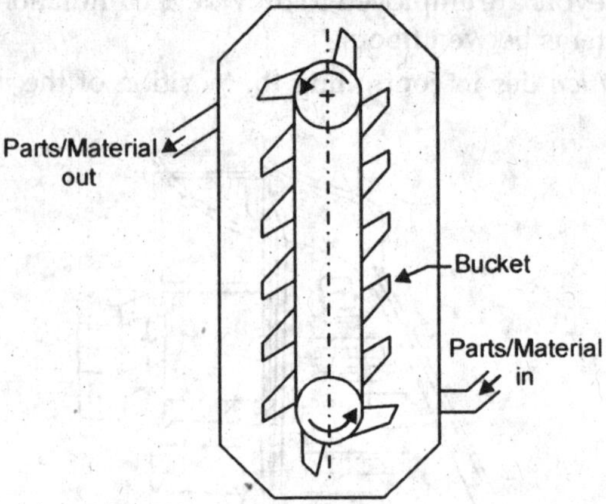

Fig. 9.7. Bucket elevator.

9. ***Vibrating conveyor*** : Refer to Fig. 9.8.
   — *It consists of a trough, bed, or tube.* It vibrates at a *relatively high frequency* and small amplitude in order to convey individual units of products or bulk material.
   — It can be used to convey almost all granular, free-flowing materials.

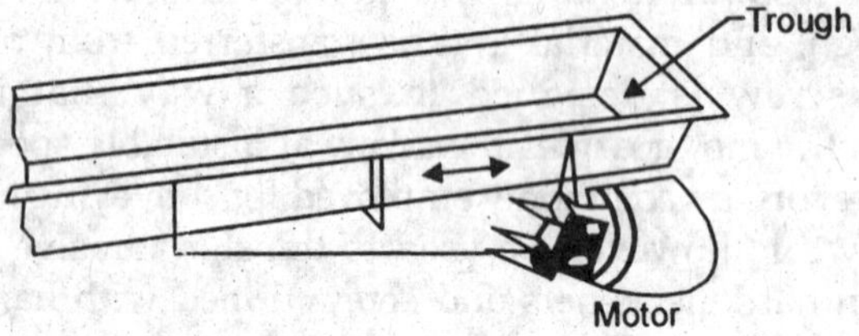

Fig. 9.8. Vibrating conveyor.

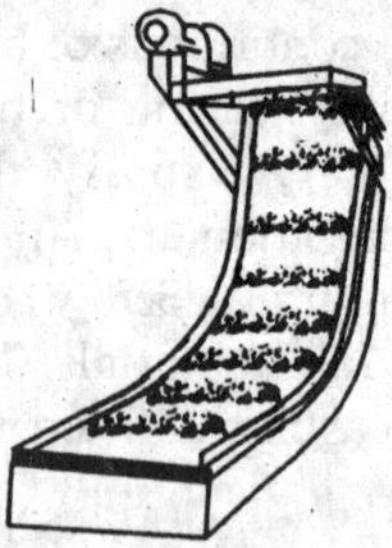

Fig. 9.9. Magnetic conveyor.

   — An *oscillating conveyor* is similar in construction, but vibrates at a *lower frequency and larger amplitude* (not as gentle) in order to convey larger objects such as hot castings.

10. ***Magnetic belt conveyor*** : *Refer to Fig. 9.9.*
   — This type of conveyor is employed to transport ferrous materials vertically, upside down, and around corners.

9.9.4. Feeding System

9.9.4.1. Introduction

— In a bulk material handling system, *a feeder is an important element because it is the means by which the rate of solids flow from a hopper or bin is controlled.*

— The feeders feed material to conveyors, processors and other equipment at a controlled rate *to maximise efficiency and production.*

— They may be considered as short conveyors that are available in multiple shapes and sizes. They differ from conveyors in that the *latter are only capable of transporting material, not modulating the flow rate.*

9.9.4.2. Characteristics of feeder

A feeder should possess the following *characteristics* :

1. The withdrawal of material through the outlet of the upstream device should be *uniform*.
2. The control of discharge rate over the necessary range should be of the *desired degree.*
3. Interface with the upstream device such that loads acting on the feeder from the upstream device are *minimal.*
4. Reliable and uninterrupted flow of material from some upstream device (say a bin or hopper).

9.9.4.3. Types of feeders

The various types of feeders/feeder devices are :

1. Screw type feeding device;
2. Vertical feeding device;
3. Cart on track feeder;
4. Tow feeder;
5. Trolley feeder device;
6. Monorail;
7. Sortation feeder device;
8. Hopper/Bowl vibratory feeding device;
9. Reciprocating feeders.

1. ***Screw type feeding device*** :
   — This device consists of a tube or U-shaped stationary trough through which a shaft-mounted helix revolves to push loose material forward in a horizontal or inclined direction.
   — It is widely used in the processing industry.
2. ***Vertical feeding device*** :
   — It is used to raise or lower a load to different levels of a facility.
   — It is employed for low-frequency intermittent vertical transfers/feeding.
3. ***Cart on track feeder*** :
   — This type of feeder is used to transport carts along a track. The carts are transported by a rotating tube.
   — To each cart a drive wheel is connected that rests on the tube which is used to vary the speed of the cart (by varying the angle of contact between the drive wheel and the tube). Carts are independently controlled.
4. ***Tow feeder*** :
   — This feeder uses towline to provide power to wheeled carriers such as trucks, dollies or carts that move along the floor. The towline can be located either overhead, flush with the floor, or in the floor.
   — It is generally employed when long distance and high frequency moves are required.
5. ***Trolley feeder device*** :
   — In this device, a series of trolleys supported from or within an overhead track is used. Trolleys are equally spaced in a closed loop path and are suspended from a chain. Carriers are used to carry multiple units of product. It does not provide for accumulation.
   — It is commonly employed in processing, assembly, packaging, and storage operations.

6. ***Monorail*:**
   - — On a monorail, one or more carriers ride. The carriers may be powered (electrically or pneumatically) or non-powered. Carriers can range from a simple hook to a hoist to an intelligent-vehicle-like device.
   - — A single-carrier, single-track monorail is similar to bridge or gantry crane. Multicarrier, track network monorail is similar to both a trolley conveyor, except that the carriers operate independently and the track need not be in a closed loop, a fixed-path automatic guided vehicle system, except that it operates overhead.
7. ***Sortation feeder device* :**
   - — The sortation conveyors are employed for *merging, identifying, inducting,* and *separating products to be conveyed to specific destinations.*
8. ***Hopper/Bowl vibratory feeding device* :**
   - — It is employed in many assembly machines to feed and orient most of *small parts.*
   - — The parts travel aıong a helical track, which passes around the inside wall of a shallow cylindrical hopper/bowl. The hopper is supported on sets of inclined leaf springs secured to heavy base. Electromagnet mounted at the base vibrates the hopper. The movement of the hopper is such that it has a torsional vibration about its vertical axis coupled with a linear vertical vibration.
9. ***Reciprocating feeders* :**
   - — Reciprocating (or plate) feeders are widely used, usually at the tail end of a conveyor or elevator to relieve pressure and drag.
   - — In this feeder, designed for feeding at a fixed rate, a plate is driven reciprocally under a head of bulk material. As the size varies the feed rate can be controlled easily. The feeders use rollers to support the belt and are engineered for maximum uptime.
   - — These feeders are employed in wet applications and for products ranging from sand and gravel to crushed stone that pass over belt feeders.
   - — Their cost is low and can handle a wide range of miscellaneous materials. However, they are not self cleaning and are unsuitable for highly abrasive materials.

9.9.5. Random Access Transport Systems (New Trends) – Automated Guided Vehicles (AGVs)

9.9.5.1. General Aspects

- — AGVs *are battery-powered vehicles that can move and transfer materials by following prescribed paths around the manufacturing floor.* They are neither physically tied to the production line nor driven by an operator like a fork lift.
- — Such vehicles have *on-board controllers* that can be programmed for complicated and varying routes as well as load and unload operations. The computer for the material handling system or the central computer provides overall control functions such as dispatching, routing, traffic control and collision avoidance.
- — AGVs usually *complement in* automated production line consisting of conveyor or transfer systems *by providing the flexibility of complex and programmable movements around the manufacturing shop.*
- — The first automated guided vehicle was developed in 1954 by A.M. Barrett.

9.9.5.2. AGVs equipment

Several types of equipment, *e.g.*, conveyors, rollers, self-powered monorails, carts, forklift trucks and various mechanical, electromagnetic, pneumatic, and hydraulic-devices and *manipulators* can be used to move materials.

— *"Manipulators"* are designed to be controlled directly by the operator, or they are automated for repeated operations, such as loading and unloading parts from machine tools, presses and furnaces. The manipulators are capable of *gripping and moving heavy parts and operating them as required between manufacturing and assembly operations.*

— Machines are often used in sequence where workpieces are transferred directly from machine to machine.

9.9.5.3. Applications of AGVs

- In Flexible Manufacturing System (FMS), industrial robots, specially designed pallets and *Automated Guided* Vehicles (AGVs) are extensively used.
- *AGVs,* which are the latest development in material movement in plants, *operate automatically along pathways with in-flow wiring or tapes for optical scanning and without any operator intervention.*
  — This transport system has *great flexibility* and is capable of *random delivery to different workstations.*
  — It *optimises the movement of material and parts in case of congestion around work-stations, machine breakdown (downtime) or failure of part of the system.*
- The movements of the AGVs are planned so that they interface with *Automated Storage/Retrieval* Systems (AS/RS), which utilise workhouse spaces efficiently and reduce labour costs.

9.9.6. Automated Storage Systems

9.9.6.1. General aspects

Industrial business firms are presently using traditional (non-automated) systems according to the personnel managing and budgetary considerations available to them. These storage systems include :

(*i*) Bulk storage system. (*ii*) Rack storage system.
(*iii*) Shelves and bins. (*iv*) Drawer storage system etc.

- The above systems require human labour to access for storage (which is *static*).
- In *highly 'automated system' loads are entered and retrieved by computer controls without any human participation except for input or control.*
- Automated systems have one *disadvantage* that they require *large investment and change in style of working.*

9.9.6.2. Objectives of automated storage systems

The automated storage systems have the following *objectives* to serve :

1. To improve stock rotation.
2. To improve security.
3. To increase capacity to store.
4. To increase storage density.
5. To recover factory or warehouse floor space which is used for storing work in process.

6. To reduce pilferage and provide safety to stores.
7. To reduce labour costs.
8. To improve customer service.

9.9.6.3. Automatic storage/retrieval systems

- This type of storage system allows the user to do store and retrieval operations with *speed and accuracy*.
- An AS/R system consists of one or more aisles and each is serviced by storage retrieval machine. The aisles have storage racks for holding the stored materials.
  — The retrieval machine can be crane or any other type.
  — Each aisle of AS/RS is provided with one or more input/output stations (also referred as pick up/delivery station) where material is delivered for storage and picked up. These input/output stations are computer controlled or manually operated or can be interfaced with any automated handling system.

Types of AS/R systems : Various types of AS/R system are enumerated and discussed as under :

(*i*) Unit load AS/R system	(*ii*) Deeplan AS/R system
(*iii*) Miniload AS/R system	(*iv*) Man on board AS/R system
(*v*) Automated item retrieval system	(*vi*) Vertical life storage/retrieval system.

(*i*) *Unit load AS/R system* :

- It is typically a large automated system to handle unit load stored on pallets or in containers.
- The system is computer controlled.

(*ii*) *Deeplan AS/R system* :

- It is high density unit load storage system which is used when large quantities of stock is to be stored but number of separate stock racks is small.
- This system stores ten or more loads in a single rack, are behind the next. Each rack is designed for flow with input on one side and output on another side.

(*iii*) *Miniload AS/R system* :

- In this storage system small loads are handled that are contained in bins or drawers.
- The S/R is specially designed to retrieve the bin and deliver it to output station at the end of aisles, so that individual item can be withdrawn from the bins. The bin or drawer must be retrieved back to its position or location.

(*iv*) *Man on board AS/R system* :

- In this system, a human operator, rides on the carriage of the machine (S/R) while the miniload system delivers an entire bin to the end of aisles pick station. The bin or container is then returned to its position in compartment.
- The man on board permits individual items to be picked directly at their storage locations.

(*v*) *Automated item retrieval system* :

- In this system, items are stored in lanes than bin or drawers.
- Any item to be retrieved is pushed from its lane and dropped into a conveyor for delivery.

(*vi*) *Vertical lift storage/retrieved system* :

- In this system a centre aisle is used to access loads in vertical position.

- Vertical lift, storage modules with heights of 10 m or more are capable of holding large inventories and help in saving floor space.

Applications of AS/R system :

Following are the *applications* of AS/R system :

1. Warehouse and distribution systems.
2. Storage of raw materials.
3. Work-in-process manufacturing :
   - — Computer control and tracking of materials;
   - — Buffer storage in production;
   - — Support in JIT;
   - — Knitting of parts for assembly.

9.10 FLEXIBLE MANUFACTURING SYSTEM (FMS)

9.10.1. Introduction

In manufacturing situations there are always conflicting requirements, the complete solutions for which are neither provided by manually operated machine tools nor by totally automated machines designed for mass production of one component. While manual machine tools are very flexible but productivity is poor. A lot of time is wasted in setting up tools, fixtures, components; loading/unloading of tools and workpiece. On the other hand the transfer line is very productive but is suitable for one or two component types. The flexible manufacturing systems (FMS) compromise for above two cases. *The FMS not only improves productivity but also provides the required flexibility enabling the factories to become more reactive to market demands.*

Flexible manufacturing system (FMS) is a manufacturing system *based on multi-operation machine tools, incorporating integrated computer control with associated support function of material handling (for automatic part handling and storage).* The system exhibits a degree of flexibility is the respect to the variation in parts (part family) which can be produced, this flexibility is not present in the transfer line concept. True FMS is possible with flexible machining centre capable of machining parts in random order, and it is integrated with the various assembly, testing and packaging operations using the appropriated software.

- *"Flexible Manufacturing System (FMS)" may be defined as a set of machines in which parts are automatically transported under computer control from one machine to another for processing.*
  - — Flexibility refers to the ability to respond effectively to changing circumstances. A *flexible system is one which is able to respond to change.*
- A FMS should be capable of coping with both *external changes* (*i.e.*, changes in the type, mix, processing requirements and quality of job allocated to the system, changes in the training and skill of operators assigned to the system), as well *internal changes or disturbances* (*i.e.*, machine and material handling system breakdowns, variability in processing time, operator absences, quality problems etc.)

9.10.2. Flexible Manufacturing Cell (FMC) and FMS

One of the most important developments in distributed numerical control (DNC) was the introduction of FMC and FMS.

— *The flexible manufacturing cell (FMC) generally has more than one CNC machine which are served by flexible material handling devices such as an "industrial robot".* For machining operations, the FMC may consist of machines like CNC lathes, milling machines, drilling machines, grinders, etc. and may even have automatic inspection stations. The job of the operator now reduces to loading the racks for the robot to pick up the parts, clamping the parts on to the machines, removing the finished parts, changing the tools and cutters, etc. (See Fig. 9.10)

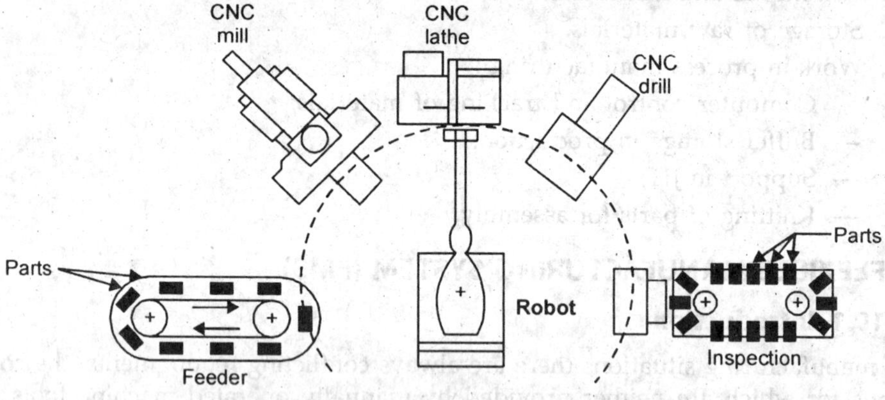

Fig. 9.10. Flexible manufacturing cell (FMC).

— FMC is generally designed for fixed process, and the parts flow sequentially between operations. *Use of FMC not only provides greater flexibility, but also reduces cycle time of the parts produced and the in-process inventory of parts.* Further, because of elimination of human error the product *quality improves.*

- When all the flexible manufacturing cells (FMCs) of a plant are interconnected, a flexible manufacturing system (FMS) is created. In fact, FMCs are small FMSs. Thus for creating a FMS, a several FMC are first installed and then interlinked.

9.10.3. Components of a FMS

The various components of a flexible manufacturing system (FMS) are :
Refer to Fig. 9.11 :

1. Machine tools.
2. Control system.
3. Handling system.
4. Planning systems.
5. CAD/CAM.

All these components have to be properly coordinated to make a functional system.

— A FMS uses various kinds of CNC machines, machining tools depending on the requirements.

— The *control system* guides the operations on the various machines to produce the required parts. It also regulates the flow of workpieces and monitors their location within the system through the material handling systems.

— The *handling system* could either be a separate unit (like *industrial robot*), or a part changer which is an integral part of the machine itself. In the latter case, cost is low as same control system and programming procedures are used for both machine and part handling. It utilizes less complicated pattern of movements. The industrial robot is best used for serving several machine in a so-called production cell.

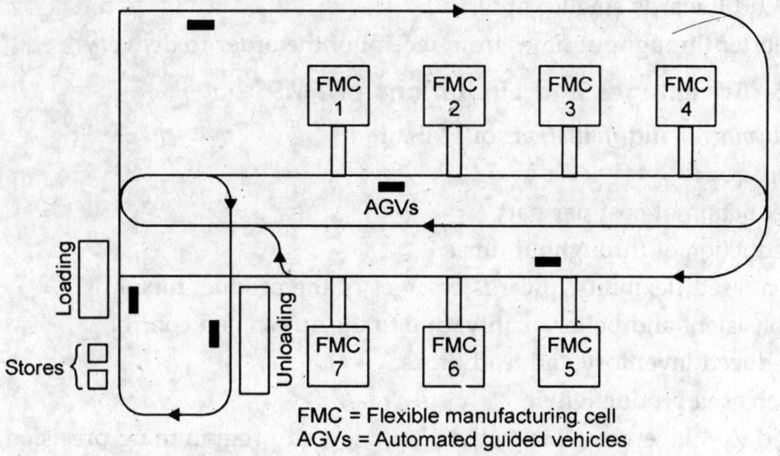

Fig. 9.11. Flexible manufacturing system (FMS).

— In principle, it is possible to build a fully automated factory where all tasks such as processing, material handling, tool changing, inspection etc. are accomplished without any assistance from operators. This however, has not yet been achieved and the human operators still play critical roles.

- *FMSs are versatile because they can produce a variety of parts and are adaptable since the system can be quickly modified to produce a completely different set of parts.*

  — A FMS consists of *several CNC machines which are served by computer controlled material handling devices and tool changing capabilities.* It can be easily *reconfigured* to produce a variety of parts because of the computer control and tool changing capability.

FMS levels : The different FMS levels are :

1. *Flexible Manufacturing Module (FMM)* :
   Example : A NC machine, a pallet changer and a part buffer.
2. *Flexible Manufacturing (Assembly) Cell [F(M/A)C]* :
   Example : Four FMMs and an AGV (Automated guided vehicle)
3. *Flexible Manufacturing Group (FMG)* :
   Example : Two FMCs a FMM and two AGVs.
4. *Flexible Production System (FPS)* :
   Example : A FMG and a FAC, two AGVs an Automated tool storage, and an Automated part/assembly storage.
5. *Flexible Manufacturing Line (FML)* :
   Example : Multiple stations in a line layout and AGVs.

9.10.4. Requirements for FMS

The various *requirements for FMS* are :

1. Better utilization of resources.
2. Shorter reaction times to market shifts.
3. Competitiveness/quality.
4. Greater variety of products and product versions.
5. Reduction is stocks.

6. Trend towards smaller orders.
7. Shorter throughput times from receipt of the order to delivery.

9.10.5. Advantages and Limitations of FMS

The *advantages* and *limitations* of FMS are :

Advantages :

1. Reduction of cost per part.
2. Reduction of throughput times.
3. Increased flexibility towards changes of the product mix.
4. Consistent and better quality, due to the automated control.
5. Reduced inventory and lead times.
6. Increased productivity.
7. Reduced inventory due to the planning and programming precision.

- The biggest *drawback* of the system is the unknown availability and reliability of the planned system.

Limitations :

1. High cost of programming.
2. Less degree of sophistication of fabrication and assembly processes.
3. Non-availability of reliable feedback devices for tool wear breakage, etc.
4. Substantial pre-planning activity.
5. Sophisticated manufacturing systems.

9.10.6. Characteristics and Features of a Typical FMS

Following are the *features and characteristics of a typical FMS :*

1. It is possible to produce at random all the variants and series of a product planned to be manufactured by a firm
2. It has the capability to quickly respond to any design changes in the product.
3. A major portion of the output of the engineering sector involves batch production and FMS offers immense cost and quality benefits for such requirements.
4. In FMS, there is no need of locking up the money in extensive stocks of finished parts due to production of a minimum batch of components in order to be economical. The work-in-progress is considerably reduced.
5. It is possible to have totally random mixes of models of a product proceeding down the line at any one time.
6. The system can be conceived in 15 to 20 minutes operations. If certain operations take longer, then multiples of similar machines can be installed in the line.
7. Components are loaded on a pallet. Means are provided to identify the exact model. Loaded pallets enter the line and wait at the start of the line until a signal that one of the first operation machines is vacant is obtained the pallets are loaded on a fixture. After the completion of the first operation, a tool in the spindle pushes in one of the coding pins and proximity switches in the line can then identify this as a first operation completed pallet. This pallet is then directed to second machine for second operation. If all the second machines are loaded this parallel may have to wait for sometime. At the end of the line the pallet is unloaded and returned via an overhead conveyor that runs above the centre track, to the start of the line.

8. The operator's intervention in the line is minimised by extensive use of touch trigger probes.
9. *Industrial robots can be used for material handling (loading/unloading), inspection activities, and assembly operations.*

9.11 COMPUTER INTEGRATED MANUFACTURING (CIM)

Computer integrated manufacturing *can be defined as the total integration of all aspects of design, planning, production, marketing and management thorough the use of computer into a single, smoothly operating system.*

— CIM is not a new manufacturing technology, it is a *new approach to manufacturing.*

— It got developed primarily in response to market demand for more product variety, better quality, and lower prices. Earlier attempts towards integration primarily attempted to lower the direct labour cost through the use of new equipment and processes.

— CIM approach can help *to improve quality, quicken response to market fluctuations and lower unit costs of production.*

— CIM is expensive and can be prohibitively expensive for small and medium-size companies, but with the *developments in microelectronics the cost is declining rapidly.* CIM plants with high level of integration in which computers and flexible automation produce the goods with very little human intervention do exist today. But plants with total integration and fully automated operations, a factory of future, is still several years away.

- The *Computer Integrated Manufacturing (CIM)* is the best context in which *industrial controls and sensors* can be understood. This concept, in some quarters, was formerly called *hierarchical control.* This concept is *applicable in both **"manufacturing"** and the **"process"** industries.* The structure is very similar, although different terms are used at certain levels, as discussed below.

Sensors and actuators. These devices provide the interface between the *controller and machine or process being controlled. **"Sensors"** translate some variables on the machine or process to a signal* (usually electrical) *that can be subsequently processed by the control equipment to obtain the current value of the variable of interest.*

Sensors can be either continuous or discrete.

Examples : *'Continuous'*— Measurement devices for temperature, pressures, flows etc.
'Discrete'—Proximity switches, limit switches, on/off level switches etc.

"Actuators" accept a signal from the control system and in some fashion impose this signal on the machine or process being controlled.

Examples : *'Continuous'*— Variable speed drives, pneumatic positioning valves etc.
'Discrete'—On/off motors, solenoid valves etc.

Machine or regulatory control. Functions at this level provide the basic regulation of the machine or process being controlled.

— For a ***machine,*** regulatory control is provided by *controllers* such as *numerical controllers or programmable logic controllers.*

— For ***processes,*** regulatory control is provided by *single-loop or multi-loop microprocessor based controllers.*

Cell control or supervisory control :

— In *"manufacturing facilities"* several machines are often grouped to form a ***cell.***

— In *"process plants"*, several unit operations are *interconnected* to form the process.

Plant control. At this level reside the various functions required for efficient operation of the entire plant. *Functions* at this level include :

— Keeping track of the raw materials and products;

— Routing production activities throughout the plant;

— Monitoring equipment utilisation etc.

The computers at this level communicate with the cell or supervisory computer in a plant Local Area Network (LAN)

Corporate :

— Normally implemented as part of the Management Information System (MIS) capabilities, functions at this level are directed to the long term or stratgic issues.

— An essential part of this function is the collection, analysis and presentation of information to corporate-level managerial personnel.

Barriers to CIM technology :

The following factors are the *main barriers to CIM technology* :

1. Shortage of expert technical manpower.
2. Shortage of expert computer power.
3. The need for a more powerful database management system (DBMS).
4. The manufacturing management who are not aware of the detailed technical benefits of automation vis-a-vis CIM.
5. The need for detailed mathematical models of physical and chemical processes.
6. The lack of standards for interfaces, communication network and programming languages.
7. A need to increase the capabilities of the majority of programmable automation devices through development of suitable sensory devices for product inspection, product identification and necessary guidance and control of CAM equipment through feedback processing.

9.12 MECHATRONICS AND CONCURRENT ENGINEERING

The *concept of "Mechatronics"* (originated by Japanese in the early 1970s) *is clearly related to the issues of* ***robotics and flexible automation.***

Mechatronics has been described as *"the union of mechanical and electrical engineering needed in producing the next generation of machines, robots and smart mechanisms for applications such as manufacturing large-scale construction and work in hazardous environment"*. It is a multi-disciplinary approach using integrated terms of designers, manufacturing engineers, purchasing, marketing, and sales personnel to design both the product and the system to manufacture it.

Complementary to the term mechatronics would be **concurrent** (*simultaneous*) **engineering,** which is a term more prominent in U.S. industries.

The focus is being given on four key areas in regard to concurrent engineering. These areas are, in many ways, fundamental to the existing discipline of industrial engineering and its future development include :

— Intelligent computer-aided design;

— Intelligent manufacturing systems;

— Human-machine interface;

— Maintenance.

The *function of "machatronics"* or in more general terms, *"concurrent engineering"*, is *to provide a national framework for the integration and implementation of the various sub-systems which comprise the disciplines and organisational functions of an engineering and manufacturing facility.*

9.13 COMPUTER AIDED PROCESS PLANNING (CAPP) SYSTEMS

— **Process planning** *involves the preparation and documentation of the plans for manufacturing the products. Computer Aided Process Planning (CAPP) represents the link between design and manufaturing in a CAD/CAM system.*

— The process planning is concerned with determining the sequence of processing and assembly steps that must be accomplished to make the product. The processing sequence is documented on a form called a *"route sheet"*. The route sheet typically lists the production operations, machine cells or work stations where each operation is performed, fixtures and tooling required, and the standard time for each task. During the last decade or so, there has been much interest in automatising the task of process planning by means of CAPP systems.

Following are the *two approaches around which the computer aided process planning systems are designed :*

1. Retrieval CAPP systems.
2. Generative CAPP systems.

1. Retrieval CAPP systems :

- These types of systems (also known as *"variant CAPP systems"* are based on the principles of *"Group technology and parts classification and coding"*.
- With these systems, a *standard process plan (route sheet) is stored in computer file and each part code number.* The standard route sheets are based on current part routing in use in the factory, or in ideal plan that is prepared for each family.

Fig. 9.12. shows the *general procedure for using the retrieval computer aided process planning system.*

- A large number of GT based variant process plans are available, such as :

  — NIPLA;

  — MICLASS;

  — CAPP etc.

2. Generative CAPP systems :

These systems *utilise an automatic computerised system consisting of decision logic, formulae, technology algorithms, and geometry based data to uniquely determine the many processing decisions for converting a part from a rough to a finished state.*

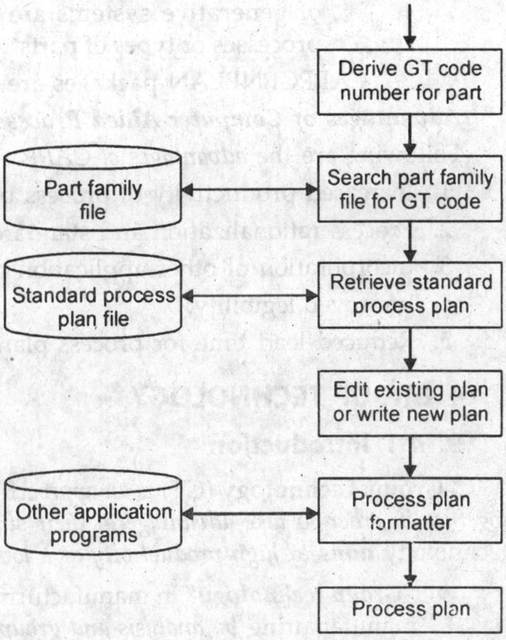

Fig. 9.12. General procedure for using one of the retrieval computer aided process planning systems.

In this approach, unlike the variant approach, *no standard manufacturing plans are predefined or stored.* Instead, the computer automatically *generates a unique operation sheet for a part every time the part is ordered and released for manufacturing.*

This system consists of the following two major components :

(*i*) ***Geometry based coding system :***

— This component *translates physical features and engineering drawing specifications into computer interpretable data.*

— The coding defines all geometric features, feature sizes and locations, and feature tolerances for all process-related surfaces. The coding scheme not only describes both the rough and finished states, but must be defined in similar terms for the part in each operation. Consequently, the coding scheme for a *generative system requires for greater detail than is required in the variant methods.*

(*ii*) ***Process knowledge :***

— This component is in *the form of decision logic and data in order to compare the part geometry requirements to manufacturing capabilities.* This logic is used to automatically determine the appropriate sequence, selecting the machine for each operation, determining cut planning or any other operation details subject to available tooling and fixturing, and calculating the setup time and cycle times for each operation.

- An additional element of the system is the *software for printing the operation sheet* and/or procedure sheet for instructing the operator according to the methods selected by the computer.
- Although a truly universal general system has yet to be developed, a number of generative systems are available today for specific manufacturing processes or types of parts.

CAPPGENPLAN packages are normally used.

Advantages of Computer Aided Process Planning (CAPP) :

Following are the *advantages of CAPP :*

1. Increased productivity of process planners.
2. Process rationalization and standardisation.
3. Incorporation of other application programs.
4. Improved legibility.
5. Reduced lead time for process planning.

9.14 GROUP TECHNOLOGY

9.14.1. Introduction

* **Group Technology** (GT) *is an approach in which similar parts are identified and grouped together in order to take advantage of their similarities in design and production.* The group technology ***aims*** at *high productivity at a low cost on short run production.*

- ***"Group technology"*** in manufacturing, is the replacing of traditional jobbing shop manufacturing *by analysis and grouping of work into families and the information of groups of machines to manufacture these families on a flow-line principle with the object of minimising setting times and throughout times.*

---

* Group technology was originated in Russia and was used during World War II.

— The traditional approach of any production unit is to use **"line layout"** where possible and **"functional layout"** in all other cases. In *"line layout"* generally the machines are laid out in a line in their sequence of usage. It is mainly used in simple process industries where all components made on the line use the same machines in the same sequence. In *functional layout* **"batch production"** is used and is based on process specialisation. In this type of factory the workers are divided into organisational units each of which specialises in a particular process or part of a process This type of layout result in *low machine utilisation and high wastage of time. The high cost of set-ups for small batches results in high manufacturing costs.*

- *Group technology is based on* **"product specialisation"**. *In this case each group of workers specialises in the production of a particular list or family of production and is equipped with all machines and equipment needed to complete these products.*
- Group technology is a new name for production system and many have used different names *viz.* **"Family grouping", "Family manufacturing"** etc. to describe this system.

— While examining the meaning of the 'Group technology' one finds that *"Group"* is a number of things classed together, and *"Technology"* is the science of industrial arts. Thus, *"group technology is the science of the industrial arts to a number of things classed together."*

- In *'group technology'* similar parts are arranged into *'part families'*. For example, a plant producing 10000 different part numbers may be able to group the vast majority of these parts into 40 or 50 distinct families. Each family would possess similar design and manufacturing characteristics. Hence the processing of each member of a given family would be similar and this *results in manufacturing efficiencies. These efficiencies are achieved by arranging the production equipment into machine groups, or cells, to facilitate workflow.*

— In product design, there are also advantages obtained by grouping parts into families. *These advantages lie in the classification and coding of parts.*

9.14.2. Advantage and Limitation of Group Technology (GT)

Following are the *advantages and disadvantages of group technology :*

Advantages :

The benefits of group technology are typically realised in the following areas :

(*i*) ***Product design benefits :***

- In the area of product design, the principle benefit is derived from the use of part classification and coding system.
- Group technology promotes *design standardisation.*

(*ii*) ***Tooling and set-ups :*** Group technology tends to promote standardisation of several areas of manufacturing. Two of these are *tooling* and *set-up.*

- In tooling, an effort is made to design jigs and fixtures that will accommodate every member of a part's family.
- The machine tools in a GT cell do not require drastic changeovers in set-up because of the similarity in the workparts processed on them. Hence, *set-up time is saved* and it becomes feasible to try to process parts in another so as to achieve a bare minimum of sets changeovers.

(*iii*) ***Materials handling*** : The group technology machine layouts lend themselves to *efficient flow of materials* through the shop. The contrast is sharpest when the flow line cell design is compared to the conventional process-type layout.

(*iv*) ***Production and inventory control*** : Production scheduling is *simplified* with group technology.

— Grouping of machines into cells reduces the number of production centres that must be scheduled.

— Grouping of parts into families reduces the complexity, and size of the parts scheduling, more attention can be devoted to the control of these parts.

Owing to the reduced set-ups and more efficient materials handling within machine cells, *manufacturing lead time and work-in-process are reduced.*

(*v*) ***Process planning*** : Proper parts classification and coding can lead to an automated process planning system.

— Even without an automated process planning system, reductions in the time and cost of process planning can still be accomplished. This is done through standardisation. New part designs are identified by their code as belonging to a certain parts family, for which the general process routing is already known.

(*vi*) ***Employee satisfaction :***

— The machine cell often allows parts to be processed from raw material to finished state by small group of workers. The workers are able to visualise their contributors to be firm more clearly. This tends to cultivate an improved worker attitude and a higher level of job satisfaction.

— Another employee-related benfit of group technology is that more attention tends to be given to product quality. Workpart quality is more easily traced to a particular machine cell in group technology. Consequently, workers are more responsible for the quality of work they accomplish.

Disadvantages/Limitations :

(*i*) The problem of identifying part families among the many components produced by a plant.

(*ii*) The expense of parts classification and coding.

(*iii*) Rearranging the machines in the plant into the appropriate machine-cells.

(*iv*) The general resistance that is commonly encountered when changeover to a new system is contemplated.

9.14.3. Part Families

A **part family** or **group** *is a collection of a parts, which either because of geometric shape and size or because of similar processing steps are required in their manufacture.* The parts within a family are different, but their similarities are close enough to merit their identification as members of the part family.

The biggest single obstacle in changing over to group technology from a traditional production shop is the *problem of grouping parts into families*. There are three general methods for solving this problems :

(*i*) Visual inspection.

(*ii*) Classification and coding by examination of design and production data.

(*iii*) Production Flow Analysis (PFA).

(*i*) ***Visual inspection :***

— This method is the *least sophisticated and least expensive.*

— It involves the classification of parts into families by looking at either their physical parts or their photographs and arranging them into similar groupings.

(*ii*) ***Classification and coding by examination of design and production data :***

— This method *involves classifying the parts into families by examining the individual design and/or manufacturing attributes of each part.* The classification results in a *code number that uniquely identifies the parts attributes.* This classification and coding may be carried out on the entire list of active parts of the firm or some sort of sampling procedure may be used to establish the part families.

Example : The parts produced in the shop during a certain given time period could be examined to identify part family categories. The trouble with any sampling procedure is the risk that sample may be unrepresentative of the entire population.

— The method of parts classification and coding *seems to be the most commonly used method today.*

(*iii*) ***Production flow Analysis (PFA) :***

— This method *makes use of information contained on route sheets rather than part drawings.*

— *Workparts with identical or similar routings are classified into part families.*

- All the above three methods are *time consuming and involves the analysis of much data by trained personnel.*

9.14.4. Formation and Establishment of Component Family

Formation of component family :

— Initially, a *pilot study* be carried out by examining simple components first, and gradually progressing to the more complex parts. However, the component family analysis should *preferably be based on the complete range of components from the products manufactured by the company.* The drawings and associated production data of the selected components are collected together, classified and sorted into code number order.

— The investigation should not be limited to the formation of potential component families, but also to access the necessary diversity in company operation. This will include the preparation of component family paper work and scheduling and control of the components into the machine group. While this investigation is in progress, the remainder of the drawing, new designs and obsolete drawings still liable to be called forward for spares replacement, should be coded.

— It is *only one-time exercise and effort is fully justified particularly, for variety control and computerisation of production planning and control.*

Establishment of component family :

— The coding provides the first stage in sorting and makes it possible to gather the components into families. If tabulated lists are visually examined, the naturally occurring families can be easily determined. These families are normally of the type *'identical in shape and function' and 'identical in shape but different in function'* and appear as blocks of near identical blocks of code number of listings. These will be the most obvious component families to begin to develop and establish machine groups.

— Finally, recheck the component tabulations and revise where necessary.

9.14.5. Collection of Production Data

— The number and sequence of machining operations, setting times and numbers of each component within a definite period of time need to be collected. By analysis of the *'machining operations and sequence'*, it is possible to *derive the types of machines required to form the machine group.* From the 'machining and setting times' and the numbers of each component, the potential load on the machine group may be established. Line balancing, however, cannot be adequately established until the component tooling requirements have been assessed. This is the *first stage* in *'family formation'*. The drawing of each component within the family is examined and the type and number of tools necessary to produce the component are determined. With the tooling analysis complete, the family is established and group layout balanced, *i.e.*, data should now include :

(*i*) Geometric shape.

(*ii*) Maximum and minimum size.

(*iii*) Material type.

(*iv*) Form and method of holding.

(*v*) Tools-type and holding.

(*vi*) Machine tools-type and capacity.

From the above information the *profile and parameters of the component family are constructed against which the acceptance or non-acceptance of new components into the family can be used.*

- Once a component family has been formed and integrated into a group layout, it does not necessarily have to remain static. Some components will become obsolete while new components will appear. The more the flexibility that can be built into the system, the more one can expect to get out of it.

QUESTIONS WITH ANSWERS

Q.9.1. Discuss briefly hierarchical levels in industrial automation systems.

Ans. The hierarchical levels in industrial automation systems may be as follows :

1. *Plant level :*
   - — It is the top level of a plant or an industrial automation system.
   - — The controller of the plant level gathers the management information from the area levels, and manages the whole automation system.
2. *Area level :*
   - — It consists of cells combined into groups.
   - — Cells are designed with an application-oriented functionality.
   - — The functions such as the setting of production targets, machine setup and shut-down, and emergency activities are prepared formulated by the area level controllers or process operators.
3. *Cell level :*
   - — The information flow, at the cell level, consists of the loading of programs, parameters and data.
   - — For the cell level operations, machine synchronisations and event handlings may require short response times on the bus. These real time requirements

are not compatible with time-excessive transfers of application programs, thus making adaptable message segmentation necessary. For achieving the communication requirements in this level, local area networks (LAN) have been used as the communication network.

4. *Field level :*

— This is the lowest level of the hierarchy of an industrial automation system and includes the field devices such as "actuators" and "sensors". These devices are also sometimes classified at the element level; their function is to transfer data between the manufactured product and the technical process.

Q.9.2. How can "flexibility" be achieved ?

Ans. *"Flexibility" is the capability of building multiple product times with the same manufacturing or assembly facility or the capability of converting the entire existing facility to produce different product lines with minimum retooling and lowest downtime. Therefore, flexibility can be achieved by organising and integrating the latest technology in he fields of :*

(*i*) Workpiece management;
(*ii*) Tool management;
(*iii*) Manufacturing equipment used in jobs order production;
(*iv*) Auxiliary equipment;
(*v*) Inspection and quality assurance;
(*vi*) Machine intelligence and sensor technology;
(*vii*) Hierarchical control system including software tools and network communications.

Q.9.3. What considerations should be taken into account while making a decision to automate a facility ?

Ans. The following parameters should be given due considerations while making a decision to automate a facility :

1. Economics;
2. The type of product;
3. The quantity and rate;
4. Skill level of workers available;
5. Reliability and maintenance;
6. Particular phase of automation;
7. Selective automation (instead of total automation).

Q.9.4. State the various reasons and limitations towards adopting 'automation' ?

Ans. ***Reasons :***

1. Lower costs;
2. Shortage of labour;
3. High cost of labour;
4. Increased productivity;
5. Competition;
6. Safety;
7. Reduced manufacturing lead-time.

Limitations :

1. Initial investment;
2. Cost of upgraded labour;
3. Labour resistance;
4. Automation is a difficult task;
5. Fluctuating demand.

Q.9.5. What strategies should be adopted for 'automation' ?

Ans. The following strategies should be adopted for automation :

1. Increased flexibility;
2. Specialisation of operations;
3. Combined operations;
4. Simultaneous operations;

5. Integration of operations;
6. On-line inspection;
7. Process control and optimisation;
8. Plant operations control;
9. Improved material handling and storage.

Q.9.6. In what ways a robot differs from a NC machine tool ?

Ans. A robot differs from a machine tools in the following respects :

1. As compared to a NC machine tool, a *robot is a lighter and more portable equipment.*
2. *Programming* of a robot is different from the *tool programming* used in NC machine tools.
3. The applications of robots are more general in nature as compared to a NC machine tool.

Q.9.7. Give the companion between 'Automated machine' and 'Robot'.

Ans. The comparison between 'Automated machine' and 'Robot' is as follows :

S. No.	*Aspects*	*Automated machine*	*Robot*
1.	*Type of job*	It performs a particular job only in the designed sequence.	In can be made to do different jobs at different times and in different sequences.
2.	*Programming*	It cannot be programmed.	It can be programmed to change the sequence of tasks.
3.	*Knowledge base, intelligence*	It has neither a knowledge base nor intelligence.	It is possible to impart intelligence base by suitable programming technique.

Q.9.8. Discuss briefly "Workcell and Interlocks".

Ans. Workcell :

- The term ***workcell*** *represents a collection of automated equipment and controls that acts as a means to coordinate all the activities within the robot workstation.* It is required because the root system by itself has a very limited utility. In order to utilise it for performing some really useful tasks, it has to be integrated with many other things like work parts, part feeders, conveyors, control devices, process machinery, tools, machine tools etc.
- A work cell need not essentially cover a single robot only, but may contain several robots. In a robot *"work station"*, some of the activities take place in a sequence and some others occur simultaneously. A controlling device, called a *"workcell controller"* or *"workstation controller"*, is, therefore suitably incorporated, which coordinates and regulates these different activities and ensures that they are performed in a proper sequence.

Interlocks :

- In *workcell control,* there has to be some means to check the continuation of sequence of the work cycle whenever needed. The means used to apply this check is known as an *"Interlock"*.
- The workcell carries *two types facilitates of interlocks* — one *"outgoing"* and the other *"incoming"*. The provision of interlocks facilitates an effective control over the workcycle by permitting the sequential activities to take place only when they should and preventing them when they should not.

Q.9.9. What are the advantages of using 'Automated Guided Vehicle System (AGVS)' in 'Flexible Manufacturing System (FMS)' ?

Ans. Following are the *advantages* of using AGVS in FMS :

1. Flexibility.
2. Real time monitoring and control.
3. Safety.

1. Flexibility :

- The route of AGVs can be easily altered and modified, simply by changing the guide path of the vehicles. This is more *cost effective* than modifying the fixed conveyor line or rail guided vehicles.
- It provides a *direct access* material handling system for loading and unloading FMS cells and accessing the automated storage and retrieval system.

2. Real time monitoring and control :

— AGVs can be monitored in real time because of the computer control. If the FMS centre system decides to change the schedule, the vehicles can be re-routed and urgent requests can be served.

— *AGVs are usually controlled through wires implanted on the factory floor using a variable frequency approach.*

3. Safety :

- AGVs can travel at a slow speed but typically operate in the range 10 to 70 m/min. They have *on-board microprocessor control* to communicate with the local zone controllers which direct the traffic and prevent the collision between vehicles as well as the vehicle and other objects.
- AGVs may also incorporate :
  — Warning lights;
  — Fire safety iterlocks;
  — Controls for safety in shops.

Q.9.10. Explain briefly how you can achieve vehicle guidance, routing and traffic control in a manufacturing plant employing AGVs.

Ans. The conventional conveyor systems are sometimes inadequate to fulfil the requirements of plants where production is carried out through interconnected workcells and where flexibility and rapid change-over times are of chief importance. Automated Guided Vehicles (AGVs) offer a variable solution for such needs.

- **AGVs** *possess intrinsic flexibility and capability of synchronous operations and easiness of integration with other automatic devices like robots, CNC, automatic storage/retrieval systems.*
- AGVs are designed for *automatic, efficient and flexible execution of transport tasks* in which a variety of loads (workpieces, fixtures, empty pallets, tool dispensers) have to be transported at regular intervals with varying frequencies and from a fairly large number of locations (input/output station set-up areas, machining centres and auxiliary machines, wait stations and storage areas).

Q.9.11. Explain briefly the 'Hierarchy of computers in manufacturing'.

Ans. Computers and computer driven devices (*e.g.*, CNC machine tools, robots), in the manufacturing firm tend to form a *"pyramidal control structure"* as shown in

Fig. 9.13. The structure represents the common structure that exists between the computers and computer driven devices in the factory.

The various computers in the hierarchy are fixed together by communication links to a distributed computer system. The individual links provide a computer communications network to forward data and information through the various level from the manufacturing operations all the way to the corporate computer level. In the opposite direction, product design, process plans, production schedules, machine commands and so on are passed down to the individual production cell. *The hierarchy allows for design engineering and manufacturing engineering function to be included within the computer network.*

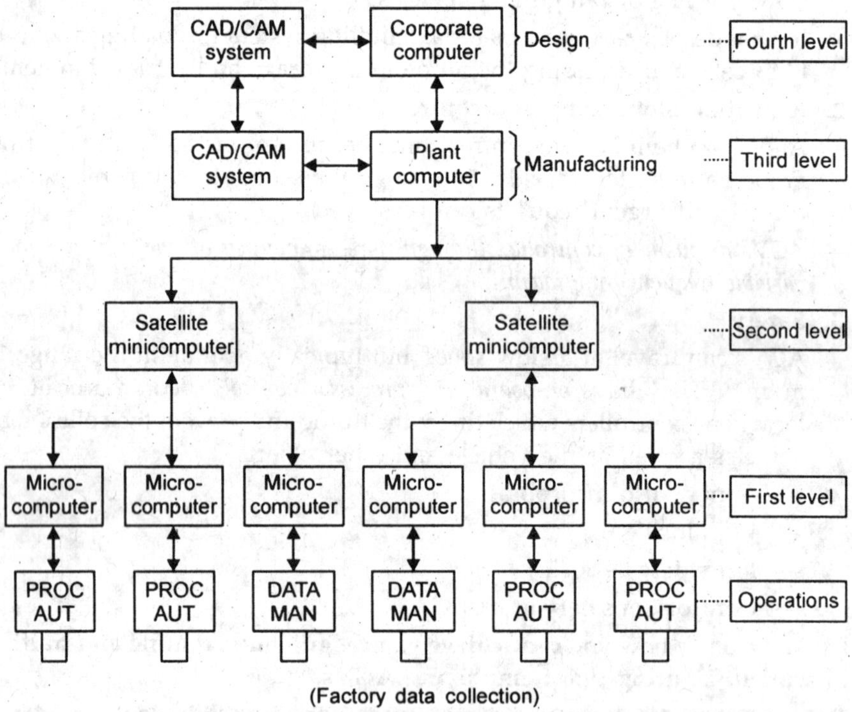

Fig. 9.13. Hierarchy of computers in manufacturing.

The following are the *benefits of the hierarchial approach to CIM* (Computer Integrated Manufacture) :

(*i*) In the hierarchial configuration, the software development can be managed more easily since computers are separated with pyramidal arrangement, programming for each project can be handled separately.

(*ii*) The hierarchial computer can be installed gradually rather than all at once.

(*iii*) This structure contains redundancy. In the event of a computer breakdown, other computers in the system are programmed to assume the critical task of the broken-down computer.

Q.9.12. What is the difference between Computer-Aided-Designed and Computer-Aided-Drafting ?

Ans. Computer-aided-design. It is used for the *initial design of the parts, sub-systems and the total system.*

This process defines all the parameters for construction of the system. Once this design is made, the numerical figures found out of calculations is presented in the form of a rough sketch to the draftsman.

Example : In the design of a beam, the dimensions are to be found out by doing a 'strength analysis'. This does not require any graphical representation. But once design is made, to convey it to the field people for actually building the beam, a *drawing* is required.

Computer-aided-drafting :

- It comes into picture *only after the design is completed.* It is *basically used to generate the working drawing to be given to the field or the shop floor to help them in production.*
- Computer-aided-drafting simply provides an *electronic drafting board* to the draftsmen and a lot of electronic tools for creation, modification, storage, retrieval, and outputting of the drawings.

Q.9.13. What is 'Computer Aided Manufacturing (CAM)' ? What are its applications ?

Ans. Computer Aided Manufacturing (CAM). *"Computer aided manufacturing is the use of computer to help in the manufacturing or production after the design process."*

- CAM basically produces instructions for the controller, a cutting or processing machine, to automatically perform a set of operations to produce that final product.
- The design outputs of a CAD application, can directly become the input to a CAM system and from the design, directly such a machine control program-called otherwise as *"Numerical Control Code"* can be generated.
  — Different machine controllers are available with every type of NC numerical control machine. Each of these controllers support one or more of these languages; ISO, APT, UNIAPT. The output of the CAM system can directly be input into these machines to perform the machining or processing required.

Applications of CAM. Following are the *applications* of CAM :

1. ***Computer monitoring and control.*** These are the *direct applications in which the computer is connected directly to the manufacturing process for the purpose of monitoring or controlling the process.*
2. ***Manufacturing support applications.*** These are the *indirect applications* in which computer is used in support of the *product operations* in the plant, but there is no direct inteface between the computer and the manufacturing process.

- Computer monitoring and control can be separated into monitoring applications and control applications. *"Computer process monitoring" involves a direct computer interface with the manufacturing process and associated equipment and collecting data from process.* The computer is not used to control the operation directly. The *control* of the process remains in the hands of *human operators,* who may be guided by the information compiled by the computer.

Q.9.14. What is CAD workstation ?

Ans. *The CAD workstation is the system interface with the outside world.* It represents a significant factor in determining how convenient and efficient it is for a design to use the CAD system. The workstation must accomplish the following *five functions :*

1. To interface with the central processing unit.
2. To generate a steady graphic image for the user.

3. To provide digital descriptions for graphic image.
4. To translate computer commands into operating functions.
5. To facilitate communications between the user and the system.

Q.9.15. List the various input and output devices used for graphics.

Ans. ***Input devices for graphics :***

- Cursor pad or cursors
- Thumb wheels
- Light pens
- Mouse
- Digitizing tablet and table
- Track ball
- Control disc
- Function keys.

Output devices :

The output devices are essentially termed as peripherals associated with the computer. Some of the output devices do not require CRT (Cathod Ray Tube) or interactive display devices. The hard copy output is printed on a paper, magnetic tape, or slide film which can be read and visualised and transported from one place to another without any distortion. Graphics hard-copy devices may be categorised in the three major types namely, *pen plotters, electrostatic printers, plotters,* and *computer-output microfilm units.* There are some other types of graphics output equipments such as *photographic devices, ink-jet* and *laser plotters, and computer controlled copies.*

Q.9.16. Discuss briefly the steps in the 'Numerical Control (NC)' in manufacturing.

Ans. Following are the steps which must be accomplished to utilise NC in manufacturing :

1. Process planning.
2. Part programming.
3. Tape preparation.
4. Tape varification.
5. Production.

1. Process planning :

- The engineering drawing of the work part must be interpreted in terms of the manufacturing processes to be used. This step is referred to as process planning and it is concerned with the preparation of a route sheet.
- The *route sheet* is a listing of the sequence of operations which must be performed on the work-part.

2. Part programming :

- A part programmer plans the process for the portions of the job to be accomplished by NC.
- There are two ways to program for NC :

(*i*) Manual part-programming.

— In *manual part programming,* the machining instructions are prepared on a form called a part *program manuscript.*

— In computer-assisted part programming much of the tedious computational work required in the manual part programming is *transferred to the computer.*

3. Tape preparation :

- A punched tape is prepared from the part programmer's NC process plan.
  - — In *manual part programming* the punched tape is prepared directly from the part program manuscript or a type writer like device equipped with tape punching capability.
  - — In *computer assisted part programming,* the computer interprets the list of part programming instructions, performs the necessary calculations to convert this into detailed set of machine tool motion commands, and then controls a tape punch device to prepare the tape for the specific NC machine.

4. Tape varification :

- After the punched tape has been prepared, a method is usually provided for *checking the occuracy of the tape.*
  - — Sometimes the tape is checked by running it through a computer program which plots the various tool movements (or table movements) on paper. In this way, major error in the tape can be discovered.
  - — The "acid test" of the tape involves trying it out on the machine tool to make the part. A foam or plastic material is sometimes used for this layout.

Programming errors are not uncommon, and it may require about three attempts before the tape is correct and ready to use.

5. Production :

- The final step in the NC procedure is to use the NC tape in production. This involves ordering the raw workparts, specifying and preparing the tooling and any special fixturing that may be required, and setting up the NC machine tool for the job.
- The machine tool operator's function during production is to load the raw workpart in the machine and establish the starting position of the cutting tool relative to the workpiece.
  - — The *NC system* then takes over and *machines the part according to the instructions on the tape.*

Q.9.17. Discuss briefly "Applications of Numerical Controls."

Ans. Now-a-days the NC systems are widely used in industry especially in the *metal working* industry. By far the most common applications of NC is for metal cutting machine tools. Within this category, NC equipment has been built to perform virtually the entire range of material removal processes including :

— Milling; — Drilling and related processes;
— Boring; — Turning;
— Grinding; and — Sawing.

Following are the *general characteristics of production jobs in metal machining for which numerical control would be most appropriate :*

(*i*) The part geometry is complex.
(*ii*) The parts are processed frequently and in small lot sizes.
(*iii*) Many operations must be performed on the part in its processing.
(*iv*) Much metal needs to be removed.
(*v*) Engineering design changes are likely.
(*vi*) Close tolerances must be held on the workpart.

(*vii*) The parts require 100 percent inspections.

(*viii*) Expensive part where mistakes in processing would be costly.

Q.9.18. Discuss briefly the steps involved in a computer aided engineering design problem.

Ans. CAD/CAM technology was initiated in the area space industry but is now widely used in all industries. It can be defined most simply as *use of computer to translate a product's specific requirements into the final physical products.* With this system a *product is designed, produced and inspected in one automatic process.*

The various steps involved in computer aided engineering design problem are shown in Fig. 9.14.

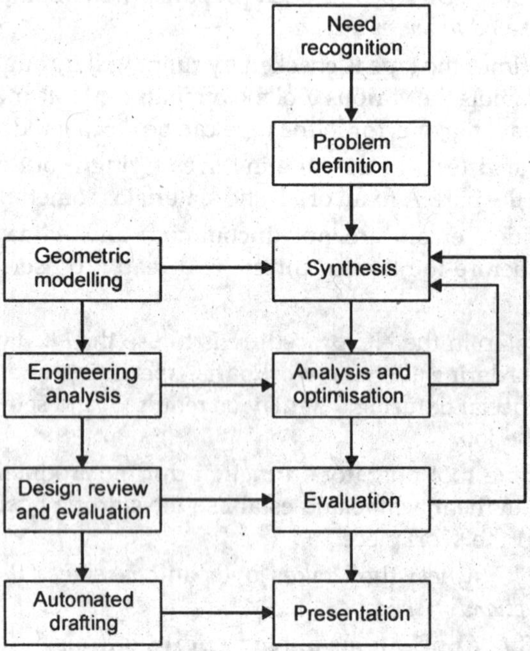

Fig. 9.14. Steps involved in computer aided engineering problems.

Q.9.19. What is CIM ? List some of the technologies of CIM.

Ans. • **Computer Integrated Manufacturing (CIM)** *is the automation of the entire manufacturing process with computers.*

— Designing a CIM system means applying systems thinking to the manufacturing enterprise ; in brief, it means viewing the organisation as a unit with certain inputs and certain desired output and designing systems that are both *computer-based and people-integrated to cause the inputs to be transformed into the outputs.*

• Some of the technologies of CIM include :

(*i*) Computer-aided design. (*ii*) Numerical control.
(*iii*) Robotics. (*iv*) Expert systems.
(*v*) Automatic guided vehicles. (*vi*) Computer-aided testing.
(*vii*) Automatic assembly. (*viii*) Communication networks.
(*ix*) Remote sensing. (*x*) Digital instrumentation.

Q.9.20. Discuss briefly the components of "Computer Integrated Manufacturing (CIM)".

Ans. The components of CIM are discussed below :

1. Computer-aided design and drafting (CADD) :

- CADD has evolved from drafting automation; automated drafting systems simplify many aspects of its creation.
- The engineering portion of a CIM organisation must use computer-aided design, not computer-aided drafting and must produce three-dimensional geometric data.
  — The three-dimensional geometric data can take three common forms : *Wire frame, surfaces* and *solids,* in order of progressively diminishing abstraction.

2. Computer-aided engineering (CAE) :

- It is the term used to describe computer-based engineering analysis of *digital models.*
- The most popular form of CAE for *mechanical engineers is finite-element analysis.* In *electronics,* there are *various forms of circuit analysis* that are applied in this manner.
- For CIM to be efficient, the model created in the computer-aided design process should be able to be digitally tested, through the design verification process that follows initial design.

3. Computer aided manufacturing (CAM) :

- It usually refers to *numerically controlled machine tools,* the means of controlling them and the associated computer-based systems within the factory.
- This category also includes other physical devices such as :
  — Flexible machining system (FMS);
  — Robots;
  — Automatic guide vehicles (AGVs); and
  — Data control systems, such as Manufacturing Requirments Planning (MRP) software.

4. Data base management systems (DBMS) :

- CIM, is in a sense, *data based manufacturing.*
- A robust CIM data base must be equally efficient with graphical and non-graphical data. Recent development in data base management technology have made the implementation of CIM easier than before.

5. Networks :

- Networking is the most important technology for CIM : The digital connectivity that allows the data to flow from design through production to field support and back is key to the "integrated" aspect of CIM.
- Salient features of networks include :
  — Topology;
  — Physical media;
  — Protocols;
  — Bandwidth;
  — Cost.
- Popular networking schemes include :
  — *Token ring and collision sensing* protocols;

— Ring and star topologies;

— *Coal and twisted-pair physical media.*

6. **Computational hierarchies :**

- **In addition to the engineering workstations and servers, the engineering data management network and accounting computers, many other computers serve** the CIM organisation. Most of them are *programmable logic controllers* or PLCs, simple computers designed for integration into industrial systems and, therefore, offering a particularly a wide range of input-output options, as well as a variety of robust installation possibilities.

Q.9.21. Discuss briefly 'Generic CIM architecture'.

Ans. CIM involves a great many functions within an organisation. The complexity of this relationship is illustrated in Fig. 9.15.

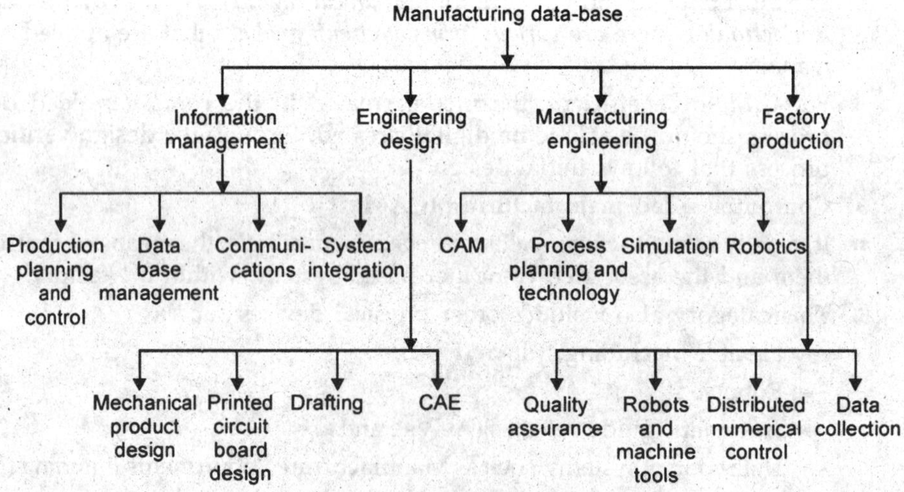

Fig. 9.15. Generic CIM architecture.

Q.9.22. What is the basic philosophy of group technology ?

Ans. Basic philosophy of group technology is to identify and bring together related or similar parts and processes, to take a advantage of the *similarities* which exist between them during all stages of design and manufacture. In this way *part design and process can be standardised and families of like parts can be produced efficiently and economically.*

It may be borne in mind that *there exists relationship between the final or finished products and the components from which they are fabricated.*

Q.9.23. What advantages can be expected by introducing group technology in an organisation ?

Ans. Following are the *advantages of group technology in manufacturing :*

1. Reliable delivery.
2. Reduction of material handling.
3. Use of robots for material handling becomes useful.
4. Space utilisation is more adequately possible.
5. Smaller varieties of tools, jigs and fixtures are required.
6. Improved quality of product and less amount of wastage is possible.

7. Improved resource utilisation results in greater amount of output.
8. Finish stock levels are also reduced.
9. Satisfactory work program.
10. Job satisfaction.
11. Standardisation of part design and minimisation of design duplication.

Q.9.24. Explain the concept of group technology and U-line in the modern production system.

Ans. Group technology :

- Group technology is a concept of grouping machines in *'Group technology cells'* so that parts having similar geometrical and processing features are processed in the same set or group of machines.
- Group technology layout is different from process type layout in which machines are grouped according to their function. Flow of parts through a factory with a group technology layout is *more streamlined* than the flow in a process layout when quantity to be produced/part vareity or (Q/P) ratio is more.
  - — Very small Q/P ratio suggests the use of process layout and very large Q/P ratio support the use of a flow shop.
  - — *Moderate Q/P ratio demands a group technology layout.*

U-line layout. Refer to Fig. 9.16.

- In flow shop type situations, various arrangements of machines are possible. The common ones are I flow or line flow; L flow when space is a problem, U flow which has feeding to and ejection from the line at the same end.

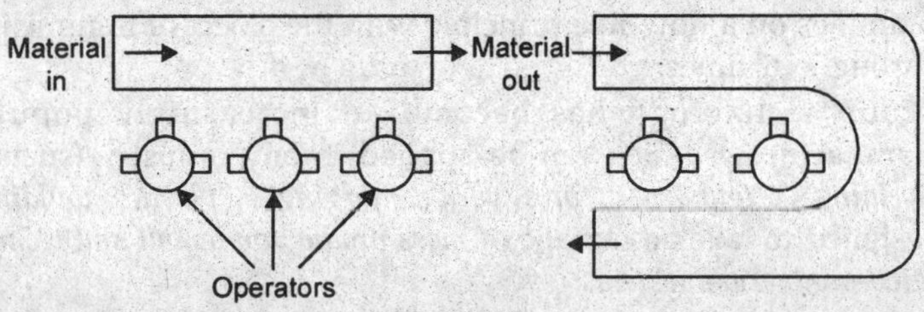

Fig. 9.16.

- A U-line is *easier to supervise.* A U-line can also adapt to different design and give different output rates through the same number of operators. When the flow rate is less, one operator can look after two machines with ease; the machine in front and the machine behind him.

Q.9.25. (*a*) What is manufacturing flexibility ?

(*b*) Give the classification of flexibility.

Ans. (*a*) The "manufacturing flexibility" is defined as *the ability to absorb various disturbances which occur in production systems, as well as the ability to incorporate and exploit new technological advances and work practices.*

(*b*) "Flexibility" can be classified as follows :

1. Machine flexibility;
2. Process flexibility;
3. Routing flexibility;
4. Product flexibility;
5. Volume flexibility;
6. Expansion flexibility;
7. Labour flexibility;
8. Material handling flexibility;
9. Operation flexibility.

Q.9.26. Explain briefly "Automated Material Handling (AMH) Systems".

Ans.
- Automated material handling systems are meant for *storing and moving materials and products on the factory floor.*
- These systems *work under "computer control".*
- They find *wide applications in shuttling workpieces between workstations, transporting work-in-process between different stages of manufacturing.*
- Material handling systems may be *simple conveyors, monorails, tow lines, carts or automated guided vahicle.* All the above material handling devices are controlled by a computer. These systems find wide applicability in flexible manufacturing systems (FMS).
- The other type of automatic material handling (AMH) is automatic storage and retrieval system (AS/RS); it is used for storing or retrieving materials in a vertical space.
- Robots are very special type of AMH and FMS and CIM need better management of AMH.

Q.9.27. What is group technology ? State its area of application.

Ans. Group technology is the realisation that many problems are similar, and that by grouping similar problems, a single solution can be found to a set of problems, thus saving time and efforts.

Group technology is a manufacturing philosophy or principle whose basic concept is to *identify and bring together related or similar parts and processes, to take advantage of similarities which exist, during all stages of design and manufacturing.* It is the replacing of traditional jobbing shop manufacture by the analysis and grouping of work into families and the formation of group of machines to manufacture these families on a flow-line principle with the object of minimising setting times and throughout times.

- Group technology has become an increasingly popular concept of manufacturing which can be applied in any industry (such as *"machining" "welding", "foundry", "presswork", "forging", "plastic moulding"* etc.) that is designed to take advantage of *mass production layout and techniques, in smaller batch-production systems.*

Features of group technology are shown in Fig. 9.17.

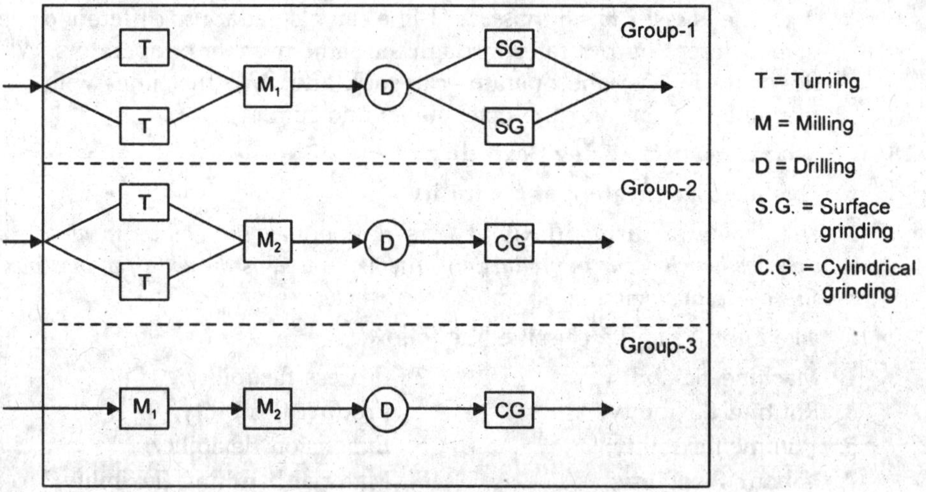

Fig. 9.17. Group layouts.

Q.9.28. Give the reasons for adopting group technology.

Ans. Modern manufacturing industries are facing a lot of challenges caused by growing international competition and fast changing market demands. These challenges which are examplified in the following list, have and can be successfully met by group technology :

1. There is an industrial trend toward the low-volume production (small lot size) of a wider variety of products in order to meet the rising demand for specifically ordered products in today's affluent societies. In other words the share of the batch-type production in industry is grouping day after day, and it is anticipated that 75 percent of all manufactured parts will be in small lot sizes.
2. As a result of the first factor, the conventional shop organisation (*i.e.,* departmentalisation by process) is becoming very inefficient and obsolete because of the wasteful routing paths of the products between the various machine tool departments.
3. There is no need to integrate the design and manufacturing phases in order to cut short the lead time, thus winning a competitive situation in the international market.

Q.9.29. What do you understand by 'Production flow analysis' ? Explain.

Ans. Production Flow Analysis (PFA) is a *method for identifying part families and associated grouping of machine tools.* It neither uses a classification and coding system, nor uses part drawings to identity families. Instead PFA is *used to analyse the operation sequence and machine routing for parts produced in the given shop. It groups parts with identical or similar routings together.* These groups can then be used *to form logical machine cells in a group technology layout.*

Since PFA uses manufacturing data rather the design data to identify part families, it can overcome the following two possible anomalies :

— *First,* parts whose geometries are *quite different* may nevertheless require similar or indentical process routings.

— *Second,* parts whose geometries are *similar* may nevertheless require process routings that are quite different.

- However, the *'disadvantage'* of using production flow analysis is that it *provides no mechanism for rationalising the manufacturing routings.* It takes the route sheets the way they, are, with no consideration being given to whether the routings are optimal or consistent or even logical.

The *procedure in production flow analysis* can be organised into the following "steps" :

(*i*) Data collection.	(*ii*) Sorting of process routings.
(*iii*) PFA chart.	(*iv*) Analysis.

Comments on PFA :

— The *weakness* of PFA is that the data used in the analysis are derived from production route sheets. The routings may contain processing steps that are non-optimal, illogical, and unnecessary. Consequently the final machine groupings that result from the analysis may be suboptimal. Notwithstanding this weakness, *PFA has the virtue of requiring less time to perform than a complete parts classification and coding procedure.* It therefore provides a technique that is attractive to many firms for making the changeover to a group technology machine layout.

Q.9.30. What is 'cluster analysis' ? Explain.

Ans. **"Cluster analysis"** *is employed for the analysis of production flow analysis chart to determine feasible group of processes and their respective packs of parts.* It makes use of algorithms for the study of similarities between objects in a quantitative manner as compared to the classification techniques, which appears to be descriptive.

"Clustering" may be defined as the *science of the classification of objects based on their possession or lack of defined characteristics.* This technique shows an approach to study the similarities between a diverse population of objects in a quantitative manner.

Clustering consists of the following *three stages.*

1. ***Preparing a post-operation matrix.*** This shows whether certain features (like a keyway on shaft) are present or absent.
2. ***Computing a similarity coefficient matrix.*** The bases of this is the extent to which the parts share common characteristics. In this case, coefficient would have a value one (1) when parts are identical and ten (10) when they have no common entity.
3. ***Performing a clustering analysis.*** In this case, the similarity between each pair of objects is examined and group of objects formed such that within each group, the objects are similar to each other according to set of rules which have been formulated previously.

Q.9.31. Discuss briefly "cellular and flexible manufacturing".

Ans. For the development of effective quality systems and automation, low volume is a major obstacle. When only a few parts will be produced little capital is made available for purchasing automated systems and up-to-date test equipment. In addition, other benefits resulting from learning curves and quality feedback may not be possible. Long lead times, high levels of inventory, and poor levels of quality usually result. *Cellular manufacturing* using the techniques of group technology and *flexible manufacturing systems* was developed to help solve these problems. **Cellular manufacturing** *increases the apparent production volumes by grouping similar parts.*

- The **Flexible Manufacturing System** (FMS) is a different approach to development of a specialised production line suitable for many different parts *using sophisticated computer control and a material handling system.*
  - **FMS** *consists of a group of manufacturing workstations connected together by an automated workpart handling system.* The system is capable of simultaneous processing of a variety of different part types at the various workstations under program control.
  - FMS system, however, is a *very expensive bridge for the gap between high-production transfer lines and lower production rates.*

Q.9.32. Explain briefly "Group technology machine cells".

Ans. The traditional view of GT includes *the concept of "GT Machine cells"—groups of machines arranged to produce similar part families.* This cellular arrangement of production equipment is *designed to achieve an efficient work flow within the cell.* It also results in *labour and machine specialisation* for the particular part families produced by the cell. This presumably raises the productivity of the cell.

Although these advantages exist in GT machine cell, it is a matter of considerable inconvenience and disruption for the shop to make the conversion from a conventional process type layout to the GT cell layout. Today many practitioners argue that it is possible to achieve a good share of the benefits of GT without physically rearranging the machining into cells.

The organisation of machines into cells can follow one of three general patterns :

1. **Single machine cell.** The single machine approach can be used for work parts, whose attributes allow them to be made on *basically one type of process*, such as turning or milling.
2. **Group machine layout :**
   - The group machine layout is a cell design in which several machines are used together, with no provision for conveyorised parts movement between the machines.
   - The cell contains the machines needed to produce a certain family of parts.
3. **Flow line design :**
   - The flow line cell design is a *group of machines connected by a conveyor system.*
   - This design *approaches the efficiency of an automated transfer line.*

Q.9.33. (*a*) Explain briefly various types of "Transfer machines".

(*b*) What are the advantages and disadvantages of transfer machines ?

Ans. (*a*) *"Transfer-type production machines"*, frequently designated as *automated machines, complete a series of matching operations at successive stations and transfer the work from one station to the next.* They are in effect a *production line of connected machines* that are synchronized in their operation so that the workpiece, after being loaded at the first station, *progresses automatically through the various stations to its completion.*

"Transfer machines" perform a variety of *machining, inspecting, and quality control functions.* They drill, mill, hone, and grind, as well as control and inspect the operations.

Types of transfer machines :

Transfer machines are of the following *three* types :

1. Rotary indexing table transfer machines.
2. In-line transfer machines.
3. Drum type transfer machines.

1. Rotary indexing table transfer machines :

In rotary transfer system the workpiece are held in fixtures on a continuous rotating table. The rotating table brings the workpieces under different machines (Fig. 9.18.)

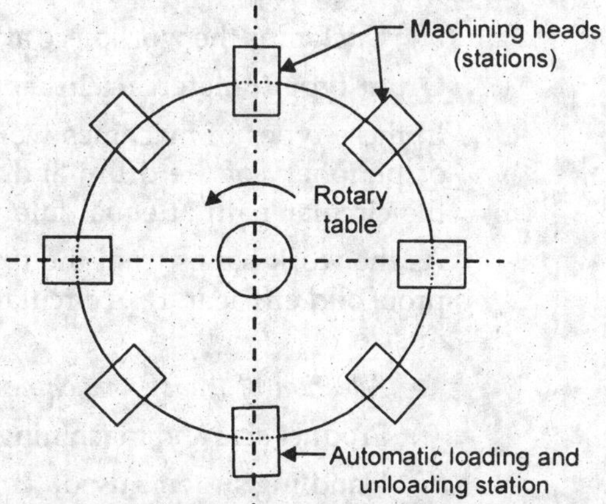

Fig. 9.18. Arrangement of rotary indexing table transfer machine.

Owing to the problems of rigidity and maintenance of proper accuracy, the table size is always limited and, hence, is the number of *stations* fixed on this type of machine. Usually, such a machine carries 6 or 8 stations around it, although for smaller workpieces this number can be as high as 16. The components size and required number of work stations are the main factors which affect the determination of table size and the number of stations.

- This method is quite compact and permits the workpiece to be loaded and unloaded at a *single location without having to interrupt the machining.*
- This type of arrangement is *best suited for automatic assembly of a product.*

2. In-line transfer machines :

In this arrangement the workpiece is held in a *fixture* or *special "pallet"*. The fixtures are located and clamped in proper position. A schematic diagram of an In-line transfer machine is shown in Fig. 9.19.

"Palletized" workholding fixtures secure the transmission during all operations. Pallets are often carried in conveyors which are indexable.

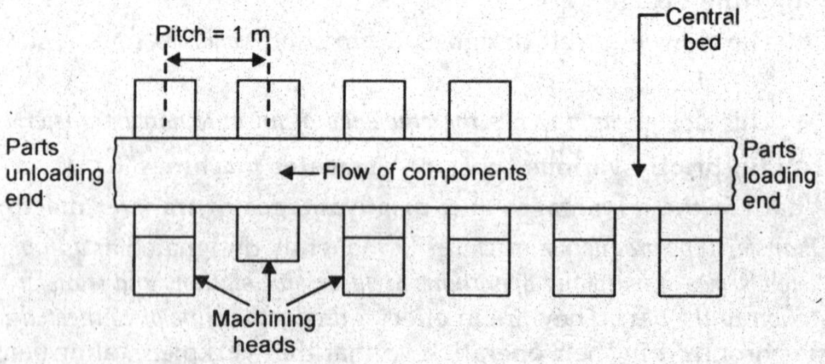

Fig. 9.19. Schematic diagram of an In-line transfer machine.

Following are the **functions** *of an in-line transfer machine :*

(*i*) Transfers workpiece to the first station and then from station to station.

(*ii*) Locates and clamps the work at each station.

(*iii*) Rapid approach of the tools to the work.

(*iv*) Feeds the tools through the cutting cycle.

(*v*) Retracts the tools clear of the work and guide bushes.

(*vi*) Unclamps the workpiece at various stations ready for further tansfer.

3. Drum type transfer machines :

In these types of machines work fixtures are fastened to the outside surface or periphery of the drum and work stations are positioned radially around the circular path at equal intervals.

As the work hangs from the fixture, the clamping arrangement must be fool proof and efficient. Like circular indexing arrangement, this too cannot be big in size.

- *Selection of transfer machines* depends on the following factors :

(*i*) Product size and machining requirements.

(*ii*) Handling systems used.

(*iii*) Floor space available.

- *Transfer machines range from comparatively small units having only two to three stations to long straight line machines with more than 100 stations.*
- These machines are *used primarily in the automobile industry.* Products processed by these machines include cylider blocks, cylinder heads, refrigeration compressor bodies and similar parts.

15.3.2. Constructional Features of a "Transfer Machine"

The principal constructional features of a 'Transfer machine' consists of the following main parts and mechanism :

1. Central bed.
2. Machining heads.
3. Automatic work holding and Transfer mechanisms.
4. Locating and clamping devices.
5. Cooling supply mechanisms.
6. Chip disposal devices.
7. Control systems.

(*b*) Advantages and disadvantages of transfer machines :

Following are the *advantages* and *disadvantages* of transfer machines :

Advantages :

1. Higher production rates are achieved.
2. Higher accuracy is obtained.
3. Less floor space is required.
4. Heavy workpieces can be easily handled.
5. The quality of products is considerably improved.
6. Complex shaped components can be conveniently machined.
7. The length of production cycle is reduced.
8. Less number of operators are required.
9. Increased tool life (resulting in further reduction in production cost).

Disadvantages :

1. High initial investment.
2. A breakdown of one machine means stoppage of whole of the production line.
3. Complex control systems are required.
4. Much time is required to change over the machine to handle a different shaped components.
5. Very high overhauling and maintenance costs of transfer lines, specifically when reshuffling is required.

Q.9.34. Explain briefly "Machining centres".

Ans. It has been observed that each machine, regardless of how highly it is automated is designed to perform basically one type of operation; furthermore, in manufacturing operations, most parts generally require a number of different machining operations on their various surfaces. None of the processes and machine tools described thus far could individually produce these parts.

Transfer lines are commonly used in high volume or mass production. There are situations and products, however, for which such lines are *not* feasible or economical particularly when the types of products to be machined change rapidly. An important concept (developed in late 1950) is *machining centres.*

A machining centre is *a computer-controlled machine tool capable of performing a variety of cutting operations on different surfaces and in different orientations on a workpiece.* In general, the *workpiece is stationary,* and the *cutting tools rotate,* such as in milling and drilling. Because of the high productivity of machining centers, large amounts of chips are produced and must be collected and disposed of properly.

— In a mahining centre, the workpiece is placed on a *"pallet"* that can be oriented in three principle direction, as well as rotated around one or more axes on the pallet. Thus, after a particular operation is completed, the workpiece does not have to be moved to another machine for subsequent operations (say, drilling, reaming, and tapping). In other words, the tools and machine are brought to the workpiece.

— After the completion of all the cutting operations, the pallet automatically moves away, carrying the finished workpiece, and another pallet containing another workpiece to be machined is brought into position by *"automatic pallet changers."*

— All movements are guided by computer control, and pallet-changing cycle times are of the order of 10-30s.

— Pallet stations are available with multiple pallets that serve the machining centre.

- Machining centres are equipped with a *"programmable automatic tool changer".* The cutting tools are automatically selected, with random access for the shortest route to the machine spindle. Tools (as many as 200 tools can be stored in a magazine, drum or chain) are identified by coded tags, bar codes, or memory chips applied directly to the tool holders).
- The maximum dimensions that the cutting tools can reach around a workpiece in a machining centre is known as the *"work envelope."*

Types of machining centres :

The basic types of designs of machine centres are :

1. Vertical-spindle machining centres.
2. Horizontal-spindle machining centres.
3. Universal machining centres.

1. *Vertical-spindle machining centres :*

- In vertical-spindle machining centres or *'vertical machining centres'* the thrust forces are directed *downward* and as such these machines have *high stiffness and produce parts with good dimensional accuracy.*
- These are generally less expensive than horizontal-spindle machines.
- These machine centres are *suitable for performing various machining operations on flat surfaces with deep cavities, such as in mould and die making.*

2. ***Horizontal-spindle machining centres :***

- These machining centres (also simply called *'horizontal machining centres'*) are suitable for *large or tall workpieces that require machining on a number of their surfaces.* The pallet can be rotated on different axes and to various angular positions.
- Another category of these machines is *"turning centres"* which are computer-controlled lathes with several features.

3. ***Universal machining centres :***

- The machining centres are equipped with both vertical and horizontal spindles.
- These are so called since they have a variety of features and are capable of machining all surfaces of a workpieces.

Characteristics of machining centres :

The major characterisitcs of machining centres are as follows :

1. *Highly automated and relatively compact* (so that one operator can attend two or more machines at the same time).
2. Significant *reduction* in the need for a variety of machine tools and a large amount of floor space (since the machines are versatile, having as many as six axes of linear and angular movements and capability of quick change over from one type of product to another).
3. These machines *can handle a variety of part sizes and shapes* efficiently, economically, and with repetitively high dimensional tolerances (of the order of $\pm$ 0.0025 mm).
4. These machines are equipped with tool-condition monitoring devices for *detecting* tool breakage and wear, as well as probes for compensation for tool wear and for tool positioning.
5. *Improved productivity, reduced labour requirements and minimised overall cost* (owing to reduction in the time required for loading and unloading workpieces, changing tools, gauging, and troubleshooting).

Selection of machining centre :

Some important factors on which the selection of a particular type and size of machining centre depends, are as follows :

1. Rate of production.
2. Type of products, their size, and their shape complexity.
3. Type of machining operations to be performed and type and number of tools required.
4. Dimensional accuracy of products.

- While selecting machining centres, though variability is the key factor, jet *these considerations must be weighted against the high capital investment required and compared with those of manufacturing the same products by using a number of the more traditional machine tools.*

HIGHLIGHTS

1. *Automation* is a technology concerned with performing a precess by means of programmed commands combined with automatic feedback control to ensure proper execution of the instructions.

2. *Elements of automation* are : (*i*) Source of power; (*ii*) Feedback controls; (*iii*) Machine programming.
3. Types of automation : (*i*) Fixed (or hard) automation; (*ii*) Programmable automation; (*ii*) Flexible or soft automation.
4. *Low cost automation* is a technology that creates some degree of automation around the existing equipment, tools, methods, people, etc. using mostly standard components available in the, market with low investment so that payback period is shorts.
5. Material handling systems are classified as : (*i*) Transportation system; (*ii*) Feeding system; (*iii*) Random access transport systems – AGVs.
6. *Automatic machines* are those machines in which both the workpiece handling and the metal cutting operations are performed automatically.
7. "Transfer machines" perform a variety of machining, inspecting, and quality control functions. These machines are of the following types : (*i*) Rotary indexing table transfer machines, (*ii*) In-line transfer machines, and (*iii*) Drum type transfer machines.
8. A *machining centre* is a computer-controlled machine tool capable of performing a variety of cutting operations on different surfaces and in different orientations on a workpiece.
9. *Group technology* is an approach in which similar parts are identified and grouped together in order to take advantages of their similarities in design and production.
10. Group technology is based on *"product specialisation."*
11. A *part family or group* is a collection of parts, which either because of geometric shape and size or because of similar processing steps are required in their manufacture.

OBJECTIVE TYPE QUESTIONS

Fill in the Blank or Say "Yes" or "No".

1. In .... machines, operations right from feeding of the stock to clamping, machining and even inspection of the workpiece are carried out automatically.
2. .... automation means replacement of human activities or involvement by automatic means partially.
3. .... means that something is done or operated by machinery and not by hand.
4. Automation increases productivity.
5. Automation does not fully ensure the human safety.
6. Palletized work-holding fixtures secure the transmission during all operations.
7. Transfer machines are used primarily in .... industry.
8. In transfer machines higher production rates are not achieved.
9. Transfer machines entail high initial investment.
10. A .... is a computer controlled machine tool capable of performing a variety of cutting operations on different surfaces and in different orientations on a workpieces.
11. Group technology is based on .... specialisation.
12. Group technology was originated in U.S.A.
13. Group technology is a new name for production system.
14. In group technology similar parts are arranged into .... families.
15. Group technology does not promote design standardization.

16. Production scheduling is .... with group technology.
17. A .... is a collection of parts, which either because of geometric shape and size or because of similar processing steps are required in their manufacture.
18. Visual inspection method is the least sophisticated and least expensive.
19. Production flow analysis method makes use of information contained on .... sheets than part drawings.
20. A .... can be a system used in information processing in which numbers or letters or a combination there of are given certain meaning.

ANSWERS

1. automatic	2. Partial	3. Mechanisation	4. Yes
5. No.	6. Yes	7. automobile	8. No
9. Yes	10. machining centre	11. product	12. No
13. Yes	14. part	15. No	16. simplified
17. part family	18. Yes	19. route	20. code.

THEORETICAL QUESTIONS

1. Define the term 'Automation'.
2. What are the advantages and limitations of automation ?
3. What are the applications of automation ?
4. Enlist the elements of automation.
5. What are the aims of automation ?
6. What is the difference between mechanisation and automation ?
7. State the various types of automation.
8. Explain briefly the following :
   (*i*) Flexible automation.
   (*ii*) Programmable automation.
9. What do you mean by 'Low cost automation'? State its advantages.
10. Give the classification of material handling systems.
11. What is a 'transfer machine' ?
12. What are the advantages and disadvantages of transfer machines ?
13. Name the various types of "Transfer machines" and explain briefly any one of them.
14. Explain briefly any two of the following machines :
    (*i*) Rotary indexing table transfer machines.
    (*ii*) In-line transfer machines.
    (*iii*) Drum type transfer machines.
15. What are the functions of an 'In-line transfer machines' ?
16. List the constructional features of a 'Transfer machine'.
17. State the advantages and disadvantages of 'Transfer machines'.
18. What is a 'Machining centre'? Explain briefly.
19. What are the various types of machining centres? Explain any one of them briefly.
20. Describe briefly any two of the following machines :
    (*i*) Vertical-spindle machining centre.
    (*ii*) Horizontal-spindle machining centre.
    (*iii*) Universal machining centre.

21. Mention the major characteristics of machining centre.
22. List some important characteristics of machining centre.
23. Explain briefly the following transfer devices :
    (*i*) Roller conveyors;
    (*ii*) Pneumatic conveyors;
    (*iii*) Chute conveyors;
    (*iv*) Magnetic belt conveyor.
24. Explain briefly any two of the following feeders :
    (*i*) Cart on track feeder;
    (*ii*) Sortation feeder device;
    (*iii*) Reciprocating feeder;
    (*iv*) Trolley feeder device.
25. What do you mean by AGVS ? Explain.
26. Discuss briefly AGVs equipment.
27. State the applications of Automated Guided Vehicles (AGVs).
28. What do you mean by "Automated storage systems" ? Explain.
29. What are the objectives of Automated storage systems ?
30. Explain briefly the various types of Automatic storage/retrieval systems.
31. List the applications of Automatic storage/retrieval systems.
32. What do you mean by 'Group technology' ? Explain.
33. List the advantages and limitations of group technology.
34. Discuss briefly "part families".

CHAPTER

10 Robot Applications

10.1 Introduction; 10.2 Advantages, disadvantages and shortcomings of robots; 10.3 Applications of robots – Welding – Spray painting – Grinding – Material handling and transfer – Machine loading and unloading applications – Assembly operations – Sorting of parts – Inspection of parts – Touch-free heart surgeries using robots; **10.4 Summary of features and applications of future industrial robots.** *Questions with Answers* – Highlights – Objective Type Questions – Theoretical Questions.

10.1 INTRODUCTION

- For more than the last two decades, robotics has rapidly moved from theory to applications and from research labs to industries. Robotics is going to be a prominent component of manufacturing industry, which will affect human labour at all levels, from managers of production to shop floor unskilled workers.
- *A programmable robot* (one of its key feature being its versatility) *with a number of degrees of freedom (DOF) and different configurations can perform specific and diverse tasks with the help of a variety of end-effectors.*
- On the industrial scene it can be reprogrammed and adopted to changes in process or production line. Robots are also finding many applications outside of the industry, in *research hospitals, space, supermarkets, services sector, farmhouses,* and even in homes as pets.

10.2 ADVANTAGES, DISADVANTAGES AND SHORTCOMINGS OF ROBOTS

The *advantages, disadvantages and shortcomings of robots* are listed below :

Advantages :

1. Greater flexibility, reprogrammability.
2. Reduced costs.
3. Elimination of hazardous tasks.
4. Improved product quality.
5. Improved productivity.
6. Greater response time to inputs than humans.
7. Accident reduction.
8. Automation less susceptible to work stoppages.
9. Maximise capital intensive equipment in multiple work shifts.

Disadvantages :

1. Greater unemployment.

2. Replacement of human labour.
3. Significant restraining costs for both unemployed and users of new technology.
4. Advertized technology does not always disclose some of the hidden disadvantages.
5. Payload to robot weight ratio is poor, often less than 5 per cent.
6. Robot structural configuration may limit joint movement.
7. Assembly dexterity does not match that of human beings, particularly where eye-hand coordination is required.

Shortcomings : Today's robots :

1. Do not think independently.
2. Do not adapt quickly to the changes in their environment.
3. Cannot react to unknown situations.
4. Are not creative or innovative.
5. Have no consciousness.
6. Cannot make complicated decisions.
7. Do not learn from mistakes.
8. Have no human feelings.

- Robots offer an excellent means of utilising technology to make a given manufacturing operation more profitable and competitive. *The main advantage offered for the industrial needs is the improved productivity and quality offered by the robots.* However, the technology is relatively new on the industrial scene. Its use in the manufacturing processes is greatly limited for multiple reasons.
- Some of the tasks which the *robots can do better* are :
  (*i*) Repetitive, back breaking and unrewarding tasks.
  (*ii*) Tasks involving danger to human or dangerous tasks.
  (*iii*) Handling dangerous materials.
  (*iv*) Spray painting.
  (*v*) Assembling products.
  (*vi*) Polishing and cutting.
  (*vii*) Inspection.

10.3 APPLICATIONS OF ROBOTS

The various applications of robots include the following :

I. Industrial applications :

1. ***Material handling :***
   (*i*) Material transfer applications.
   (*ii*) Machine loading and unloading applications.
   (*iii*) Pelletizing application.
2. ***Processing applications :***
   (*i*) Arc welding. (*ii*) Spot-welding.
   (*iii*) Spray painting. (*iv*) Paint scraping.
3. ***Assembly applications :***
   (*i*) The assembly task. (*ii*) Peg-in-hole assembly.

4. ***Inspection applications :***
   (*i*) Sensor based inspection. (*ii*) Vision based inspection.
   (*iii*) Testing.

II. Non-industrial applications :

1. Home sector.
2. Health care.
3. Service sector.
4. Agriculture and farms.
5. Research and exploration

Some of the above applications are discussed in the following text.

10.3.1. Welding

"Welding" is one of the major uses for an industrial robot. Actually, the following two distinct types of welding operations are readily and economically performed by robots :

1. Spot welding;
2. Arc welding.

1. Spot welding :

— In this type of welding, the *robot is taught a series of distinct points.*

— Since the metal parts that are to be joined may be quite irregular (in three dimensions), a wrist with good dexterity is often required (*e.g.*, three degrees of freedom). This permits the welding tool to be aligined properly at the desired weld point without the gun coming into contact with other portions of the part.

— Typically, the welding tools carried by these robots are *large, and reasonably heavy.* Also, it is usually necessary for the manupulator to have a *long reach.* As a consequence, *large point-to-point servo-controlled robots (either hydraulically or electrically actuated) are normally used for this purpose;* these robots are widely used in *"automobile industry"*.

— Because the weld points are pre-taught, sensory information is generally *not* required in order to energize the welding gun. It is, however, possible to utilize the increased motor current that results when the tool makes contact with the part to initiate the welding operation.

2. Arc welding :

— In arc welding, an irregularly shaped seam or wide joint must be made.

— In this case, *a continuous-path servo-controlled robot that is often specifically designed for this single application is most usually the choice.*

— If the parts to be welded can be accurately positioned and held in place, the complex *three-dimensional path can be pre-taught and no external sensors may be necessary.* At present, a number of manufacturers include a position sensor that is placed in front of the welding tool and can therefore provide information concerning irregularities in the weld path.

— *Where a wide joint is to be handled, th robot can be programmed to produce a weave type of motion.* This ensures that weld covers the entire gap.

— The workspace of the welding robot should be large enough to accommodate size of the parts to be welded. A *five* degrees of freedom (DOF) manipulator can weld parts in a plane while *six* DOFs are required for *welding complex contours.*

- *The "major advantage" of a robotic welder is that arc time (a critical parameter in determining the weld's strength) can be carefully controlled.*
- The arc welding is also extensively used in the *auto-industry.*

Advantages of robotic welding :

1. Higher productivity.
2. Improved safety of worker.
3. Good quality of product.
4. Better control over the welding process.

10.3.2. Spray Painting

— In robot spray painting operations, the spray gun is the robot's end-effector. (Fig. 10.1)

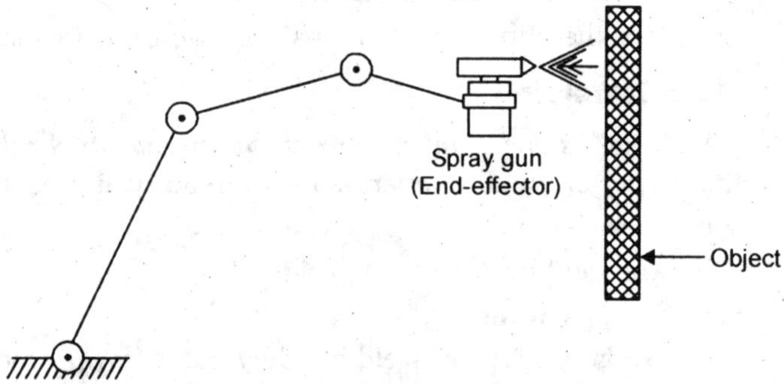

Fig. 10.1. Spray painting.

— The use of industrial robots is increasing in spray painting operations due to hazard conditions prevalent in such operations like fumes, and mist in atmosphere, high noise from the spray nozzle, fire hazards due to flammable paint and breathing difficulties in painting area.

— Robots employed for this purpose are usually capable of performing *both straightline and continuous-path motions.*

— Programming a spray-painting robot is usually performed by the best human operator. His actions are then mimicked by one or more robots.

— The spray-painting application generally does *not* require the use of external sensors. However, it is necessary that the part to be painted be accurately presented to the manipulator.

Advnatages : The use of robots in spray painting leads to the following *advantages :*

1. Removal of operators from the hazardous environment.
2. Consistency in operation.
3. High product quality.
4. Increased productivity.
5. Less painting material usage.
6. Reduced direct labour costs.
7. Saving in energy consumed.

10.3.3. Grinding

- When two metal pieces are welded using arc welding, a bead is formed at the seam. Where a smooth surface is required for *appearance* sake (such as on auto bodies) for *functionality*, it is usually necessary to perform a grinding operation. This is also a natural task for a robot since the manipulator can use the same program that was employed in the arc welding operation. All that must be done is to remove the welding tool and replace it with a rotary grinder.
- Another important grinding task is on "*metal castings*". Here the *robot is taught the correct shape of the casting using continuous-path programming*. The grinder then

removes any undersired high spots and corrects areas of the casting that are too large.

- Still another robotic grinding application is that of *"deburring"*. Here the unwanted material that remains around the back side of a drilled hole is ground away to leave a smooth surface.

In the above grinding applications, there is always some uncertainty in the dimensions of the part being worked on. Consequently, *sensory information is often needed* to permit the robot to more accurately *"feel"* the account of the part. This is especially important in case of smoothing of the arc weld bead.

☞ Other applications involving a rotary tool :

Besides the rotary grinding or deburring applications robots are also currently used for :

(*i*) *Drilling holes;*
(*ii*) *Routing;*
(*iii*) *Polishing;*
(*iv*) *Nut running;*
(*v*) *Driving of screws.*

— In cases (*i*) and (*ii*), preprogramming of either points or paths can be performed when extreme accuracy is not required. However, where exact placement of drilled holes is needed, it may be necessary to utilize a *template*. The difficulty with doing this is that unless the robot wrist has some *"give"* (*i.e.*, compliance), any misalignment of either the part or the robot itself will result in a damaged template and/or an inaccurately placed hole. This problem is overcome by means of a compliant wrist which permits the drill bit to be aligned in the template hole even if there is a positional error; RCC (remote-centred compliance) has been used for this purpose.

10.3.4. Material Handling and Transfer

Robots are commonly used for shifting an object from one location to another. The prominent areas of application for this purpose are :

(*i*) Transfer of parts from one conveyor to the other.
(*ii*) Pelletizing and depalletizing.
(*iii*) Transfer of blanks from an incoming conveyor to the machine tool for further processing.

A material handling operation in which the primary objective is to move a part from one location to another without any complex constraints is the simplest operation for a robot. The application usually requires a relatively *simple robot with few degrees of freedom (DOF) and a simple controller.* These operations are commonly called *pick-and-place (or pick-n-place)* operations (Fig. 10.2)

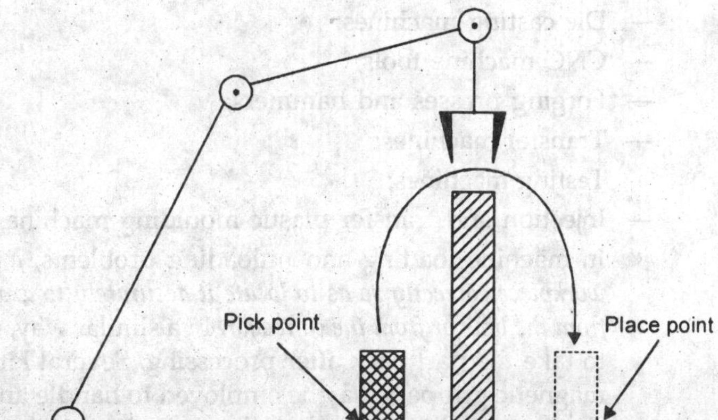

Fig. 10.2. Pick-and-place task.

- Often, it is necessary to acquire a part from a remote location and to place it in a compartmentalised (*e.g.*, a rectangular away) box or carton. Once all the compartments are filled, the box is moved (by either the same robot or a larger one) to another location within the workcell, where it is sealed and stocked for future use. Such an operation is referred to as **pelletization**. The inverse operation of unloading an array of objects and placing them in another place within the work space (*e.g.*, on a conveyor belt) is called **depalletization**. In depalletizing, the pickup point is different for each part while the delivery point is fixed.

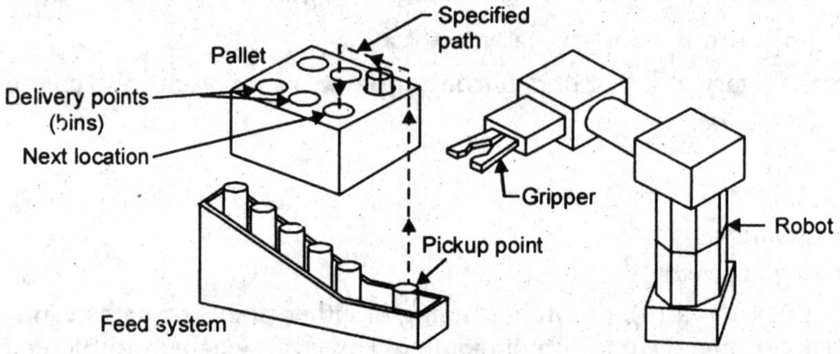

Fig. 10.3. A palletizing operation.

- Other important part-handling applications involve the *acquiring of blank or unfinished parts and feeding them into some type of machine tool for finishing (e.g., a punch press)*. This application is often a dangerous one for human beings and so is ideally suited for a robot.
- Similarly, in the *metal working industries*, a common task is *to produce finished castings or extrusions*. Such work can be dangerous since it is necessary for human being to work in the vicinity of hot furnaces, punch presses, lathes, and/or drill press. *Robots are ideally suited for this type of job because they can resist the high temperature environment and can be programmed to avoid collision with the various other machine tools present in the workcell.*

10.3.5. Machine Loading and Unloading Applications

Robots are commonly used for loading of stock parts and unloading of finished parts on :

— Die casting machines;
— CNC machine tools;
— Forging presses and hammers;
— Transfer machines;
— Testing machines;
— Injection and transfer plastic moulding machines, etc.

- In machine loading and unloading problems, *a robot should be able to orient the workpiece correctly so as to locate it accurately to a particular machine after picking up from the bins or from the conveyor*. In a similar way, the robot gripper should be able to take the part back after processing. Several kinds of machanical, vacuum and magnetic grippers may be employed to handle and manipulate objects of varying shapes, weights and materials.

Fig. 10.4. shows a loading and unloading a die casting machine by a robot.

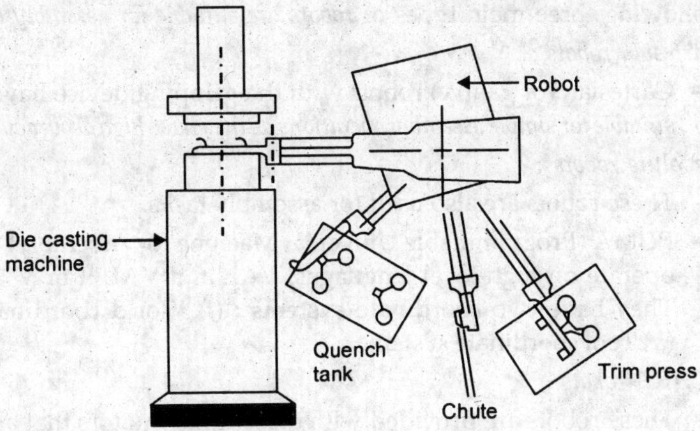

Fig. 10.4. Loading and unloading a die casting machine by a robot.

10.3.6. Assembly Operations

Robots find applications in assembly areas involving :

(*i*) Screwing of studs and screws in threaded holes.

(*ii*) Screwing and unscrewing of nuts.

(*iii*) Insertions of shafts in holes;

(*iv*) Insertions of electronic components in electronic assemblies.

(*v*) Assemblies of small electric motors, plugs, switches, etc.

So far maximum potential for robot application is observed only in *back type assembly work*.

- Human beings are capable of assembling a group of diverse parts to produce either a finished product or a subassembly because of their ability to utilise good eye-hand coordination in conjunction with the important sense of touch. However, these jobs may be extremely tedious because of their repetitious nature. Therefore, there is a need to *automate* the assembly process.
- However, not every final product may qualify for automated assembly. *The requirements for the improved, more efficient "robotic assembly" are :*

  (*i*) Faster robots;

  (*ii*) Limited capability cheap robots;

  (*iii*) Versatile and inexpensive grippers;

  (*iv*) Identification of assembly families;

  (*v*) Improved assembly efficiency;

  (*vi*) Low cost feeders.
- The *design for robotic assembly* should evolve from the design of the product. The important points to be considered for design of robotic assembly are :

  (*i*) The components for the assembled product should be selected for ease of assembly.

  (*ii*) Parts should be designed for feeding and orienting for automated assembly for which product simplification and if necessary, redesigning the products are necessitated.

— *The compliance of the gripper is the most important property in the assembly.*

Robots for assembly work :

The following three main types of *robots are suitable for assembly operations :*

(*i*) *Cartesian robots :*

— Cartesian (or gantry) robots with the gripping device having 2-3 DOF are *suitable for simple assembly operations as they have high accuracy and repeatability.*

(*ii*) *Revolute robots :*

— These robots are also used for assembly tasks.

— PUMA (Programmable Universal Machine for Assembly) type robots that operate on high level languages, VAL and VAL II or V^+ have six DOFs. They have two coordinate systems : (*i*) World coordinate system, and (*ii*) Tool coordinate system.

(*iii*) *SCARA robots :*

— These robots are provided with direct drive motors that allow high speed with accelerations and backlash-free, fast and accurate motions.

— The accuracy is around ± 0.076 mm and repeatability is ± 0.025.

10.3.7. Sorting of Parts

Robots are used to sort out correct parts from a lot of finished parts and place them in proper locations in respective bin.

- Fig. 10.5. shows *gauging and sorting of washers :*

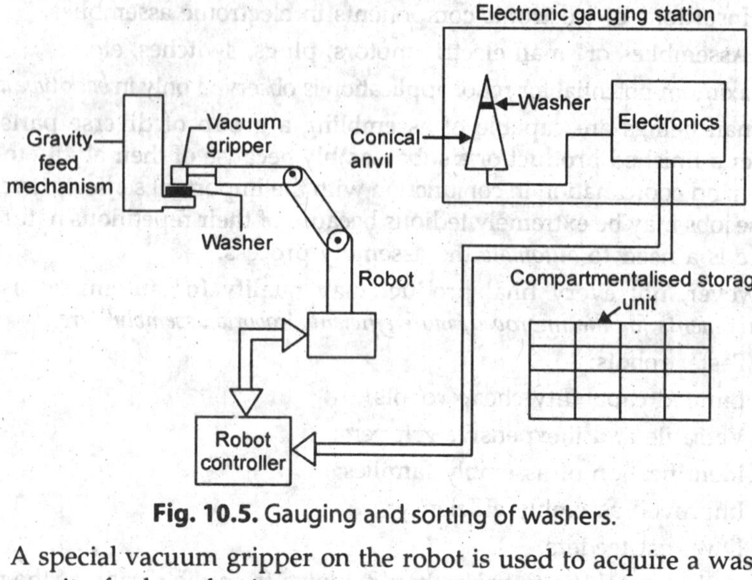

Fig. 10.5. Gauging and sorting of washers.

— A special vacuum gripper on the robot is used to acquire a washer from the gravity feed mechanism.

— The washer is then placed on the conical anvil at the electronic gauging station where the diameter of the washer's hole is determined. This information is sent to the robot's controller which causes the washer to be reacquired and then placed in the proper bin of the compartmentalised unit.

- Another less common sorting application involves the *use of a "vision system"* :

— A group of parts is randomly placed on a conveyor belt.

— The vision system is placed upstream of the robot and determines the type and orientation of the part passing within the field of view of the comera. This information is passed to the robot controller, which then directs the manipulator to move to the correct location with the appropriate gripper orientation. The robot is then able to place the acquired object into the correct bin (or on to another conveyor).

— This type of application is found in factories where many different castings must be handled (*e.g.*, in the automobile industry).

10.3.8. Inspection of Parts

- Robots have been employed to *inspect finished parts or subassemblies in order to increase product quality*. The *"automobile industry"* is an example of a group of companies that is striving to upgrade its product by automating the inspection process.
- Inspection of *"electronic devices"* has also been performed by robots. For example, a *printed circuit (PC) board* often must be checked for missing or improperly drilled holes before the components are placed on the board. Two techniques can be used to accomplish this :

— In the *first*, the robot picks up the board and puts it onto a special-purpose jig. Probes extend through properly drilled holes and make contact with electrodes on the other side of the board. If all contacts are made, the part is good and the robot can place it in the appropriate bin. If there is even one missing contact, the robot's program branches and causes the PC board to be placed in the reject bin.

— The *second technique* utilises a *"vision system"* that consists of one or more video cameras. The robot places the PC board on a light table. The pattern of light passing through the drilled holes is sensed by the vision system and compared to a stored pattern for a good board. The vision system then commands the robot to place the part in the correct bin.

Fig. 10.6 illustrates an *inspection task*.

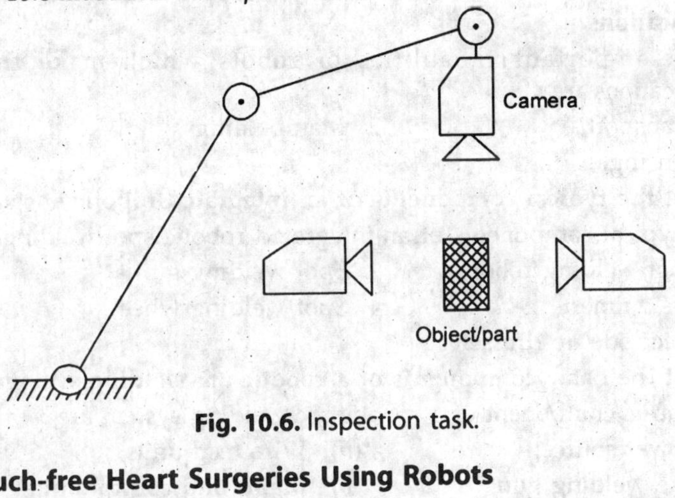

Fig. 10.6. Inspection task.

10.3.9. Touch-free Heart Surgeries Using Robots

The use of robots in Touch-free heart surgeries is becoming more popular these days.

- In robotics surgery, the incision made is quite small; there is no need to cut any bones; the patient does not need to be put on a heart-lung machine; and the entire surgery can be performed with just two or three small holes in the chest.

- In robotic surgery, the robotic cardiac surgeon sits away from the patient on a computer console where he sees the patient's heart magnified on the screen and moves highly-sensitive instruments that direct the camera-equipped, three armed robot. The robot precisely watches the doctor's hand and wrist movements, translating them to tiny instruments placed inside the patient through the small holes.

Advantages :

(*i*) Less blood loss.
(*ii*) Less post-operative pain or discomfort.
(*iii*) Decreased risk of infection.
(*iv*) Significantly shortened hospital stay.
(*v*) Much less scarring.
(*vi*) Faster recovery.

10.4 SUMMARY OF THE FEATURES AND APPLICATIONS OF FUTURE INDUSTRIAL ROBOTS

The summary of the features and applications of future industrial robots is given in Table 10.1 on page no. 453.

QUESTIONS WITH ANSWERS

Q. 10.1. State the reasons for using robots in industries.

Ans. The robots are used in industries because of the following *reasons* :

1. Better quality of the product.
2. Elimination of hazardous tasks.
3. Improved productivity.
4. Reduced costs.

Q. 10.2. Name three important capabilities of robots which make them useful for applications.

Ans. Three important capablities of robots which make them useful for applications are :

1. Transport
2. Manipulation
3. Sensing.

Q. 10.3. Enlist the typical components of an integrated robotic spot welding cell.

Ans. The typical components of an integrated robotic spot welding cell are :

1. Spot welding robot.
2. Spot welding gun.
3. Weld timer.
4. Spot welding wheel.
5. Electrode tip dresser.

Q. 10.4. Enlist the basic components of a robotic arc welding system.

Ans. The basic component of a robotic arc welding system are :

(*i*) Power source;
(*ii*) Wire feed units;
(*iii*) The welding gun;
(*iv*) Mains and control unit;
(*v*) Gas supply unit;
(*vi*) Robot main body;
(*vii*) Controller;
(*viii*) Teach box;
(*ix*) Terminal;
(*x*) Floppy disc;
(*xi*) Input-output modules.

Table 1.1 : Summary of Features and Applications of Future Industrial Robots

S.No.	Features	Applications					
		Part loading and unloading	*Material handling*	*Spot (track) welding*	*Arc welding*	*Spray coating*	*Electronic assembly*
1.	*Structure*	Polar, cylindrical jointed arm (adaptable)	Jointed adaptable robot arm	Polar, jointed adaptable robot arm	Polar, modular cartesian with adaptable jointed arm	Jointed arm with adaptable gun	Jointed adaptable cartesian modular robotic arm
2.	*Degrees of freedom (DoF)*	4 – 5 multiple arms	3 – 5	5 – 6	5 – 6	6 or more	3 – 6
3.	*Drive system*	Electronic servo motors (for heavy pay loads)	Servo motors	Electronic stepper motors	Direct drive servo motors	Hydraulic actuators	Stepper motors and direct drives
4.	*Program*	Programmable Automation Control (PAC)	Programmable Automation Control (PAC)	Programmable Logic Controllers (PLCs)	Programmable Logic Control (PAC)	Programmable Logic Controllers (PLCs)	Programmable Automation with Controller Area Network (CAN)
5.	*Control system*	Microcontollers and motion controllers with vision	Motion cont-rollers with sensor technology	Micro-controllers with changeable functions	Continuous path motion controllers with sensor technology	Continuous path motion controllers	Microcontrollers, nodes with sensors and end-effectors with vision
6.	*Nature of task*	Complicated and safe environments	Safe/hazardous complicated	Simple and safe	Complicated and unsafe	Simple and unsafe	Complicated and safe

Q. 10.5. Enlist the sequences of handling process.

Ans. Following are the *sequences of a handling process* :

— Transfer of robot arm up the workspiece.

— Fine motion approaching the workspiece.

— Grasping.

— Fine motion uprising the workspiece.

— Transfer the workspiece to the desired position.

— Fine motion down to the destination position.

— Release the workspiece.

— Fine motion upward.

Q. 10.6. Give five examples of material handling applications.

Ans. Five examples of material handling applications are listed below :

1. Transfer of parts from a machine to an overhead conveyor.
2. Transferring parts from one conveyor to another.
3. Transferring parts from a processing line to a conveyor.
4. Loading bins and fixtures for subsequent processing.
5. Moving parts from a warehouse to a machine.

Q. 10.7. Explain briefly peg-in-hole assembly task.

Ans. Fig. 10.7. shows an elementary assembly operation involving assembly of mating a peg in a hole :

— For human beings putting a peg in a hole is an extremely simple and trivial task. However, surprisingly it is very difficult to have peg-in-hole assembly being carried out by a robot manipulator. If for some reasons the hole is not exactly in the location expected or peg is grasped wrongly, a manipulator would be most of the time unable to perform the assembly successfully without any human help. The uncertainties embedded in the physical world make the task most difficult.

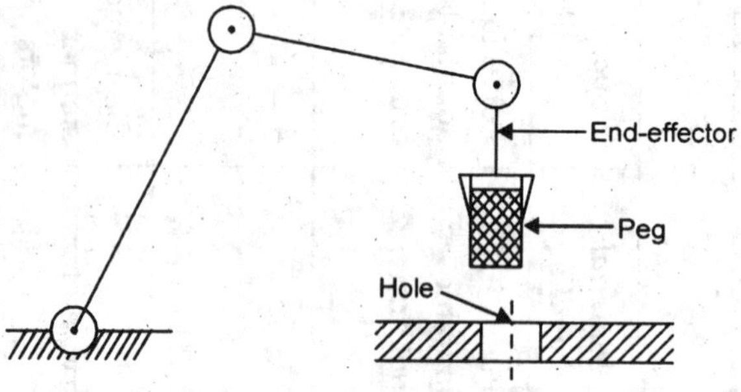

Fig. 10.7. Peg-in-hole assembly task.

— The peg-in-hole assembly can have different difficulty levels. More complex peg-in-hole assembly *involves assembly of multiple pegs into multiple holes simultaneously. In multiple peg-in-hole assembly, all the pegs must be aligned with the corresponding holes for the assembly to take place.*

Q. 10.8. Explain briefly "Sensor based" and "Vision based" inspection.

Ans. Sensor based inspection :

— The specific physical dimensions can be determined by the robot with the help of *sensor placed on the gripper fingers of a material handling robot.*

— In material handling each part is grasped by the gripper and moved from one place to the other. The specific physical size information can be obtained by the sensors fitted on the fingers of the gripper and the robot can determine whether the size is within tolerance limit or not. *If the size is correct part is placed on the desired place and if the size is incorrect, the part is dropped into a waste bin.*

Vision based inspection. A typical robotic vision system is capable of analyzing a two-dimensional scene. The two-dimensional view of the scene is obtained from a single camera. If more features of the object are required to be inspected, the camera is maneuvered into different positions by the robot manipulator. Ultimately, the robot manipulator can be used to present the part to a stationary vision system (camera) in different orientations. Thus robot's task is a sort of material handling task.

— The following factors must be considered while designing a robot vision inspection system :

(*i*) Robot should have sufficient degrees of freedom to manipulate the camera or the part, as the case may be.

(*ii*) Proper lighting of the workspace.

(*iii*) The field of view should be large enough to accommodate the part.

(*iv*) Cameras with required resolution to extract dimensions and other features to the desired accuracy.

Q. 10.9. Enlist the various types of justification of robots.

Ans. The various types of justification of robots are :

1. *Technical justification for potential applications.*
2. *Quantitative justifications.*
3. *Qualitative justification :*
   - Quality
   - Productivity
   - Hazards
   - Flexibility.

Q.10.10. How are assembly systems classified ? Explain them briefly.

Ans. Assembly system may *classified* as follows :

1. Manual assembly.
2. Semi-automatic assembly.
3. Automatic assembly.
4. Flexible assembly.

1. Manual assembly :

— It is carried out by the assembler who has simple and mostly passive auxiliary equipment at his disposal, such as tables, fixtures, component boxes etc.

— It is the *most flexible system but it lacks consistency due to the fatigue of the assemblers.*

2. Semi-automatic assembly :

— The assembly is carried out with automatic machines in which some manual operations are also adopted.

3. **Automatic assembly :**
   - It is a mechanical assembly comprising machine systems which follow a program.
   - The system accomplishes the task based on the pre-planned program.
4. **Flexible assembly :**
   - *Flexibility* refers to the system's ability to accept variations in different modes of the assembly tasks accommodating the design characteristics of the components and production variations without any considerable charge in the toolings.
   - *"Flexibility assembly" includes versatility and adaptability to cope with the new demands.* It takes decisions depending on the system conditions.
   - It is *suitable for multistation assembly tasks.*

Q.10.11. List the main R & D challenges of future industrial robot systems.

Ans. The main *R & D challenges of future industrial robot systems* are :

1. Fine manipulation/High precision
2. Human-robot-collaborative work cells.
3. Cooperating robots.
4. Hyper flexible manufacturing systems.
5. Micro-and Nan-manufacturing.

Q.10.12. Discuss briefly "Robot safety".

Ans. A robotic system is an integration of robots, machines, computerised information channels and human beings, where no element can be considered perfect or immune from eventual failure and malfunction. Since the humans work rather close to the robots this increases the risk of mutual damage. All these factors necessitate the formulation of safety guidelines that indicate how the condition of conflict can be minimised. In a practical sense, safety procedures and devices allow the authorised entry of humans into a robot's working envelope with minimum risk of injury. Hardware devices and sensors would monitor all anticipated reasonable access to a robot's work envelope.

Safety devices make use of physical safeguards available in a variety of forms; some of these are :

(*i*) Pressure mats.
(*ii*) Infrared light beams or light barriers.
(*iii*) Vision systems.
(*iv*) Simple contact micro-switches.
(*v*) Restrained keys to the access doors.

- The *flashing red lights often seen within a work zone indicate that an apparently stationary robot is activated, but awaiting an input, or performing a time-delayed operation.*

Q.10.13. Explain briefly the following robots :
(*i*) Walking robots; (*ii*) Undersea robots.

Ans. (*i*) Walking robots :

- Most of the industrial robots designed so far are mostly stationary and some of them are mobile in the sense that they have been placed either on automated guided vehicle (AGV) or on the rails. None of them have really any legs or similar structures. Such robots are useful for transportation inside the plants.

— The walking robots can prove quite useful in *military* since they can move on loose sand or gravel, muddy ground, across the slope of rough train and may replace soldiers and save valuable human lives. These robots will also prove useful in handling dangerous radioactive materials in the area of *nuclear engineering*. Telechiric manipulators are already in use for doing such work. Other promising areas of walking robot are : *Research and exploration of planets; Prosthetics and orthetics; Rescuing operation in case of fire hazards etc.*

— In walking robots, mobility may be provided by legs or wheels. Legs are nothing but levers with some joints and may be short or long. All the legs should be properly designed with suitable joints to have several DoFs as well as stability.

(*ii*) Undersea robots :

— For undersea remote work *submersible robots* are used. This type of robot is a *reprogrammed manipulator that has articulated hands or grippers.* It has many subsystems such as *heavy duty slave arms, actuator drives, computer control, multiplex communication, bilateral force feedback, servocontrol, digital cassette recording units, vision and other sensors.*

— An undersea robot works on the *principle of teleoperated system,* in which a *master arm is actuated to manipulate the slave arm that works in a remote place.* The master and slave have several DoFs.

— Undersea teleoperated robotic manipulators are capable of working at various sea depths (ranging from 100 m to 6000 m). They are very useful for *remote maintenance; oil and mineral exploration; vehicle control; inspection; remote handling.*

HIGHLIGHTS

1. Some of the tasks which the robots can do better are :

   (*i*) Repetitive, back breaking, and unrewarding tasks; (*ii*) Tasks involving danger to human or dangerous tasks; (*iii*) Handling dangerous materials; (*iv*) Spray painting; (*v*) Assembling products; (*vi*) Polishing and cutting; (*vii*) Inspection.

2. The various applications of robots include the following :

   (*i*) Industrial applicatons; (*ii*) Processing applications; (*iii*) Assembly applications; (*iv*) Inspection applications.

OBJECTIVE TYPE QUESTIONS

Fill in the Blank or Say "Yes" or "No".

1. One of the key features of a robot is its ....
2. Robots are not creative ....
3. Robots can make complicated decisions.
4. Robots do not think independently.
5. .... is the process of stacking or storing the material, parts or cartons on a pallet in a specified manner.
6. In depalletizing, the pick up point is different for each part while the delivery point is fixed.
7. Welding application is the least common process applications for industrial robots.

8. .... is the final stage of manufacturing and it is manual labour intensive.
9. Assembly means fitting two or more discrete parts together to form a new product or subassembly.
10. To perform the inspection tasks robot requires various sensors or .... systems.

ANSWERS

1. versatility.	2. Yes.	3. No.	4. Yes.	5. Palletizing.
6. Yes.	7. No.	8. Assembly.	9. Yes.	10. vision.

THEORETICAL QUESTIONS

1. What are the reasons for using robots in industries ?
2. State the advantages and disadvantages of robots.
3. What are the shortcomings of robots ?
4. List some of the tasks which robots can do better.
5. List the main applications of robots.
6. Why robots are commonly used in hazardous environment ?
7. Why it is difficult to employ robots where variability in workpiece exists.
8. Explain briefly the following robot applications :
   (*i*) Welding;
   (*ii*) Spray painting;
   (*iii*) Material handling and transfer.
9. Describe briefly the any two of the following applications of robots :
   (*i*) Machine loading and unloading application.
   (*ii*) Assembly operations.
   (*iii*) Sorting of parts.
   (*iv*) Inspection of parts.

SHORT ANSWER QUESTIONS

SHORT ANSWER QUESTIONS

Q. 1. *What is an industrial robot ?*

Ans. A ***robot*** may be defined as follows :

The definition adopted by International Standards Organisation (ISO) and agreed upon by most of the users and manufacturers is :

"*An industrial robot is an automatic, servo-controlled, freely programmable, multipurpose manipulator, with several areas for the handling of workpieces, tools or special devices. Variably programmed operations make the execution of a multiplicity of tasks possible.*"

Or

According to the Robotics Industries Association (formerly the Robotics Institute of America):

"*An industrial **robot** is a reprogrammable multifunctional manipulator designed to move materials, parts, tools or specialised devices through various programmed motions for the performance of a variety of tasks*".

This definition is compatible with the classification of programmable automation, but robots are also used in flexible automation and fixed automation systems. *Example*. An automation line using many robots to perform spot welds in which robot programs are downloaded from a computer to a specific robot controller could be considered a high-production flexible automation system.

Q. 2. *What are the functions of a robot?*

Ans. The functions of a robot can be classified into *three* areas :

1. "*Sensing*" the environment by external sensors.

   Example : Vision, voice, touch, proximity and so on.

2. "*Decision making*" based on the information received from the sensors.
3. "*Performing*" the task decided.

Q. 3. *What are non-tactile sensors?*

Ans. These are ***contactless sensors** which sense the signals remotely*, but only within the specified range of distance from the object.

— They detect and measure magnetic fields, infrared and ultraviolet light, X-rays, electrical fields, ultrasonic sound waves or electromagnetic waves.

— Typical non-contact robotic sensors include:

(*a*) Proximity sensors;

(*b*) Electro-optical sensors;

(*c*) Range imaging sensors.

Q. 4. *What are non-servo and servo-controlled robots?*

Ans. (*i*) ***Non-servo controlled robots*:**

— Non-servo-controlled robots are also called *limited-sequence robots*, *end-point robots*, *pick and place robots* or *bang-bang robots*.

— Such a robot is controlled by setting mechanical stops or limit switches to establish end points of travel of each joint. The mechanical set-up to give the proper position and sequence of stops serves as a rudimentary programming approach rather than a computer-intensive robot programming language.

(*ii*) ***Servo-controlled robots***:

— These robots are normally sub-divided into either 'continuous-path' (CP) or 'point-to-point' (PTP) devices. In either case, however, information about the position and velocity (and perhaps other physical quantities) is continuously monitored and fed back to the control system associated with each of the joints of the robot. Consequently, each axis loop is "*closed*".

Q. 5. *What do you mean by 'Intelligent robots'?*

Ans. ***Intelligent robots:***

— These are very advanced state of art robots and *possess adequate artificial or machine intelligence*, somewhat analogous to the sensory perception of the neuro-muscular coordination that human beings are capable of.

— They can not only explore the environment on their own machine perceptions and evaluate them in real time, but also execute the necessary motor functions matching the action of their sensory inputs.

— Advanced robots have been built with mobility to not only move over floors but also to climb, ability to avoid obstacles, high power-to-weight ratios, compactly assembled, with on board sensors, instruments and power supplies.

Q. 6. *Which factors should be considered while selecting a robot for a particular application?*

Ans. The following *factors* should be considered while *selecting a robot for a particular application*:

1. Load capacity required.
2. Adequate number of degrees of freedom.
3. Speed of movement.
4. Reach.

- The robots may be powered electrically, hydraulically or pneumatically. The electrically powered robots have the advantage of accurate movements and quiet operation.

Q. 7. *List the qualities which are expected in robots.*

Ans. The following *qualities* are expected in robots:

(*i*) Tactile sensing ; (*ii*) Vision ;

(*iii*) Mobility ;

(*iv*) Other important qualities:

— General purpose hands ;
— Computer interpretation of the visual and tactile data ;
— Multiple appendage hand-to-hand coordination ;
— Minimised spatial intrusion ;
— Man-robot voice communication ;
— Inherent safety ;
— Interaction with other technologies ;
— Total self-diagnostic fault tracing etc.

Q. 8. *Define the term 'Repeatability'.*

Ans. ***Repeatability is the measure of robot's ability to position an object at a previously taught point in the work envelope.***

— **Owing to inherent errors present, particularly due to mechanical sources, the robot will not be able to return to exact programmed point.**

Q. 9. *Explain briefly 'Control resolution'.*

Ans. It relates to the *system's capability* (both controller and the positioning device) *to divide the range of the total movement into closely spaced points that can be identified.* Thus it would represent the minimum noticeable movement achievable.

It may be added that the controller can generate pulses of very small duration but the positioning device should be able to respond and change its position accordingly. In such a case :

$$\textit{Control resolution} = \frac{\text{Range of movement}}{2^n}$$

where, n = Number of bits devoted to a joint, and

2^n = Number of addressable points.

Q. 10. *Enumerate the ways by which the robots can be controlled.*

Ans. The robots can be *controlled* in the following ways:

1. Non-servo control.
2. Servo-control.
   (*i*) Point-to-point control
   (*ii*) Continuous path control:
      (*a*) Servo control approach
      (*b*) Preview control or feed forward control
      (*c*) Path planning or trajectory calculation approach.

Q. 11. *What are the main characteristic features of 'servo-controlled robots'?*

Ans. Servo-controlled robots incorporate *feedback devices* on the joints or actuators of the manipulator which continuously measure the position of each axis. *These have much more manipulative quality and can position the end-effector anywhere within the total work envelope.*

Q. 12. *List the basic methods of improving robot accuracy.*

Ans. The basic methods of improving robot accuracy are:

1. Joint calibration.
2. Open-loop calibration.
3. Closed-loop calibration.

Q. 13. *What are the two principal categories of robot sensors?*

Ans. The two principal categories of robot sensors are :

(*i*) "*Internal state sensers*" :

— These sensors deal with the detection of variables such as arm joint position, which are used for robot control.

(*ii*) "*External state sensors*" :

— These sensors deal with the detection of variables such as range, proximity and touch.

— External sensing is used for robot guidance, as well as for object identification and handling.

Q. 14. *What is 'Robot vision'?*

Ans. Although proximity, touch, and force sensing play a significant role in the improvement of robot performance, '*vision*' is recognised as the most powerful of robot sensory capabilities.

Robot vision *may be defined as a process of extracting, characterising, and interpreting information from images of a three-dimensional world.* This process, also commonly referred to as *machine or computer vision,* may be subdivided into following six areas :

(*i*) Sensing ; (*ii*) Preprocessing ;
(*iii*) Segmentation ; (*iv*) Description ;
(*v*) Recognition ; (*vi*) Interpretation.

Q. 15. *What is an 'End-effector'?*

Ans. ***End-effector*** *is any object attached to the robot flange ("wrist") that serves a function.* This includes:

— Robotic grippers;
— Robotic tool changers;
— Robotic collision sensors;
— Robotic rotary joint;
— Robotic press tooling;
— Compliance device;
— Robotic point gun;
— Robotic deburring tool etc.

Q. 16. *Explain briefly vacuum grippers?*

Ans. It is often difficult to grasp large flat objects; however, it can be done by the use of ***vacuum gripper***.

- These grippers are mainly "*suction grippers*" that use a *vacuum pump or a venturi jet to generate vacuum* between a suction cup and the gripping plane.
  — The suction cups are made of rubber like materials (silicone, neoprene). Based on the design and the material, suction cups are suitable for workpiece temperatures upto 200°C. These cups are really important because the efficiency depends on the air-pressure source.
  — *Number of grippers/cups determines the size and weight of object to be grasped.*
  — *For handling soft materials cups made of harder material are used.*
- The vacuum gripper is tolerant of errors in the orientation of the part and is *especially suited for pick-and-place work*

Uses. Vacuum grippers are used for picking up *metal plates, pans of glass, or large light weight boxes.*

Q. 17. *State the requirements for gripper's design.*

Ans. The design of gripper should satisfy the following *requirements* :

1. To match the abilities of the arm and controller.
2. To grasp and hold the object securely.
3. To complete the task.

- Different grippers are needed for different applications, which are of highly varied types. These may be : *To hold complicated shapes which may be fragile, assembling parts and fitting them into clamps/fixtures etc.* (Once a part is picked up it must be securely held in such a way that its position and orientation are as desired).

Q. 18. *List the characteristics of a sensing device.*

Ans. A sensory device should possess the following *characteristics*:

1. Range:

- It refers to the ***minimum and maximum change in input signal to which the sensor can respond.***
- The sensor should have a ***wide operating range.***

2. Response:

- The sensor should be capable of responding to changes in the sensed variable in ***minimum time.***
- Ideally, the response should be ***instantaneous.***

3. **Accuracy:**
   - The measurement accuracy should be *as high as possible.*
   - The sensor's output should properly reflect the input quantity being measured or sensed.
4. **Sensitivity:**
   - The sensitivity relates to the *change in the output exhibited by the sensor for the change in input.*
   - It should be *as high as possible.*
5. **Linearity:**
   - The sensory device should exhibit the same sensitivity over its entire operating range.
6. It should be *suitable for the environment* in which it is to be employed.
7. It should have a suitable physical size, cost and ease of operation.
8. It should *not disturb* or have any effect upon the quantity it senses or measures.
9. It should *include isolation* from receiving excess signals or electrical noise that could give rise to the possibility of misoperation or damage of the sensor, circuit or system.

Q. 19. *What are touch or tactile sensors?*

Ans. ***In robotics, touch or contact sensors are used to obtain information associated with the contact between a manipulator hand and objects in the workspace.***

- Several physical principles have been exploited in the development of tactile sensors. In most cases, the developments in tactile sensing technologies are *application driven.* It should be recognised that the operation of a touch or tactile senor is *amply dependent on the material of the object being gripped.* Although the sensors are capable of working with rigid objects, yet work has shown that conventional sensors can be *modified* to operate with non-rigid materials.

  Touch/Contact sensors can be subdivided into the following two principal categories :

  1. Binary sensors.
  2. Analog sensors.

Q. 20. *Name some common devices which are used as position sensors.*

Ans. Some common devices which are used as position sensors are :

1. Potentiometers (or Pot)
2. Encoders (optical)
3. Linear variable differential transformer (LVDT).

Q. 21. *Name the main components of absolute encoders.*

Ans. ***Absolute encoders*** usually consist of the following *three components* :

(*i*) A multiple-track (or channel) light source.

(*ii*) A multiple-channel light receiver.

(*iii*) A multiple-track rotary disc.

— Normally, light emanating from a linear, N-element source (*e.g.* LEDs) is made to pass at right angles through the disc and is received or collected by a corresponding linear array of N light sensors (*e.g.* phototransistors) mounted on opposite side of the disc. The disc is divided into circumferential *tracks* and radial *sectors.*

— Absolute rotational information is obtained by utilising one of several possible code formats.

Q. 22. *What are proximity sensors?*

Ans. ***Proximity sensors*** *(sometimes referred to as external state sensors) generally have a binary output which indicates the presence of an object within a specified distance interval.* "Range sensors," in contrast, yield an estimate of the distance between a sensor and a reflecting object.

These sensors, typically, are used in robotics for near-field work in connection with object grasping or avoidance.

Q. 23. *What is the working principle of an eddy current sensor?*

Ans. *An* **eddy current sensor** *is used as a proximity switch and for determining the accuracy and repeatability of commercial robotic manipulators.*

***Working principle*:**

When a coil is supplied with an alternating current an alternating magnetic field is produced. If there is a metal object in close proximity to this alternating magnetic field, then eddy currents are induced in it. The eddy currents themselves produce a magnetic field which *distorts* the magnetic field responsible for their production. Consequently, the *impedance of the coil changes and so the amplitude of the alternating current. This change, at some preset level, can be used to trigger a switch.*

Q. 24. *What are capacitive proximity sensors?*

Ans. These sensors are potentially capable of *detecting all solid and liquid materials,* with various degrees of sensitivity (unlike inductive and Hall-effect sensors which can detect only ferromagnetic materials).

The capacitive sensors are *based on detecting a change in capacitance induced by a surface that is brought near the sensing element.*

Q. 25. *Name the three systems which utilise the 'fibre optics technology" in the robotics and for the manufacturing field.*

Ans. Following are the three systems which utilise the "fibre optics technology" in the robotics and/or the manufacturing field. :

1. Opposed configuration.
2. Retroreflective configuration.
3. Diffuse configuration.

Q. 26. *What is the function of a range sensor?*

Ans. *The function of a* ***range sensor*** *is to measure the distance from a reference point (normally on the sensor itself) to objects in the field of operation of the sensor.*

- Range sensors are employed for robot *navigation and obstacle avoidence*, where interest lies in estimating the distance to the closest objects, to more detailed applications in which the location and general shape characteristics of objects in the workspace of a robot are desired.

Q. 27. *What do you mean by 'Robotic vision'?*

Ans. ***Robotic vision*** *may be defined as the process of extracting, characterising and interpreting information from images of a 3-dimensional world.*

- Robotic vision (also termed as "*Computer vision*" or "*Machine vision*") is an important sensor technology with potential applications in many industrial operations.
- Robotic vision is primarily targetted at *manipulation and interpretation of image and use of this information in robot operation control.* Robotic vision requires *two aspects to be addressed :*

  (*i*) Provision for visual input;

(*ii*) Processing required to productively utilise the visual information in a computer-based system.

- The greatest virtues of a vision system are :
  - (*i*) Repeatability;
  - (*ii*) Accuracy;
  - (*iii*) Ability to produce approximately the same results when given approximately the same inputs.

Q. 28. *Enumerate the methods which are used to program the work cycle of a robot.*

Ans. In order to programme the work cycle of a robot the following *methods* are used :

1. Manual programming method.
2. Walk through programming method.
3. Teach pendant or lead through programming method.
4. Off-line programming method.

Q. 29. *What are 'Teach pendants'?*

Ans. ***Teach pendants*** *are hand held boxes that allow control of each manipulator joint or of each cartesian degree of freedom.*

— Some such controllers allow testing and branching, so that simple programs involving logic can be entered.

— Some teach pendants have alphanumeric displays and approaching hand-held terminals in complexity.

- *The primary functions of a teach pendant are :*
  - (*i*) To guide the robot or motion device, while teaching locations.
  - (*ii*) To serve as the primary point of control for initiating and monitoring operations.
  - (*iii*) To support application programs.

Q. 30. *What is 'fluid power'?*

Ans. *The term **fluid power** refers to the energy that is transmitted via a fluid under pressure.* With "*hydraulics*," that fluid is a liquid such as *oil* or *water*; while in "*pneumatics*", the fluid is typically compressed air or nitrogen gas (inert gas).

"*Fluid power technology*" *is a means to convert, transmit, control and apply fluid energy to perform useful work.*

Applications *of fluid power* include the following areas:

(*i*) Agriculture;	(*ii*) Construction;
(*iii*) Ships;	(*iv*) Aviation;
(*v*) Defence;	(*vi*) Transportation;
(*vii*) Fabrication;	(*viii*) Material handling;
(*ix*) Automation.	

Q. 31. *What are 'hydraulic valves'? How are they classified?*

Ans. ***Hydraulic valves*** *are mechanical devices used for controlling hydraulic parameters in hydraulic systems.*

***Classification of valves*:**

In hydraulic system various types of valves are used to control or regulate flow medium, but can be *broadly* classified into two types:

1. *Infinite position valves.* These valves allow any position between fully open and closed to modulate flow pressure.

— These valves may also be called *analog* position valves.

Examples: Water tap.

2. *Finite position valves.* This type of valve has discrete positions, usually just opened and closed, each providing a different pressure and flow condition.
   — These valves are commonly described by *x*/*y* designation, where *x* represents number of paths or ports and *y* number of positions; For example a 4/3 valve means the valve has 4 ports/ways and 3 positions.

An analogy between these two types of valves is the comparison between an *electric light dimmer* and a *simple on/off switch.*

Q. 32. *What is a 'linear actuator'?*

Ans. *A **fluid power hydraulic cylinder** is a 'linear actuator' which is most useful and effective in converting fluid energy to an output force in a linear direction for performing work such as pulling or pushing in a variety of engineering applications such as in machine tools and other industrial machinery, earth moving equipment, construction equipment and space applications.*

A hydraulic cylinder usually consists of a *movable element*, a *piston* and a *piston rod* operating within a cylindrical bore.

Q. 33. *What are 'rotary actuators'?*

Ans. ***Rotary actuators** are the hydraulic pneumatic equivalents of electric motors.* For a given torque, or power, a rotary converter :

— is more compact than an equivalent motor;

— cannot be damaged by an indefinite stall;

— can be safely used in an explosive atmosphere.

For variable speed applications, the complexity and maintenance requirements of a rotary actuator are similar to a thyristor-controlled D.C. drive, but for fixed applications, the A.C. induction motor is simpler to install and maintain.

Q. 34. *What are 'hydraulic motors'?*

Ans. An ***hydraulic motor** is a rotary actuator wherein hydraulic energy is converted into mechanical energy in the form of rotary motion and torque which may be used for doing work.* Thus, its function is just opposite to that of an hydraulic pump.

These motors very closely resemble pumps in construction. In an hydraulic motor high pressure oil pushes a freely moving element, thereby developing torque and continuous rotating motion. Since both inlet and outlet ports at times be pressurised, therefore, most hydraulic motors are drained externally.

- The maximum performance of a motor in terms of pressure, flow, torque, speed, efficiency, expected life and physical configuration is determined by the following factors:
  (*i*) Leakage characteristics;
  (*ii*) Efficiency of the means used to convert pressure surface and output shaft;
  (*iii*) Ability of the pressure surface to withstand hydraulic force.
- Hydraulic motors are *specified or rated* by:
  (*i*) Maximum operating pressure;
  (*ii*) Displacement (volume size);
  (*iii*) Speed;
  (*iv*) Torque capacity.

Q. 35. *What is 'fluid circuit'?*

Ans. ***A hydraulic or pneumatic circuit*** *(designed to perform a particular job) consisting of a large number of equipments (like power unit, service unit, accessories, valves actuators etc.) combined in logical sequence to get desired output in the form of motion of actuators and shown diagrammatically by using certain standard symbols of components, constitutes a* ***fluid circuit****.*

- Circuit diagrams are prepared by arranging various components into a sequence according to their functions. In case of very complex fluid circuits, where it is difficult to understand the 'start' and 'end' of the circuit, the designation systems are classified as : (*i*) *Systems using digits*, and (*ii*) *Systems using alphabets*.
- *The major difference between hydraulic and pneumatic circuits is that hydraulic elements are designed for high pressures (20 bar to 400 bar) and pneumatic circuits are designed for low pressures (5 bar to 20 bar).*

Q. 36. *State the advantages of 'pneumatic logic circuits'?*

Ans. ***Pneumatic logic controls*** offer the following *advantages* :

1. Systems are simple in design.
2. The life expectancy of such systems is longer than their electrical counterparts.
3. The response of pneumatic switching elements is faster than electrical systems
4. Employed in specialised applications, *e.g.*, explosive environments.

Q. 37. *Define the term 'Automation'?*

Ans. ***Automation*** may be *defined* as follows :

"*It is the process of following a predetermined sequence of operations with little or no human labour, using specialised equipment and devices that perform and control manufacturing processes.*"

Or

"*It is a system in which many or all of the processes in the production, movement, and inspection of parts and materials are automatically performed or controlled by self-operating devices.*

Or

"*It is technology concerned with performing a process by means of programmed commands combined with automatic feedback control to ensure proper execution of the instructions.*"

— ***Partial automation*** means replacement of human activities or involvement by automatic means only *partially.*

— In ***full automation***, the human involvement is totally eliminated and the process is entirely carried out and controlled through automatic means alongwith a proper feedback system.

Q. 38. *What are the types of automation?*

Ans. The various types of "automation" are:

1. Fixed automation (Hard automation).
2. Programmable automation.
3. Flexible automation (Soft automation).

Q. 39. *What is 'low cost automation'?*

Ans. ***Low cost automation*** *is a technology that creates some degree of automation around the existing equipment, tools, methods, people, etc., using mostly standard components available in the market with low investment, so that payback period is 'short'.*

- Low cost automation is one of the very practical, safe, economical and rewarding strategies in the present regime of globalisation and liberalisation to enhance productivity through *quality enhancement* and *cost reduction*.
- Low cost automation system may be adopted for the following *activities*:

— Loading;	— Feeding;
— Clamping;	— Machining;
— Forming;	— Welding;
— Gauging;	— Assembly and packing etc.

Q. 40. *What are 'Automated Guided Vehicles (AGVs)?*

Ans. ***AGVs** are battery-powered vehicles that can move and transfer materials by following prescribed paths around the manufacturing floor*. They are neither physically tied to the production line nor driven by an operator like a fork lift.

Such vehicles have *on-board controllers* that can be programmed for complicated and varying routes as well as load and unload operations. The computer for the material handling system or the central computer provides overall control functions such as dispatching, routing, traffic control and collision avoidance.

AGVs usually *complement in* automated production line consisting of conveyor or transfer systems *by providing the flexibility of complex and programmable movements around the manufacturing shop.*

The first automated guided vehicle was developed in 1954 by A.M. Barrett.

Q. 41. *What do you mean by 'Flexible Manufacturing System (FMS)'?*

Ans.
- ***Flexible Manufacturing System (FMS)** may be defined as a set of machines in which parts are automatically transported under computer control from one machine to another for processing.*

— "*Flexibility*" refers to the ability to respond effectively to changing circumstances. A *flexible system is one which is able to respond to change.*

- A FMS should be capable of coping with both *external changes* (*i.e.*, changes in the type, mix, processing requirements and quality of job allocated to the system, changes in the training and skill of operators assigned to the system), as well *internal changes or disturbances* (*i.e.*, machine and material handling system breakdowns, variability in processing time, operator absences, quality problems etc.)

Q. 42. *What is 'Group Technology (GT)'?*

Ans. ***Group Technology (GT)** is an approach in which simuar parts are identified and grouped together in order to take advantage of their similarities in design and production.* The group technology ***aims*** at *high productivity at a low cost on short run production.*

- ***"Group technology"*** in manufacturing, is the replacing of traditional jobbing shop manufacturing *by analysis and grouping of work into families and the information of groups of machines to manufacture these families on a flow-line principle with the object of minimising setting times and throughout times.*

INDEX

T

NOTES

NOTES

NOTES

NOTES